ON THE WINGS OF CHECKERSPOTS

ON THE WINGS OF CHECKERSPOTS
A Model System for Population Biology

Edited by

Paul R. Ehrlich
Ilkka Hanski

2004

OXFORD
UNIVERSITY PRESS

Oxford New York
Auckland Bangkok Buenos Aires Cape Town Chennai
Dar es Salaam Delhi Hong Kong Istanbul Karachi Kolkata
Kuala Lumpur Madrid Melbourne Mexico City Mumbai Nairobi
São Paulo Shanghai Taipei Tokyo Toronto

Published by Oxford University Press, Inc.
198 Madison Avenue, New York, New York, 10016

www.oup.com

Library of Congress Cataloging-in-Publication Data
On the wings of checkerspots : a model system for population biology / edited by Paul R.
Ehrlich, Ilkka Hanski.
p. cm.
Includes bibliographical references and index.
ISBN 0-19-515827-X
1. Nymphalidae. 2. Population biology. I. Ehrlich, Paul R. II. Hanski, Ilkka.
QL561.N9O5 2004
577.8'8—dc21 2003051740

9 8 7 6 5 4 3 2 1

Printed in the United States of America
on acid-free paper

To Anne and Eeva

Preface

Bay Checkerspots

I clearly remember the first time I saw a Bay checkerspot butterfly. It was not in a sunlit patch of serpentine grassland overlooking San Francisco Bay, but in my bedroom in suburban New Jersey in 1948. There, as a 16-year-old devoted butterfly collector, I opened the first box of butterflies I had ever received in exchange from another collector. Bill Hammer in California had mailed me a generous lot of specimens, which he traded to me for some rather common eastern butterflies. In that pile of specimens in neatly folded paper triangles was a series of *Euphydryas editha bayensis*. Little did I know then that I was to spend most of my life in California and devote much of it to the study of that very butterfly.

About a decade later I arrived at Stanford University, thrilled to have been hired as an assistant professor of biology to teach evolution and entomology, replacing the eminent morphologist and student of scale insects, Gordon Floyd Ferris. Jobs were hard to get, and I considered myself especially lucky to land one where the opportunities to do field work stretched through much of the year. I was tired of always waiting until May or June for interesting butterflies to start flying. The Bay area and Stanford seemed like paradise on earth to one fresh from graduate school and postdoctoral training in Kansas and Chicago. Indeed, almost four decades later I still feel fortunate to live in the land of milk and honey.

In the course of my training with Charles Michener, Robert Sokal, and Joseph Camin, I developed a strong interest in understanding evolutionary processes in natural populations. Joe Camin and I had been fortunate enough to elucidate one of the few examples documenting natural selection in the field, a task made possible by an unusual combination of strong selection balanced by migration in a population of water snakes (Camin and Ehrlich 1958, Ehrlich and Camin 1960).[1] My early experience with that project greatly shaped my career. Other investigations of selection in nature also have been focused on unusual situations. One classic example is the reaction of moth populations to increased industrial pollution, first studied in the Midlands of Great Britain and later in Europe and North America (Kettlewell 1955, 1973, Owen 1997, Majerus 1998, Brakesfield and Liebert 2000). Another was strong selection on land snails detectable

1. Water snakes (*Nerodea sipedon*) on the islands of Lake Erie, unlike those in the swamps of most of North America, tend to have reduced banding. This is a result of selection by visual predators for bandlessness against the uniform gray limestone rocks of the islands. This selection pressure is balanced by individuals migrating from banded lakeshore swamp populations. It is these opposing forces that allowed selection to be detected by comparing the amount of banding in juvenile individuals with that in adults. At birth there were many banded individuals, but relatively few of them survived to adulthood. More recent work on the *N. sipedon* system has confirmed and expanded our early observations (King 1993, King and Lawson 1995).

because thrushes that prey on them break their shells on traditional rocks, permitting the color patterns of the snails being eaten to be compared with those in the entire population (Sheppard (1951, 1952, Cain and Sheppard 1954).[2] All those studies showed that predation could be a potent factor in determining the color patterns of both moths and snails by favoring genotypes whose phenotypes were better camouflaged against the backgrounds on which they lived. I was (and still am) concerned with whether these and the handful of other studies of natural selection were a biased sample of selection pressures in nature. Other sampling problems have hindered understanding the dynamics of natural populations. People have been mostly concerned with changes in the size of large populations. That is because of our interest in pests and in organisms that we want to harvest. With trivial exceptions, these have large populations. But most natural populations are small, and if we want to understand how the world works, understanding the dynamics of small populations is important. Indeed, it is essential in a world entering an unprecedented era of extinctions (see, e.g., Myers 1979, Ehrlich and Ehrlich 1981, Ehrlich and Wilson 1991, Wilson 1988, 1992, Heywood 1995). Central questions of conservation biology deal with predicting the fates of small populations and finding ways to increase their chances of persistence.

When I arrived at Stanford in the fall of 1959, I was determined to make at least a small start at correcting the sampling bias of studies of natural populations. I decided to put a portion of my research effort into teasing apart the population dynamics, general ecology, and population genetics of some populations that were not doing anything spectacular—just apparently going along with no vast outbreaks of numbers or dramatic changes in the proportions of different kinds of individuals because of strong selection. My long-term interest in butterflies (my dissertation had been on their morphology and higher classification) and the pioneering work of E. B. Ford (e.g., Ford and Ford 1930) on their ecological genetics made choosing a butterfly population, especially a checkerspot population, as a model system a natural decision.

2. Detecting selection coefficients in the vicinity of .01 (one genotype reproducing itself 99% as well as another) is virtually impossible in the field. One of the triumphs of theoretical population genetics has been to show that such selection pressures would have been sufficient, given the enormity of the time available, to account for evolution from coacervate droplets to people. Thus we do not know whether the high levels of selection that have been observed are due simply to our ability to observe them or whether they are indeed typical of natural situations.

Fortunately, just after my arrival at Stanford, I decided to go to the 1959 meeting of the Lepidopterists' Society in Santa Barbara and drove down with two local butterfly collectors, who have since passed away: Bill Tilden and Elton Sette. I discussed my plans with them, and Elton suggested that I study the Bay checkerspots on Jasper Ridge, a partly undeveloped area of Stanford campus used for grazing, recreation, and occasionally for teaching and research by the Department of Biological Sciences. Elton said that there was a population in the grassland on the ridge and said he would be glad to give me samples of specimens that he had taken on the ridge at different times going back to 1936.

The die was cast. In the fall of 1959, I mapped the Jasper Ridge area where Elton said the butterflies flew and divided it into arbitrary areas, from A to H (figure 1.2). In the spring of 1960 I began to visit the area with the intention of starting a mark–release–recapture study. I was accompanied by a graduate student, Susan Davidson, now Susan Thomas, wife of my late Stanford colleague, botanist John Thomas. I saw no butterflies on the first two visits, but on March 31 at 10 AM I saw and caught my first live Bay checkerspot. My ancient mimeographed data sheet says it was a male, taken in area G, in excellent condition. It was marked with a coded number one, a felt-pen smear on the underside of the apex of the right forewing, and released. Then Susan and I moved off in search of number two, little realizing the numbers marked on Jasper Ridge would eventually reach many thousands in a research program that would span more than four decades and would be central in a long battle to prevent Jasper Ridge from being converted into a housing development (figures 1–3).

So the circle was closed. The butterflies that I fondly remember arriving dead in little triangular envelopes, smelling faintly of moth balls, were now living entities that would help me understand more about how the world worked. I have never regretted the decision to work with *Euphydryas*, and one of the greatest pleasures of doing so has been the number of graduate and postdoctoral students and colleagues that I have gotten to know as friends as we struggled together to solve the mysteries of the biology of their populations. Much of the *Euphydryas* work reported here was done by those colleagues (as the citations will show). Several of them have authored chapters in this volume.

We are an inbred lot. Mike Singer, now a professor at the University of Texas, is one of the most knowledgeable *Euphydryas* biologists. He and his

Figure 1. Susan Davidson Thomas and Duncan Porter marking *Euphydryas editha* at Jasper Ridge, spring 1960. Photo by Paul R. Ehrlich.

students are unsurpassed as investigators of *Euphydryas* oviposition behavior and the evolution of host plant choice, and he is married to another *Euphydryas* biologist, Camille Parmesan, known for her important work on the response of checkerspots and other butterflies to global climate change. Carol Boggs, director of Stanford's Center for Conservation Biology (CCB), has worked on various problems in the reproductive biology of *Euphydryas* and has broad experience with other butterflies. She is one of my academic grandchildren, having received her Ph.D. with another distinguished *Euphydryas* biologist at Texas, my former student Larry Gilbert (who deserted *Euphydryas* for *Heliconius*). Carol is also the wife of my Stanford colleague Ward Watt, a leading butterfly evolutionist who has been continually helpful to me as I have struggled to understand the *Euphydryas* system.

Alan Launer, when he was an undergraduate, worked with my student Dennis Murphy (now too busy saving the biota of the western United States from runaway development to devote much time to his first love, *Euphydryas*) and me on the most intensive study ever done of a Jasper Ridge *E. editha* population. After getting his doctorate at Harvard doing research on fishes, Alan returned to Stanford to work more on *Euphydryas* and to become the university's official conservation biologist. Dennis Murphy, both as a graduate student and as a former director of the CCB, mentored Stu Weiss. Stu has devoted decades of his life to understanding the dynamics of Bay area *E. editha* populations, both as a researcher at the CCB, during the course of his doctoral work, and as a biological consultant. John McLaughlin got his Stanford doctorate in ecological theory with Joan Roughgarden, returned to the CCB as a postdoc to work on *Euphydryas*, and now collaborates with us as a faculty member at Western Washington University. The youngest member of our *Euphydryas* team is Jessica Hellmann, who got her doctorate in our group working on *E.*

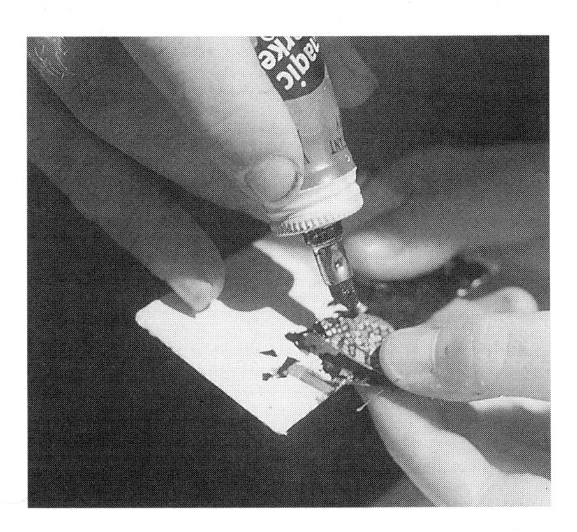

Figure 2. Using a felt-tipped marker to number (1-2-4-7 system; see Ehrlich and Davidson 1960) a *Euphydryas editha* at Jasper Ridge in spring 1960. With these markers it was best to support the wing with a piece of cardboard, making it necessary to have two people in a marking team. Photo by Paul R. Ehrlich.

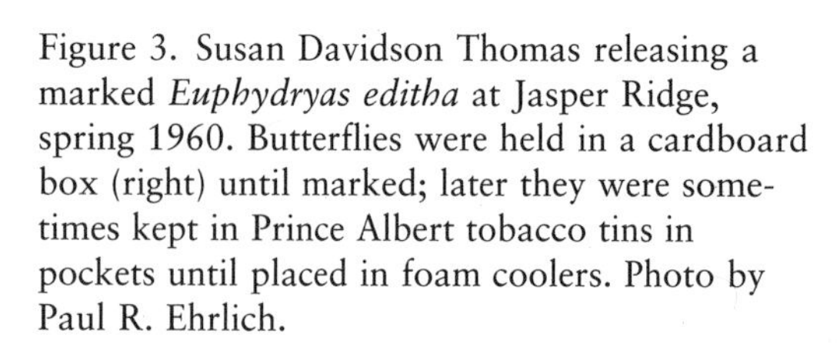

Figure 3. Susan Davidson Thomas releasing a marked *Euphydryas editha* at Jasper Ridge, spring 1960. Butterflies were held in a cardboard box (right) until marked; later they were sometimes kept in Prince Albert tobacco tins in pockets until placed in foam coolers. Photo by Paul R. Ehrlich.

editha, is now a faculty member at the University of Notre Dame and is pursuing her burning passion to understand the community dimensions of population dynamics. We all share broad interests in butterflies, the principles of population biology (ecology, evolution, and behavior) and protecting biological diversity, as do Ilkka Hanski and our other European checkerspot biologist friends and colleagues with whom we've had the great pleasure of writing this volume. Thus we have used the basic theme of the population biology of checkerspots as a springboard for discussing a variety of related issues. It has been my privilege and pleasure to work with them and with all the many other friends who have participated in our checkerspot adventures.

Mickey Suen, an undergraduate Stanford student who worked on *Euphydryas editha* with Mike Singer in 1970–71, helped interest his sister, Leslie, in the Jasper Ridge Preserve. Leslie eventually became a docent in the Preserve's growing public outreach program. More than three decades later, a few years after Leslie's tragic death, Mickey and I met on Jasper Ridge. It was June 2, 2002, a lovely day with freshly emerged *Euphydryas chalcedona* flying in abundance in the chaparral, nectaring on *Aesculus* and *Eriodictyon*. It was a day of nostalgia for me, and the *E. chalcedona* reminded me of my former student Irene Brown, who has dedicated a quarter of a century to studying them, allowing our group to contrast their behavior with that of the other Jasper Ridge *Euphydryas*, *E. editha*. Susan Thomas, my wife, Anne, and I enjoyed watching *chalcedona*'s mating behavior as we hiked up to the serpentine grassland on the Ridge's crest, where so long ago Susan and I had started our research. We then descended to a knoll where about 200 people were gathered, including Stanford's Provost, its Dean of Humanities and Sciences, and Mickey. We were there to dedicate a marvelous new laboratory building to Leslie's memory. Many people from the local community, including Leslie's husband, Tony Sun, old friends Bill and Jean Lane, and Eff and Patricia Martin, had worked hard to help Stanford (and, especially, the Preserves' far-sighted Administrative Director Philippe Cohen) make that building a reality. It is a critically needed structure dedicated to research and teaching, and a model solar "green" building that puts energy into California's electricity grid.

The ceremony represented, more than 42 years after marking that first *Euphydryas editha*, the final securing of the Jasper Ridge Biological Preserve as one of Stanford's premier facilities—and it's best known one. The Ridge has been the site of hundreds of research projects, including now a critical experiment on the effects of global climate change. It is by far the best such research area on the campus of any major university—one which, not coincidentally, also has the best environmental science faculty in the world. Jasper Ridge also has been involved in the education of tens of thousands of Stanford students, pupils from Bay area schools, and local people interested in science and nature. The Preserve is a monument to the very best of town–gown cooperation, showing that open space can be preserved even in the midst of a badly overdeveloped area with incredibly high land prices. One never can tell where smearing a checkerspot's wing with a felt-tip pen can lead.

Paul R. Ehrlich
Stanford, California

Glanville Fritillaries

Writing this book has been a thrilling experience. Never could I have expected the number of parallel yet complementary results that have been obtained in the two long-term checkerspot projects, the one conducted in Stanford in sunny California, the other one on the not-always-so-sunny Åland Islands in Finland. Striking similarities go back to days before the research began. Like Paul Ehrlich, I was a keen collector of butterflies and moths in my youth. Three species of checkerspot butterflies were familiar to me from the southeastern corner of Finland, where I had the luxury of spending all my summers as a schoolboy. I, too, traded a few butterflies, in my case some eastern Finnish butterflies for western ones, and the latter included two specimens of the Glanville fritillary from the Åland Islands. This happened some time in the late 1960s, when I was 15 or 16 years old. The two butterflies had been collected in 1941 and 1955 from two different localities in Åland, one of which continues to be a stronghold of the species even today, while the other one has succumbed to habitat destruction and is now practically devoid of this species.

My experiences as a boy catapulted me to a career in population and community ecology. I have been excited by questions about population regulation and causes of rarity and commonness of species. My research has been in the interface between theory and field studies. The organisms that I know best include dung beetles, blowflies, pine sawflies, shrews, and voles. Throughout my career, I have been particularly intrigued by the spatial structure of populations and its consequences for population dynamics. The many years that I spent chasing butterflies and moths left notions of patchy distribution of species in heterogeneous landscapes hardwired in my brain. The work on blowflies and dung beetles was primarily concerned with small-scale patchiness of populations exploiting highly ephemeral resources. One of the main questions was the mechanisms of coexistence of dozens of ecologically similar species dependent on just one or a few distinct resources.[3] For anyone delighted by the sight of a multitude of species scrambling in a small space, I can recommend nothing better than a leisurely examination of cow pats in a sunny pasture! Later on my interests moved to larger spatial scales, and ever since I discovered Richard Levins's metapopulation concept in the mid-1970s, I must have been one of his most committed apostles.

From 1972 until 1991 I paid little attention to butterflies, but this was to change in winter 1991. At the time I had started what turned out to be long-lasting and most enjoyable collaboration with Mats Gyllenberg, a mathematician specializing in structured population models. Mats had previously worked on populations of cancer cells, but I managed to convince him that spatially structured "real" populations are so much more interesting. Our modeling efforts produced predictions about the possibility of alternative stable states in metapopulation dynamics, which were so interesting that I wanted to test them (Gyllenberg and Hanski 1992, Hanski and Gyllenberg 1993; for the result of the empirical work, see Hanski et al. 1995b). My old love for butterflies reemerged, and I began searching for a species to study. I knew, of course, of Paul Ehrlich's work on *Euphydryas editha*, which was another reason for narrowing down the search to one of its Old World relatives, the Glanville fritillary (*Melitaea cinxia*). I had never seen it alive, but my inquiries suggested that it might be the right species for the job and that the Åland Islands might be the right landscape (no choice really here, because the species had gone extinct from the rest of Finland in the 1970s). I have not had a reason to regret the decision.

3. The mechanisms that allow two or more species that compete with each other nonetheless to coexist in the same community pose one of the classic problems in ecology, first formulated in mathematical terms by Lotka (1925) and Volterra (1926) and explored experimentally in hugely influential research by Gause (1934). Briefly, the most likely mechanism facilitating coexistence is a difference in the kinds of resources used by the competing species, which puts less pressure on individuals of different species (the competitor) than on individuals of one's own species. I realized, during my doctoral studies on dung beetles in Oxford, that the small-scale aggregated spatial structure so characteristic of populations of insects and many other creatures is likely to shift the balance in the strengths of intraspecific and interspecific competition toward the former (because individuals tend to be most aggregated with others of their own species). I showed with a simple model that this would automatically help coexistence (Hanski 1981). I later found, to my great embarrassment, that Charles Elton, whom I had the privilege to meet and talk to in Oxford in the late 1970s, had clearly seen this possibility in his visionary paper published in 1949 (Elton 1949). It may also be added that the combination of my studies on dung beetles and Levins's metapopulation model led to the concept of core and satellite species (Hanski 1982), which involves processes that may generate alternative stable states in metapopulation dynamics (Hanski and Gyllenberg 1993)—the phenomenon that stimulated the birth of the Glanville fritillary project. Here, hence, is a direct line from my early studies on dung beetles to the metapopulation studies on checkerspots and other species.

Two of my students, Mikko Kuussaari and Marko Nieminen, left for Åland on April 24, 1991, to search for a network of habitat patches and a metapopulation of the Glanville fritillary that would live in it. The following day brought encouraging news: they had found a large local population with tens of larval groups in a landscape that might well be suitable for our studies (figure 4). The habitat patches (dry meadows) were mapped within an area of some 20 km^2, and Mikko and Marko started an intensive mark–release–recapture study on June 6 when the first male was seen flying. I joined them on June 17 and caught my first Glanville fritillary at 9:40 AM the following morning: a male number 148, marked 3 days earlier in the same population. Three minutes later I had marked my first unmarked butterfly, a very fresh female, number 291 (figures 5 and 6). Our efforts in that first summer led to a research paper (Hanski et al. 1994), in which we concluded that 'the *Melitaea cinxia* metapopulation we analyzed provides a contrasting example to the *Euphydryas editha* metapopulation reported by Harrison et al. (1988). Unlike the latter case, there is no large "mainland" population in the *M. cinxia* metapopulation, and its long-term persistence appears to depend on genuine extinction-colonization dynamics.'

Mikko Kuussaari went on to complete his doctorate on the larval biology of the Glanville fritillary. Since then he has become a senior researcher in the Finnish Environment Institute, working on biodiversity conservation in agricultural landscapes. Marko Nieminen's doctoral project was on spatial dynamics of moth populations, but later he returned to the Glanville fritillary and is now a senior researcher in our Metapopulation Research Group (MRG). Marko is responsible, among other things, for the smooth running of the annual surveys of the hundreds of local populations, leading a small army of some 40 undergraduate field assistants employed for a few busy weeks in the spring and late summer. Before Marko, the annual surveys were run by Juha Pöyry, another keen lepidopterist, who later became a doctoral student in Mikko Kuussaari's project. Another doctoral student of mine, Niklas Wahlberg, was until recently an active member of the MRG, and he greatly expanded our vision through his molecular phylogeny of checkerspot butterflies and more detailed comparative studies of five of the Finnish species.

Figure 4. The 2-ha dry meadow in Finström, northern Åland, where larvae of *Melitaea cinxia* were first recorded in the spring 1991 (patch 1 in Hanski et al. 1994). The 72 larval groups with which the experimental metapopulation on the island of Sottunga (Figure 13.7) was started in the fall 1991 were collected from this meadow. Photo by Ilkka Hanski.

Figure 5. Butterflies have been marked by writing a number on the underside of the hindwing with a fine-tipped felt pen. Photo by Niklas Wahlberg.

At present, he pursues molecular systematics of butterflies in Stockholm.

The modeling work that I started with Mats Gyllenberg has been an essential part of the Glanville fritillary project, to the extent that I have stopped commenting on whether butterfly biology is driving modeling or vice versa. In his doctoral project, Atte Moilanen much increased the rigor with which we can apply models to real metapopulations, including the Glanville fritillary metapopulation in Åland. Atte continues to play a key role in the MRG, and so does Otso Ovaskainen, a young mathematician who joined us only three years ago but who has already successfully injected a huge amount of real mathematics into our modeling work and is on his way to a career in mathematical biology.

For the first five years the research on the Glanville fritillary was focused rather narrowly on metapopulation ecology and dynamics. Four researchers in particular have helped widen our horizon. Ilik Saccheri, now in the School of Biological Sciences in the University of Liverpool, added a population genetic component to our work and started the ongoing work on the role of inbreeding in the metapopulation dynamics of the Glanville fritillary.

Figure 6. Ilkka Hanski releasing *Melitaea cinxia* on the island of Kyrkogårds in June 1992. Photo by Mikko Kuussaari.

This line of research is currently pursued by Sari Haikola as her doctoral project. Guangchun Lei, now a professor at Beijing University, pioneered studies into host–parasitoid interactions involving the Glanville fritillary, a line of research that has greatly expanded over the past few years under the supervision of Saskya van Nouhuys, who has shared her time between Cornell University in winter and the Åland Islands in summer. Finally, since 1996, Michael Singer has been a most valued collaborator on our project, sharing with us his unsurpassed knowledge of checkerspot population biology, especially anything and everything related to host plant preference by female butterflies. We started the collaboration with the idea of merging the ecological colonization–extinction dynamics, which has been my speciality, with the evolutionary dynamics of female host-plant choice, Mike's speciality. Several chapters in this volume illustrate the success of this collaboration. Mike Singer also forms another bridge between the two long-term butterfly projects to which this volume is devoted, as he was one of the first students of Paul Ehrlich and knows both *M. cinxia* and *E. editha* so well.

The large Glanville fritillary metapopulation in the Åland Islands may be the best studied example of classic metapopulations. While we have been busy documenting the biology of this species living in its naturally highly fragmented landscape, conservation biologists have become increasingly aware of the massive threat that habitat loss and fragmentation pose to populations and species worldwide. The Glanville fritillary has become an important model, almost an icon, of species struggling for survival in fragmented landscapes, and I have attempted to follow in the steps of Paul Ehrlich and others in voicing the broader implications of our research results. This volume demonstrates what wonderful model systems the checkerspots have been in advancing our conceptual understanding of many basic issues in population biology over the past four decades, in bridging the still-too-wide gap between theory and field studies, and in bringing into sharper focus some of the most pressing (and depressing) environmental issues that we cannot be silent about.

Ilkka Hanski
Helsinki, Finland

Acknowledgments

This book, being the collaborative project of two large research groups and many researchers, is the product of the efforts of many more people than the authors. We would like to express our deep appreciation to all of those listed below—without them, *On the Wings of Checkerspots* would never have been completed. We beg their indulgence for not specifying each precise form of help; if we did this it would be the longest chapter of the book. But we think our list underlines what a group enterprise science truly is.

We have not acknowledged the many internal reviews done by authors of one or more chapters who read those of other authors. Many other colleagues were kind enough to review and provide insightful comments upon parts of the manuscript. They include Mike Begon, Arjen Biere, David Boughton, Deane Bowers, Erica Fleishman, Peter Grant, Niklas Janz, Lukas Keller, Gary Luck, Sören Nylin, Otso Ovaskainen, Juha Pöyry, Ron Rutowski, Mark Shaw, Jamie Smith, Constantí Stefanescu, Jorma Tahvanainen, Chris Thomas, Juha Tuomi, Dirk van Nouhuys, Peter Vitousek, Ray White, and Christer Wiklund. Two colleagues and two of our authors read the entire manuscript and contributed a series of extraordinarily helpful comments: Susan Harrison, Hanna Kokko, Dennis Murphy, and Mike Singer. The following supplied us with information or made helpful suggestions on technical issues: Christian Anton, Brendan Bohannan, Irene Brown, Mar Cabeza, Gretchen Daily, Jared Diamond, Johanna Ehrnsten, Marc Feldman, Liz Hadly, Sari Haikola, Jeff Harvey, Aaron Hirsh, Cheri Holdren, Maaria Kankare, Anna-Liisa Laine, Simon Levin, Leif Lindgren, Evgeniy Meyke, Atte Moilanen, Kim O'Keefe, Richard Moore, Judith Myers, Otso Ovaskainen, Camille Parmesan, Hanna Paulomäki, Eeva Punju, Ron Pulliam, Jane Reid, Bob Ricklefs, Terry Root, Tomas Roslin, Mark Shaw, Hans-Peter Tschorsnig, Kata-Riina Valosaari, Peter Vitousek, and Ward Watt. For all of their help we are extremely grateful.

Nona Chiariello, Philippe Cohen, Alan Grundman, numerous docents, and many other people associated with Stanford University's Jasper Ridge Biological Reserve have been very supportive of the Ehrlich group's checkerspot work there over many decades. Waste Management Inc. provided the group access to the Kirby Canyon butterfly reserve, and many other landowners were kind enough to put up with butterfly chasing on their properties. The Rocky Mountain Biological Laboratory served as an excellent and cooperative base for checkerspot work in Colorado. And the U.S. Fish and Wildlife Service worked with the Ehrlich group to get *Euphydryas editha bayensis* listed as "threatened" and to provide permits for subsequent work on its populations. In Finland, the Nåtö Biological Station has been the center for *Melitaea cinxia* studies in Åland since the early 1990s, and we thank Carl-Adam and Eva Hæggström for all the help they have provided. Husö Biological Station also provided us with accommodation and lab space when needed. A very special thank you goes to the hundreds of landowners in the Åland, who have allowed us to work on their land. In mainland Finland, the staff of the Tvärminne Zoological Station has supported our experimental work for many years.

We all owe debts to our colleagues who are not primarily "Melitaeaologists" but who have discussed and debated issues relating to checkerspot biology with us over the years. A representative, but far from exhaustive, list would include (D indicates now deceased): Juha Alho, Charles Birch, Paul Brakefield, Mark Camara, Theodosius Dobzhansky (D), Marc Feldman, E. B. Ford, Wille Fortelius, Karin Frank, Larry Gilbert, Mikko Heino, Dick Holm (D), Tad Kawecki, Guangchun Lei, Ernst Mayr, Charles Michener, Hal Mooney, Isabelle Olivieri, Otso Ovaskainen, Peter Raven, Charles Remington, Jim Rohlf, Ophélie Ronce, Joan Roughgarden, Elton Sette (D), Bob Sokal, John Thomas (D), Ward Watt, Peter Vitousek, and Darryl Wheye.

Tapio Gustafsson, Evgeniy Meyke, Ann McMillan, Marjo Saastamoinen, Joan Schwan and Anu Väisänen helped with many production tasks. Ann and Joan did yeoman work in preparing the line art, and Zdravko Kolev provided beautiful drawings. George Austin and Tapio Gustafsson oversaw the production of the plates, and John Fay was enormously helpful in working with digital imaging problems. The staff of Falconer Biology Library, especially Jill Otto, performed with their usual skill and alacrity in running down references, and Pat Browne and Steve Masely once again handled copying chores at the Stanford end rapidly and accurately. Others who provided various forms of important technical help include Anders Albrecht, Mathijs Doets, Elizabeth Garcia, Janne Heliölä, Larry Huldén, and Jaakko Kullberg.

We are indebted to the U.S. National Science Foundation for a series of grants to Paul Ehrlich, Michael Singer, and Carol Boggs and to the Academy of Finland for long-term support to Ilkka Hanski and the Metapopulation Research Group. The Ehrlich group also benefited greatly from support from Peter and Helen Bing, John and Susan Boething, Stanley and Marion Herzstein, the late LuEsther T. Mertz, and the Hewlett, Joyce Mertz-Gilmore, and Winslow foundations. The Hanski group acknowledges the support for the *Fragland* network from the European Union's Training and Mobility of Researchers' Programme. Ward Watt (Carol Boggs's husband) put up with many late-night and weekend phone calls from Paul Ehrlich to Carol, remaining civil the whole time. The editors' wives, Anne Ehrlich and Eeva Furman, patiently endured more than we'd like to describe in the course of this work.

Contents

Contributors

Carol L. Boggs
Center for Conservation Biology
Department of Biological Sciences
Stanford University
Stanford, California 94305, USA

Paul R. Ehrlich
Center for Conservation Biology
Department of Biological Sciences
Stanford University
Stanford, California 94305, USA

Ilkka Hanski
Department of Ecology and Systematics
University of Helsinki
FIN-00014 Helsinki, Finland

Jessica J. Hellmann
Department of Biological Sciences
University of Notre Dame
Notre Dame, Indiana 46556, USA

Mikko Kuussaari
Finnish Environment Institute
FIN-00251 Helsinki, Finland

Alan E. Launer
Center for Conservation Biology
Department of Biological Sciences
Stanford University
Stanford, California 94305, USA

John F. McLaughlin
Department of Environmental Sciences
Western Washington University
Bellingham, Washington 98225, USA

Dennis D. Murphy
Department of Biology
University of Nevada
Reno, Nevada 89557, USA

Marko Nieminen
Department of Ecology and Systematics
University of Helsinki
FIN-00014 Helsinki, Finland

Ilik J. Saccheri
School of Biological Sciences
University of Liverpool
Liverpool L69 7ZB, UK

Mika Siljander
Department of Ecology and Systematics
University of Helsinki
FIN-00014 Helsinki, Finland

Michael C. Singer
Department of Integrative Biology
University of Texas
Austin, Texas 78712, USA

Saskya van Nouhuys
Department of Ecology and Systematics
University of Helsinki
FIN-00014 Helsinki, Finland
and
Department of Ecology and Evolutionary Biology
Cornell University
Ithaca, New York 14853, USA

Niklas Wahlberg
Department of Zoology
Stockholm University
S-106 91 Stockholm, Sweden

Stuart B. Weiss
Palo Alto, California, USA

ON THE WINGS OF CHECKERSPOTS

Euphydryas editha

Melitaea cinxia

Vignettes of *Euphydryas editha* and *Melitaea cinxia*, used to help identify figures in this book that deal with these species. Drawings by Zdravko Kolev.

1

Checkerspot Research

Background and Origins

PAUL R. EHRLICH AND ILKKA HANSKI

As the laws of Nature must be the same for all beings, the conclusions furnished by this group of insects must be applicable to the whole organic world; therefore, the study of butterflies—creatures selected as the types of airiness and frivolity—instead of being despised, will some day be valued as one of the most important branches of Biological science.

Henry Walter Bates

This book is a story about butterflies, whose beauty and variety fascinated us as children and fascinates us still. It tells how we and our colleagues developed one group of butterflies, the checkerspots, into a model study system with which to do research and answer basic questions in ecology and evolution. We've tried to create one population biological analogue to the well-known model systems in other biological disciplines, such as the fruit flies of classical genetics and the chick embryos of early embryology. This book describes what has been discovered while studying the checkerspot system and how those findings relate to what is known about other organisms. It is a story of increasing understanding both about how the living world works and about how to preserve humanity's natural capital, biodiversity (Daily 1997).

The primary goal of the research begun on checkerspots (the tribe Melitaeini) more than four decades ago was to uncover processes at the population level of biological organization (Ehrlich and Holm 1962). What were the spatial patterns in which populations exist? What controlled those patterns? Were most populations open—that is, freely exchanging individuals with other populations—or were they relatively closed? What patterns could be found in the changes of the size of checkerspot populations, and what were the factors driving those changes? Were they density dependent or density independent? How did the dynamics of populations interact with their genetics? In short, we wanted to make the natural populations of checkerspots into a tool for generating and testing ideas that would further the understanding of population biology. Many of the questions posed above, and many others, have been answered for a set of checkerspot populations, and the answers have implications for other organisms that are less easily studied in the field, as you will see. In the course of doing this research (although it was not an original goal), we learned a great deal about the costs and benefits of focusing research efforts on model systems. Understanding the functioning of natural populations—how they are distributed and structured, how and why their sizes change, and how they evolve—is the major intellectual challenge of population biology. It is at this foundation level of ecology and evolution that the dynamics and evolution of single-species populations and metapopulations of checkerspots, and the behavior of the individuals of which they consist, have been investigated. Apart from its scientific significance, knowledge of ecology and evolution of populations is of great practical importance. For instance, humans want to harvest economically valuable organisms sustainably and to suppress populations of other organisms that compete with people and are designated as "pests." Conservationists want to find ways to avert the disappearance of populations (Ehrlich and Daily 1993, Hughes et al. 1997, Ceballos and Ehrlich 2002) and of the species of which they are components (Myers 1979, Ehrlich and Ehrlich 1981). And demographers, ecologists, and evolutionists seek to estimate the size and genetic attributes of future human populations and the pathogens and vectors that will threaten them (Daily and Ehrlich 1996a, 1996b).

Our use of checkerspots as a model system developed in a period of extraordinarily rapid progress in population biology. To help address the questions above, an impressive body of theory has been developed in both ecology and population genetics (e.g., MacArthur and Wilson 1963, Kimura and Ohta 1971b, Lewontin 1974, Roughgarden 1979, Emlen 1984, Tilman and Kareiva 1997, Hanski 1999b, Turchin 2002; for a less technical summary, see Ehrlich and Roughgarden 1987). But the difficulties of testing that set of theories in the field have remained severe, and considerable uncertainty and controversy still surround both the demographic (Cappuccino and Price 1995, Rhodes et al. 1996, McLaughlin et al. 2002a) and genetic dynamics (e.g., Hartl and Clark 1997, Watt and Dean 2000) of natural populations, and detection of evolutionary forces in nature remains in its infancy (Endler 1986). Perhaps most important, the connections between demography and genetics largely remain to be investigated in wild populations.

1.1 Model Systems in Population Biology

The rate of disappearance of natural habitats is so high that many species will go extinct before their most basic biological characteristics have been recorded. Many if not most species will vanish before they can even be given a latinized name (Ehrlich 1964, Myers 1979, Ehrlich and Ehrlich 1981, Wilson 1992). With the exception of a few species, the distribution of populations has not even been preliminarily mapped, while many thousands of the billions of populations, essential to maintenance of ecosystem services (Daily 1997, Luck et al. 2003), disappear daily under human onslaught (Hughes et al. 1997, 1998). The majority of earth's millions of species (Ehrlich and Wilson 1991) will be exterminated before we gain even a partial understanding of their biology, let alone the knowledge we now have for intensively studied organisms such as the nematode *Caenorhabditis elegans* (e.g., Wood 1988, Sternberg 1990, de Bono and Bargmann 1998), *Drosophila* fruit flies (e.g., Morgan 1911, Dobzhansky 1947, Konopka and Benzer 1971, Schaeffer and Miller 1993), *Bicyclus* butterflies (Brakefield et al. 1998, Wijngaarden 2000), white-crowned sparrows (*Zonotrichia leucophrys*) (Baptista 1975, Baptista and Petrinovich 1986), and chimpanzees (*Pan troglodytes*) (Goodall et al. 1979, Goodall 1986, Ehrlich 2000).

In the face of this biological holocaust, what the scientific community primarily concerned with biodiversity should have been doing has long been clear: taxonomists, ecologists, and evolutionists should have been emulating the success of geneticists by concentrating their efforts on carefully chosen model systems (Ehrlich et al. 1975), the biodiversity equivalents of gut bacteria (*Escherichia*), *Drosophila*, and laboratory mice, and the model systems that have greatly aided progress in other biological disciplines as well (Krogh 1929). But instead of developing their own model systems, population biologists have by and large taken a "shotgun" approach to nature and have paid a high price in lack of progress and of prestige within the scientific community for their folly (Ehrlich 1997). While it is now too late to add much to the existing crude overview of the vast panoply of biological diversity, it is not too late to develop a substantially more detailed and useful understanding of a limited number of taxa. We need comprehensive pictures of the diversity, distribution, and ecological relationships of a suite of organisms that could provide grist for the mills of future evolutionists and ecologists. The way things are going now, in a few centuries some (or even most) of those groups will be primarily the subject of study by paleontologists. Think of how wonderful it would be today if we had records of such comprehensive studies of Cretaceous dinosaurs, birds, and plants—or of the primates of 6–15 million years ago (when the common ancestors of chimps and people were alive)—with sample systems scattered through them in which population dynamics and microevolution had been investigated.

Our research groups working with checkerspot butterflies have attempted to establish a model system that would contribute to a baseline of information equivalent to those developed in laboratory-based disciplines. We are taking a broad-scale approach to sets of butterfly populations that can be relatively easily observed in nature, as opposed, for example, to fruit flies, which are most readily studied in milk-bottle microcosms. Our goal has been to develop a multivalent approach that would make checkerspots one of the insect equivalents of the classic evolutionary–ecological model system, the Geospizinae—Darwin's finches (e.g., Lack 1947, Bowman 1961, Abbott et al. 1977, Grant 1986, 1999, Sato et al. 1999). At another level, birds as a group could be viewed as a model system (they are certainly a "well-studied system" in the sense of Vitousek 2002), and so could butterflies (Boggs et al. 2003).

There is still time to reap great rewards from expanding the number of model systems under investigation in population biology, especially if the sample of systems is taxonomically and geographically stratified. Some of these systems should be drawn from diverse, already well-known, higher taxa to maximize the chances for inducing generalities. But in the face of the disappearance of much of what they study, most professional taxonomists and ecologists are not switching to working on carefully selected model systems (one important exception is provided by plant population biologists working on *Arabidopsis*; e.g., Mitchell-Olds 2001). One does not need to search far for the reasons. The training of professional taxonomists largely produces workers who are taxon-bound, many of whom persist in doing alpha (species description) and beta (simple classificatory "revisions") taxonomic studies of groups they are interested in (or were assigned to by their major professors), or working out hypothetical phylogenies. No sense of urgency is usually transmitted to students or among professionals, so it is not surprising that taxonomic studies are undertaken as if there were many centuries to "complete the job." The training of professional ecologists does not usually emphasize the importance of consistent work with model systems; hence the literature is (like that of systematics) clogged with isolated bits of information on a vast variety of organisms and communities (but see Matson et al. 1987, Vitousek 2002, 2003). And taxonomists and ecologists have not been able to collaborate on even the most basic work, such as all-taxa inventories of a few dozen selected 1-ha sample plots, that would give science a reasonable picture of how ratios of abundance of different kinds of organisms vary geographically and provide a baseline for observing different vulnerabilities of those groups to environmental change (May 1988, Ehrlich 1997, 2002).

Taxonomic groups that might be chosen for "completion" today (i.e., working out the relationships, distribution, and natural history for virtually all species) include most vertebrates, vascular plants, and, of course, butterflies. The true butterflies (Papilionoidea: the families Papilionidae, Pieridae, Nymphalidae, and Lycaenidae) are the most obvious invertebrate taxon from which to select model systems for detailed taxonomic, ecological (including ecophysiological), evolutionary, and behavioral study because of the abundant knowledge of their natural history (assembled in large part by amateurs) and the ease with which they can be identified and marked in the field. Other candidates among terrestrial arthropods are ants, because of their importance in ecosystem functioning, especially in the tropics (e.g., Schneirla 1971, Beattie 1985, Hölldobler and Wilson 1990, Gordon 1999); bees, because of their key roles in pollination (e.g., Michener 1974, 2000, Heinrich 1979); mosquitoes and ticks, because of their public health importance (e.g. Gillett 1972, Smith 1973, Pfadt 1985, Oaks et al. 1991, Sonenshine 1991, Berenbaum 1995a); and tiger beetles and ground beetles, which have many attributes that make them good indicators of biodiversity (e.g., Pearson and Cassola 1992, Niemelä et al. 1993).

But the Papilionoidea are closer to completion than those other candidates; and good field guides are already available for many regions. With modest funding, guides for virtually all areas of the world could be rapidly assembled. The amount of basic distributional information is staggering for the best studied areas, such as the British Isles (e.g., J. Thomas and Webb 1984, Asher et al. 2001; figure 1.1). As this volume shows, data on changes in population structures and dynamics of Papilionoidea, while inadequate, are more extensive than for any other major group of invertebrates. As model sys-

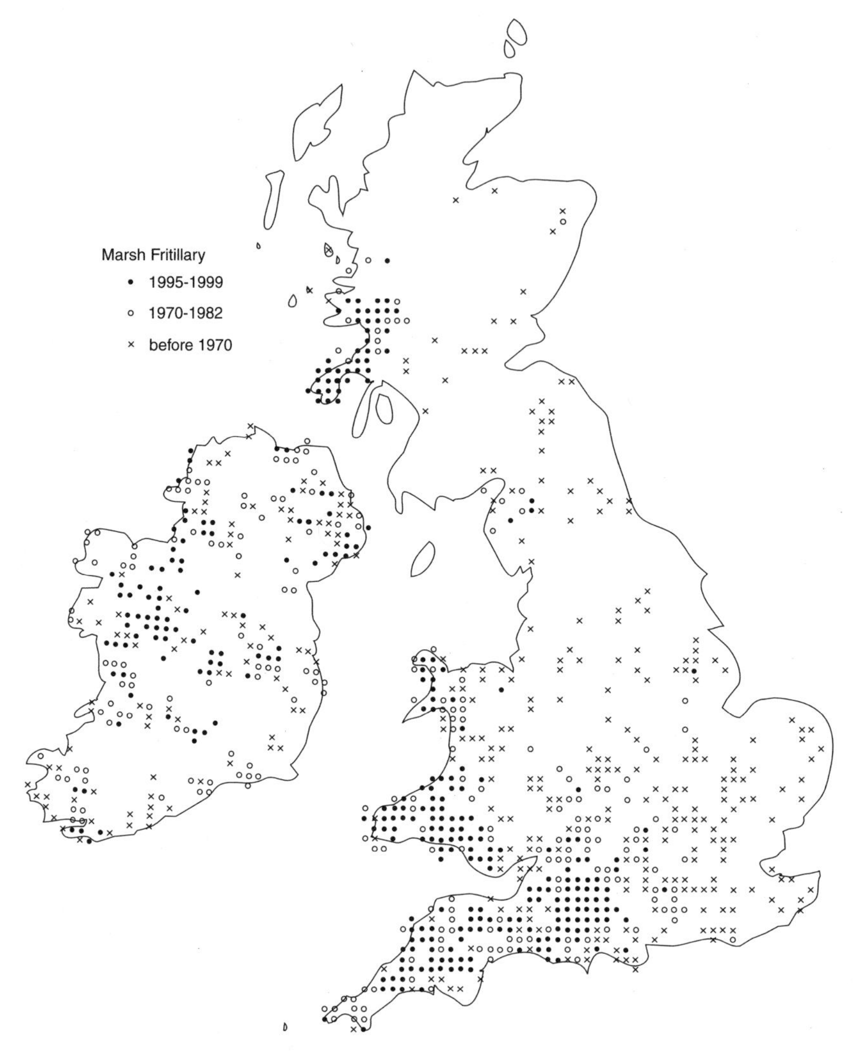

Figure 1.1. Distribution of *Euphydryas aurinia* in the United Kingdom in 10-km^2 squares. The dots represent the distribution in the survey in 1995–99, the open circles represent 1970–82, and the crosses show the pre-1970 distribution (based on Asher et al. 2001).

tems in population biology, butterflies have the advantages of being readily identifiable in the field, easily marked without serious disturbance (Ehrlich and Davidson 1960, Orive and Baughman 1989), showing easily measured phenotypic characteristics on their colorful wings, and often being relatively easy to propagate in the laboratory. Butterflies frequently have well-delimited local populations, which greatly facilitates demographic and genetic field studies. It was for these reasons, as well as for their natural beauty, that our groups focused their attention on checkerspots. We wanted to simultaneously study the dynamics, genetics, and behavior of suites of populations to obtain as comprehensive and integrated view of their biology as possible.

1.2 Checkerspot Populations

In the middle of the last century, the spatial structure of populations was not an important topic of research. At that time Allee and his colleagues (1949, p. 265), in their classic textbook, gave two definitions of populations:

1. The organisms, collectively, inhabiting an area or region.
2. A group of living individuals set in a frame that is limited and defined in respect to both time and space.

Using these sorts of definitions, much discussion and research had been devoted to the issue of whether population sizes were "regulated" and if so, how. The issue gelled long ago around concepts that were eventually labeled "density dependence" and "density independence" (e.g., Howard and Fiske 1911, Nicholson 1933, Smith 1935, Thompson 1939, Elton 1949, Andrewartha and Birch 1954, May 1979). Subsequent research in population biology paid some attention to the spatial limits of populations, as when Dowdeswell and colleagues (1949) recorded the isolation of three small populations of the satyrine butterfly *Maniola jurtina* separated by only a few hundred meters on the small island of Tean in the Scilly Isles. The idea of spatial structure was also inherent in the notion of a "Mendelian population," put forward by Dobzhansky (1950) as "a reproductive community of sexual and cross-fertilizing individuals that share in a common gene pool." Sharing in a common gene pool at least implies spatial proximity. But in dividing the serpentine grassland (one based on soils with an unusual mineral mix because they are derived from serpentine rock) occupied by Edith's checkerspot *Euphydryas editha* on Jasper Ridge on Stanford Campus into areas A to H (figure 1.2), and in deciding to follow the fates of butterflies in those areas over the long term, Ehrlich's research group started a new era in the study of natural animal populations in 1960.

That research has gradually shifted the focus in population ecological studies from broad generalities about regulatory factors based on data gathered with little attention to spatial structure to studies focused on that structure itself. The early results led to the realization that investigation of the detailed structure of local populations was required to understand the dynamics of natural populations. The checkerspot research also led quickly to another conclusion—namely, that "the influence on population size of various components of environment varies with population density, among species, among local populations, and through time" (Ehrlich and Birch 1967). Our later research on checkerspots has abundantly supported that conclusion, suggesting that for many, if not most, species the entire concept of a "species ecology" is questionable. Populations of the same checkerspot species often use different larval host plants (including secondary hosts), have different patterns of adult nectaring, vary in their spatial population structure and responses to environmental perturbations, have different relationships with the parasites that attack them, and so on. In *Euphydryas editha,* differences in the topography of habitats (and thus of topoclimates) can lead to population dynamic differences in local populations situated just a few hundred meters from one another. In short, the studies have demonstrated in practice that population biology should be taken literally: it is the biology of populations rather than of species.

We have been less successful at answering the population-genetic questions that originally interested Ehrlich—for example, whether selection favoring heterozygotes would counter the decay of genetic variability caused by small population sizes. Attempting to understand the evolution of *Euphydryas* populations using quantitative genetic techniques proved disappointing, and although the results of early allozyme studies threw some light on the "neutrality" controversy, they were not promising enough to merit much further effort with the techniques available at that time. Ehrlich and his group therefore shifted their research onto more tractable problems in population ecology. Molecular genetic techniques are evolving rapidly and now allow issues ranging from microevolutionary forces operating within populations to population differentiation and phylogenetic relationships to be addressed with more powerful tools. The checkerspot system is well poised to permit further progress on a wide range of evolutionary issues.

Pioneering work on *Euphydryas editha* brought the issue of population extinction to the fore in population biology and later in conservation biology, where understanding the mechanisms of population extinction is now recognized as a major problem (e.g., Soulé 1987, Gaston 1994, Lawton and May 1995, Meffe et al. 1997, Landweber and Dobson 1999). This was accomplished by long-term comparative studies that showed that local populations frequently went extinct (e.g., Ehrlich et al. 1980, McLaughlin et al. 2002a). Extensive studies of *E. editha* in the San Francisco Bay area have shown how a complex interaction between macroclimate and spatial heterogeneity in habitat patches produces microclimates that control the delicate phenological relationship of developing *E. editha* larvae and senescing larval host plants. This relationship dominates the dynamics of California

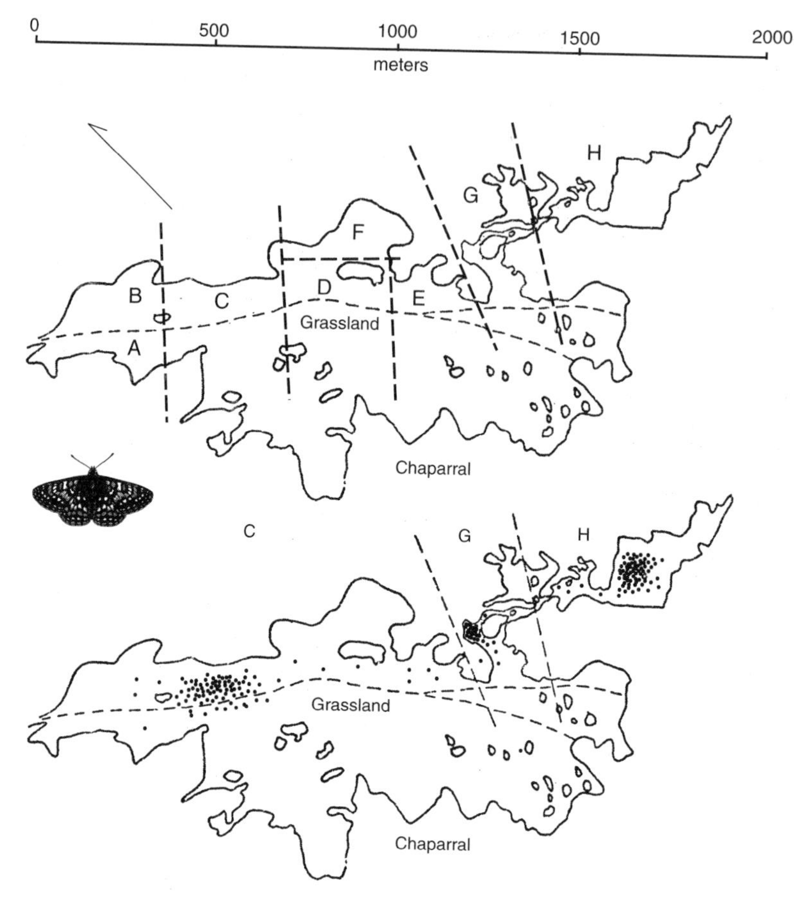

Figure 1.2. Divisions of habitat at Stanford's Jasper Ridge Preserve were designated A–H before Ehrlich ever saw butterflies flying, as shown in the upper figure (after Ehrlich 1961b). Mark–release–recapture studies then showed that what was once thought to be one population of *Euphydryas editha* was actually three distinct populations or "demographic units." The demographic units turned out to be centered in C, G, and H; A–F were later condensed to "C," so that the three units became known as C, G, and H, as shown in the lower figure where each dot represents the position of first capture of an individual in 1960.

checkerspot populations, although endogenous factors and availability of nectar resources also play roles. Unraveling these interactions made it possible to produce predictive models of the dynamics of the populations and to explain the observed time courses to extinction (McLaughlin et al. 2002a).

Discoveries of the key interactions between weather and topography influencing the extinction of local populations and the existence of "reservoir" and "satellite" populations, began to move studies on *Euphydryas editha* in California more into the realm of metapopulation biology (Ehrlich and Murphy 1987a, Harrison et al. 1988). This research laid some of the empirical groundwork for the rapid development of metapopulation studies that was being ignited about the same time, including the research by Hanski (1999b) investigating *Melitaea cinxia* in Finland.

1.3 From Populations to Metapopulations and Beyond

Isolated populations are convenient subjects for field studies, and such populations correspond to what much of the theory in population biology has traditionally addressed. In these situations, the mechanisms hypothesized to influence the dynamics of populations and their evolution can be efficiently dissected either by statistical analysis of observational data or by experiments. Unfortunately, natural populations do not often come in such neatly delineated packages. Rather, individual populations under study are likely to be coupled via migration and gene flow with other populations. Ignoring this coupling can lead to grossly misleading conclusions. Thus a population that appears relatively stable may in fact be a sink population, whose persistence is entirely dependent on migration from other (source) populations (Pulliam 1988, 1996). Theoretical predictions about the influence of migration on local dynamics have challenged our intuition and ability to conduct relevant experiments. For instance, models predict that complex dynamics in local populations in the absence of migration may become transformed into simpler dynamics in coupled populations (Gyllenberg et al. 1993, Ruxton et al. 1997) or, conversely, migration may induce cyclic or even chaotic dynamics where isolated populations would exhibit numerical stability (Ruxton 1993, Rohani et al. 1996). A population may not be locally well adapted if gene flow from other populations living under different environmental conditions is substantial (Haldane 1931, Barton 1987, Dias and Blondel 1996, Barton and Whitlock 1997). On the other hand, gene flow may enhance local adaptation by providing fresh genetic variation. Sets of coupled populations may be adapted to conditions at multiple localities in heterogeneous landscapes, in which case it would be misleading to focus on what is happening in just one local population.

Our checkerspot research has evolved to cover many of the above issues of metapopulation biology, but the starting point in the work on *Melitaea cinxia* by Hanski's research group was the goal to understand the rules of persistence of species living in highly fragmented landscapes. The underlying concept has been Richard Levins's (1969) vision of a metapopulation as a "population" of local populations linked by migration. *Melitaea cinxia* in the Åland Islands in Finland closely agrees with Levins's vision. A major ongoing challenge is to understand in quantitative terms how habitat fragmentation influences metapopulation persistence. And not just habitat fragmentation in an abstract (spatially implicit) sense, as in the original Levins model, but taking into account the actual spatial configuration of the habitat, whether the habitat patches are clustered in space or are more dispersed, how much variation there is in patch areas, and so forth. These questions are closely related to a major concern in conservation: how should limited resources be used to ameliorate the effects of habitat loss and fragmentation on endangered species? *Melitaea cinxia*, like scores of butterflies and other organisms, has become endangered due to changes in land use, pushing questions about persistence of species living in fragmented landscapes high on the agenda of research councils and funding agencies.

There is a vast amount of knowledge of butterfly natural history that greatly facilitates any kind of population biological study. But butterflies also possess traits making it likely that they exhibit classic metapopulation structures and dynamics in many landscapes. Butterflies have small body sizes, high rates of population increase, short generation times, and specific resource requirements (Murphy et al. 1990). Many habitat types used by butterflies are naturally fragmented or fragmented by humans, forcing species dependent upon them to have similarly fragmented populations. Small body size means that the number of individuals in even small habitat fragments may be large enough to constitute a local breeding population, especially because small species tend to have lower migration rates than large ones. A high population growth rate implies that, following population establishment, local populations grow quickly to the local carrying capacity, unless they go extinct. Short generation time means, among other things, that stochastic events are not buffered by great longevity of individuals, which would reduce the risk of population extinction. Finally, habitat specialists with short generation times are likely to be greatly affected by interactions between large-scale weather perturbations and local environmental conditions such as variation in slope exposures and moisture conditions. Spatial variation in local environmental conditions decreases the level of spatial synchrony in population dynamics due to large-scale weather effects, and spatial variation thereby facilitates metapopulation persistence. In summary, we are not in favor of rigid classifications of different types of (meta)population structures (figure 1.3), but rather emphasize the relevant processes (Hanski 1999b). Whichever way one sees it, butterflies present a great model system for investigating issues that are fundamental to metapopulation biology.

Figure 1.3. Rampal Etienne's (2002) perspective on the problem of deciding whether a particular species has a "metapopulation structure" or not. Drawing by Mathijs Doets.

Conducting field studies at large spatial scales—indeed, just estimating the magnitude of migration and gene flow, and spatial and temporal variation in them—pose daunting challenges for empirical research (e.g., Ehrlich 1965). Nonetheless, research on *Melitaea cinxia* has demonstrated that working even with hundreds of local populations is sometimes feasible. One immediate advantage that comes with a study simultaneously concerned with many small populations is the opportunity to accumulate much data on the key metapopulation processes of extinction and recolonization. The *Melitaea cinxia* project in the Åland Islands has so far recorded on the order of 1000 population extinctions and establishments. As is reported in detail in this volume, the actual mechanisms and causes of local extinction are many, including deterioration of habitat quality either due to natural (successional) or human-generated processes, demographic and environmental stochasticities, natural enemies, and inbreeding depression. Given the critical role that population extinction plays in ongoing environmental deterioration and in the loss of biodiversity and ecosystem functioning, it is helpful to have well-understood model systems that allow the study of recurrent extinctions. Another advantage that accrues from studying systems with high turnover rate is the opportunity to contrast populations of known but different histories. For instance, a comparison of the migration behavior of individuals originating from newly established versus old populations showed that females from new populations are significantly more dispersive than females from old populations (Hanski et al. 2002). This finding strongly suggests that recurrent colonizations select for increased migration rate, in agreement with modeling results for the same system (Heino and Hanski 2001). Had nature not sorted out for us the offspring of colonists versus residents to study and compare, it would have been much more difficult to investigate the significance of individual variation in traits that might influence migration rate and other life-history characteristics and their evolution.

Our focus in this volume is on empirical findings and their conservation implications, but it is noteworthy that the long-term research on *Melitaea cinxia* has also been instrumental in the development of the spatially realistic metapopulation theory (Hanski 2001). This new approach to metapopulation theory synthesizes the classic metapopulation theory of Levins (1969, 1970) and the dynamic theory of island biogeography of MacArthur and Wilson (1963, 1967). The general setting is the same as in Levins's classic metapopulation model, but the spatially realistic theory takes explicitly into account the effects of landscape structure on metapopulation processes, an element shared with the MacArthur-Wilson island theory. The theoretical advances are reported in a series of recent publications, including Hanski (1992, 1994b, 1999b) for the foundations, Ovaskainen and Hanski (2001, 2002) and Ovaskainen (2002a) for rigorous mathematical theory, and Hanski (2001) and Hanski and Ovaskainen (2003) for up-to-date reviews.

Spatially realistic metapopulation theory also brings the fields of metapopulation ecology and landscape ecology closer together. Merging these two fields has been a goal for a long time (Hanski and Gilpin 1991), but in practice not much has yet happened. The two disciplines continue to largely adhere to their own research traditions and literatures, even to the extent that shared key concepts such as landscape connectivity are used in a different manner (Tischendorf and Fahrig 2000, Moilanen and Hanski 2001). Spatially realistic metapopulation theory adds the structure of the landscape to metapopulation models, an element lacking in previous theory and the element that particularly interests landscape ecologists. Admittedly, even this model is a simplification both from the perspective of metapopulation biology and landscape ecology. For instance, the spatially realistic metapopulation theory retains the presence/absence description of local populations, a major shortcoming for many metapopulation biologists, and it takes no explicit account of the structure of the intervening (matrix) habitat, a major shortcoming for landscape ecologists. Further theoretical development is necessary in many directions. The emerging field of "countryside biogeography," conceived in the intellectual home of the research on *Euphydryas editha*, represents one promising new approach (Daily et al. 2001, Ricketts et al. 2001, Hughes et al. 2002). For conservation purposes, the development of methods to assess rapidly how entire communities of species, as opposed to just single species, are distributed across spatially heterogeneous and increasingly disturbed landscapes and to determine how those landscapes can be modified to make them more hospitable to biodiversity is essential. Our research on the natural enemy complexes associated with checkerspot metapopulations makes a contribution here, but many theoretical and empirical challenges remain to be answered.

1.4 Long-term Investigations in Population Biology

Focusing on particular model systems involves commitment to long-term studies, but there are major disincentives to carrying out such studies in population biology. First, long-term studies are considered too risky. Graduate students do not ordinarily pursue them because they hope to complete their doctoral degrees in less than a decade. Junior faculty members want numerous and quick publications to satisfy criteria for promotion to tenure and to ensure renewal of grants given in cycles of a few years. Second, the prospects that research sites will remain undisturbed tend to be dim, especially in the all-important tropics. One of us (PRE) has already had long-term research sites destroyed or made inaccessible in California, Trinidad, and Tanzania. In California, a key population of *Euphydryas editha* near Jasper Ridge was bisected and greatly damaged by freeway construction and another utterly destroyed by a combination of that construction and the building of a housing development (figure 1.4). In Trinidad, a population study of the long-wing butterfly, *Heliconius ethilla* (Ehrlich and Gilbert 1973), was ended when squatters burned the forest site. At Gombe Stream in Tanzania the kidnapping of several Stanford students by Zairean rebels ended an intended long-term study of the dynamics of a butterfly mimicry complex when it had just begun. And even without outright destruction of study sites, things tend to change in the course of time, in response to either local or global environmental changes, which is an inevitable complication, as well as one of the opportunities, with long-term studies. Finally, long-term studies may turn, like anything else human beings do, to repeating what has already been done beyond the point where real gains continue to accrue.

One of the earliest long-term studies of population size through time was carried out by a scientist once described by Niko Tinbergen as "an American housewife [who] was the greatest scholar of them all"—Margaret Morse Nice. Nice (1937, 1943) studied a population of song sparrows (*Melospiza melodia*), mapping the territories of males on a section of the flood plain of the Olentangy River in Columbus, Ohio, between 1930 and 1935. The male population fluctuated in size from a high of 44 to a low of 17. At the highest sparrow density, most of the available habitat was occupied and the territories were compressed in size, apparently close to the limit that can provide adequate resources.

Nice's song sparrows and many other bird studies showed changes in population size that were small relative to those demonstrated by a population of the Palaearctic checkerspot *Euphydryas aurinia* in Cumberland County in western England (figure 1.5). That population was tracked for more than 50 years, and reports of its fluctuations in classic work by E. B. Ford, a pioneer in ecological genetics, led Ehrlich to decide to begin his work on *E. editha* at Jasper Ridge. Adults of *E. aurinia* were abundant when records were first kept in 1881, and the population exploded in the mid-1890s, when larvae were reported as swarming, pupae were seen all over the place, and adults occurred in "countless numbers," or "in clouds" (Ford and Ford 1930). In peak years, "larvae could be taken by gallons" (Wilkinson 1907). The host plant, *Succisa pratensis* (Dipsicaceae), suffered great defoliation, and caterpillars were recorded as switching to feeding on *Lonicera* (Caprifoliaceae, a closely related, chemically similar family), which presaged the extensive

Figure 1.4. Destruction of the Woodside Colony, one of the "stepping stones" important to the maintenance of the Jasper Ridge populations of *Euphydryas editha*. Photo by Paul R. Ehrlich.

studies of food-plant relationships of checkerspots that would be ongoing a century later (chapter 6). The outbreak years were followed by a decline and two decades of small population sizes, possibly due to heavy parasitism from braconid wasps. In some years only a few adults were found, and in 1920 a "diligent search" uncovered only 16 caterpillars. Then there was another explosion between 1920 and 1924, and stabilization at high numbers until observations ceased in 1935 (Ford and Ford 1930, Ford 1975). This long-term record raised intriguing questions about population dynamics in checkerspots, their interaction with natural enemies, and their population genetics (chapter 10). Unfortunately, since Ford's time, *E. aurinia* has declined dramatically in the United Kingdom (figure 1.1),

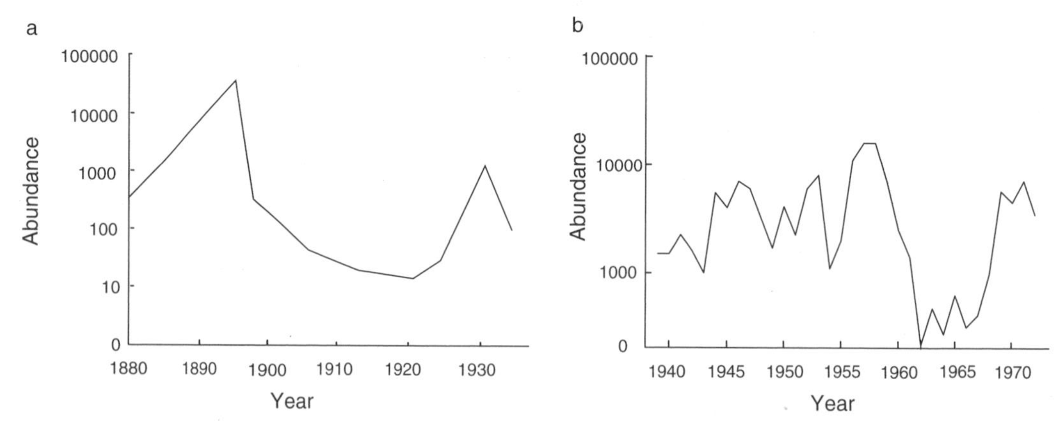

Figure 1.5. Records of changing population sizes in *Euphydryas aurinia* (a) and *Panaxia dominula* (b) in England (from Ford 1975).

and it now belongs to the large set of habitat specialist species that are threatened by habitat loss and fragmentation (Asher et al. 2001).

The early research of British evolutionist Phillip Sheppard also helped set the stage for the development of the checkerspot system by demonstrating that the population biology of *Euphydryas aurinia* was similar in key aspects to that of another lepidopteran. Sheppard studied a population of the day-flying scarlet tiger moth *Panaxia dominula* at Cothill in Berkshire, England, which fluctuated in abundance some 75-fold between 1941 and 1972 (e.g., Sheppard 1956, Ford 1975; figure 1.5). This population occupied a marsh of approximately 6 ha and had no significant exchange of individuals with other populations. The moths are subject to high mortality as late-instar larvae and pupae, partly from cuckoos and possibly other birds and shrews, but primarily from an unknown cause, possibly disease. Females lay about 200 eggs, and the population thus has a much greater potential for rapid increase than those of birds. The fluctuations in the size of the Cothill colony were largely independent of those at the Sheepstead Hurst colony 2 km away. The two are clearly separate demographic units, leading to the conclusion that their dynamics were "probably controlled to a considerable extent by local changes in the environment" (Ford 1975).

The stability of most bird populations can be attributed to their territoriality, while the checkerspots and *Panaxia* are not similarly territorial, though some male butterflies defend small areas where they wait for receptive females to arrive or fly by. More fundamentally, endothermic birds and mammals have relatively stable populations and are often limited by space. This is because individual birds and mammals are generally long lived, have relatively large body sizes, and their population sizes are thus less influenced by various environmental fluctuations that constantly affect those of butterflies, moths and other invertebrates. The overall outcome of these differences is that terrestrial invertebrate populations in general vary in size much more than those of vertebrates (Hanski 1990). One major exception to this rule is cyclic population dynamics seen in lynx and hares in the subarctic (Krebs and Myers 1974, Ford 1975) and in voles and lemmings in northern Europe and elsewhere at high latitudes (Elton 1924, Hansson and Henttonen 1985, Hanski et al. 1991, 2001, Hanski and Henttonen 2002). It is clear that there are fundamental population dynamic and other biological differences between most invertebrates and most vertebrates; hence research on a butterfly model system will not necessarily enhance the understanding of all aspects of a vertebrate population.

Work on checkerspots has helped answer the seemingly simple question of the relationship between population size in generation $t + 1$ to population size in generation t, which has been the subject of great controversy in population ecology. Andrewartha and Birch (1954) debated the issue vigorously with A. J. Nicholson (1933, Nicholson and Bailey 1935). Andrewartha and Birch emphasized the role of weather conditions in influencing population fluctuations, and they tended to deemphasize density-dependent population regulation, while Nicholson was the chief protagonist of the view that populations are strongly regulated by density-dependent processes (e.g., Elton 1949). Though to a large extent the heat of the argument has now dissipated and a consensus has been emerging (Sinclair 1989, Hanski 1990, Turchin 1995, 2002), many important questions about population regulation still remain. We return to these issues later in chapter 3.

As noted earlier, much of the research in population dynamics suffers from the deficiency that the demographic units are not properly delimited. Research has not been focused on populations that are either so isolated from others that their dynamics could not be significantly influenced by migration or on populations in which the role of migration could be rigorously assessed. Classic long-term studies of insect populations typically share this deficit—for instance, Schwerdtfeger's (1941) study of forest insects and the otherwise fabulous Rothamsted survey of moths and aphids in the United Kingdom (e.g., Taylor 1974, 1986). This is not to say that long-term studies would be worthless unless focused on species with clear-cut population structure or unless based on methods that allow resolution at the population level. For instance, analysis of the data collected in the Rothamsted insect survey has produced unequivocal evidence for populations of moths and aphids being regulated by density-dependent processes (Woiwood and Hanski 1992). However, analysis of overall density dependence in poorly defined samples typically leaves open questions about the actual mechanisms that generate the density-dependent feedback in population dynamics and the spatial scale at which these processes operate.

The failure to delimit populations can have practical consequences. For example, thorough knowledge of the number of populations ("stocks") of the Peruvian anchoveta could have led to harvesting strategies that might have prevented the catastrophic crash of the fishery in the 1970s (e.g., Boerema and Bulland 1973, Idyll 1973). Focus on the spatial structure of populations and metapopulations has been one of the unifying themes in our long-term research on checkerspots, as evidenced by most of the chapters in this volume. Different population dynamic patterns are observed at different spatial scales, reflecting the operation of diverse mixes of population dynamic processes (figure 1.6). Thorough understanding of population dynamics is not possible without attention to all essential processes operating at different spatial scales.

We hope that checkerspots will contribute increasingly to our understanding of the mechanisms of evolution in sexual organisms. Work on Lepidoptera has already shown the way. Philip Sheppard, A. J. Cain, E. B. Ford, and their British colleagues pioneered work on evolution with their studies on natural populations of butterflies and moths. Sheppard's classic research on *Panaxia* traced the frequency of the "*medionigra*" gene in the population at Cothill between 1939 and 1972 (summarized in Ford 1975). This work led to a substantial disagreement between Ford and Sewall Wright (Fisher and Ford 1947, Wright 1948) on the roles of selection and drift, a debate that echoes to this day in the form of the slowly dying "neutrality controversy" (e.g., Lewontin and Hubby 1966, Kimura 1968, Kimura and Ohta 1971b, Lewontin and Krakauer 1973, Lewontin 1974, Gillespie 1991, Koskinen et al. 2002). Meanwhile, the fine research of Ward Watt and his collaborators at Stanford has shown the utility of *Colias* butterflies as a model system for understanding the operation of selection at the molecular and physiological levels.

1.5 Checkerspots and Conservation

In 1960, when Ehrlich caught his first Bay checkerspot on Jasper Ridge, the issue of the preservation of biodiversity, humanity's "natural capital," was not even on the radar screen of the scientific community, let alone that of the general public. Rachel Carson's book *Silent Spring* (1962), which launched public concern about the environment in the United States, was still two years in the future. Ehrlich's conservation concerns were focused on two issues, both rooted in his boyhood experiences while collecting butterflies in the eastern United States. The first was development, as he had seen many of his favorite collecting sites disappear under suburban sprawl. The second was overuse of pesticides. So much DDT had been sprayed in New Jersey that it was difficult to find food plants that were safe to feed caterpillars that he was attempting to rear. That concern had been heightened by his graduate assistantship under Robert Sokal at the University of Kansas, studying the evolution of DDT resistance in fruit flies. But Jasper Ridge was not yet threatened by development in 1960, and there was no pesticide use at the site. That, and the seemingly large size of Jasper Ridge as a home for a small insect, meant that the conservation status of the population there was the furthest thing from Ehrlich's mind when he used a felt-tipped pen (Ehrlich and Davidson 1960) to mark *E. editha* number one.

But as the ephemeral status of the local populations at Jasper Ridge and elsewhere became clear

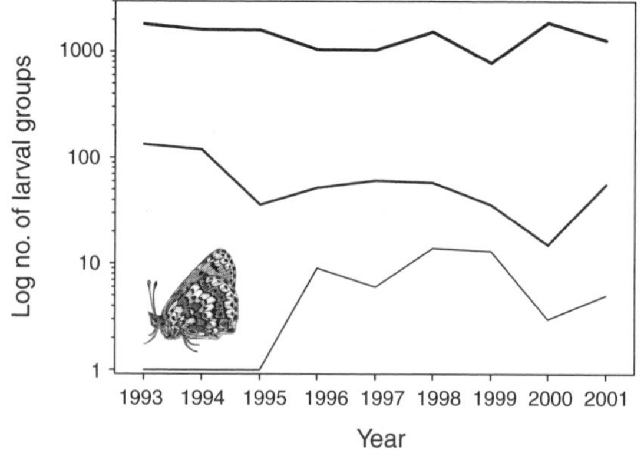

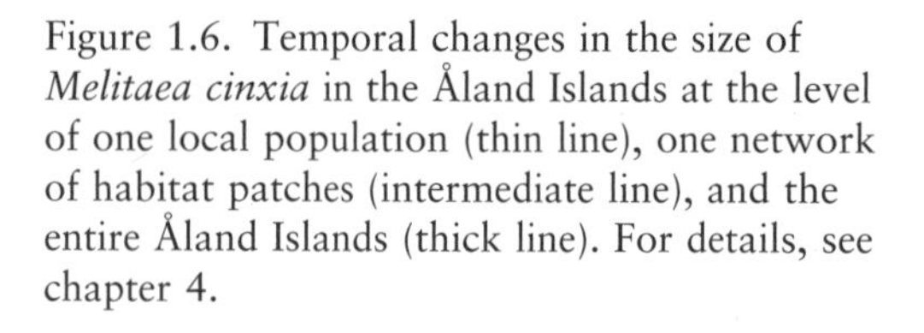
Figure 1.6. Temporal changes in the size of *Melitaea cinxia* in the Åland Islands at the level of one local population (thin line), one network of habitat patches (intermediate line), and the entire Åland Islands (thick line). For details, see chapter 4.

(Ehrlich et al. 1980), and the global plight of biodiversity was elucidated (Myers 1979, Ehrlich and Ehrlich 1981), the focus of *Euphydryas* research turned increasingly to issues related to conservation. Documented extinctions engendered an interest in the mechanisms and significance of extinctions that had previously been little pursued by population biologists. The connection of climate to the dynamics of *E. editha* and *E. chalcedona* populations (Ehrlich et al. 1980, McLaughlin et al. 2002a) laid the groundwork for examining the impact of climate change using butterflies as a model system (Parmesan 1996, Parmesan et al. 1999).

In 1991, at the time when Hanski started his research on *Melitaea cinxia* in the Åland Islands, conservation biologists were already turning to metapopulation ecology. Merriam (1991) went as far as claiming that "metapopulation models have largely replaced equilibrium island biogeography as a way of thinking about terrestrial habitat islands, fragmented habitats, and heterogeneous terrestrial environments in general." Hanski and Simberloff (1997) examined in some detail the apparent "paradigm shift" from the island theory to metapopulation theories. In hindsight, it is curious that the "shift" was perceived to be so fundamental. With the perspective of the spatially realistic metapopulation theory, which unites island theory and classic metapopulation theory (section 1.3), we now realize that no fundamental change occurred at all. Instead, we have learned to apply the core ideas in the two previous theories in the same framework and have learned to apply them to ordinary fragmented landscapes rather than to sets of oceanic islands (Hanski and Simberloff 1997). It is noteworthy that whereas birds provided the bulk of early examples for the island theory, butterflies have gained a somewhat similar status in current metapopulation studies, especially in Europe (C. D. Thomas 1994b, Hanski and Kuussaari 1995, C. D. Thomas and Hanski 1997, Hanski 2002b). Further research is needed to see how far results from birds and butterflies can be extrapolated to less well-studied organisms.

Since the early 1990s, metapopulation ideas have appeared with increasing frequency not only in the conservation biology literature but also in management plans and discussions. A telling example is how the *Melitaea cinxia* project has been featured in the context of national debate on forest conservation in Finland. Only about 1% of the natural forests remain in southern Finland, and what remains occurs in small fragments. Metapopulation ideas have naturally been invoked in the analyses and discussions of the persistence of species that are adapted to natural forests and are currently threatened by extinction. Applying the conceptual understanding of metapopulation dynamics obtained in the *Melitaea cinxia* project to forest species living in highly fragmented landscapes has been helpful but has also been challenged by those who oppose conservation (Hanski 2002a). The *Melitaea cinxia* model system has attained a role in conservation that could not have been anticipated in 1991.

1.6 About This Book

The structure of this volume reflects our conviction that an effective model system in population biology has to be studied at sufficient depth and breadth to provide the biological context in which particular research questions can be meaningfully addressed. For instance, knowledge of the population structure and life history of *Euphydryas editha* at Jasper Ridge allowed an in-depth analysis of the gradual extinction process that has been documented for those populations over the past decades and can be related to the global changes currently affecting most populations on earth (McLaughlin et al. 2002b, McLaughlin et al. 2002a). The extensive knowledge of the causes of population extinction in *Melitaea cinxia* has allowed a more balanced assessment of the role of inbreeding depression in the dynamics of small local populations than would have been otherwise possible (Saccheri et al. 1998, Haikola et al. 2001, Nieminen et al. 2001). The long-term records that we have assembled for local populations and metapopulations gives an opportunity to document not only the occurrence of particular phenomena or processes, but also to evaluate the conditions for their occurrence in space and time. Thus the interaction between *M. cinxia* and one of its specialist parasitoids, the braconid wasp *Cotesia melitaearum*, was initially thought to drive much of the spatial variation in host density, as has been predicted by theoretical models (Hassell et al. 1991, 1994, Nee et al. 1997). We now know, however, that the coupling of the host and the parasitoid is dynamically strong only in certain parts of the study area and has been strong in only some periods of time. Once again, a short-term study in one small area might have led to misleading conclusions.

We start by describing in chapter 2 our two focal checkerspot species in a broader ecological and evolutionary context: what sort of butterflies are they, how many related species are there, what is their taxonomic position and evolutionary history, where do they occur, what is peculiar in their biology, how does research on checkerspots relate to general issues in ecology and evolution? Chapters 3 and 4 focus on the population ecology—population structures and dynamics—of the two species and the populations on which most of the research has been conducted; our research on them was originally initiated because of the population structure of each species. This volume presents not only an overview of the past research but also provides us an opportunity, for the first time, to relate the two complementary research projects to each other. Chapters 5–7 deal with the reproductive biology of adult checkerspot butterflies and with their larval biology. These are key elements in all holometabolous insects, and reproductive biology is a key to the population biology of all sexually reproducing organisms; research on checkerspots and other butterflies has contributed many fundamental results and insights. Decisions made by ovipositing females relate to another critical stage in the life of oligophagous insects. Checkerspots provide some of the best studied examples of local host plant specialization, including demonstrations of rapid evolution in host plant preference. Checkerspot females typically lay eggs in clutches, and the larvae stay in groups, typically consisting of full sibs, for a varying length of time; some like *M. cinxia* stay in groups until the end of larval development. Group-living poses another series of ecological and evolutionary questions and puzzles.

Chapter 8 discusses the interactions between the host butterflies and their parasitoids and other natural enemies. Here is a good demonstration of how field-based research on model species naturally and inevitably expands to cover interactions among several species and how such more extended studies may greatly benefit of the model system context. The small community consisting of *M. cinxia,* its two host plants, its two specific larval parasitoids and their hyperparasitoids, has been studied in the naturally highly fragmented landscape in the Åland Islands, and it represents one of the best studied small metacommunities. Chapter 9, on dispersal behavior, is placed after the chapters on population ecology, as patterns of dispersal are hardly possible to understand without knowing the spatial population structure of the focal species in sufficient detail. It turns out that interspecific interactions are critical here also, as the oviposition host-plant preference of females significantly influences migration behavior in heterogeneous fragmented landscapes. Our description of the population biology of checkerspot butterflies, especially as represented by the most thoroughly investigated populations and metapopulations of *E. editha* and *M. cinxia*, is completed by chapter 10 on population genetics, a topic on which much remains to be done.

The remaining four chapters expand the contents of the book to broader issues and to wider implications of our research. Chapter 11 asks to what extent our results are representative of other populations of *E. editha* and *M. cinxia*, of other checkerspot species, of other butterflies, and ultimately of other organisms in general. Chapter 12 examines in greater detail the concept of "model system;" that is what we designed the two checkerspot systems to be, and that concept has been in our minds throughout our studies. What is a model system in population biology, and how successful have the checkerspot systems been as model systems? This chapter takes the view that a successful model system is not only a vehicle for testing existing models and ideas, but it also facilitates further development of the conceptual framework within which population biological research is being conducted. Chapter 13 turns to conservation biological implications of the research results. Like so many other population biologists, we feel that it is our responsibility to bring our contribution to both the scientific and public debates on conservation. Chapter 14 looks back to the previous chapters and to the many years devoted to checkerspot studies and asks what have we learned and what more is there to be learned.

Our two checkerspot projects were started because the prospective study systems were thought to provide prime research opportunities for two purposes. Ehrlich selected *Euphydryas editha* in the autumn of 1959 for research on an "ordinary" species, to contrast with the then-prevailing tendency to investigate the population biology of species that are either constantly abundant or reach high densities during dramatic population oscillations. Thirty years later, in 1991, Hanski was searching for a species that would not only have well-delimited small local populations, but would have many of them located sufficiently close to each other to form a metapopulation. In hindsight, we take much pleasure in observing how well the studies of the two model systems complement one another.

2

Introducing Checkerspots

Taxonomy and Ecology

DENNIS D. MURPHY, NIKLAS WAHLBERG, ILKKA HANSKI, AND PAUL R. EHRLICH

2.1 Introduction

In this chapter we set the stage for the rest of the book. First we examine checkerspots as a group, reviewing what we can infer of their evolutionary past and what we know of their current distribution, characteristic biology, and how they illuminate general issues in ecology, evolution, and conservation biology. Then we introduce our central characters, Edith's checkerspot (*Euphydryas editha*) and the Glanville fritillary (*Melitaea cinxia*). We concentrate on the aspects of their ecology, especially their relationships to their larval host plants and their spatial population structure, that have been pivotal to their use as model systems. We examine the range of ecological variation exhibited by our focal species and also by their closest relatives.

Checkerspots are representatives of one superfamily, the Papilionoidea, of the scaly-winged insects the Lepidoptera. The Papilionoidea contains four major families: Nymphalidae, Lycaenidae, Pieridae, and Papilionidae (Ehrlich 1958a, Kristensen 1976, Scoble 1992). The Lepidoptera, which also include moths and skippers, are morphologically the most uniform of the great insect orders and are relatively consistent in their resource use and physiology as well. Almost all adult lepidopteras feed on nectar, honeydew, or other liquids (some adult moths do not feed at all), and the larvae of the majority of species eat plant foliage, flowers, or fruits. The wing scales, which bear the pigments and structural colors that often make butterflies so attractive, separate easily from the wings. Scales probably evolved to increase insulation and to help generate lift in flight (Scoble 1992). In addition, because scales make adult butterflies slippery, they presumably aid butterflies in eluding predators such as lizards and spiders.

The Papilionoidea, characterized by clubbed but not recurved antennae and a special overlapping mechanism to coordinate the beating of the fore- and hindwings (Ehrlich 1958a, 1958b), are widely distributed on all continents save Antarctica and on most oceanic islands. They fly from tropical lowland forests and barren deserts to alpine and arctic tundra. The superfamily comprises about 15,000 species, most of which are diurnal. Their closest relatives, which often are also called butterflies, are the largely diurnal skippers (Hesperioidea), a superfamily with more than 3000 species. Skippers are usually placed in a single family, Hesperiidae, and they are almost as widely distributed as the Papilionoidea. The most recently described superfamily of butterflies is the Hedyloidea, a nocturnal taxon classified as part of a monophyletic group with the Papilionoidea and Hesperioidea (Scoble 1986). Hedyloidea consist of a single family, Hedylidae, with 35 species in tropical America. The remaining 100,000 or so described species of the Lepidoptera, collectively called moths, have been divided in a recent comprehensive work into 38 superfamilies and 101 families (Scoble 1992). The Papilionoidea thus represent just a small fraction of the species diversity, as well as the morphological, ecological, and behavioral diversity of the Lepidoptera. This fraction is almost certain to shrink because most of the Papilionoidea have already been scientifically described, but many more species of moths remain unknown. Butterflies are the largest lepidopteran superfamily to have penetrated the diurnal niche. Skippers and a number of species and larger taxa of moths are also diurnal.

Euphydryas editha and *Melitaea cinxia* are closely related members of the family Nymphalidae, the "four-footed" butterflies. Nymphalids are united by a suite of morphological characteristics (Ehrlich 1958a), but the taxonomic structure within the family remains unresolved (DeVries et al. 1985, Harvey 1991, Martin and Pashley 1992, Brower 2000), likely in part due to the evolutionary success of the family. The Nymphalidae and the Lycaenidae (blues, hairstreaks, and metalmarks) are the two most species-rich butterfly families, each comprising about 6,000 species (Ackery 1984). Neither seem to have experienced the extinctions of intermediate (parental) taxa that make taxonomic subdivision relatively easy in other groups. The most recent classification (Harvey 1991) of Nymphalidae places 494 species in the Nymphalinae, the subfamily that contains the checkerspot tribe Melitaeini, as well as the tribes Nymphalini, Coeini, and Kallimini (table 2.1).

The nymphalids are characterized by greatly reduced front (prothoracic) legs and (with a single exception) by their lack of the full-sized tarsal claws found on all three pairs of legs in most insects. Instead, the forelegs of nymphalids have been converted into organs that, in the female, carry the sensory apparatus used in selection of appropriate plants on which to deposit eggs (Calvert 1974). Exactly why checkerspots and many other groups of butterflies have, through evolution, lost the stable six-legged gait characteristic of insects is not well understood. It has been hypothesized that the smaller forelegs in the Nymphalidae protect the delicate tarsal receptors from the wear of walking (Myers 1969). However, in the Nymphalidae and in the other butterfly groups that show prothoracic leg reduction, the Lycaenidae and Hedyloidea, the forelegs of males are more reduced than the forelegs of the females, even though females have more use for the receptors (e.g., Ehrlich 1958b). In addition, in the lycaenid subfamily Lycaeninae, the somewhat reduced forelegs are still used in walking (Ehrlich 1958a, Scoble 1986, Robbins 1988).

2.2 Terminology and Systematics

The term "checkerspot" refers to the repeated pattern elements that are characteristic of the colored marking of the wings of many nymphalid butterflies (Nijhout 1991). "Fritillary" derives from the same root as that of the plant genus *Fritillaria*, in the lily family, which has checkered markings on the petals of its drooping flowers. The common names, as is often the case, tend to confuse communication (Murphy and Ehrlich 1983). In Europe, "fritillary" is widely applied to butterflies of the subfamilies Nymphalinae and Heliconiinae with checkered markings; in North America, "fritillary" is generally reserved for members assigned to genera called variously *Argynnis* (including *Speyeria, Mesoacidalia*, and *Fabriciana*), *Boloria*, *Agraulis, Dione,* and *Dryas* (all Heliconiinae). In this book our primary focus will be on a subset of genera in the tribe Melitaeini of the subfamily Nymphalinae (table 2.2), for which we will use the term "checkerspots." Otherwise, we largely will dispense with common names. A sample of checkerspots and related nymphalid butterflies is shown in plates I–IV.

Throughout our work, we have had to make decisions about what latinized nomenclature to use. We follow the rule of obligatory categories and treat conservatively ("lump") the ranks in the hierarchic classification that must be specified for every animal (species, genus, family, as opposed to subgenus, superfamily, etc.; Ehrlich 1983). Lumping at those key ranks provides an antidote to the substantial oversplitting of butterfly taxa that has resulted, among other things, in a multiplication of genera (Ehrlich and Murphy 1981a), including genera in the Melitaeini (e.g., Higgins 1978). Use of broadly

Table 2.1. The classification of the subfamily Nymphalinae.

Family	Subfamily	Tribe	No. of Species
Nymphalidae	Nymphalinae	Nymphalini	91
		Coeini	12
		Kallimini	113
		Melitaeini	278

Table 2.2. The current generic classification of the tribe Melitaeini.

Tribe	Subtribe	Genus	Synonyms
Melitaeini Tutt, 1896	Euphydryina Higgins, 1978	*Euphydryas* Scudder, 1872	*Occidryas* Higgins, 1978
			Eurodryas Higgins, 1978
			Hypodryas Higgins, 1978
	Melitaeina Tutt, 1896	*Melitaea* Fabricius, 1807	*Lucina* Rafinesque, 1815
			Schoenis Hübner, 1818
			Cinclidia Hübner, 1818
			Mellicta Billberg, 1820
			Melinaea Sodoffsky, 1837
			Didymaeformia Verity, 1950
			Athaliaeformia Verity, 1950
		Chlosyne Butler, 1870	*Morpheis* Geyer, 1833
			Anemeca Kirby, 1871
			Coatlantona Kirby, 1871
			Limnaecia Scudder, 1872
			Charidryas Scudder, 1872
			Thessalia Scudder, 1875
		Microtia Bates, 1864	*Texola* Higgins, 1959
			Dymasia Higgins, 1960
		Antillea Higgins, 1959	
		Poladryas Bauer, 1961	
		Higginsius Hemming, 1964	*Fulvia* Higgins, 1959
		Atlantea Higgins, 1959	
		Gnathotriche Felder & Felder, 1862	*Gnathotrusia* Higgins, 1981
	Phyciodina Higgins, 1981	*Phyciodes* Hübner, 1819	
		Phystis Higgins, 1981	
		Anthanassa Scudder, 1875	*Tritanassa* Forbes, 1945
		Dagon Higgins, 1981	
		Telenassa Higgins, 1981	
		Ortilia Higgins, 1981	
		Tisonia Higgins, 1981	
		Tegosa Higgins, 1981	
		Eresia Boisduval, 1836	
		Castilia Higgins, 1981	
		Janatella Higgins, 1981	
		Mazia Higgins, 1981	

inclusive taxa provides a relatively stable nomenclature for communicating with nonspecialists and amateur lepidopterists. We believe that nonobligatory categories are best used for communicating taxonomic details of interest primarily to specialists. For example, Wahlberg and Zimmerman (2000) retained a conservative generic classification and used nonobligatory categories (tribe, subtribe) to illuminate the complex relationships displayed by the nymphalids they studied.

The tribe Melitaeini is a group of some 270 species of butterflies that form a distinct group within the subfamily Nymphalinae (sensu Harvey 1991). Analysis of several independent data sets, both morphological (Kons 2000) and molecular (Wahlberg and Zimmermann 2000, Zimmermann et al. 2000, Wahlberg unpubl. data), has demonstrated that the tribe is a monophyletic clade. The taxonomy of this group of butterflies owes much to the efforts of the British entomologist Lionel Higgins (e.g., Higgins 1978, 1981). Higgins distinguished three different groups of checkerspots, the Euphydryina, Melitaeina, and Phyciodina, which later were recognized as subtribes of the Melitaeini by Wahlberg and Zimmermann (2000; table 2.2). Relationships among the three subtribes have been studied using several data sets based on morphological characters (Kons 2000), mitochondrial DNA sequences (Wahlberg and Zimmermann 2000, Zimmermann et al. 2000), and two nuclear DNA sequences from different chromosomes (Wahlberg unpub. data; figure 2.1). Cladistic analyses of all of these data place

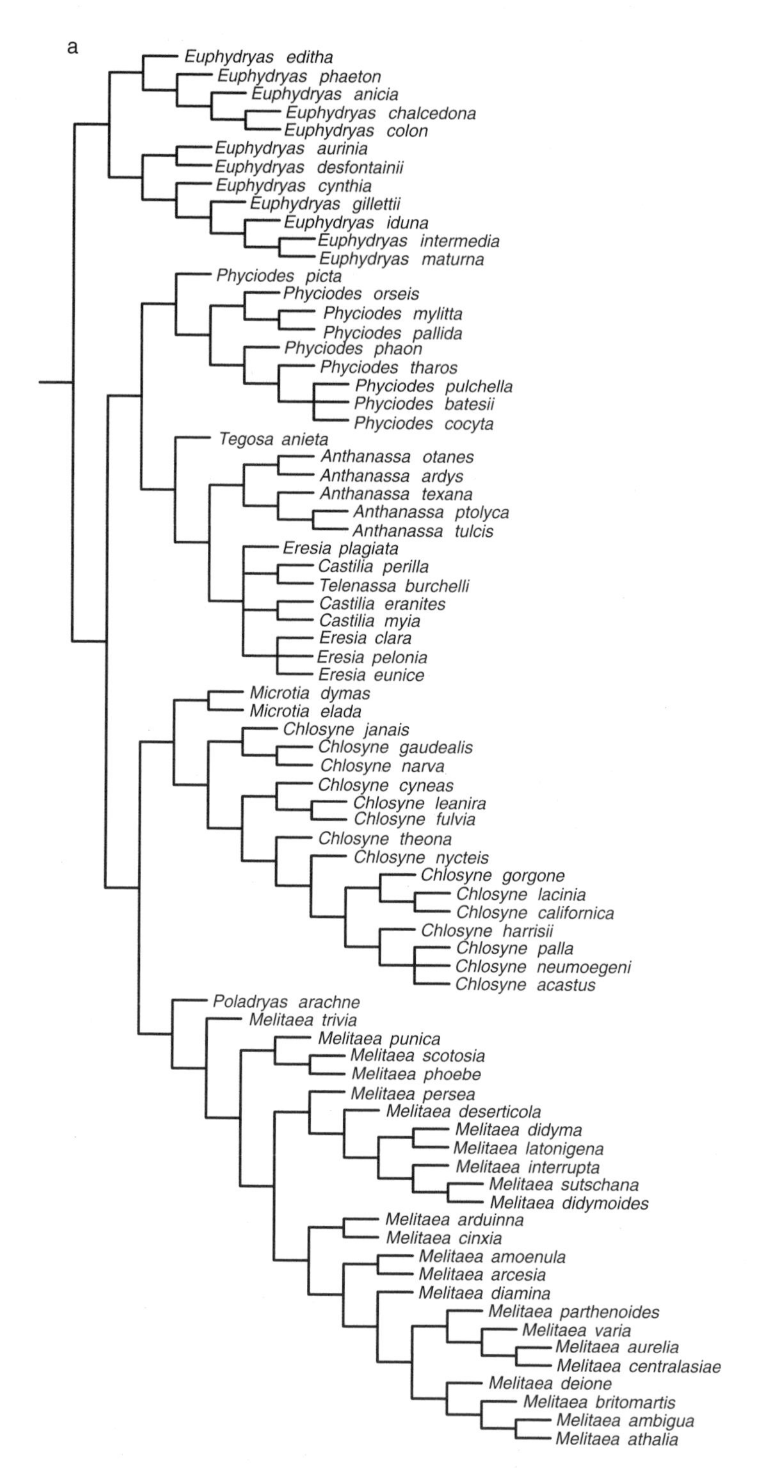

Figure 2.1. Three phylogenetic hypotheses for species in the tribe Melitaeini. (a) Strict consensus of 12 equally parsimonius trees for a data set of two mitochondrial gene (COI and 16S) sequences (Wahlberg and Zimmermann 2000). (b) The single most parsimonious tree found for a data set of one mitochondrial gene (COI) and two nuclear gene (EF-1α and *wingless*) sequences (Wahlberg, unpub. data). (c) Strict consensus of 510 equally parsimonious trees found for a morphological data set with 123 characters (Kons 2000).

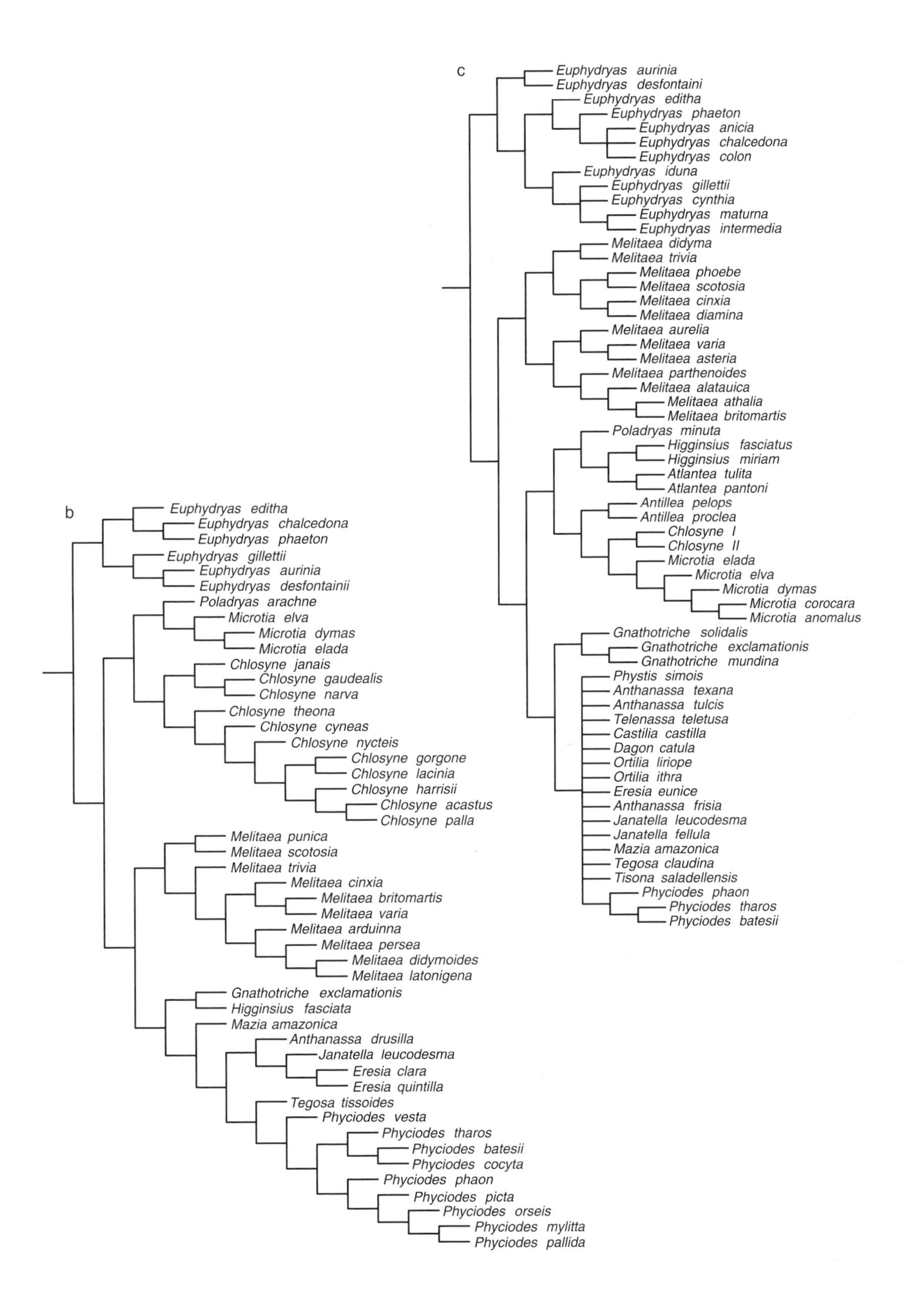
c
Euphydryas aurinia
Euphydryas desfontaini
Euphydryas editha
Euphydryas phaeton
Euphydryas anicia
Euphydryas chalcedona
Euphydryas colon
Euphydryas iduna
Euphydryas gillettii
Euphydryas cynthia
Euphydryas maturna
Euphydryas intermedia
Melitaea didyma
Melitaea trivia
Melitaea phoebe
Melitaea scotosia
Melitaea cinxia
Melitaea diamina
Melitaea aurelia
Melitaea varia
Melitaea asteria
Melitaea parthenoides
Melitaea alatauica
Melitaea athalia
Melitaea britomartis
Poladryas minuta
Higginsius fasciatus
Higginsius miriam
Atlantea tulita
Atlantea pantoni
Antillea pelops
Antillea proclea
Chlosyne I
Chlosyne II
Microtia elada
Microtia elva
Microtia dymas
Microtia corocara
Microtia anomalus
Gnathotriche solidalis
Gnathotriche exclamationis
Gnathotriche mundina
Phystis simois
Anthanassa texana
Anthanassa tulcis
Telenassa teletusa
Castilia castilla
Dagon catula
Ortilia liriope
Ortilia ithra
Eresia eunice
Anthanassa frisia
Janatella leucodesma
Janatella fellula
Mazia amazonica
Tegosa claudina
Tisona saladellensis
Phyciodes phaon
Phyciodes tharos
Phyciodes batesii
b
Euphydryas editha
Euphydryas chalcedona
Euphydryas phaeton
Euphydryas gillettii
Euphydryas aurinia
Euphydryas desfontainii
Poladryas arachne
Microtia elva
Microtia dymas
Microtia elada
Chlosyne janais
Chlosyne gaudealis
Chlosyne narva
Chlosyne theona
Chlosyne cyneas
Chlosyne nycteis
Chlosyne gorgone
Chlosyne lacinia
Chlosyne harrisii
Chlosyne acastus
Chlosyne palla
Melitaea punica
Melitaea scotosia
Melitaea trivia
Melitaea cinxia
Melitaea britomartis
Melitaea varia
Melitaea arduinna
Melitaea persea
Melitaea didymoides
Melitaea latonigena
Gnathotriche exclamationis
Higginsius fasciata
Mazia amazonica
Anthanassa drusilla
Janatella leucodesma
Eresia clara
Eresia quintilla
Tegosa tissoides
Phyciodes vesta
Phyciodes tharos
Phyciodes batesii
Phyciodes cocyta
Phyciodes phaon
Phyciodes picta
Phyciodes orseis
Phyciodes mylitta
Phyciodes pallida

Euphydryina as the most basal group of checkerspots, but the relationships within and between the other two subtribes are in conflict among the data sets. All the data sets suggest that the subtribe Melitaeina should be split into two subtribes, the Melitaeina (comprising the Palaearctic genus *Melitaea*) and the "Chlosynina," which has not yet been formally described (including the New World genera *Chlosyne*, *Microtia*, *Texola*, *Dymasia*, and most likely *Poladryas*).

Morphology and Development

Adult checkerspots are medium-sized butterflies. Their forewings are 1.5–3 cm long, and their dominant colors are red, orange, and black or brown (plates I–IV). Several tropical checkerspots with bright, contrasting coloration are involved in mimicry complexes. Checkerspots usually fly in a low, straight line or zig-zag, and their variegated wing patterns may make it difficult for avian predators to follow them in flight. All adults feed on nectar from flowers, often keeping their wings spread while feeding. Little is known about the natural enemies of adults, although heavy predation by birds occurs in some circumstances (e.g., Bowers et al. 1985).

Mating behavior varies from species to species. Males either search actively for females or watch for them from perches. Many species have two or more generations (broods) annually. *Melitaea cinxia* has obligate diapause and one generation per year in most of Europe, but facultative diapause and from one to three generations per year in parts of southern Europe (chapter 7). *Melitaea didyma* and *M. deserticola* may have up to three generations (Higgins and Riley 1983). Nearctic *Euphydryas*, in contrast, are nearly always univoltine. *Euphydryas editha* has a facultative partial second generation after late summer or early autumn rains in southern California and northern Baja California. In southern locations, *Euphydryas* larvae will frequently remain in diapause or return to diapause after briefly feeding to avoid adult flight and reproductive failure in drought years when resources are scarce.

Development from egg to adult in the Palaearctic sister species *Euphydryas intermedia* and *E. maturna* may often take two years (Luckens 1985, Wahlberg 2001a), and the arctic *E. iduna* most likely has a development time of several years, though no conclusive evidence for this yet exists. The Nearctic *Euphydryas gillettii* also may take more than a year to go from egg to adult (E. H. Williams et al. 1984). In general, all or most nontropical checkerspot larvae have the capacity for repeated diapause (M. Singer pers. comm.), though in natural populations development practically always takes only one year.

Larval checkerspots feed on a taxonomically restricted suite of plants in the subclass Asteridae. They are known to utilize plants in 16 families, although the majority of records are from just four families: Acanthaceae, Asteraceae (Compositae), Scrophulariaceae, and Plantaginaceae (Wahlberg 2001b). Like many members of the Asteridae, the Scrophulariaceae and Plantaginaceae, which serve as the primary hosts of *Euphydryas editha* and *Melitaea cinxia*, contain secondary chemicals called iridoid glycosides (e.g., Bowers 1983b, 1988, 1991, Bowers and Puttick 1986, Wahlberg 2001b). Host plants in the Acanthaceae and Asteraceae, however, do not contain iridoid glycosides. The latter two host plant families appear to have been repeatedly and independently colonized by ancestral checkerspot species that originally fed on iridoid-containing plants (Wahlberg 2001b). As we discuss in chapter 8, the iridoids that are sequestered by checkerspot larvae may make adult butterflies sufficiently distasteful to predators to afford them roles as models in mimicry complexes (Bowers 1991).

Checkerspots tend to lay their eggs in clusters, and young larvae often live colonially, typically under a silken web. In most species older larvae are dark-colored, making them efficient "solar collectors" in climates in which development time may be thermally limited (see chapter 7). The black and orange coloration may be aposematic, again possibly related to the iridoids in their food. Larvae of many species have branched, conical spines on the body (plate XI). Young larvae spend the winter (or summer in Mediterranean climates) in diapause.

2.3 Evolution, Infraspecific Variation, and Conservation

The Melitaeini are thought to have evolved in the Nearctic region (North America north of Mexico) and then repeatedly radiated into the Neotropical (Latin America) and Palaearctic (Eurasian) regions (Wahlberg and Zimmermann 2000). *Euphydryas* appears to have invaded the Palaearctic once and evolved a complex of species there; the ancestor of *E. gillettii* then reinvaded North America. The genus *Melitaea* seems to have evolved entirely within the Palaearctic, although most species in the tribe Melitaeini occur in the Western Hemisphere. Wahlberg and Zimmerman have suggested that repeated glaciations at the start of the Pliocene (some 5 million years ago) created habitats that were invaded by ancestral Melitaeini. The Melitaeini subsequently speciated, perhaps in response to repeated episodes of habitat fragmentation and isolation in glacial refugia. Molecular data suggest that in Europe glacial refugia have played a decisive role in the evolution of many insect species (Hewitt 1996, 1999).

Like most other butterflies, checkerspots tend to be rather sedentary, and as a result they show substantial geographic variation as they respond genetically and phenotypically to local conditions (Ehrlich 1984a). As Ford and Ford (1930) pointed out in a classic paper on *Euphydryas aurinia*, even within populations there tends to be substantial variation in size and wing color patterns, both among individuals in the same generation and between generations (e.g., Ehrlich and Mason 1966, Mason et al. 1967). Much of this variation has been described taxonomically in the form of named subspecies. But subspecies usually have little biological significance because they are based on arbitrarily selected characters that are not consistently correlated with other characters (Wilson and Brown 1953, Gillham 1965). Some named subspecies, which we refer to as "ecotypes," comprise suites of populations that occur in ecologically similar circumstances and exhibit similar patterns of habitat choice and oviposition host plant use. Other subspecies may contain several ecotypes. Conversely, where subspecies have been named from wing pattern only, a single ecotype may contain populations assigned to several subspecies.

Current infraspecific taxonomy gives only a limited and often distorted picture of the genetic variation in species across large geographic areas. Luckily, biologists now have access to information that much more reliably reflects the evolutionary history and relationships among populations. A recent molecular phylogeographic study of *Melitaea cinxia* has revealed three distinct clades in Europe, apparently reflecting the locations of glacial refugia and subsequent geographic expansion of the species (chapter 10). Populations in the Åland Islands of Finland belong to the eastern clade, which extends from the Stockholm archipelago in Sweden to the eastern shore of Lake Baikal in Siberia and the Tian Shan range in China. Members of the central clade are distributed from Lebanon through Turkey, Greece, Bulgaria, and Hungary to Denmark and southern Sweden. The southern clade includes haplotypes in southwestern Europe, including Spain, France, and Italy.

In some circumstances, geographic variation is evidence of allopatric speciation in progress. Checkerspot species, like those of most other groups of butterflies, span a wide range of geographic divergence. Although *Euphydryas phaeton* and the narrowly distributed *Euphydryas gillettii* are relatively constant in phenotype across their ranges, the remaining Nearctic *Euphydryas* exhibit dramatic variation. There are 21 described subspecies of *Euphydryas editha* in California alone. The highly variable *Euphydryas editha* frequently co-occurs across western North America with *Euphydryas chalcedona* or the closely related *E. anicia* and *E. colon*, flying in sympatry and often in synchrony without any observed hybridization. But *Euphydryas chalcedona*, *E. anicia*, and *E. colon* have complicated relationships. Many dozens of subspecies have been described based on variation in wing colors and markings and in the structure of male genitalia. In intermountain western North America, diverse patterns of geographic variation occur among these latter "species." Across the Mojave Desert and down the eastern slope of the Sierra Nevada, *Euphydryas chalcedona* shows clinal variation. In the western desert, most butterflies have red wing markings, historically assigned to *E. chalcedona*. Moving eastward, there is a classic step cline toward darker butterflies that historically were assigned to *E. anicia*. In the Sierra Nevada, most butterflies at the highest elevations are red. Along a descending elevational gradient, the proportion of dark butterflies increases; all are considered *E. chalcedona*. At the eastern base of the Sierra Nevada, darker forms shift abruptly to bright red butterflies, which have traditionally been treated as *E. anicia*.

The red *Euphydryas anicia* extend east and north across the entire Great Basin, where on its northeastern boundary they fly in the Humboldt River drainage in usually brief temporal overlap with almost completely black checkerspot butterflies that have been referred to as *Euphydryas colon*. The red butterflies fly early in the summer, with females usually ovipositing on *Castilleja* on warm slope exposures. The black butterflies fly later, ovipositing on *Symphoricarpus* in cooler valley-bottom microclimates. As the flight season for the red butterflies draws to a close and the earliest black individuals emerge, mating between individuals of these well-differentiated putative species can occur at the ecotone. A small number of butterflies with both red and black markings can be found in a much larger pool of all red and all black individuals. This zone of contact between *Euphydryas anicia* and *E. colon* is not unique, but the introgression may be. On the northwestern edge of the Great Basin, black-dominated phenotypes of both checkerspots co-occur in similar ecological and phenological circumstances with no sign of mixing. Austin and Murphy's (1998) analysis of these complicated circumstances led them to conclude that *Euphydryas chalcedona*, *E. anicia*, and *E. colon* are best treated as a single highly polymorphic species, which occurs as a ring species (Futuyma 1998) of sorts across montane and coastal western North America. Such a treatment is consistent with Scott's (1978) assessment of the three taxa based on genitalic characters and with genetic analyses (Brussard et al. 1989, Wahlberg and Zimmermann 2000, Zimmermann et al. 2000).

The genus *Melitaea* includes about 70 species with a center of diversity in the mountains of Central Asia. There appear to be three major clades within *Melitaea*: the subgenus *Mellicta*, the *didyma*-species group, and the *phoebe*-species group. *Melitaea cinxia* does not belong to any of these groups and seems to be basal to the *Mellicta* clade (figure 2.1). The sister species of *M. cinxia* is not yet known. Two species (*M. diamina* and *M. arcesia*), considered to be allied to *M. cinxia* by Higgins (1981), actually are more closely related to the *Mellicta* assemblage (Wahlberg and Zimmermann 2000). The closest relative of *M. cinxia* that has been analyzed in molecular studies appears to be *Melitaea arduinna*, an enigmatic species from southeast Europe and Asia whose larvae feed on *Centaurea* (Asteraceae). Higgins (1941) originally placed *M. arduinna* in the *didyma*-group. With respect to infraspecific geographic variation across the entire range of *Melitaea cinxia* in Europe and Asia, there are several described subspecies, some of which, like *M. cinxia atlantis* in the Atlas Mountains in North Africa, are quite distinct (Higgins and Riley 1983). Tennent (1996) claims that the North African lowland *M. cinxia* feeds on *Centaurea*.

Overall, then, the checkerspots appear to be a relatively young group of butterflies that diversified from a common ancestor roughly 5 million years ago. Many checkerspot lineages, including *Euphydryas, Melitaea, Phyciodes*, and the subgenus *Mellicta* (Wahlberg and Zimmermann 2000), have not differentiated enough to make the taxonomic delineation of species easy. Indeed, molecular studies of the sequences of mtDNA genes suggest that entities considered species within these groups have diverged very recently (N. Wahlberg, unpub. data), perhaps within tens of thousands to hundreds of thousands of years, rather than millions of years. If this is true, the impact of glacial cycles on speciation could have been great.

While the biological information conferred by conventional infraspecific taxonomy may not be especially reliable in an evolutionary context, it has taken on great meaning to conservation. In the United States, vertebrate species can be protected under the Endangered Species Act at the level of "distinct population segments," that is, populations that are morphologically or ecologically unique. Protection is not conferred to populations of invertebrates, which, if properly delimited, would be meaningful biological entities. However, subspecies of invertebrates are eligible for protection if the entire subspecies is at risk. As a result, coastal populations of *Euphydryas editha,* recognized as the subspecies *E. editha editha* (previously in the literature as *E. e. bayensis*) and *E. editha quino* have been beneficiaries of subspecific recognition. The U.S. federal government prohibits the direct harming (taking) of individuals of these butterflies, including their eggs, larvae and pupae, and, importantly, protects their habitats from modification that would jeopardize the continued existence of the butterflies. Likely to soon join those *E. editha* subspecies on the endangered species list are two additional candidates for federal protection, *E. editha taylori*, known from north-coastal Washington, and *Euphydryas anicia cloudcrofti*, restricted to the Sacramento Mountains of New Mexico.

Subspecies of *Melitaea cinxia* in Europe do not have legal protection, partly because the range of infraspecific variation is limited within any one European country and partly because *M. cinxia* is not threatened in the European Union as a whole (van Swaay and Warren 1999). In Europe, a few butterfly subspecies are protected under the Species Directive, including *Hesperia comma catena*, *Agriades glandon aquilo*, *Clossiana improba improbula*, and *Erebia medusa polaris*. Somewhat like the Endangered Species Act in the United States, European Union legislation stipulates that the habitat of species and subspecies listed in the Species Directive cannot be altered in a manner that would threaten the persistence of populations. The conservation status of *M. cinxia* in Europe varies from country to country. For example, it is a species of "conservation concern" in the United Kingdom (Asher et al. 2001), and "vulnerable" in Finland (Rassi et al. 2001). *Melitaea cinxia* is still widespread and common in much of southern and eastern Europe (Hanski and Kuussaari 1995).

2.4 Life Histories of Edith's Checkerspot and the Glanville Fritillary

Edith's Checkerspot (Euphydryas editha)

The life histories of *Euphydryas editha* in the San Francisco Bay area of California and *Melitaea cinxia* in the Åland Islands in southwestern Finland are broadly representative of those of the Melitaeinae in general. Bay area *E. editha* adults fly in March and April in patches of grassland on serpentine-based soils that are surrounded by chaparral, oak woodlands, nonserpentine grasslands, and human development (figure 2.2). Adults take nectar from the abundant flowers of a variety of annual and perennial plants (Murphy et al. 1983), which in the spring make the landscape attractive to both butterflies and humans. When Paul Ehrlich first arrived at Stanford University in 1959, he sought a butterfly species to study that was relatively stable and occurred in distinct colonies. Local butterfly collectors described a colony of *E. editha* on Stanford's Jasper Ridge Biological Preserve, about 8 km from the main campus, and told him that the larval host plant was an annual plantain, *Plantago erecta* (Plantaginaceae).

Basic ecology of Bay area *Euphydryas editha* was quickly uncovered. Females lay masses of 20–350 eggs (average about 50) on or near *P. erecta* (Labine 1968, Singer 1972). The eggs hatch in 13–15 days, and young larvae from each egg mass often live communally in a silken tent. Larvae grow through three instars, and then, as the summer drought commences and their host plants senesce, they molt into a fourth instar and enter diapause. First and second instar larvae cannot enter diapause. When host

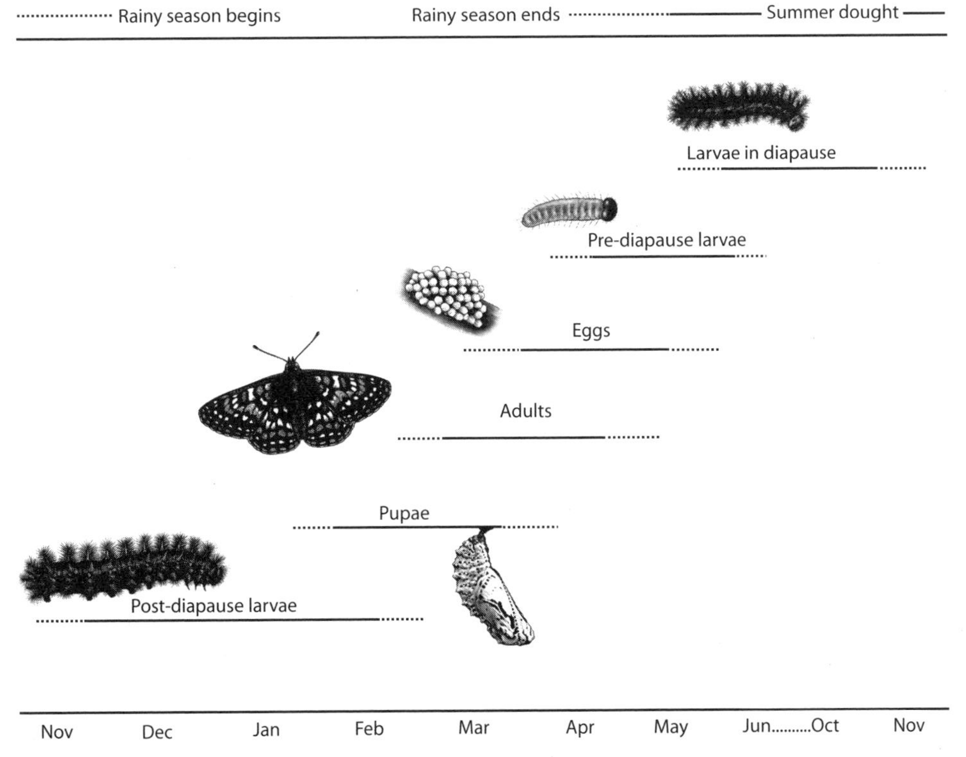

Figure 2.2. Life cycle diagram of the Bay checkerspot butterfly (*Euphydryas editha* populations near San Francisco, California). Eggs are laid by adult females in the spring and hatch into prediapause larvae that feed until the late spring and early summer. Larvae then diapause through the summer drought and emerge with fall and winter rains as postdiapause larvae until pupation. Drawings by Zdravko Kolev.

plants germinate at the start of the late autumn rainy season, larvae break diapause and feed to maturity as solitary individuals (figure 2.2).

Prediapause larvae are in a race to reach diapause before their host plants senesce (Singer 1972). The grasslands that support the Bay checkerspot and its larval host plants dry rapidly in late spring. Mortality of prediapause larvae commonly exceeds 99% (White 1974), with survival rates depending on the availability of green and edible host plants. Detailed studies of the ecology of prediapause larvae show that survival is facilitated in several ways. Eggs laid early in the flight season on *P. erecta* have a high probability of producing larvae that can mature to diapause on their oviposition plant or adjacent plants (Cushman et al. 1994, Hellmann 2002a). Survival probability is enhanced for larvae on *Plantago* on cooler slope exposures or deeper soil, where *Plantago* senesces more slowly. At Jasper Ridge, where gophers are common, gopher-tilled soils support large, long-lived, deeply rooted *Plantago*, providing favorable circumstances for larval growth and survival. Alternatively, eggs may be laid on plants in the family Scrophulariaceae, *Castilleja densiflora* and *C. exserta* (= *Orthocarpus densiflorus* and *O. purpurascens*, respectively) (Hickman 1993). Those eggs, deposited nearly any time after the plants germinate, have a strong chance of producing larvae that reach diapause. Although *Castilleja* germinate later than *Plantago*, they senesce later as well. Singer's discovery that some prediapause larvae at Jasper Ridge disperse from senescing *Plantago* to still green *Castilleja* suggested that co-occurence of the two plants may improve habitat quality for *Euphydryas editha* in the San Francisco Bay area. At Jasper Ridge and some other locations, there is not enough *Castilleja* in many years to support a considerable proportion of larvae, so use of *Plantago* is essential. *Castilleja* presumably is not used exclusively because it germinates later than *Plantago* and therefore is not immediately available to postdiapause larvae.

The race between growing larvae and senescing plants may imply a poor fit between *Euphydryas editha* in the Bay area and its host plants and habitat. However, it is possible that a better fit existed until quite recently, although this is far from a settled issue, even among those who do research on *Euphydryas*. Virtually all of California's native grasslands (e.g., Yogi 1996), once covering several million hectares, have been converted to a landscape dominated by non-native plants. The conversion was initiated by the first Spanish explorers and missionaries, who both purposefully and accidentally introduced aggressive Mediterranean grasses and forbs to California. These plants provided better forage for livestock but rapidly outcompeted and replaced most native grassland vegetation. The exotics subsequently spread across a landscape already impacted by drought and overgrazing in the late 1800s. Since the early 1900s, non-native species have become dominant except in a few refugia with shallow or inhospitable soils.

One of these less readily invaded soils is derived from serpentine rock. Serpentine soils are poor in essential nutrients, including nitrogen and phosphorous, and rich in toxic elements such as nickel, chromium, and magnesium. This substrate supports an extensive assemblage of endemic plant species, as well as native, formerly widespread grasses and forbs. *Euphydryas editha* now inhabits a network of serpentine soil-based habitat patches that range in size from just a fraction of a hectare to several thousand hectares. Serpentine-based soil drains readily, drying weeks earlier than other soil types and promoting early senescence of vegetation. Thus, the butterfly's larval host plants are available for less time on serpentine soils than they would have been on other soils. This shortened growing season has become the dominant source of larval mortality and the key factor in *E. editha*'s local persistence (Singer 1972).

Plantago erecta, the two *Castilleja* species, and *Euphydryas editha* undoubtedly had a more extensive distribution across nonserpentine soils two centuries ago (Ehrlich and Murphy 1987b). But the rather sedentary behavior of *E. editha* today suggests that even before invasion of non-native plants, the species may have been colonial. Concentrations of forbs that serve as larval host plants and adult nectar sources for *E. editha* were likely intermittent and associated with disturbance, occuring as early successional patches separated by bunchgrasses. Sedentary behavior at sites of eclosion may have been as adaptive then as it is now (Ehrlich 1961b, 1965, Ehrlich et al. 1984, Murphy et al. 1986), although genetic evidence suggests that gene flow among *E. editha* populations was much greater in the past than it is today (Slatkin 1987; chapter 10).

The Glanville Fritillary (Melitaea cinxia)

Melitaea cinxia flies in June in the Åland Islands in the northern Baltic, at the northern range limit of its European distribution (figure 2.3). The butter-

Figure 2.3. Maps of the Åland Islands showing (a) the islands from which there are historical records of *M. cinxia,* (b) the present distribution, (c) islands to which *M. cinxia* has been introduced (Sottunga in 1991, still persisting; see figure 13.7; Husö in 1992, failed in two years—this introduction was done in the context of a large-scale experiment on migration; see Kuussaari et al. 1996), and (d) the islands where habitat suitable for *M. cinxia* has been mapped.

fly is found on the main Åland Island (1500 km^2) as well as on a few nearby islands that are larger than 10 km^2, but it is absent from the hundreds of smaller islands and islets in the Åland archipelago that mostly lack suitable habitat for the butterfly. Across the Åland landscape, small fields and pastures are dispersed among rocky outcrops and stretches of coniferous forest. The human community in Åland is fiercely independent and takes pride in protecting the historic landscape, including the small, dry meadows used by *M. cinxia* (plate XII). Many of these meadows occur on outcrops of granite bedrock and are therefore fairly permanent features in the landscape. *Melitaea cinxia* went extinct on mainland Finland by the 1980s (Marttila et al. 1990, Kuussaari et al. 1995). The number of dry meadows on mainland Finland and in many other parts of northern Europe has been dramatically reduced, apparently below a threshold that would allow long-term persistence of *M. cinxia* and many other butterflies. Characterizing this extinction threshold is one of the key foci of research on *M. cinxia* in Åland (chapters 4 and 13).

The initiation of the *Melitaea cinxia* project in the Åland Islands and research on *Euphydryas editha* at Jasper Ridge have close parallels. Ilkka Hanski originally was aware of one particular butterfly population and its larval host plant, a perennial plantain, *Plantago lanceolata*. As it turned out, only one other host plant is used by *M. cinxia* in Åland, *Veronica spicata*. Both *E. editha* at Jasper Ridge and *M. cinxia* in Åland feed on plants in the family Plantaginaceae. A recently developed molecular phylogeny suggests that *Plantago* and *Veronica* are sister genera (Olmstead et al. 1993, Angiosperm Phylogeny Group 1998, Olmstead et al. 2001). The distribution patterns of the two host plants in Åland are different: *V. spicata* occurs primarily in the northwestern part of the main Åland Island, whereas *P. lanceolata* is widespread (Kuussaari et al. 2000; see figure 7.2). Much of the habitat for both plant species has been created by human activity, and both appear to have been unintentionally introduced to Finland at least 500 years ago (Kukkonen 1992, Hämet-Ahti et al. 1998). Some records of *V. spicata* in mainland Finland suggest the possibility of post-glacial relict populations (Kukkonen 1992), but these are isolated and small at present.

The life history of *Melitaea cinxia* is generally similar to that of *E. editha* (Kuussaari 1998), though there is an obvious seasonal difference in larval growth, reflecting the difference in climate. Butterflies fly in June and early July (figure 2.4). Females oviposit groups of 100–200 eggs, preferably on *V. spicata* in western Åland and on *P. lanceolata* in eastern Åland (Kuussaari et al. 2000; chapter 4). Thus, despite the relatively small size of Åland (< 50 km across), there has been geographic differentiation in a key life-history trait: oviposition host plant preference. In contrast to *E. editha*, larvae of *M. cinxia* from a single egg cluster (full sibs) live communally past diapause in a silken web that they spin around the host plant (in high elevation populations in the Alps *M. cinxia* larvae disperse after diapause; M. Singer pers. comm.). Before diapause, the larval group typically must move at least once to a new host plant after consuming the original plant. Larval groups often split into two subgroups. This behavior increases probability of survival when food is scarce but reduces the probability of surviving the winter because relatively large larval groups are necessary to spin the special "winter nest" (figure 2.5) required for successful diapause (chapter 7). Even before diapause, there is substantial mortality from predation and parasitism (chapter 8) and from starvation if the host plant senesces in the middle of the summer. Summer droughts are not typical in Åland, but dry periods have been un-

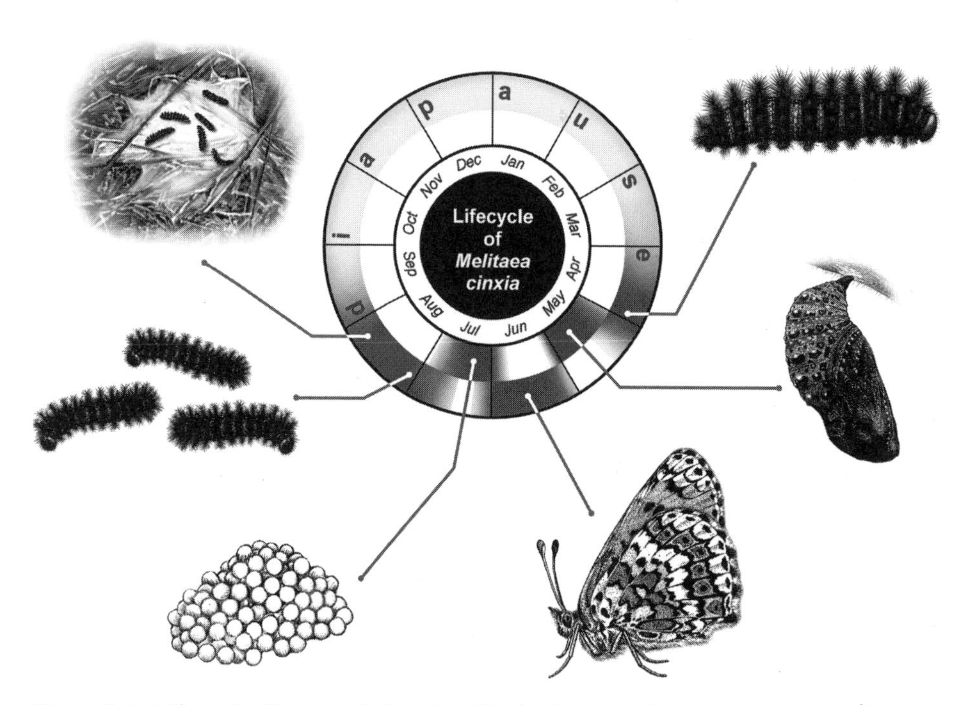

Figure 2.4. Life cycle diagram of the Glanville fritillary (*Melitaea cinxia*) in the Åland Islands in southwestern Finland. Eggs are laid by females in early summer (June to early July) and hatch in two to three weeks time. The prediapause larvae live gregariously in a silken web. In late August the larvae spin a dense winter web inside which they diapause. Diapause is broken after snow melt in March–April, and larvae continue to feed gregariously until pupation in late April–early May. Drawings by Zdravko Kolev.

Figure 2.5. Winter nest of *Melitaea cinxia* larvae. Drawing by Zdravko Kolev.

usually frequent in the past decade, perhaps due to global climate change. Because *V. spicata* and *P. lanceolata* occur on naturally dry, rocky outcrops, the plants wither quickly during dry periods.

Melitaea cinxia larvae remain in groups until late August, when they spin the initially white and conspicuous winter nest (plate IX) and then enter diapause. This has important practical consequences for research. The backbone of the research on *M. cinxia* in Åland is an annual survey of population size on approximately 4,000 small meadows. The time window for the survey is limited because by mid-September the winter nests turn gray and are more cryptic. Even during late August and early September, when the winter nests are easy to spot at the base of larval host plants, about 40 people are needed to complete the survey. When the snow melts in late March or early April, larvae break diapause and start basking in the sun in small groups. Larvae are able to increase their body temperature far beyond ambient temperature, and they start feeding and growing as soon as new leaves of the host plants are available. In early spring, larval growth is often food limited, but food availability becomes less limiting as the season advances.

Much of the work on checkerspot model systems has focused on relationships between the butterflies and their larval host plants and on the interactions between population dynamics and spatial distribution of populations. Next, we introduce the basics of this research to provide orientation for navigating the more detailed treatment in the rest of the book.

2.5 Ecology of Host-Plant Associations

Although many results from the early research on populations of *Euphydryas editha* at Jasper Ridge seemed applicable to other herbivorous insects, it was also clear that the Jasper Ridge populations had ecological characteristics unique to their isolation on patches of serpentine soil-based grassland. After all, populations of the species are scattered from northern Mexico to southern Canada and from the Pacific coast to the Rocky Mountains. Within that far-flung distribution, *E. editha* can be found among cactus in arid gullies at the edge of the Mojave Desert, in forest openings in foothill oak woodlands, in sagebrush-dominated subalpine meadows in Colorado, and on windswept ridgelines on the crest of the Sierra Nevada—all locations far beyond the distribution of *Plantago erecta* and the checkerspot's favored nectar resources in California's coastal grasslands.

Populations of *E. editha* in California show complex patterns of host and habitat use (see figure 12.3). For example, Jasper Ridge *E. editha* lays eggs on *Plantago erecta*, *Castilleja exserta*, and *C. purpurescens* and takes nectar from nearly a dozen species of plants in March and April. Populations less than 100 km to the east have similar black, red, and yellow markings, but they fly among dense chaparral on complex terrain more than a month later in the spring, lay their eggs on *Pedicularis densiflora*, and take nectar from the flowers of the shrub *Eriodictyon*. Also in April and May, but on the other side of the broad Central Valley, at low elevations in the Sierra Nevada, a distinctive, bright red *E. editha* flies in oak forest openings, ovipositing exclusively on *Collinsia tinctoria*. Several months later, dark and diminutive *E. editha* are found high on glacial tarns and alpine peaks up to 4000 m in elevation, depositing eggs on *Penstemon davidsoni* and *Castilleja nana*.

The mechanisms that underlie these complex differences partly involve oviposition host-plant preferences varying in space (Singer and Parmesan 1993) and time (Singer et al. 1993)—across habitats and from flight season to flight season. Virtually all polyphagous populations show differences among years in patterns of host use, as described in chapter 6. Distinct populations of individuals that look similar and fly in ecologically similar habitats nonetheless exhibit local adaptations to resource quality and availability. The availability of resources is often so unpredictable and sometimes so fleeting that many *E. editha* populations are at constant risk of extirpation. The same situation applies to many other checkerspot species, although not to *M. cinxia* in Åland. In the case of *M. cinxia*, variation in resource availability is only one of several reasons that small local populations are so ephemeral. Oviposition host-plant preference in *M. cinxia* influences the rate of colonization of currently unoccupied meadows (Hanski and Singer 2001; chapter 9). Despite the presence of just two host plant species in the Åland Islands, there is distinct geographic differentiation in oviposition host-plant preference even within the main Åland Island (Kuussaari et al. 2000).

2.6 Spatial Population Structures

The spatial structure of populations—the patterns in the distribution of individuals that affect reproduction, distribution of offspring, and thereby interactions among developing individuals—has been another focal point of our research on checkerspots. The isolated colony of *Euphydryas editha* that Ehrlich originally selected for his research consisted of three well-delimited local populations living in discrete fragments of serpentine vegetation at Jasper Ridge. Migration among the three populations was limited, and exchange of individuals with other populations was virtually nonexistent. But even in the San Francisco Bay area of central California, other populations of *E. editha* have different spatial structures. For example, a population at Del Puerto Canyon in the Inner Coast Range uses *Pedicularis densiflora* as a larval host and uses shrubs that tend not to occur near the larval hosts as adult nectar sources. In this population, individual movements average more than an order of magnitude longer than those at Jasper Ridge (Gilbert and Singer 1973; figure 2.6). This difference hints at the complexity of spatial population structures discovered subsequently in other populations of checkerspots, including those of *E. editha* and *Melitaea cinxia*.

The prominent features of the *Melitaea cinxia* system in Åland that fascinated Hanski are highly fragmented spatial structure, frequent turnover of small local populations, and long-term persistence of the metapopulation as a whole in a balance between stochastic extinctions and recolonization of temporarily unoccupied meadows. Indeed, the *M. cinxia* system in Åland is a real-world approximation of the classic metapopulation concept, originally introduced by Richard Levins in 1969. In 1991, however, all that was known was that the species had occurred at least at some sites in Åland in the past. During the first few years, understanding of the spatial population structure sharpened. We now know that there are around 4000 small habitat patches, including patches just hundreds of square meters in area and patches of marginal quality. The largest patches are only a few hectares in area; thus there are no "mainland" patches nor mainland populations. Consequently, all local populations are small enough to have a significant risk of extinction. At any one time, the number of occupied meadows is 300–500, and only about 10% of the suitable meadows are simultaneously occupied. Uncovering the processes that determine the rate of change in the spatial distribution of the butterfly has been one of our principal research tasks.

Given the small size of most habitat patches for *M. cinxia* in the Åland Islands, it is not surprising that many populations consist of just one group of larvae. We know that larvae in this situation are usually full sibs because the majority of females only mate once, typically soon after emergence. In such a case, we still refer to a single group of larvae as a population, because a population (demographic unit) is defined as a group of individuals that mate among themselves and have relatively independent

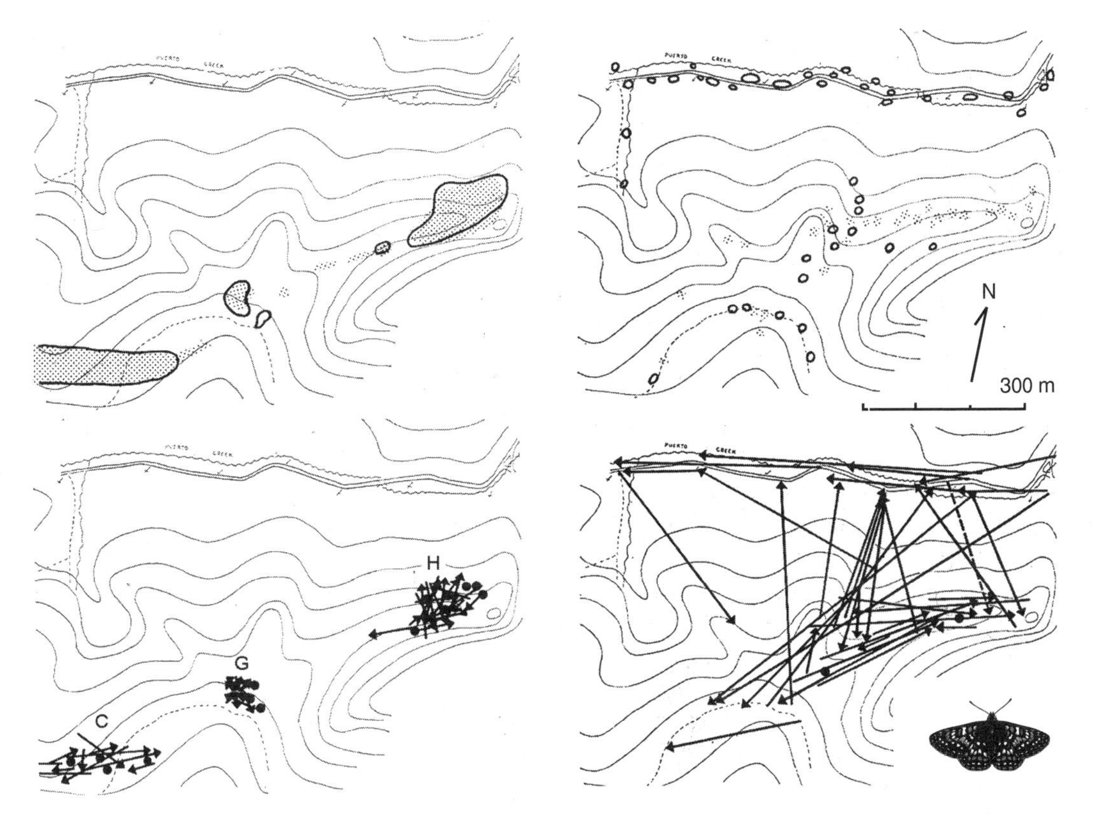

Figure 2.6. Movements of marked individuals of *Euphydryas editha* at Jasper Ridge (left) and in the Del Puerto Canyon (right) compared, both shown on the map of Del Puerto Canyon. The upper panels show the distribution of larval (stippled) and adult resources (encircled by bold line). At Jasper Ridge, the larval resources are *Plantago erecta* and *Castilleja densiflora*; at Del Puerto Canyon larvae feed on *Pedicularis densiflora* and adults on *Eriodictyon* and *Achillea*. The lower panels show typical pattern of adult movements between capture events. Arrows represent movements with 1 or more days between capture events; large dots represent individuals marked on some previous day and recaptured at original point of marking. Based on Gilbert and Singer (1973).

dynamics. Patches of habitat are so widely scattered that individuals could not "sample" many patches even if they flew away from their natal meadow. Although there is substantial migration of individuals, there is also high probability that in the smallest populations mating takes place among full sibs or other close relatives. This, of course, leads to inbreeding, which occurs regularly in the *Melitaea cinxia* metapopulation and lowers the fitness of the offspring (Saccheri et al. 1998, Haikola et al. 2001, Nieminen et al. 2001). Given the detrimental consequences of inbreeding, it is perplexing why females do not avoid mating with brothers (Haikola et al. 2004) and why selection has not "purged" deleterious genes. These issues are discussed in chapter 10.

Inbreeding depression is severe enough in the smallest *Melitaea cinxia* populations to elevate their risk of extinction. Inbreeding, however, is not the only process that dramatically shortens the expected lifetime of the small populations. One of the major lessons from research on *M. cinxia* is that many factors influence the risk of population extinction (Hanski 1998b), including demographic stochasticity, environmental stochasticity, deterministic environmental change, and interactions with natural enemies (parasitoids). Timing of host plant senescence, which is so relevant to persistence of *Euphydryas editha* populations in California, is also influential.

Three decades of studies of populations of *Euphydryas editha* in the San Francisco Bay Area have amended initial inferences regarding their structure. *Euphydryas editha* was believed to be extraordinarily sedentary, limited by "intrinsic barriers" to dispersal between suitable habitat at Jasper Ridge (Ehrlich 1961b). We now know that although most individuals do have high site fidelity, with typical migration distances of hundreds of meters or less (Ehrlich 1965, Brussard et al. 1974), some move substantial distances, undoubtedly facilitating gene flow and founder events. For example, at Edgewood Park, 10 km from Jasper Ridge, it was discovered that *E. editha* moved freely across grassland areas substantially larger than those at Jasper Ridge. Subsequently at Morgan Hill, 50 km from Jasper Ridge, a small portion of the population was found to move across several kilometers of intermittent habitat (Harrison et al. 1988).

After initial studies at Jasper Ridge, there was little reason to doubt that larger habitat patches would offer resident butterflies more resources and reduce the probability of extinction. Years later, however, it became obvious from work at Morgan Hill that not only area but also topographic heterogeneity contributes to habitat quality for *Euphydyas editha*. During drought, butterflies survive best on cool, moist slope exposures. During extremely wet years, in contrast, butterflies persist on warm, dry slope exposures. Accordingly, the population on Jasper Ridge's smaller but topographically diverse habitat, patch H, survived years longer than the population on larger, but more topographically homogenous patch, C (chapter 3). One early action to conserve *E. editha* on Jasper Ridge was to eliminate cattle grazing. After all, what could be less compatible with natural processes in a native grassland ecosystem than intensive grazing by exotic livestock? Cessation of grazing, however, may have been a crucial step down the path toward extinction of *Euphydryas editha* at Jasper Ridge. We now know that selective grazing of grasslands can slow the invasion of non-native, weedy plant species that may otherwise outcompete members of the native flora. Similarly, moderate levels of cattle and sheep grazing help prevent grasses from invading meadows in the Åland Islands, although intensive grazing increases larval mortality of *Melitaea cinxia* (Hanski 1999b). The competitive ability of non-native plants has been dramatically demonstrated by studies at Morgan Hill, downwind from the urban centers of San Francisco and San Jose. Atmospheric deposition of nitrogen at Morgan Hill is increasing the rate of invasion of non-native plants relative to Jasper Ridge, which receives far less atmospheric nitrogen (chapter 3).

In brief, our checkerspot studies have produced evidence for both classic metapopulation structure, in which a large number of small habitat patches are occupied by relatively ephemeral local populations (*Melitaea cinxia* in Åland); mainland–island metapopulation structure with a large and permanent "mainland" population (the Morgan Hill system with the surrounding smaller patches); as well as the patchy population structure of Harrison (1991), with so much movement among small nearby "populations" that they do not comprise independent demographic units (the Morgan Hill system itself). Furthermore, long-term research by Singer and associates in the Sierra Nevada has produced fascinating results showing source-sink metapopulation structures in *E. editha*, including robust understanding of the mechanisms at the individual level that underpin the population dynamics (chapter 9). The generally sedentary behavior of checkerspots, in comparison with many other butterflies (chapter 11) and many other taxa predisposes them to spatially structured populations, but it is also clear that the primary determinant of the more fine-grained spatial structure is the structure of the landscape and the distribution of suitable habitat, which may show dramatic temporal changes (chapters 6 and 9). Much of our work on the biology of checkerspots should be explicitly related, and has been related, to the nature of the landscape in which the populations live.

It is impossible to verify the prehistoric spatial population structures of checkerspots or any other organisms, but some informed guesses can be made. In Asia, *Melitaea cinxia* and several other steppe-inhabiting checkerspots appear to be associated with *Veronica incana*, a common species closely related to *V. spicata*, the preferred host plant in the Åland Islands and northern Europe. The present discontinuous distribution of *M. cinxia* in northern Europe reflects the naturally patchy habitat, which has become increasingly fragmented because of human activities. *Euphydryas editha* has adapted to diverse habitats and food plants in western North America (e.g., Singer 1971b, Ehrlich et al. 1980, Ehrlich and White 1980, Holdren and Ehrlich 1982) and tends to have a colonial distribution even in seemingly continuous habitats like the sagebrush flats of subalpine Colorado (Ehrlich and Wheye 1984). The patchy distribution pattern of *E. editha* in Colorado supports the speculation that the distribution of San Francisco Bay area *E. editha* before the invasion of European weeds, when *Plantago* was much more widespread, was already somewhat discontinuous. Human activities have, of course, greatly increased habitat fragmentation for *E. editha* as they have that for *M cinxia*.

2.7 Concluding Remarks

The modern, well-delimited populations of *Euphydryas editha* in California and *Melitaea cinxia* in the Åland Islands have facilitated our research by permitting us to ask a wide range of questions about the dynamics and biology of individual demographic units, including the biological consequences of spatial population and metapopulation structures. These questions are examined in detail in most of the chapters in this volume. Checkerspots provide an excellent model system to study the biology of small populations, including phenomena such as density dependence (Andrewartha and Birch 1954), the rescue effect (Brown and Kodric-Brown 1977), the Allee effect (Allee et al. 1949), and inbreeding. Research on *M. cinxia* in particular has addressed genetic and demographic processes related to spatial population structure (Hanski 1999b).

Even so, the checkerspot story is incomplete. For example, we have not yet exploited the opportunity to study a sample of populations and metapopulations spanning the structural continuum from highly fragmented to extensive uniform habitats. Such comparisons would, for example, allow us to determine the extent to which populations have adapted morphologically or behaviorally to the structure of their landscape. That information might allow us to better predict the response of checkerspots, other invertebrates, and even small vertebrates to further human fragmentation of habitat. Results on *Melitaea cinxia* suggest that even subtle differences in the level of habitat fragmentation can have noticable consequences for the dispersal behavior of female butterflies (Heino and Hanski 2001, Hanski et al. 2002). Resolving whether there are differences in the biology of species that have evolved in patchy landscapes versus species whose habitat has been artificially fragmented is highly relevant for conservation.

We hope this chapter has shown that checkerspots not only offer us beauty and fascinating variety, they also offer us windows of opportunity to understand important aspects of how the biological world functions. We'll now take a longer look at the view through those windows.

3

Structure and Dynamics of *Euphydryas editha* Populations

JESSICA J. HELLMANN, STUART B. WEISS, JOHN F. McLAUGHLIN,
PAUL R. EHRLICH, DENNIS D. MURPHY, AND ALAN E. LAUNER

3.1 Introduction

The research reported in this volume began more than 40 years ago with studies of the Bay checkerspot butterfly, a subspecies of the widely distributed Edith's checkerspot butterfly. Working initially with the populations on Stanford University's Jasper Ridge Biological Preserve, Ehrlich and colleagues asked some fundamental questions: why does the Bay checkerspot butterfly inhabit the area that it does? Why does it not occur elsewhere? Why is it sometimes abundant and sometimes rare? How does population size relate to genetic variability? Are these populations evolving at a rate that can be detected on an ecological time scale? Answers to such questions from checkerspots and other well-selected systems have helped produce a comprehensive picture of how the world works at the population level. As research on the Bay checkerspot proceeded, however, many questions took on new applied and conservation significance. Work at Jasper Ridge and other sites established that Edith's checkerspot butterfly populations were quite unstable and that many habitats were threatened by human development and habitat degradation. Most recently, the Jasper Ridge populations that were the focus of the Ehrlich group's long-term study have become extinct.

We review here the current understanding of population structure and dynamics that has emerged from four decades of work on Edith's checkerspot butterfly, *Euphydryas editha*. We focus primarily on the Bay checkerspot, which has been studied from 1959 to the present. This butterfly is typically assigned to the subspecies *bayensis*, a name we will continue to use to avoid confusion. The first descriptions of *E. editha* were apparently based on specimens from the San Francisco Bay area; hence technically the proper name is *E. editha editha*. Other studies exploring the dynamics of *E. editha* have been pursued at Rabbit Meadow in the Sierra Nevada mountain range since 1979 (chapter 9) and opportunistically at other sites (e.g., Parmesan 1996). As this chapter and subsequent chapters illustrate, *E. editha* displays a diversity of population dynamic patterns. The dynamics of *E. editha* populations in the San Francisco Bay area provide a foundation for exploring this diversity and its relationship to factors such as habitat size and isolation, disturbance history, position within the species' range, and adaptation to local food resources. A major conclusion emerging from this research is that the interplay between habitat quality and climate is a critical determinant of the dynamics of local populations.

We begin with a description of the Bay checkerspot and the characteristics that make it distinct from other *E. editha*. Second, we turn to the factors that play a role in its population dynamics. Third, we discuss the factors that may lead to extinction of Bay checkerspot populations, particularly those factors responsible for the documented disappearance at Jasper Ridge. Fourth, we discuss our understanding of the dynamics of *E. editha* in the San Francisco Bay region, knowledge that developed as research expanded to include an increasing number of populations at different sites. Finally, we briefly contrast the Bay checkerspot with other *E. editha* ecotypes. In this chapter and others, we use the term "ecotype" to refer to groups of populations with similar ecological characteristics such as larval food plants, timing of life cycle, and climatic zone of their habitat. Ecotype can be used in either a general or specific sense. For example, populations of the Bay checkerspot form one specific ecotype found in one small region of California. The Bay checkerspot joins with other coastal populations, however, to form a more general ecotype living in coastal grasslands and typically feeding on annual hosts. These populations contrast with ecotypes using predominantly perennial hosts and/or found in very different habitat types such as at higher elevations in the Sierra Nevada.

3.2 Habitat and Life History of *Euphydryas editha bayensis*

Euphydryas editha is distributed from British Columbia to Baja California and from the coast of California east to Colorado, Wyoming, and Alberta, Canada (White and Singer 1974) (chapter 2). Across this range, populations are typically scattered, discrete, and isolated (Scott 1986). Figure 3.1 shows the locations of the best studied populations in California and western Nevada. Groups of populations often differ from each other in host plant usage and timing of life cycle events (e.g., summer versus winter diapause). In coastal areas, *E. editha* larvae feed on *Plantago* and occasionally on *Castilleja* and *Collinsia*. Noncoastal populations feed on *Pedicularis*, *Penstemon*, *Collinsia*, *Castilleja*, and, less commonly, on other scrophulariacious plants (chapters 6 and 7).

The Bay checkerspot butterfly is confined to patches of native grassland in the San Francisco Bay area, California, which occur almost exclusively on serpentine-based soils (figure 3.2). These patches are surrounded by chaparral, oak woodlands, and nonnative grassland on nonserpentine substrates. Serpentine soils maintain a native flora because their unusual chemical composition helps prevent domination by invasive grasses (Tadros 1957). These chemicals include high concentrations of nickel, chromium, and magnesium and relatively low levels of major nutrients, including phosphorous and nitrogen (Walker 1954).

The largest habitat patch occupied by *E. editha bayensis* is Morgan Hill, a >1500 ha site near the southern edge of the subspecies range just south of San Jose (figure 3.2). (In various publications, Morgan Hill has been called Kirby Canyon, East Hills, and Coyote Ridge, and it contains the Kirby Canyon Butterfly Reserve.) A smaller number of more isolated habitat patches, including Jasper Ridge, are found on the San Francisco peninsula and on the east side of San Francisco Bay. In general, the grassland patches inhabited by *E. editha* are larger and more isolated than patches used by *Melitaea cinxia* in the Åland Islands, as described in chapter 4 (see figure 4.3). Before the introduction of Eurasian grasses approximately 200 years ago, the habitat of the Bay checkerspot butterfly may have been much more widespread (Ehrlich and Murphy 1987a, Cushman et al. 1994). The degree to which this more extensive habitat was continuous versus patchy is unknown.

Populations of *E. editha bayensis* feed on the annual forbs, *Plantago erecta*, *Castilleja densiflora*, and *C. exserta* (figure 3.3). (These *Castilleja* species were formerly in the genus *Orthocarpus*.) Key events of the butterfly's annual life cycle coincide with the growing season of these and other native, annual grassland plants. Prediapause larvae hatch in the late spring and feed on presenescent plants until they are sufficiently mature to enter an obligatory diapause during the summer drought (figure 2.2). Of the prediapause hosts, *P. erecta* is typically the most widespread and consistently abundant from year to year, while *Castilleja* species tend to be spatially and temporally more variable (figure 3.4). Larvae that survive to diapause emerge as winter rains begin and continue feeding (largely on *P. erecta*) until they pupate in the early spring. Adults emerge after approximately two weeks, and the adult flight season lasts four to five weeks, depending on weather conditions and population size (Hellmann et al. 2003). Adults feed on nectar from a variety of native flowers, most notably *Muilla maritima*, *Layia platyglossa*, *Lasthenia californica*, *Allium serratum*, and *Lomatium macrocarpum*.

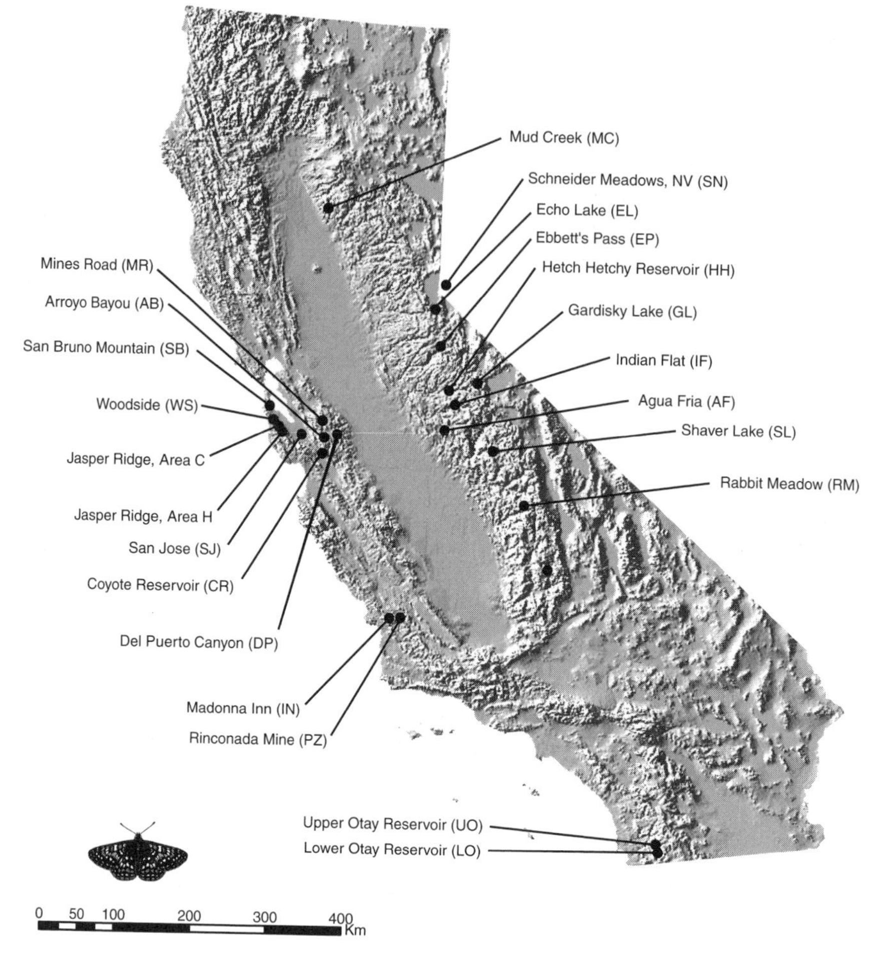

Figure 3.1. Map of the locations of the best studied *Euphydryas* populations in California and in far western Nevada.

3.3 Population Sizes and Spatial Structures

Traditional methods of analyzing mark–recapture data work well for univoltine organisms like the Bay checkerspot that have closed, discrete populations and in which all individuals are of reproductive age (for more complicated systems, see Schwarz and Seber 1999). Research on the dynamics of *E. editha bayensis* began with an intensive mark–recapture experiment in 1960 at the Jasper Ridge Biological Preserve (Ehrlich and Davidson 1960; figure 3.5). Mark–recapture experiments have been repeated at Jasper Ridge most years since, producing one of the most comprehensive data sets on the dynamics of a suite of invertebrate populations (NERC Centre for Population Biology 1999, Hellmann et al. 2003). Estimates of population size are calculated from mark–recapture data using the ratio of marked to unmarked adult butterflies captured on a particular sampling day (Jolly 1965). Daily estimates are then summed over an entire flight season using Scott's (1973) method to estimate the total number of adults flying in a single season. It appears that handling by researchers in these experiments has minimal impact on the behavior and longevity of the butterflies (e.g., Orive and Baughman 1989).

The intensity of mark–recapture surveys at Jasper Ridge varied over the years. Years with intensive surveys (57% of study years in the habitat patch labeled "area C" and 50% in the habitat labeled "area H"; figure 1.2) had sample sizes of as many

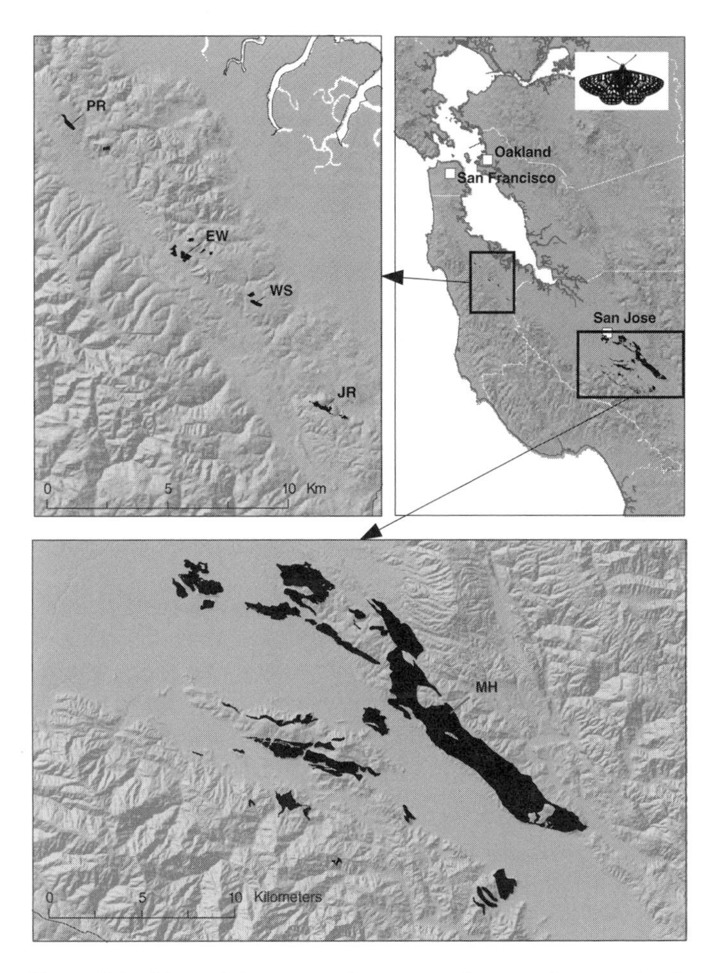

Figure 3.2. Map of the two main groups of serpentine grassland habitats in the San Francisco Bay Area. One cluster of *E. editha* populations is at Morgan Hill (MH), south of San Jose (bottom panel). The other cluster is in the vicinity of the Jasper Ridge Biological Preserve (JR, left panel). Of these latter patches, only the habitat at Edgewood County Park (EW) was extant in 2000. PR is Pulgas Ridge. WS was the site of the Woodside population.

as 500–2000 individuals when populations were large, recapture rates > 20%, and sampling spread over the entire season. Remaining years had lower rates of recaptures or data for only portions of the flight season were collected (eight years). The most striking finding from the first year of mark–recapture study at Jasper Ridge was the discovery that adult butterflies were confined to serpentine grassland. These areas are dominated by native plants (total about 7 ha) and are located within a larger expanse of grassland dominated by non-native plants (figure 2.6). Early recapture results (figure 3.6) clearly showed that these circumscribed areas contained three separate populations that rarely exchanged individuals; migration rates between two of the populations averaged < 2.6% of 1048 adults recaptured in 1960–63 (Ehrlich et al. 1975; table 3.1, figure 2.6). Because populations were largely independent, they have been deemed "demographic units" (Brown and Ehrlich 1980), synonymous to "local populations" in the *Melitaea cinxia* metapopulation in the Åland islands, Finland (chap-

Figure 3.3. The larval host plants of the Bay checkerspot butterfly, *Plantago erecta*, *Castilleja exserta*, and *Castilleja densiflora*. The two *Castilleja* species are typically grouped into one host type in checkerspot studies (e.g., Hellman 2002). Drawings by Zdravko Kolev.

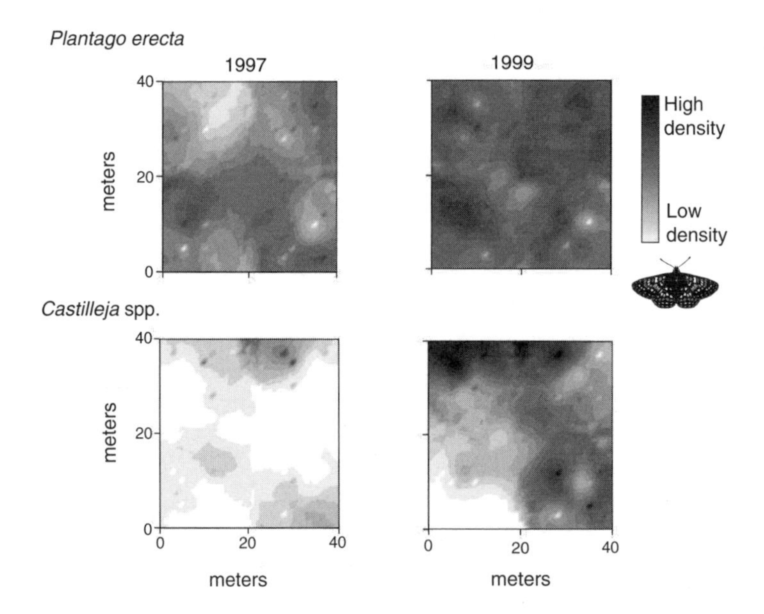

Figure 3.4. Contour diagram of the abundance and distribution of *E. editha bayensis* host plants on representative slopes in two survey years at Morgan Hill (Hellmann 2000). Dark shades represent high density; light shades represent low density or absence. In general, *Plantago erecta* is spread relatively evenly across slopes and throughout the study area, whereas the occurrence of *Castilleja* is spatially and temporally variable.

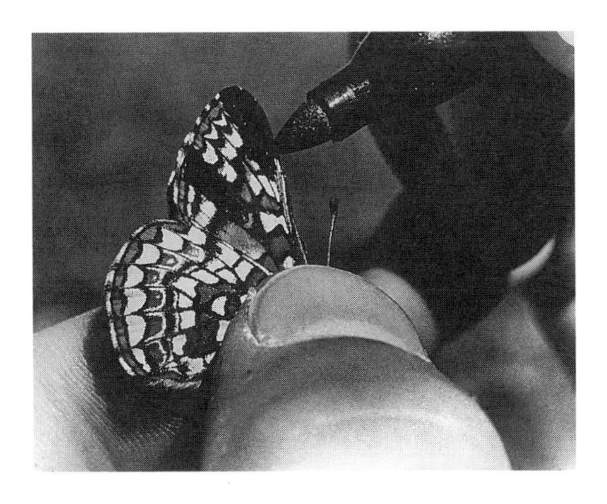

Figure 3.5. Marking *Euphydryas editha* number 30, using a modern felt-tip pen with a flow rate not requiring wing support (see figure 2). Photo by Paul R. Ehrlich.

ter 4). The areas occupied by these demographic units are partially separated from one another by chaparral and oak woodland, but a butterfly could easily fly from one area to another (distance from area C to area H is 700 m). The rarity of inter-area dispersal shows that *E. editha* is fairly sedentary. Recaptures outside of areas of original capture continued to be infrequent in subsequent years of mark–recapture sampling (Ehrlich 1961b, 1965).

The long-term data from the three habitat patches at Jasper Ridge give a retrospective view of the population dynamic similarities and differences among closely neighboring populations. Figure 3.7 shows the estimated size of each population at Jasper Ridge until its extinction (Hellmann et al. 2003). The population in area G (between areas C and H; figure 1.2, see also figure 12.1) was ephemeral. At its peak, this patch had approximately 200 adults, and over the study period the population was extirpated twice and recolonized once (Ehrlich et al. 1975). Just one checkerspot butterfly has been observed in area G since 1973. Populations in areas C and H fluctuated in the range of hundreds to thousands of adults until approximately 1983 when the population in area C began a relatively rapid decline and then disappeared by 1991. The decline of the population in area H was less abrupt; its fluctuations continued around a long-term decline that began in the mid-1970s and accelerated in the late 1980s until its extinction in 1997 (McGarrahan 1997). Recognition of the distinctiveness of these three demographic units was critical to subsequent study of their dynamics.

Table 3.1. Data reproduced from Ehrlich (1965) of early mark–recapture data, reflecting the movement of individuals among habitat patches at Jasper Ridge.

Year	Recapture in Area of Previous Capture	Total Recaptures	Percent in Area of Previous Capture
1960	216	224	96.4
1961	411	425	96.7
1962	159	164	97.0
1963	235	235	100.0

The populations at Jasper Ridge show variability in several parameters in addition to population size (Hellmann et al. 2003). Considering years with recapture rates >20%, the average sex ratio of adult males to females was 1.08 in area C ($\sigma = 0.13$; $n = 14$) and 2.02 in area H ($\sigma = 1.12$; $n = 17$; Ehrlich et al. 1984; chapter 5). The number of adult females was greater than the number of adult males in six years in area C, whereas female-biased sex ratios were never observed in area H. These results suggest that mortality or dispersal rates by sex differed in the two habitat patches, given that the primary sex ratio is close to 1:1. Data also suggest that the adult sex ratio may have been skewed toward males when total population size was low (Hellmann et al. 2003). The population in area C almost always flew later than the population in area H, and females almost always flew later than males in both populations (Iwasa et al. 1983). Loss rate, a parameter that includes mortality and dispersal, differed qualitatively between the sexes and populations. In general, females showed higher rates of mortality and/or dispersal from each population than did males, but because interpatch dispersal was rare, differential mortality by sex appears to have occurred in some years.

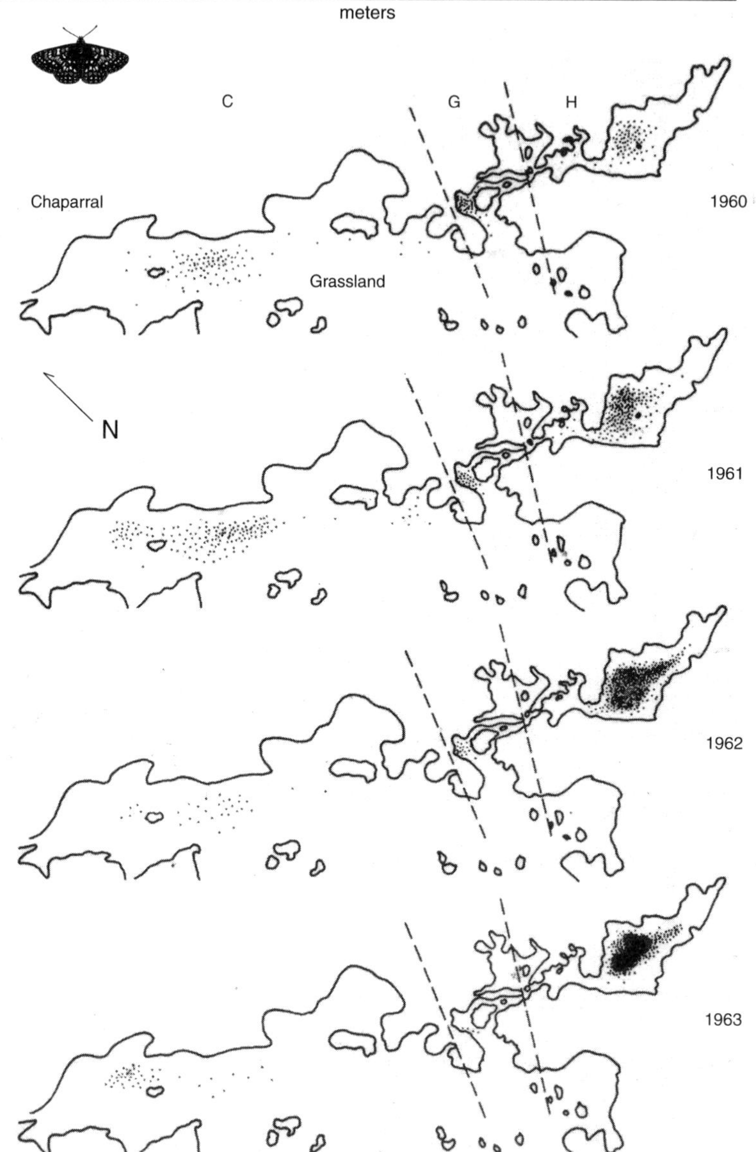

Figure 3.6. Geographic representation of the relatively independent dynamics of populations in areas C, G, and H at Jasper Ridge. Dots represent the position of first capture of an individual. Note the decline of the population in G while the population in H was exploding (more than 10-fold increase) and the population in C fluctuated (after Ehrlich 1965).

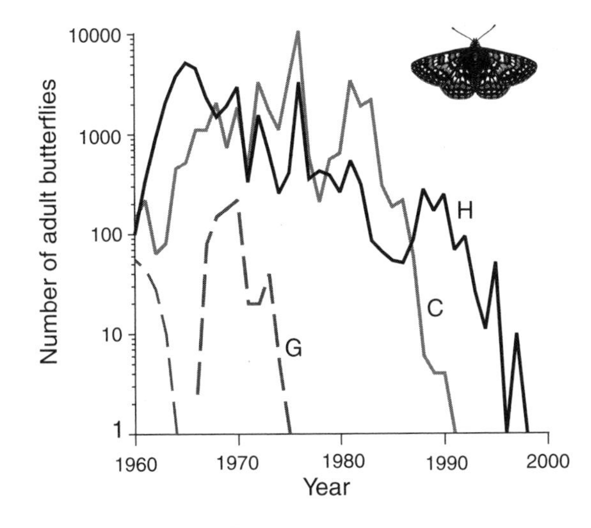

Figure 3.7. Estimated adult population sizes at Jasper Ridge calculated from results of mark–release–recapture surveys (Hellmann et al. 2003).

At Bay checkerspot butterfly sites beyond Jasper Ridge, the primary method of estimating abundance has involved counting postdiapause larvae. Large postdiapause larvae are relatively easy to detect in the short grassland vegetation, and their numbers are estimated using a timed search technique that is extrapolated to an entire habitat patch (Murphy and Weiss 1988b). For postdiapause larval sampling, the habitat is stratified into "thermal strata" based on potential March 21 clear-sky insolation, or the amount of in-coming solar radiation. Insolation is determined by latitude, date, time of day, slope, aspect, topographic shading, and cloud cover; thermal strata within a habitat primarily reflect topography (section 3.4). Multiple samples in each stratum allow errors to be estimated, and recently the method has been extended to cover > 1000 ha of habitats. This method is preferred over mark–recapture of adults in large, well-mixed populations.

Very few populations of Bay checkerspot remain. Those that have been monitored, apart from Jasper Ridge, are at Morgan Hill and Edgewood Park (figure 3.2). Morgan Hill has fluctuated from 12,000 to 500,000 adults, never reaching very low numbers (figure 3.8), while Edgewood Park declined from 100,000 adults in 1981 to near-extinction in 2002.

3.4 Factors Affecting Population Dynamics

With this quick summary of the abundance and demography of the Bay area *E. editha* populations, we now ask: what causes checkerspot populations to change in size and distribution across space? One notable aspect of checkerspots is that adult females lay a large number of eggs. This reproductive output represents a huge potential for growth, but in most populations and in most years, nearly all immature individuals die before reaching reproductive stage (chapter 7). Determining the causes of this mortality is key to understanding and predicting the dynamics of this butterfly. Research at Jasper Ridge and other sites now suggests that prediapause larval host plant use, weather, and topographic diversity combine to determine the dynamics of Bay checkerspot populations.

The Relationship between Larvae and Plants

Singer (1971b, 1972) was the first to ask how populations of *E. editha bayensis* use food resources and how they are affected by food availability over time. Singer found that larvae from eggs laid at Jasper Ridge in 1969 and 1970 had difficulty reaching diapause. Although the butterflies chose nonsenescent host plants for oviposition, senescence often occurred during the two weeks before eggs hatch. This caused approximately 80% of larvae to starve soon after hatching. In 1970 this figure rose to approximately 99%, when hosts underwent senescence before the larvae had reached a size at which they could respond to lack of food by entering diapause. Field observations of larvae that did survive this critical period indicated that they could do so in three ways. First, they came from eggs laid near

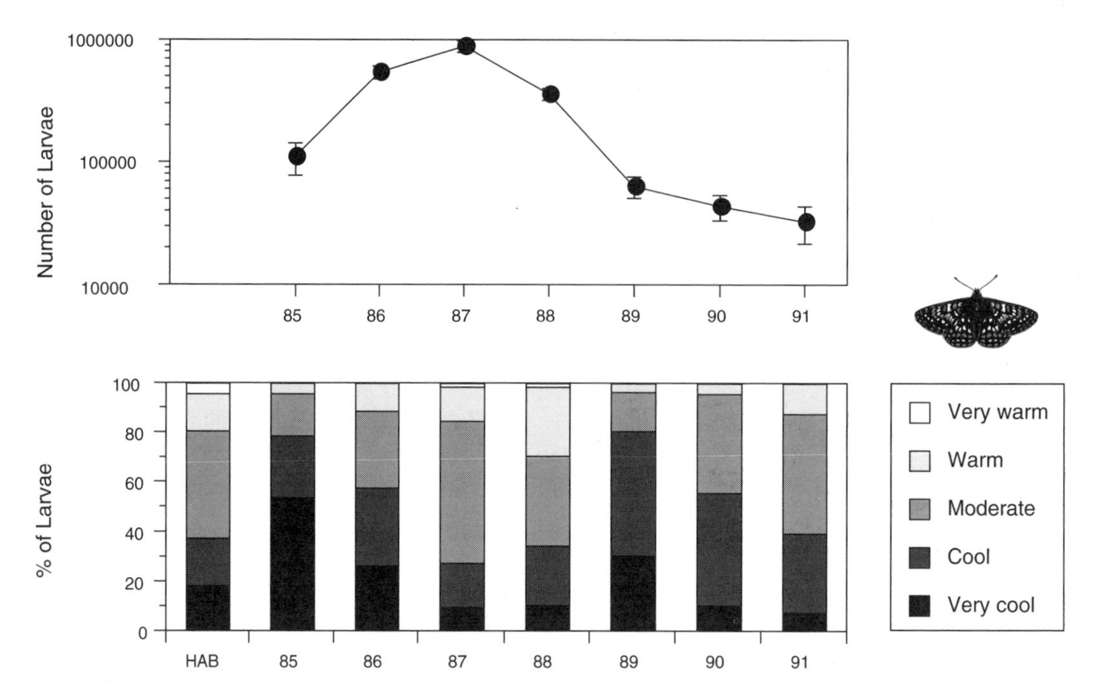

Figure 3.8. Changes in population size of *E. editha* in the Kirby Canyon Preserve at Morgan Hill in 1985–91 (see also Hellmann et al. 2003). Error bars show 95% confidence intervals. The lower panel shows proportional representation of larvae in different thermal strata. "HAB" refers to the proportion of the different slope types in the habitat as a whole.

Plantago erecta on gopher mounds, where *P. erecta* stayed green longer than plants on soil untilled by gophers. Second, they came from eggs laid near *P. erecta* that lasted for sufficient time without gopher activity, usually on a cooler, north-facing slope. And third, they came from eggs laid on or near *Castilleja*, a host that tended to remain edible longer than *P. erecta*. These three situations involved three different resource types for prediapause larvae. Singer concluded that the timing of host senescence relative to the timing of larval growth was the primary determinant of the number of larvae that survived to adulthood and that prediapause larval survivorship was the primary factor responsible for changes in population size.

More recent work by Hellmann (2002c) on the interaction between larvae and plants built upon Singer's observations between prediapause larval survival and host plant use and explores how variation in host plant phenology, host availability, and host use affects larval growth and survivorship. Using field enclosures (0.75 × 0.75 m), Hellmann observed how larvae forage in the serpentine grassland. She found that prediapause larvae dispersed widely among host plants and shifted from approximately equal use of the two host species toward exclusive use of *Castilleja* as the growing season progressed (figure 3.9a). This shift and subsequent survivorship to diapause was independent of the host on which larvae originally hatched. These results suggest that larvae select and modify their diet over time, and what they choose to eat correlates with changing quality of plants.

Hellmann also quantified how environmental conditions and the host plant species available to foraging larvae affect their growth and survivorship (Hellmann 2002c). Using the controlled conditions of a greenhouse, she manipulated the surface temperature of containers holding either both *Castilleja* and *P. erecta* or only *P. erecta*, into which she introduced prediapause larvae. These treatments replicated dissimilar microclimates occurring in the field (moderate vs. extreme surface temperatures) and the two types of host plant environments that larvae experience (one vs. two hosts). The experiments confirmed that temperature affects the rate of host plant senescence; high temperature accelerates the loss of water, nitrogen, and phosphorous

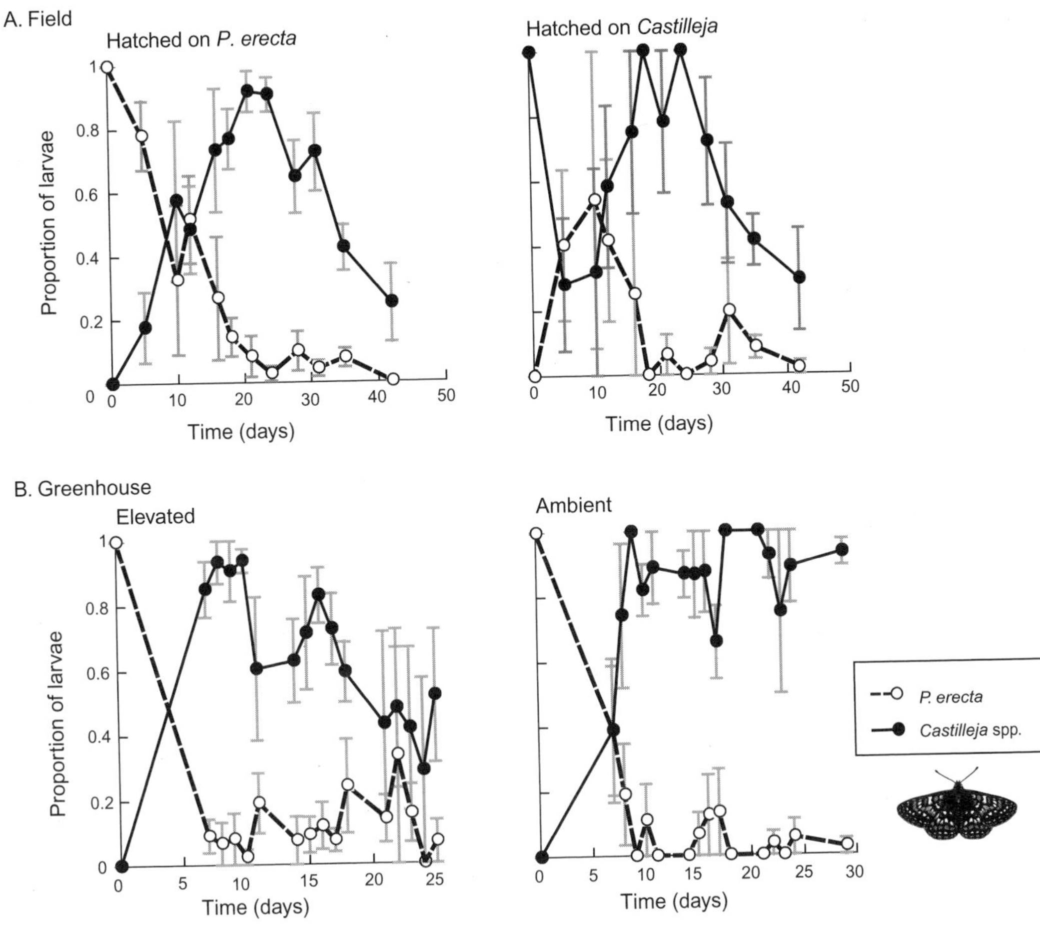

Figure 3.9. Fraction of larvae on different substrates through time in (A) field and (B) greenhouse studies (Hellmann 2002c). In (A), two types of data are shown: containers with larvae that all hatched on *Plantago erecta* and those with larvae that all hatched on *Castilleja*. In (B), two types of data are also shown: containers with larvae under elevated temperature (surface temperature = 30°C) and under ambient temperature (surface temperature = 20°C). In (B), all larvae hatched on *P. erecta*.

and does so more strongly in *P. erecta* than in *Castilleja*. *Castilleja* had a higher average concentration of nitrogen and phosphorous but was more variable in nutrient content than was *P. erecta*. Larvae feeding on plants in the greenhouse did not show a strong preference for *P. erecta* at any point in their development, and those that could dispersed to *Castilleja* as the season progressed (figure 3.9b). This shift happened earlier under elevated temperature because plants senesced quickly. Survivorship in containers with only *P. erecta* was lower than in containers with both hosts, and elevated temperature increased larval survival in containers with both hosts. This is because higher temperatures accelerate larval growth, enabling larvae to shorten their development. Elevated temperature seems to allow larvae to stay in phase with *Castilleja*, but not with *P. erecta*.

In general, growth rates of ectothermic animals increase rapidly with increasing temperature to an optimum and subsequently decrease until a lethal temperature is reached (Sharpe and DeMichele 1977, Ratte 1984). Reflecting this pattern, Fleishman et al. (2000b) found a relationship between stage of prediapause larval development and microclimate (topographic position) in the field; *E. editha bayensis* larvae on warm slopes tended to grow faster than larvae on cool slopes. Environmental conditions, such as temperature, directly affect larval growth and indirectly affect larval feeding via host plant senescence. The balance of these effects on the number of larvae surviving to diapause seems to depend at least in part, on the host species that larvae consume (Hellmann et al. in review).

The apparent inferiority of *P. erecta* as a larval host raises the question: why do adult females oviposit on this species at all? A partial answer may be in the relative consistency of *P. erecta* abundance among years (figure 3.4) and varying nutrient content among individual plants. Laying eggs on *Castilleja* is an option for females in only some years. The dispersal abilities of larvae suggest that adults may choose an optimal neighborhood, rather than an individual plant, in which to lay eggs (Hellmann et al. in review). The most desirable location to lay an egg is an area where both hosts are available. Singer (1971a) suggested that females may restrict their search to areas rich in flowers, enhancing opportunities for larvae to diversify their diet. An interesting aspect of diet evolution in this and other small foragers is the potential influence of spatial scale on natural selection. Plasticity of larval behavior may be selected in environments where individuals can make their own resource decisions, when resources (host plants or slope exposures) are fine grained. In environments where alternative resources are spread far apart or coarse grained, selection may act primarily on the oviposition choice of adults.

Another key factor in the relationship of larvae and host plants is the timing of adult emergence, a factor that largely determines the start of larval feeding. Murphy et al. (1983) and Cushman et al. (1994) estimated how the reproductive success of females changes over time due to the constraints of host plant senescence (see also Murphy et al. 1983). They posited that reproductive success drops sharply as the season progresses, based on an estimation of daily fecundity (number of eggs produced by a female as a function of size and age) and the mortality rate of prediapause larvae from eggs laid at different times in the season (as a function of the time larvae have to feed until the host plant senesces). They concluded that under conditions of relatively high larval mortality, few females (as low as 11%) have enough surviving offspring to replace themselves, and because the earliest emerging females contribute the majority of individuals to the next generation, effective population size may be substantially smaller than census population size in *E. editha bayensis*. On the other hand, previous studies suggested that by selecting hosts that had not senesced, which is a smaller and smaller subset of plants as time goes on, reproductive success may hold constant for some portion of the flight season. In 1970, for example, Singer and Ehrlich (1979) reported that groups of eggs laid two and three weeks after the beginning of the flight season had a nearly equal chance of surviving at egg hatch (71% and 74% of groups, respectively, died). Including *Castilleja* in the larval diet also extends the foraging period and may boost larval survival relative to feeding only on *P. erecta*, particularly for eggs laid later in the season (Singer 1972, Hellmann 2002a). Cushman et al.'s (1994) analysis considers effects of senescence on larvae, not on oviposition, and considers only the dynamics of *P. erecta*; therefore the pressures for early emergence described in their analysis are likely to be mitigated somewhat by adaptive strategies of host choice during the flight season (chapter 6).

The Significance of Weather

Studies of Bay checkerspot larvae and their host plants indicated that butterfly abundance is strongly mediated by abiotic conditions. Ehrlich (1965) first suggested that the population dynamics of Jasper Ridge *E. editha* might be climate-sensitive, but year-to-year changes in population size were not correlated among the three principal populations at Jasper Ridge (figure 3.6), subject to the same yearly fluctuations in weather. Comparisons of distribution maps of larval survival with maps of their resources showed that populations in areas C and H depended on different resources (Singer 1971a). Spring survival of larvae in area C was correlated with abundance of *P. erecta,* which had its growth period extended by the digging of gophers. Survival in area H was correlated with the abundance of late-senescing *Castilleja* and of *P. erecta.* From 1969 to 1971, this difference appeared to cause suitable habitat patches to differ in area C in each year, while the larval distribution in area H was more stable. Singer (1971b) suggested that a year without suitable habitat patches was more likely to occur in area C than in area H. Therefore, he predicted that area C should be the less stable of the two populations, and its persistence might depend on intermittent recolonization from area H. Over the period 1961–

69, the population in area H fared better after dry winters, while the population in area C did the reverse, performing better following wet winters. Thus the data began to indicate a dissimilar response in the two populations to a major environmental factor (rainfall) driving population dynamics (Singer 1971b; figure 3.10). Taking advantage of the much more extensive data set that has by now been accumulated, McLaughlin et al. (2002a, 2002b) have shown that weather conditions play a dominant role in the long-term dynamics and extinction of the populations at Jasper Ridge, and indeed that the two populations C and H exhibited dissimilar responses to weather events (for details, see section 3.5).

Further evidence of the sensitivity of *E. editha bayensis* populations to weather conditions can be seen in their gross responses to weather extremes. Ehrlich et al. (1980) found that the California drought of 1975–77 was "correlated with unusual size changes" in a number of *E. editha* populations in distinct habitat types. Each of five Bay checkerspot populations included in their analysis declined over the drought period (figure 3.11). Dobkin et al. (1987) showed that several years of extremely heavy rainfall caused population declines in area H at Jasper Ridge. Both drought and deluge are hypothesized to alter the timing of the insect–plant interaction, much like temperature, but in different ways. Drought deleteriously affects populations by reducing the period of host plant availability, but deluge slows larval growth. Further work is needed to understand how temperature and precipitation combine to affect the timing of larval and plant development; work by Hellmann (2002c) suggests that both temperature and moisture availability are critical determinants of plant senescence and larval growth rates.

Historical sensitivity to climate and weather suggests that checkerspot populations will be strongly affected by shifts in climate (Murphy and Weiss 1992, Parmesan 1996, Parmesan et al. 1999). Mechanistic models of the relationship between larvae and plants suggest that the dynamics of Bay checkerspot populations in particular will be markedly affected by climate change (Hellmann et al. 2004). Small changes in the temporal overlap of larvae and host plants, as could be caused by a shift in mean regional temperature, may result in significant changes in population growth rates. One model of host plant senescence and larval movement between host plants indicates that if senescence relative to diapause occurs 10% sooner, a currently viable population could be extirpated in fewer than 60 years (Hellmann et al. 2004). Populations of mobile larvae that are able to take advantage of longer-lasting hosts (*Castilleja*) will fare better under such a scenario, suggesting the importance of polyphagy in buffering populations against environmental variation and change. More recent evidence suggests that climate change in the form of increased climatic variability may already be affecting checkerspot butterfly populations (McLaughlin et al. 2002a; section 3.5).

Microclimate and Spatial Variation of Environmental Factors

Weather varies across time and space, of course, and weather and topography interact to create a spatial component in the dynamics of local checkerspot

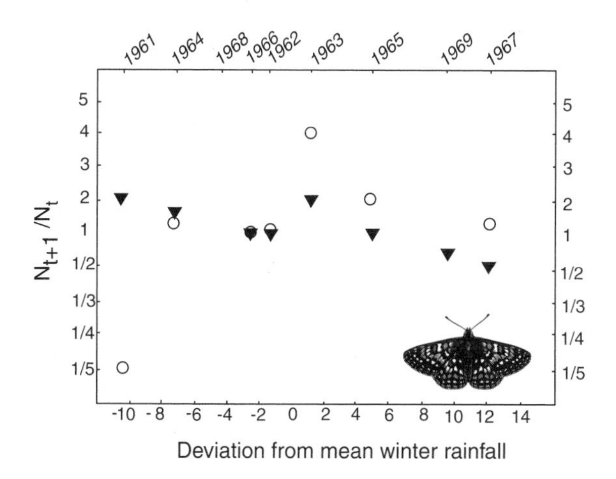

Figure 3.10. Early illustration of the relationship between rainfall (deviation from mean of winter rainfall, November–March of year t in inches) and proportional change in population size between years t and $t + 1$ in area C (open circles) and area H (filled triangles) at Jasper Ridge, as based on Singer (1971a). Correlation between the difference of the N_{t+1}/N_t values for the two habitats and the corresponding rainfall value is significant at $p = .017$ (Spearman rank correlation), suggesting a dissimilar response in the two populations.

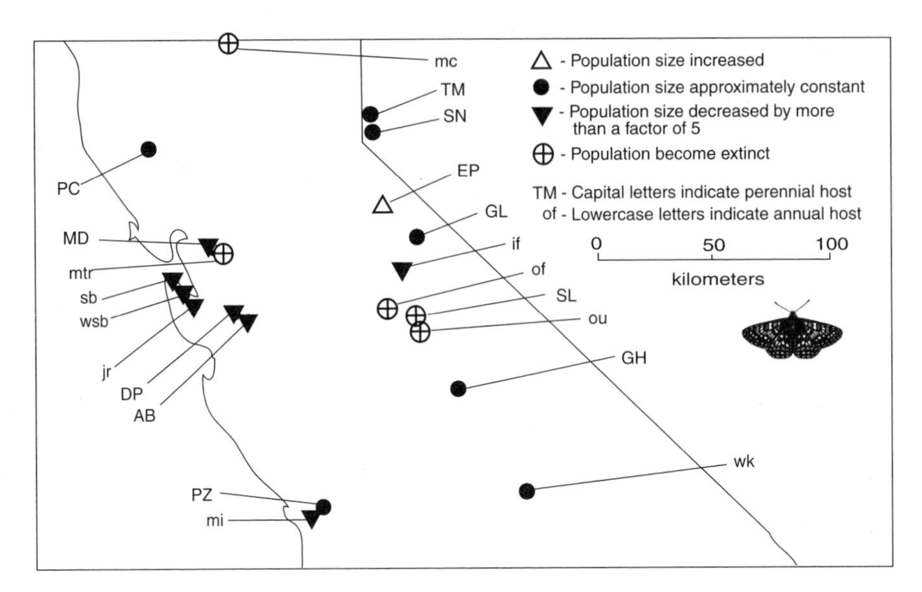

Figure 3.11. Responses of *Euphydryas editha* populations to the California drought of 1975–77 (after Ehrlich et al. 1980).

populations. Using a combination of biophysical modeling, empirical field studies, and population modeling, Weiss et al. (1988) examined the complex relationships among weather, topography, and dynamics of the Bay checkerspot butterfly at Morgan Hill. Topoclimates, or topographically mediated microclimates, strongly affect the distribution and abundance of larvae and butterflies within a habitat (Geiger 1965). We commonly think of weather as a mean condition that applies over an entire region at a given time, but for many small animals and plants, weather is a local phenomenon. Weather experienced by individual organisms is affected by factors such as surrounding vegetation and soil type and, strongly, by topography.

Initial site surveys at Morgan Hill indicated that the size of the population and the habitat that supported it were at least two orders of magnitude larger than at other known sites (figure 3.2). Approximately 1500 ha of serpentine grassland exist along a main ridge, and adjacent smaller patches of habitat are found on numerous side ridges and in canyons across an elevational range of 300 m (figure 3.12, plate V). Slopes facing all compass directions reach 40°. Surveys in 1984 and 1985 indicated that adults, host plants, and nectar sources were widespread across the habitat but that postdiapause larvae were found primarily on cooler north- and east-facing slopes. Abundance and survival appeared to vary by slope also at Jasper Ridge (Singer 1972), but the Morgan Hill site offers much more topographic diversity, habitat area, and denser populations in which to study this phenomenon.

Using insolation models for tilted surfaces (measurements of in-coming solar radiation) and field measurements of temperature, Weiss and colleagues translated the complex terrain of the Morgan Hill into a thermal gradient. Weiss et al. (1988) observed striking spatial variation in surface temperature across the grassland in relation to insolation (figure 3.13), similar to that observed at Jasper Ridge by Dobkin et al. (1987). Surface temperatures in short grassland vegetation with a high proportion of bare soil are strongly influenced by insolation. Noontime surface temperatures on south-facing slopes can exceed air temperature by 20–30°C through the growing season, and those on flat areas exceed air temperatures by 5–12°C. In contrast, surface temperatures on north-facing slopes can be at or below air temperature. Within a slope there can be as much as a 15°C difference between the warmest and coolest spots. For small foraging caterpillars, this variation can be the difference between successful foraging and overheating, finding senescent or nonsenescent host plants, or growing slowly or quickly. We briefly discuss the effects of topography on several life stages and aspects of *E. editha bayensis* life history.

Topography influences postdiapause larval growth. Postdiapause larvae must grow from about 3 mg to 300–500 mg to reach a size at which they

Figure 3.12. Aerial photograph of the Morgan Hill, showing topographic heterogeneity. Photo by Paul R. Ehrlich.

can pupate. After larvae break diapause and begin feeding in the winter, maximum air temperatures are well below the body temperatures required for larval growth. Postdiapause larvae are efficient solar collectors; their black color (plate VIII.2) maximizes absorption of insolation, and their setae (the hairy structures protruding from the body) minimize convective heat loss (Weiss et al. 1988). When basking, last instar larvae reach temperatures 10–12°C above air temperatures near the ground. Because insolation varies across slopes, ground-level temperatures, larval body temperature, and larval growth rates also vary spatially. Field collections of larvae demonstrated that those on warmer slopes were several weeks ahead in development compared to those on cooler slopes. Weiss and colleagues (1987) weighed cohorts of postdiapause larvae, marked them, and released them on various slopes through the growing season. Gains in larval mass were a linear function of slope-specific insolation, with a maximum growth rate of about 50 mg/day for last instars. Postdiapause larvae also could disperse up to 10–20 m/day, which allowed them to transfer from cooler to warmer slopes and speed their development by a week or more.

Topography also modifies the development of pupae. Placement of pupae on various slopes in the field showed that pupae, like larvae, develop faster on warmer slopes than on cooler slopes and that development rate is proportional to insolation (Weiss et al. 1988, Weiss and Murphy, 1993). Extreme heat experienced in some portions of the habitat also appears to cause pupal mortality. Ninety percent of one sample of pupae placed on a south-facing slope during a March heat wave ($T_{air} > 30°C$) were killed by surface temperatures that were well over 40°C (Weiss, unpub. data). Pupae are likely to be better protected than other life stages, however, in most field situations, because they are normally in sheltered positions and protected by a small amount of silk webbing (plate VIII.5).

Differences in larval and pupal development across slopes lead to differences in adult emergence times across the topoclimatic gradient. Empirical relationships between insolation, air temperature, and the growth and development rates of postdiapause larvae and pupae have been integrated into a climate-based computer model (Weiss et al. 1993) (also implemented in a Geographic Information System [Weiss and Weiss 1998]) that predicts dates of adult emergence across complex topography. The models predict the course of development by summing slope-specific insolation over time, with adult emergence requiring a total of 630 MJ/m^2 of inso-

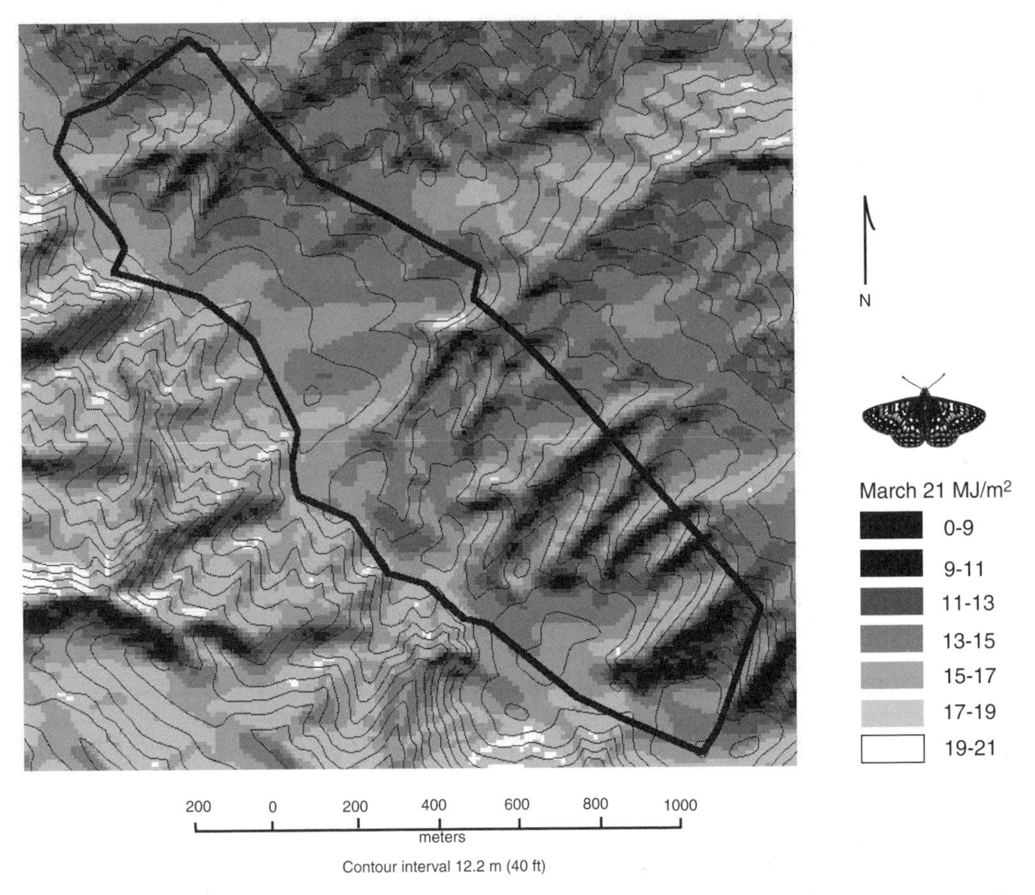

Figure 3.13. Topographic map of the Kirby Canyon Butterfly Preserve (~100 ha) at Morgan Hill showing March 21 insolation. Morgan Hill is the largest and most topographically diverse of the Bay checkerspot habitats.

lation accumulated over sunny days (larvae grow slowly under cloudy and rainy conditions). The model also allows for simulation of adult emergence curves for any given weather sequence, topographic configuration, and distribution of larvae. The results of simulation for the years 1976–89 indicate that the spatial variation in adult emergence (the difference between extreme north- and south-facing slopes) is about 43 days, greater than the amount of year-to-year variation in emergence (28 days) on the same slope. The emergence curve is also affected by the distribution of postdiapause larvae on different slopes. Observed changes in larval distribution were shown to alter mean emergence date in this model by 10–12 days. These results suggest that the size and shape of the emergence curve for adult butterflies is determined by macroclimate, topoclimate, and historical larval distributions and varies among sites due to local topography and differences in larval distributions.

Just as development and emergence vary in space with topography and topoclimate, so does host plant phenology. Recall that larvae experience different host plant conditions in different locations within a given habitat area just as they do in different years (Fleishman et al. 2000b, Hellmann 2002c). To estimate when the host plants undergo the various stages of their life cycle, Weiss et al. (1988) set up transects in the field to assess visually their dates of initial flowering, peak flowering, average flowering, and senescence. *Plantago erecta* phenology at Morgan Hill exhibited a four-week difference in peak flowering between years (1990 and 1991) and a two to four week difference between slope extremes within the same year (south 22°; north 30°; figure 3.14). Data for *Castilleja densiflorus* and *C.*

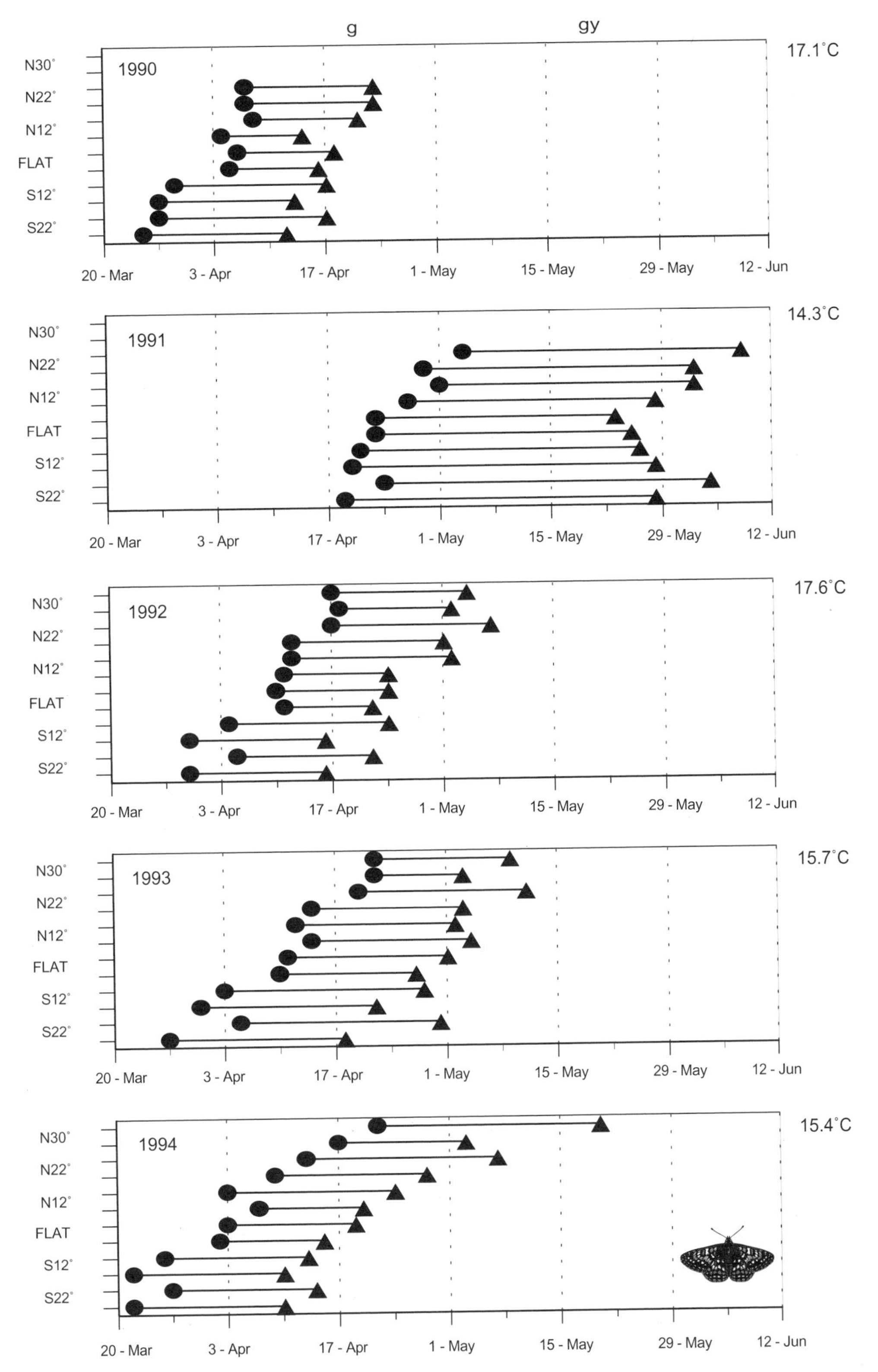

Figure 3.14. Timing of peak flowering (circles) and senescence (triangles) assessed visually across slopes at Morgan Hill from 1990 to 1994 for *Plantago erecta*. Mean temperature in April is given on the right.

exserta were similar, but those species flowered, on average, 10 and 17 days later than *P. erecta*. Warmer temperatures in April shortened the time to senescence, and cooler temperatures prolonged it. Nectar sources also showed the same phenological patterns, so that topographic diversity assures availability of nectar through the flight season. These observations of senescence are similar to other visual assessments made at Jasper Ridge (Dobkin et al. 1987) and are confirmed by nutrient analysis of plants grown in the greenhouse (Hellmann 2002c).

Integration over Space and Time

The distribution of postdiapause larvae across habitats and over time reflects the significance of larval–plant interactions, weather, and topography to Bay checkerspot populations. Singer's early study of spatial patterns of larval survival at Jasper Ridge (Singer 1971b, 1972) and later studies at the larger Morgan Hill site showed that the distribution of postdiapause larvae generally expands and contracts over the topoclimatic gradient. The hypothesis is that the interaction of climate and topography drives this process: favorable weather leads to relatively high survival on a range of slope types, but unfavorable weather precludes survival on all but the coolest of slopes (assuming that host plant resources are equally abundant on slopes of all exposures). At Morgan Hill, for example, postdiapause larvae cover a wide range of slope types when abundance is high, but when population size is small, a greater proportion of larvae are found on moderate to cool slopes (Weiss et al. 1988, 1993). This phenomenon is illustrated for the period 1985–91 in figure 3.8. From 1985 to 1987, the Morgan Hill population grew from 100,000 to nearly 1 million larvae, while at the same time the proportion of larvae on warmer slopes increased markedly. During the large decline from 1988 to 1989, the larval population retreated back to cooler slopes. Postdiapause larvae indicate where prediapause larvae survived to diapause in the previous spring, reflecting patterns of host senescence that vary with slope exposure and gross environmental conditions.

Effect of Other Factors on Population Dynamics

Factors other than prediapause mortality from starvation seem to play a secondary role in the dynamics of *E. editha bayensis* populations. Estimates based on field-collected larvae suggest that parasitism is low in most extant populations in the Bay area (White 1986; chapter 8). Birds and other vertebrate predators are rarely observed feeding on adults or larvae. Spider predation may occur, as does parasitism or predation of pupae (White 1986) and some egg predation. Habitat-wide resource limitation due to larval host plant defoliation has not been observed in populations of this ecotype. Finally, nectar availability may be important in extending adult fecundity and the number of eggs laid (Murphy et al. 1983), but like predation, it does not appear to affect dynamics to the degree that host senescence does. In other populations of *E. editha*, the influence of these factors can be quite different, however. For example, Moore (1989a) found a high level of parasitism, and David Boughton (pers. comm.) reports intense resource competition at Rabbit Meadow in the Sierra Nevada mountains.

3.5 Population Extinctions

Theoretically oriented ecologists have historically been preoccupied with the necessity of some density-dependent factors operating in population dynamics to keep populations from reaching excessive densities. Empiricists have often been more preoccupied with the question of which factors are mostly responsible for the observed population dynamics. Controversies have occurred over how to interpret empirical results in which no evidence for density dependence could be found. These views can be largely reconciled if one assumes that density-dependent factors can act either constantly or so infrequently that they have little observable effect on short-term population behavior. It is in this perspective that we should view the results on *Euphydryas editha bayensis* population dynamics, with the apparent dominance of weather-related density-independent processes.

In this section, we examine the causes of population extinction in *E. editha bayensis*. By analyzing actual extinction events, well documented for the Jasper Ridge populations (figure 3.7), we may uncover which factors affect *E. editha* populations most strongly. Drawing upon the population processes outlined above, we explore five possible explanations for the population losses we have seen at Jasper Ridge (table 3.2).

Table 3.2. Hypotheses about the causes of extinctions in *Euphydryas editha bayensis* populations at Jasper Ridge Biological Preserve.

Hypothesis	Predictions[a]	Evaluation
1. Habitat loss	(a) Extinctions below threshold in habitat number and isolation	(a) Yes
	(b) Spatial synchrony within large habitats	(b) No
2. Weather	(a) Extinctions follow severe or protracted droughts	(a) Yes: JRC; no: JRH
	(b) Extinction in JRC before JRH	(b) Yes
	(c) Weather variables forecast population fluctuations	(c) Yes
	(d) Declining long-term trend in growing season rainfall	(d) No; increasing trend
	(e) Long-term increased variation in rainfall or temperature	(e) Yes; increased variability in rainfall
3. Plant declines	(a) Correlation between plant cover (year $t - 1$) and butterfly population size, N_t	(a) No: *Lasthenia*; yes: *Plantago* and JRH
	(b) Similar responses in JRC and JRH to plant declines	(b) No; inverse responses
4. Parasitoids	(a) Strong or complex endogenous population dynamics	(a) No; endogenous component weak, both populations
	(b) Generalist parasitoid: density-independent mortality	(b) No; asynchronous dynamics, driven by weather
5. Human impact	(a) Detectable effects of destructive sampling	(a) Maybe (Harrison et al. 1991)
	(b) Large population decreases following intensive study	(b) No

[a]JRC, Jasper Ridge habitat area C; JRH, Jasper Ridge habitat area H.

Unfavorable Weather

As outlined in section 3.4, populations of the Bay-area ecotype of *Euphydryas editha* decline after years with extreme weather (hot and dry or cold and wet) because these extremes reduce the temporal overlap of larvae and plants (Singer 1972, Cushman et al. 1994, Hellmann 2002c). Based on this observation, extinctions should follow a sequence of several years of unfavorable weather, conditions that push population size steadily lower. Our understanding of topographic effects on *E. editha bayensis* also suggests that when extinctions do occur, they should happen sooner in habitats with relatively little topographic heterogeneity because topographic diversity reduces the impact of extreme weather (Singer and Ehrlich 1979, Dobkin et al. 1987, Ehrlich and Murphy 1987a, Weiss et al. 1988).

Unfavorable weather, such as droughts in 1976–77 and the late 1980s, and extremely wet winters of 1996 and 1998, did precede large declines in abundance at Jasper Ridge. Extinctions of populations in areas C (1991) and H (1998) followed such events (figure 3.7). Furthermore, extinction occurred first in area C, the larger but more topographically homogeneous patch. More refined predictions have recently been made by McLaughlin et al. (2002b), who modeled the population dynamics that would result from various mechanisms acting on the Jasper Ridge populations, including endogenous (density-dependent) factors, exogenous effects of weather, and combined influences of both endogenous and exogenous factors. The structure of the model is explained in box 3.1. In both areas C and H, McLaughlin et al. found that population fluctuations were largely driven by climatic variability: rainfall in year t predicted population size in year $t + 1$. The timing and the form of the weather effects differed between the two areas. In area C, where the topography is more homogenous and thus microclimates are more uniform, the effect of weather was severe. In area H, where topography and microclimates are more variable, the population showed damped oscillations over time and weaker effects of weather on population size. Differences in the dynamics of the two populations are evident in the dissimilar shapes of the response surfaces produced by the models (figure 3.15).

Historical data indicate that extreme weather became more frequent toward the demise of the populations' at Jasper Ridge. Seasonal rainfall data from San Jose, California (National Climatic Data Center [NCDC] 2001), the longest climate data set available in the region, show an upward but not statistically significant trend in the total seasonal rainfall, but, more important, they show an increase

BOX 3.1 Model of Population Dynamics at Jasper Ridge

Population modeling provides a powerful approach for testing hypotheses about population dynamics (Kendall et al. 1999). Population dynamics can be reconstructed from time-series data using response surface methodology (RSM; Box and Draper 1987, Turchin and Taylor 1992, Turchin 1996). RSM predicts population growth rates using generalized polynomial regression of transformed independent variables. The method is highly flexible: it evaluates diverse functional forms, considers several model dimensions (number of variables included), and can detect nonlinearities and time lags in population dynamics (Turchin 1995). Other methods sharing these advantages (Sugihara and May 1990, Ellner et al. 1991) require longer time series than are available for *E. editha bayensis.*

Changes in population size can be caused by endogenous (density-dependent) feedback, exogenous factors (e.g., weather events), or interactions between endogenous and exogenous factors. These three kinds of dynamics can be modeled using time-lagged population data, weather data, and combinations of population and weather data (Lewellen and Vessey 1998). Details about the modeling method and results for *E. editha bayensis* are in McLaughlin et al. (2002b).

Modeling Endogenous Dynamics To model endogenous effects of multiple factors when only population data are available, RSM uses an equation containing multiple time lags. Equation 3.1 represents per capita rate of population change, N_t/N_{t-1}, as a function of population size in d previous years, N_{t-i}, and exogenous noise, ϵ_t:

$$N_t/N_{t-1} = f(N_{t-1}, N_{t-2}, \ldots, N_{t-d}, \epsilon_t). \tag{3.1}$$

The response surface (figure 3.15) is constructed by approximating f, using polynomial combinations of transformed predictor variables ($X \equiv N_{t-i}^{\theta_1}$) and r log-transformed rate of population change, $r_t \equiv \log_e (N_t/N_{t-1})$. Equation 3.1 reduces to equation 3.2 for a two-dimensional model ($d = 2$) with quadratic dependence ($q = 2$) on transformed predictor variables $X \equiv N_{t-1}^{\theta_1}$, $X \equiv N_{t-2}^{\theta_2}$:

$$r_t = a_0 + a_1X + a_2Y + a_{11}X^2 + a_{22}Y^2 + a_{12}XY + \epsilon_t. \tag{3.2}$$

Model dimension (d), polynomial degree (q), and variable transformations (θ_i) are determined in different steps to avoid selecting spuriously complex models. The best model is selected with cross-validation, which identifies the model with greatest prediction accuracy (Efron and Tibshirani 1993, Turchin 1996). Prediction accuracy is measured with the prediction coefficient of determination, R_p^2 (Turchin 1996), for log-transformed population size ($L_t = \log N_t$), which depends on the ratio of the mean square prediction error to population variance,

$$R_p^2 = 1 - \frac{\sum_i (L_i - \hat{L}_i)^2}{\sum_i (L_i - \bar{L})^2}, \tag{3.3}$$

where $\hat{L}_i$ is abundance predicted in year i, and $\bar{L}$ is the mean of the log-transformed abundances L_i. Potential values for R_p^2 range from 1, indicating perfect prediction, to $-\infty$. RSM includes a given variable only if that variable increases prediction accuracy.

Modeling Exogenous Dynamics Models for exogenously driven changes in *E. editha bayensis* populations contained total growing season (October–April) rainfall

(*continued*)

for each year as a predictor variable. Exogenous models included rainfall data from either the current growing season (W_t), the growing season in the year preceding adult flight (W_{t-1}), or both:

$$N_t/N_{t-1} = f(W_t, W_{t-1}). \qquad (3.4)$$

Mixed Endogenous-Exogenous Models A third class of models includes both lagged population size and total growing season rainfall as predictor variables. Equation 3.5 is a two-variable mixed model containing the endogenous variable N_{t-1} and the exogenous variable, W_{t-1}:

$$N_t/N_{t-1} = f(N_{t-1}, W_{t-1}). \qquad (3.5)$$

In practice, equation 3.5 is transformed to a form analogous to equation 3.2.

Simulating Population Dynamics Simulating population trajectories reveals dynamical properties extracted by each kind of model. Simulations with endogenous models run freely from a starting population size. Simulations with exogenous or mixed endogenous–exogenous models can be run with the observed sequence of precipitation data (McLaughlin et al. 2002b) or with precipitation data representing various climate scenarios (McLaughlin et al. 2002a). Simulation results in figure 3.17 were obtained using random samples of precipitation data (figure 3.16) from decades with low (figure 3.17a, b) and high (figure 3.17c, d) variability.

in variance over time (figure 3.16). McLaughlin et al. (2002a) found that variance in seasonal precipitation was significantly higher in intervals just before the extinction of the two populations than in the early part of the time series. Simulation of population responses confirmed that increased variability can cause greater population fluctuations and increased extinction risk. Based on a shift in the variance of the precipitation record, McLaughlin et al. (2002b) divided the precipitation data into two periods corresponding to the past (pre-1971) and the present climate (post-1971). Using the model described in box 3.1, they simulated replicate trajectories for each population using the data, adjusted to reflect conditions at Jasper Ridge before and after 1971. They found that under the historical regime, conditions prevailing before 1971, simulated populations in area C persisted an average of 444 years and populations in area H an average of 162 years (figure 3.17). Under the current regime (post-1971), however, the average persistence time of the population shifted significantly downward to an average of 19 years for the population in area C and 52 years for the population in area H. This strongly suggests that it was not just unlucky sequences of weather, but shifts in the underlying climate that caused the demise of the Jasper Ridge populations.

Loss of Habitat Connectivity

Numerous factors may have set the stage for climate to drive population extinctions in *Euphydryas editha*. In particular, widespread declines in the area and connectivity of native grasslands likely reduced the available habitat for this butterfly, possibly affecting its ability to withstand weather extremes. Any reduction in population size or rates of butterfly immigration from neighboring populations may have decreased the life span of the Jasper Ridge populations and be the ultimate cause for population losses. Unfortunately, the process of invasion by Eurasian grasses into California was essentially complete before *E. editha* came under scientific study (Murphy and Ehrlich 1989). However, several indirect pieces of evidence support the influence of habitat loss on population sustainability and time to extinction.

Because currently suitable habitat is distributed in discrete patches (figure 3.2), we can use metapopulation theory to examine the likelihood of regional persistence of the Bay checkerspot butterfly (see box

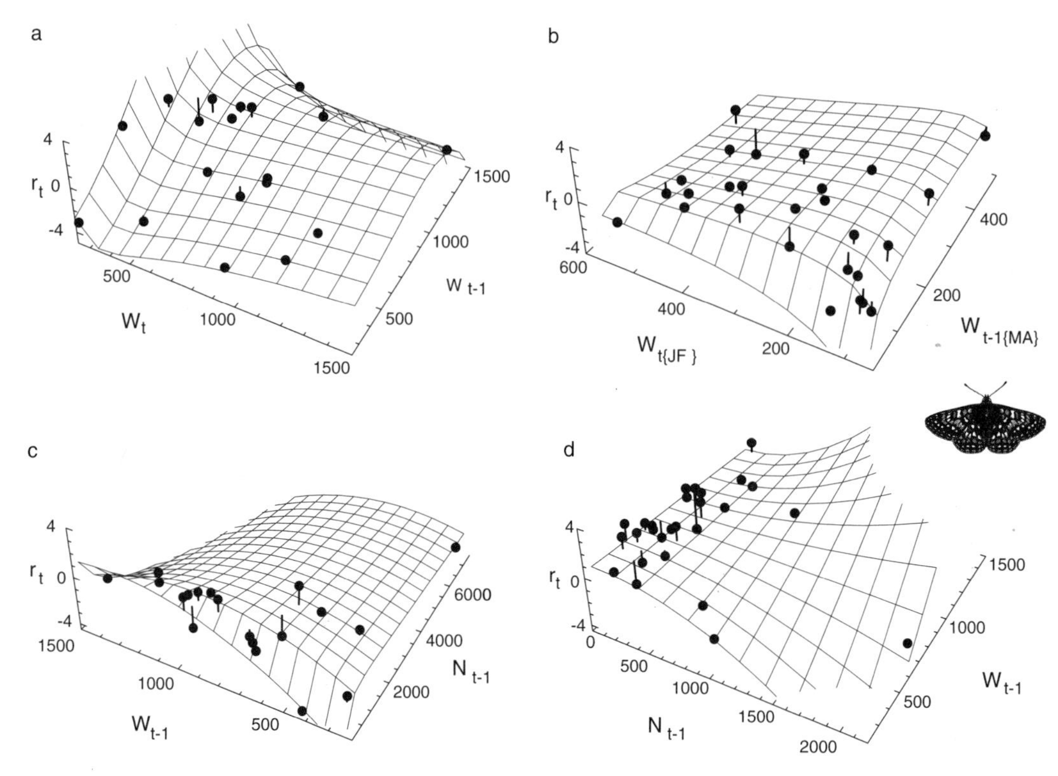

Figure 3.15. Response surfaces for population growth of *Euphydryas editha* at Jasper Ridge. Estimates of growth rate (r_t) are plotted against population size in the previous year (N_{t-1}) and/or rainfall (W) using the exogenous and mixed endogenous–exogenous models of box 3.1 (McLaughlin et al. 2002b). Points are the observed growth rate. The lines from these symbols to the surface illustrate the magnitude of residual variation. The data for the population in area C are shown in (a) using an exogenous model with total growing season rainfall in the current and preceding year as predictor variables and in (c) using a mixed endogenous–exogenous model. The data for the population in area H are shown in (b) using an exogenous model with winter (January–February) rainfall in the current year and spring (March–April) rainfall in the preceding year as predictor variables and in (d) using a mixed endogenous–exogenous model.

12.1). As a rule of thumb, a meta-population consisting of extinction-prone but well-connected local populations should inhabit a network of at least 15–20 patches to be viable (Gurney and Nisbet 1978, Hanski et al. 1995b, Hanski and Ovaskainen 2000). Viability here means that the lifetime of the metapopulation is much longer than the lifetime of an average local population. Empirical data on butterfly metapopulations in Europe support the existence of a threshold value of around 20 patches (Thomas 1994a, Hanski et al. 1995b). This value exceeds the current number of Bay area habitat patches by a factor of at least 2 (Murphy and Weiss 1988a). Furthermore, some of these patches are currently so isolated that it is doubtful whether any migration occurs among them (figure 3.2). A caveat here is that some remaining habitat patches may be sufficiently large (such as Morgan Hill) that their expected time to extinction of their populations is very long. The small and isolated habitats at Jasper Ridge, however, do not likely fall into this category. For the Bay checkerspot butterfly to persist to modern times at Jasper Ridge, the populations in areas C and H must have been linked to large, high-quality "reservoir" habitats that were close by. No such sites have occurred since the 1970s, suggesting that the loss of neighbors or larger habitat areas was the ultimate reason that populations like those at Jasper Ridge could no longer persist. Before the 1970s, the Woodside population, less than 5 km away (figure 3.2), may have filled this reservoir role. This population was reported to have had a very high population density.

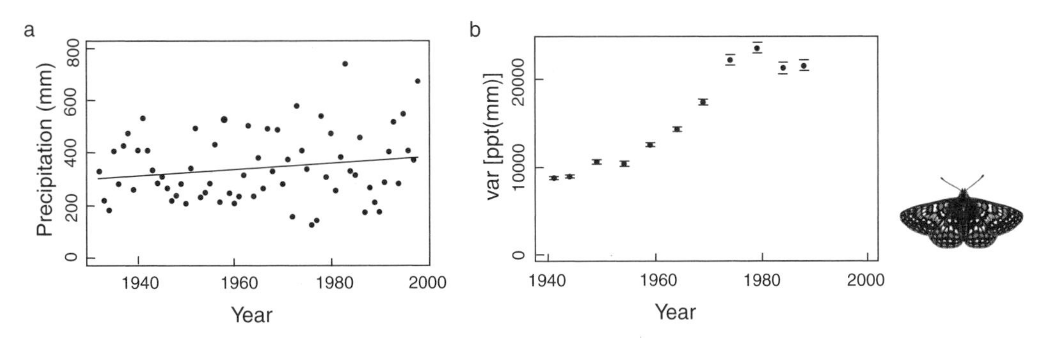

Figure 3.16. Historical precipitation records for San Jose, California, the longest sequence of local weather data on record (McLaughlin et al. 2002a). The total rainfall accumulated during the growing season from 1932 to 2000 is given in (a), where the regression line shows a nonsignificant increase in rainfall over this period ($p = 0.15$). The variance in rainfall, shown in (b), is calculated as bootstrap means and standard deviations in 20-year intervals. Values are plotted at the 11th year of each interval. Variance is significantly higher at the end (just before extinction of *E. editha* at Jasper Ridge) than at the beginning of the time series ($p = .001$).

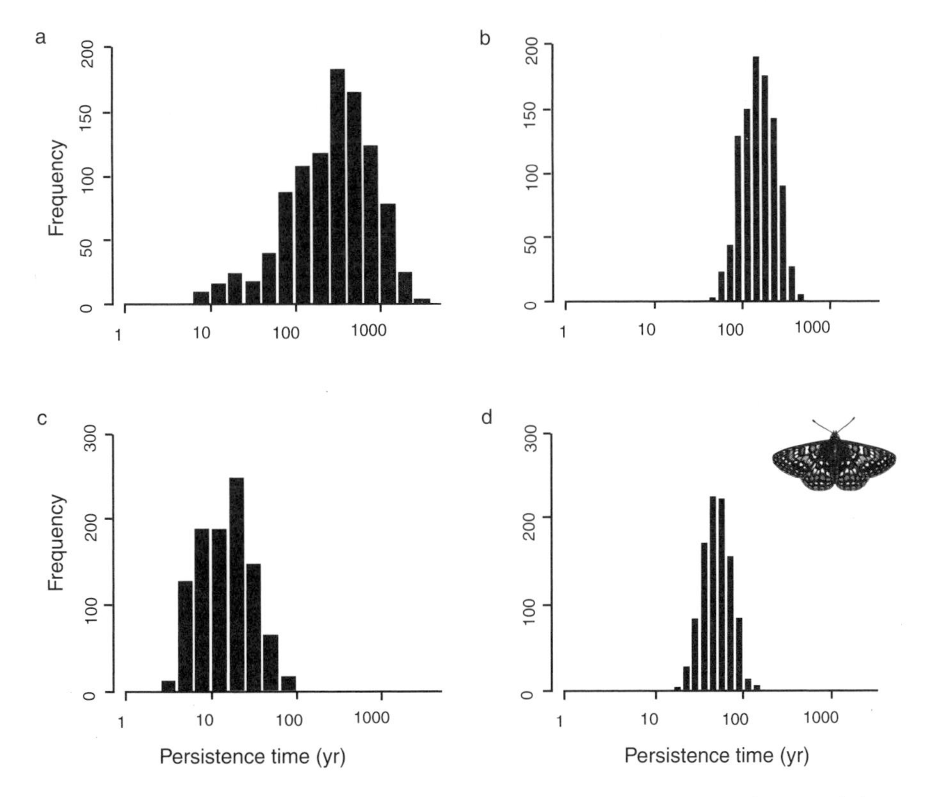

Figure 3.17. Distributions of persistence times in many replicate simulations of the two Jasper Ridge populations under historical and present climate. Simulations were performed using the bootstrapped precipitation data from San Jose (figure 3.16) before and after 1971 and the population models built for Jasper Ridge (McLaughlin et al. 2002a, 2002b; box 3.1). The distribution of times to extinction for simulated populations in area C are given on the left (a and c); the distribution for area H are on the right side (b and d). The mean of the distribution under the historical climate (1932–70), top panels (a and b), is significantly higher than the mean persistence time under the present climate (1971–2000), bottom panels (c and d; $p < .001$).

Loss of serpentine grasslands outside the preserve in more recent decades is not likely to be the proximate cause of declines at Jasper Ridge. The distance between Jasper Ridge and its nearest serpentine habitat exceeds by 100-fold the radius within which adults have been observed to detect habitat and lies at the upper limit of observed adult dispersal (Harrison 1989). Therefore, it is unlikely that the population dynamics at Jasper Ridge were significantly affected by immigration from other populations (or changes in those populations due to any recent habitat losses). Populations at Jasper Ridge may have been rescued by neighboring populations historically, but most recently, the influence of neighbors on the dynamics of the populations at Jasper Ridge was small (see also section 3.6).

Declines in Larval Host Plants or Adult Nectar Sources

Checkerspot butterflies require plants for larval foraging and adult nectaring. Changes in these resources could have caused the extinctions if they became limiting sometime during the lifetime of the Jasper Ridge populations. A decrease in larval host plants one year would reduce adult abundance the following year because adult abundance depends strongly on prediapause larval survival. A decrease in nectar plants also would reduce adult abundance in the following year because nectar abundance limits egg production by adult females, although females emerge with a full or nearly full complement of eggs, and nectar feeding only increases late-season laying (Murphy et al. 1983, Cushman et al. 1994; but see Boggs 1997b; section 3.4). Recall that the timing of food resources is a prominent feature of the ecology of *E. editha bayensis*, but this factor is not necessarily correlated with total plant abundance.

We can use plant data collected in area C from 1983 to 1993 (Hobbs and Mooney 1995; figure 3.18) to determine whether any changes in plant abundance occurred in the period preceding population extinction. Assuming that changes in plant cover were similar in areas C and H, regression analysis of butterfly abundance against the abundance of *Lasthenia californica*, a major nectar source, and *Plantago erecta*, a larval food plant, indicates that plant declines were not responsible for the population losses (table 3.3). Relationships between butterfly abundance and plant cover were weak, except for a correlation between population size in area H and *P. erecta* cover. In addition, regression slopes for areas C and H were opposite in sign, and there is no reason to expect this difference because most factors affecting plants, such as rainfall and nitrogen deposition, would affect both habitats (differences in gopher abundance could affect plant abundance and would be site-specific, however). Further, even in years of lowest cover, *L. californica* and *P. erecta* grew in excess of potential checkerspot consumption. A final possibility concerning plants is that the removal of cattle from Jasper Ridge in 1960 led to the extinctions. If changes in plants resulted from cattle removal, however, they appear to have been completed by 1983,

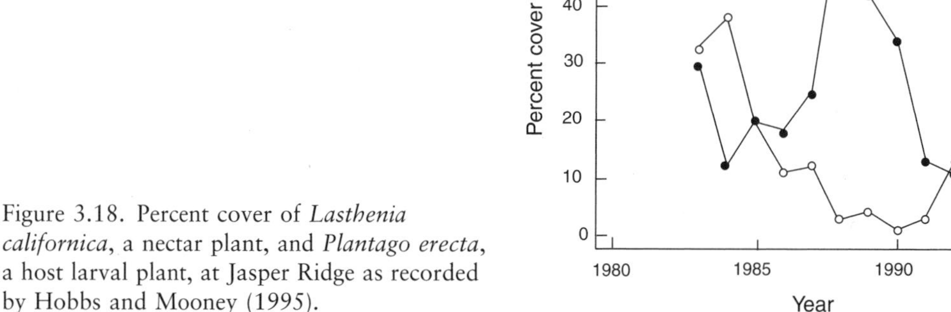

Figure 3.18. Percent cover of *Lasthenia californica*, a nectar plant, and *Plantago erecta*, a host larval plant, at Jasper Ridge as recorded by Hobbs and Mooney (1995).

Table 3.3. Results of simple linear regression of *E. editha bayensis* adult male abundance versus preceding year percent cover of plants used by *E. editha bayensis* adults (*Lasthenia californica*) and larvae (*Plantago erecta*).

Population	Plant Species	n	b	$p\{b = 0\}$	r^2
JRC	*Lasthenia californica*	5	0.15	.271	.38
JRH	*Lasthenia californica*	11	−0.02	.549	.04
JRC	*Plantago erecta*	5	−0.08	.762	.04
JRH	*Plantago erecta*	11	0.06	.015	.50

To normalize distributions, abundance data for males were log-transformed and plant percent cover data were arcsine transformed. Plant cover data were recorded from plots in area C during 1983–93, by Hobbs and Mooney (1995). b = slope of regression line. For area C, regressions include data from 1983–88 only because the abundance of *E. editha bayensis* males reached zero in 1988 and never exceeded four individuals thereafter (see figure 3.18).

and this happened well before population extinction. Nevertheless, we cannot rule out the possibility that cattle removal caused a subtle change in the resources of the butterfly population that took many years to lead to extinction.

Predation

Long-term research with *E. editha bayensis* has failed to find a strong influence of predators or parasitoids on butterfly population dynamics (section 3.4; chapter 8). Nevertheless, there are a number of parasitoids that potentially could have contributed to population losses at Jasper Ridge. In the model built for the Jasper Ridge populations (box 3.1), the effect of three specialist parasitoids should appear in the density-dependent (endogenous) component of the butterfly's dynamics (Ehrlich 1965, McLaughlin et al. 2002b). Extinctions would result either from undamped oscillations or from parasitoid-induced oscillations compounded by density-independent events. However, the endogenous component in the dynamics of both Jasper Ridge populations was weak to nonexistent, suggesting that specialist parasitoids had little effect on population declines. Alternatively, the effect of generalist parasitoids (which attack other hosts in addition to *E. editha*) should appear in the density-independent (exogenous) component of the butterfly's dynamics. Exogenous factors strongly influenced changes in size of the populations of both areas C and H, but a generalist also would have caused different dynamics in areas C and H because the populations were not in close synchrony. Furthermore, such a parasitoid also would have been strongly affected by weather because rainfall explained most of the fluctuations in the Jasper Ridge population models. This combination seems unlikely, however.

Human Disturbance

A final possible explanation for extinction at Jasper Ridge is the destruction of individual butterflies or habitat disturbance caused by people, including researchers. Adults were routinely removed from the populations (Harrison 1991), decreasing reproductive success, and eggs, pupae, or larvae must have been trampled by researchers or visitors. Other impacts could have included trampled plants and compacted soils (Murphy and Ehrlich 1989). If sampling or handling of individuals were responsible for population losses, population decreases should have been particularly severe in years after destructive sampling or intensive study: notably 1964 when all butterflies that were captured were removed from the populations, and 1981 when the population in area H was sampled intensively with mark–recapture methods.

Populations persisted long after the 1964 sampling season—in fact, growing soon thereafter (Hellmann et al. 2003). Using the 1981 data, Harrison et al. (1991) estimated the effects of adult removals from areas C and H. They concluded that population reductions due to destructive sampling were small relative to variability caused by environmental conditions (effects of sampling on population sizes were not statistically detectable). Nevertheless, they estimated that sampling increased extinction risk between 0 and 15%, depending on female egg production before sampling. The effects of intensive sampling can also be evaluated qualitatively by examining population growth rates between 1981 and 1982. For both populations, the growth rate [$r_t = \log(N_t/N_{t-1})$] in 1982 was less than the mean growth rate over all years but was within half of a standard error of the mean (table 3.4; Hellmann et al. 2003). This means that there were not unusually large decreases in population size immediately after 1981. We also can use the models outlined in box 3.1 to predict the change in population size that would have occurred in area H if it were not sampled intensively in 1981. In this case, the model underestimates the observed decline between 1981 and 1982 (estimated: $r_{1982} = -0.237$; observed: $r_{1982} = -0.340$), but the magnitude of the underestimate is within the model's prediction uncertainty (table 3.4; see McLaughlin et al. 2002b). Although the 1981–82 decline in area H was greater than would be predicted from weather conditions alone, it was not unusually large.

Assuming that the difference between predicted and observed population declines in 1982 was due to direct human impact and that other human impacts in other years had similar effects, the difference between observed and predicted growth rates (table 3.4) provides an estimate of the direct impact of human activities. After subtracting this effect, mean growth rates since 1969 remain negative for both the C and H area populations (adjusted means: $r_{JRC} = -0.047$; $r_{JRH} = -0.152$). Hence, research and other direct human factors may have reduced persistence time marginally at Jasper Ridge, but impacts of other factors were considerably more important.

We conclude that the likely dominant factors in population extinction at Jasper Ridge were climate change, as the proximate cause of extinction, and habitat loss in the surroundings, as the ultimate trigger for regional decline of the species.

3.6 Regional Dynamics

Although the Jasper Ridge populations have disappeared, other populations of *Euphydryas editha bayensis* still persist. Widespread events, like drought or heavy rain, affect multiple sites, but local events cause widely separated populations to have qualitatively distinct dynamics. Small amounts of dispersal do link habitats and populations together, though not strongly enough to synchronize their dynamics. Dispersal of even a few individuals can also link distant populations genetically, preventing genetic divergence over time at neutral loci (Roughgarden 1979; chapter 10).

A number of researchers have investigated adult movements of *E. editha* with some conflicting results. Brussard et al. (1974) showed that Bay checkerspot adults fly relatively freely within their local habitat and that sexes showed relatively little difference in movement patterns. They also found no apparent pattern that restricted intrapopulation gene flow; thus local populations should be relatively well-mixed. When a butterfly "decides" to leave, albeit infrequently, it typically moves steadily across nonhabitat in a directed fashion. Work by Harrison (1989) showed that butterflies released in nonhabitat areas do not orient toward distant habitant patchs.

As for the propensity to move, Brussard et al. (1974) observed higher rates of movement at the beginning and end of the flight season in *E. editha bayensis*. Similarly, Murphy and White (1984) suggested that populations of *E. editha* in southern

Table 3.4. Population declines from 1981 to 1982 in Jasper Ridge *E. editha bayensis* populations and declines predicted by models of population responses to weather conditions (McLaughlin et al. 2002b).

Population	r_{1982}	$\bar{r}_t$	SE($\bar{r}_t$)	$\hat{r}_{>1982}$	Prediction Error ($r_{1982} - \hat{r}_{1982}$)	SE ($\overline{r_t - \hat{r}_t}$)
JRC	−0.463	−0.297	0.381	−0.213	−0.250	0.156
JRH	−0.340	−0.255	0.233	−0.237	−0.103	0.158

Differences between observed and predicted declines provide an upper estimate of the impact of intensive mark–recapture sampling conducted in 1981. Growth rates were calculated as $r_t = \log(N_t/N_{t-1})$. r_{1982} is the growth rate observed in $\hat{r}_{1982}$ is the growth rate predicted in 1982 using population response models. $\bar{r}_t$ is the mean observed growth rate over all years. SE ($\overline{r_t - \hat{r}_t}$) is the standard error of the mean difference between observed and predicted growth rates. For JRH, the prediction error in 1982 is within one standard error of the mean prediction error (mean = 0).

California show variable dispersal rates and changes in vagility across generations. In these southern populations, changes were largely driven by changes in resource availability: when resources were scarce, vagility increased. Work by Sisk et al. (2004), in contrast, suggested that dispersal rates of *E. editha bayensis* do not increase when resources are limiting at the end of a flight season. They also did not find a seasonal, locational, or interannual difference in dispersal rates. The relative constancy of low interpatch dispersal at Jasper Ridge over time also suggests that dispersal in *E. editha bayensis* is not strongly correlated with population size.

Evidence for Metapopulations

A metapopulation is a set of local populations connected by migration (Harrison et al. 1988, Hanski 1999b). In metapopulation theory, the risk of extinction of a local population often is assumed to be a function of the area of the respective habitat patch (box 12.1). The probability that a currently unoccupied site will receive colonists from existing populations is assumed to decline with distance. Because populations of *E. editha bayensis* appear to go extinct relatively frequently but exist in a shifting mosaic of distinct populations, some groups of habitat patches may support metapopulations.

To test this hypothesis, Harrison et al. (1988) examined the occupancy and characteristics of small habitat patches surrounding the Morgan Hill colony (figure 3.2). Initial surveys suggested that distance from Morgan Hill was a key factor in whether a patch was occupied by checkerspots. In 1986, 8 small patches within 4.4 km of Morgan Hill were occupied, but none of a possible 15 patches outside this radius were occupied. Assuming that the small habitat patches have ephemeral local populations, these results suggest that the maximum colonization distances are around 4–5 km, which is the same as for *Melitaea cinxia* (van Nouhuys and Hanski 2002a; chapter 4). Harrison (1989) found that butterflies released into nonhabitat could fly up to several kilometers and not locate suitable habitat. She also estimated, however, that the probability that a dispersing female successfully reproduces in her new habitat is small. Transplants of postdiapause larvae in 38 patches produced adults in only 6 sites 1 year later. Harrison et al. (1988) found that a logistic regression including distance from Morgan Hill, topographic composition, and abundance of larval and nectar plants significantly explained the presence or absence of butterflies in 59 potential patches in the neighborhood of Morgan Hill. This implies that both distance from the source population and habitat quality (topography and plant abundance) determine patch occupancy. Because of this distance-dependent occupancy, local populations in the Morgan Hill area appear to constitute a metapopulation. In particular, they conform to the mainland–island model of metapopulations—a metapopulation with one large source of colonists surrounded by smaller, extinction-prone satellites. The other metapopulation study of *E. editha*, at Rabbit Meadow (chapter 9), exhibits a structure more akin to the classic model (box 12.1) than to the mainland–island model (Hanski 1999b), where all local populations have a significant risk of extinction (C. D. Thomas et al. 1996).

It is important to realize that the total abundance and number of sites occupied by *E. editha bayensis* appears to be steadily declining over time. Recent human development and the invasion of historical habitats by grasses has likely increased the isolation between the sites (considering both actual distances and the hospitality of interpatch matrix to movement), which would decrease the number of occupied sites. As we argued above, there currently are too few habitat patches on the San Francisco Peninsula, including Jasper Ridge, to sustain a metapopulation over the long-term. It also seems unlikely that remaining sites on the peninsula are close enough for much interpatch colonization to occur (figure 3.2). Neighboring sites, which were fairly distant by the 1990s and with relatively small population sizes, did not rescue the populations at Jasper Ridge from extinction. For that matter, the long-lasting population in area H did not rescue its nearest-neighbor populations from extinction (the populations in areas C and G).

Genetic Similarities among Local Populations

Dispersal affects the genetic composition of populations as well as their abundances. Baughman et al. (1990a) examined the genetic differentiation among *E. editha* populations (including *bayensis*) to determine the degree to which populations and subspecies shared either a recent common ancestor or current gene flow. Examining allozyme variation at 19 loci, they concluded that either a significant amount of gene flow occurs among populations or there is strong stabilizing selection. These patterns may reflect historical connectivity, rather than current gene flow, however (Slatkin 1987). Before the invasion of non-native grasses and/or before the last glaciation (8,000–10,000 years ago), checkerspot populations may have had greater dispersal among habitat patches (section 3.5). Alternatively, the amount of dispersal that could be inferred from Harrison et al.'s (1988) results may be sufficient to prevent significant genetic differentiation among *E. editha bayensis* populations over time. It is not possible at present to estimate which of these explanations is more likely.

Other characters such as morphology and host plant use tell a very different story of population differentiation. Groups of *E. editha* populations defined on the basis of allozyme data do not correlate well with groups delineated by morphology (subspecies classifications) or host plant use (Baughman et al. 1990a). This suggests that host choice is not a good indicator of genetic relatedness or gene flow via dispersal among populations; diet seems to be a relatively plastic trait, and ecotypes using new hosts seem to evolve relatively quickly (Singer et al. 1992a, Radtkey and Singer 1995, C. D. Thomas et al. 1996). This lack of correlation also suggests that the subspecies that have been described for *E. editha* are not natural genetic units.

3.7 Other *Euphydryas editha* Ecotypes

To put the work described above in a broader context, we briefly contrast the qualities of Bay area populations with other *E. editha* populations. One of the early results of extending research beyond Jasper Ridge was that *E. editha* did not have an "ecology" per se but rather that populations and groups of populations had distinctive sets of ecological relationships (Gilbert and Singer 1973, Ehrlich et al. 1975). One of the simplest ways to contrast these relationships is to examine population responses to the abiotic environment, notably drought. We already mentioned (section 3.4) that *E. editha bayensis* populations respond negatively to drought, presumably via effects on the timing of host senescence relative to prediapause larval growth. Populations of *E. editha* in the inner coast range (*E. editha luestherae*) show more variable responses to drought than do *E. editha bayensis*. Two of three study populations remained constant after the 1975–77 drought, for example, but one, at Del Puerto Canyon (Figs. 3.1, 3.11), plunged to a low of three individuals in 1978 before rebounding the next year. Del Puerto Canyon is one *E. editha* population that appears to show strong density dependence in its dynamics. Intensive competition and overexploitation of food resources is responsible for larval starvation at this site, and drought appears to decrease the abundance of the dominant host, *Pedicularis*, thereby exacerbating resource competition.

In southern California, *E. editha quino* populations are similar to *E. editha bayensis* in the ephemeral nature of their hosts and the effect of weather on phenology, but dissimilar in the frequency with which individuals disperse from their natal habitat in response to weather events. The drought of 1975–77 was less severe in southern than in central California, and populations actually experienced great population explosions during this period. At this time, dispersal out of habitat patches also was observed on a number of occasions (Murphy and White 1984). Populations remain sedentary when food resources are plentiful, but dispersal increases after extreme scarcity of larval host plants arising from high population densities (White 1974). Populations of this ecotype are notorious for disappearing from sites and reappearing, presumably because they are easily extirpated from small or otherwise marginal habitats. Between outbreak years, the ecotype presumably persists in only the largest "reservoir" habitats, from which migrants reoccupy many temporarily vacant sites.

In populations of *E. editha* that feed on *Collinsia tinctoria* between 150 and 1800 m elevation on the western slopes of the Sierra Nevada, adults are able to respond adaptively to host phenology and to oviposit on plants that will last long enough to support larvae to diapause (Singer 1971b). As a consequence, mortality of prediapause larvae from host senescence only occurs in severe drought years. However, these insects are also susceptible to climate in the early postdiapause stage. In the drought year of 1977, most *Collinsia* at these sites remained in the seed bank. At one site, Agua Fria, host density was reduced by two orders of magnitude, and only 23 plants could be found (Singer and Ehrlich 1979). Larvae that emerged from diapause would have been unable to find sufficient food to fuel their return to diapause until the following year. Three of the four studied populations of this ecotype became extinct at this time. The sole survivor of this group of populations was the one (Indian Flat) that covered the largest area and had the greatest diversity of host senescence times due to its many different slope aspects.

In populations at moderate elevations (2000 m), annual hosts undergo senescence but perennial hosts do not, while at the highest elevations (above 3000 m), frequent thunderstorms keep vegetation edible all summer even when low elevations are experiencing drought (figure 3.19). One example of a metapopulation relying on both an annual and a perennial host plant occurs at Rabbit Meadow (2300–2400 m; Singer and Thomas 1996, C. D. Thomas et al. 1996, Boughton 1999). Cleared patches where the annual host was used for oviposition occurred interspersed with undisturbed patches where the (traditional) perennial host was used. Insects in the clearings developed fast and were able to oviposit on phenologically suitable annual hosts. Slower developing insects in the undisturbed patches did migrate to clearings but arrived too late to produce surviving offspring (Boughton 1999). When clearings were unoccupied, the phenological differences between the patches generated a set of sources and sinks, with clearings acting as sinks. When clearings were occupied, these source–sink relationships were reversed (this system is described in greater detail in chapter 9). It is interesting that larval survival in adjacent populations with different hosts was negatively correlated in this system, so that conditions that were favorable in one habitat were unfavorable in the other. This situation is analogous to that in *E. editha bayensis* populations, where different habitats and slopes with dissimilar topography and host plants buffer populations against fluctuations in the abiotic environment.

Outside of California, we know relatively little about the mechanisms controlling *E. editha* population size. In Colorado, the occurrence of low-rainfall years seems to play a role (Holdren and Ehrlich 1982), and drought may have been influential in recent extinctions of coastal populations in British Columbia (subspecies *taylorii*) (Guppy and Fischer 2001). For each of the *E. editha* ecotypes, the response of a population to extreme weather appears to depend on the details of the relationships between the insect and its host plants. We can generalize from *E. editha bayensis* that ac-

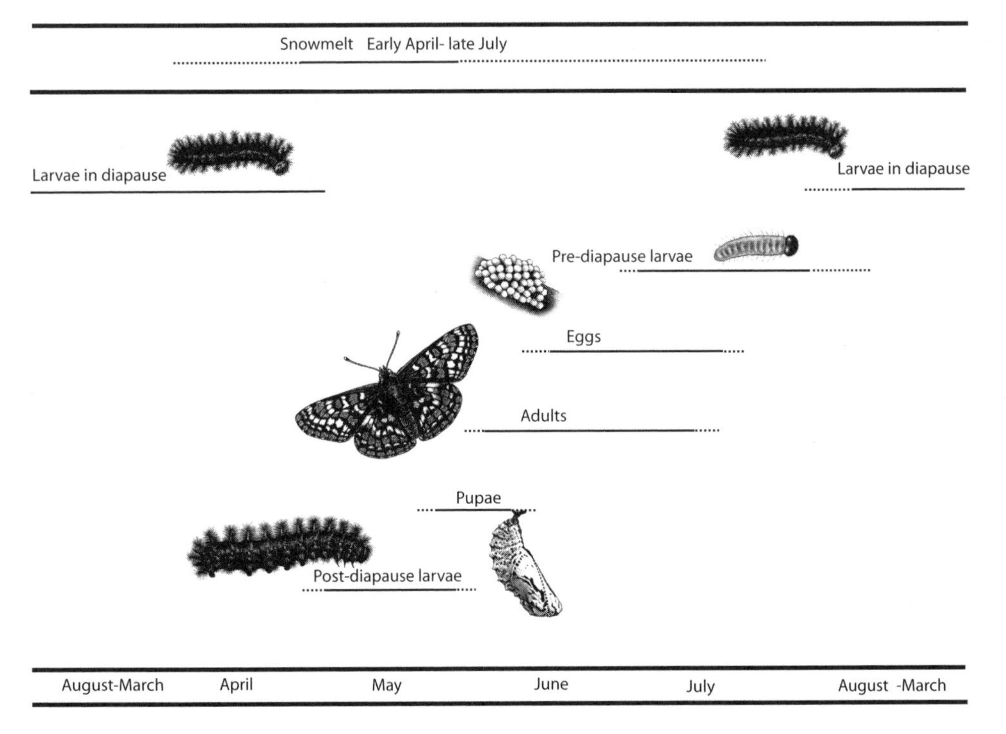

Figure 3.19. Life cycle diagram of noncoastal *E. editha* populations. This life cycle is representative of populations residing at high elevations in the Sierra Nevada. These populations are univoltine, like *E. editha bayensis* (figure 2.2), but diapause during the winter months. Drawings by Zdravko Kolev.

cess (either temporal or sheer abundance) to host resources is an essential, and often the principle, factor limiting population size.

3.8 Conclusion

The type of detailed understanding of local populations outlined in this chapter is essential for managing and conserving checkerspots and other insect species (chapter 13). In the control of pests, for example, an understanding of population abundance, location, and timing, as well as variation in these factors over time, are useful in improving efficiency and effectiveness of pesticide application. To conserve endangered insects, it is important to understand which factors buffer against population declines and which portions of a landscape serve as refugia. All factors taken together, timing of life-history events and the synchrony of butterflies and their resources, appears particularly important in *E. editha bayensis* and possibly other *E. editha* ecotypes and insect taxa. A major theme emerging from studies of this species is the significance of resource diversification in terms of habitat heterogeneity and resource use. This concept has been referred to as a portfolio effect (Boughton and Singer 2004), much like managing stocks on the stock market: diversification buffers against fluctuating conditions.

The long history of research on *E. editha* has provided a number of particular lessons for population ecology. First is the realization that it is critical to identify demographic units if sensible conclusions are to be drawn about the dynamics of populations. Second is the recognition that demographic units commonly go extinct. Third, conserving large, topographically diverse habitats with multiple plant resources helps populations persist through extreme environmental events (Ehrlich and Murphy 1987a, Weiss and Murphy 1993, Hellmann 2002c, Hellmann et al. 2004). Fourth, the finding that climate change can cause population losses confirms that regional changes in the environment will threaten native biota (Murphy and Weiss 1992, Hill et al. 1999b, Parmesan et al. 1999, C. D. Thomas and Lennon 1999, Hughes 2000, McCarty 2001, McLaughlin et al. 2002a). Phenology and indirect impacts of climate change through host plants may be important ways that taxa are susceptible to climate change (Hellmann 2002c). The ultimate product of research on the Bay checkerspot is the mechanistic understanding of factors controlling population dynamics, as well as a grasp of the complex interactions of these factors across space and time.

4

Structure and Dynamics of *Melitaea cinxia* Metapopulations

MARKO NIEMINEN, MIKA SILJANDER, AND ILKKA HANSKI

4.1 Introduction

To place our research on the Glanville fritillary butterfly (*Melitaea cinxia*) in a historical perspective, consider the state of metapopulation ecology in 1990. The earlier history, reviewed by Hanski (1999b), includes outstanding conceptual contributions by Wright (1940) in population genetics and evolution and by Nicholson (1933) and Andrewartha and Birch (1954) in population ecology. Several significant though somewhat isolated studies appeared in the 1950s and 1960s. Huffaker (1958) showed with a laboratory system of herbivorous mites and predatory mites that patchy habitat led to sufficient spatial asynchrony in population densities to allow the two species to coexist when they could not coexist in a more uniform habitat (the predator would cause the extinction of the prey). Den Boer (1968) argued on the basis of his studies on ground beetles that assemblages of local populations without strong density dependence but connected by some migration would persist by virtue of being influenced by different sequences of environmental perturbations—the spreading-of-risk idea. Gadgil (1971) showed for the first time how habitat patchiness would influence the evolution of a key life-history trait, migration rate. And then there was the illustrious work of Levins (1969), who proposed a simple mathematical model to demonstrate the possibility of metapopulation-level persistence of a set of extinction-prone local populations, essentially because he assumed that different populations have independent dynamics, and hence there was time for recolonizations to compensate for local extinctions before the entire metapopulation would go extinct. But in spite of these studies, the field was still wide open in 1990, as is shown by the few citations to the key word "metapopulation" before 1990 and the exponential growth of such citations afterwards (Hanski and Simberloff 1997).

In terms of metapopulation theory, the pre-1990 days had seen an upsurge of simple metapopulation models following Levins (1969, 1970) for competing species (Levins and Culver 1971, Horn and MacArthur 1972, Slatkin 1974, Hanski 1983) and for predator–prey metapopulations (Maynard Smith 1974, Hilborn 1975, Nachman 1987, Taylor 1988). Empirical metapopulation studies were just starting to appear. Harrison (1988, 1989, 1991) described how *Euphydryas editha* had a mainland–island metapopulation structure in Morgan Hill in California (chapter 3): one large habitat patch supported a large population that functioned as a source of colonists to outlying smaller patches of habitat (figure 3.2) with smaller ephemeral populations. Rolstad and Wegge (1987) and Soulé et al. (1988) demonstrated how habitat fragmentation influenced the occurrence of bird species in boreal forests and California chaparral vegetation, respectively. Smith (1980) had established a long-term study of the American pika in California, in a man-made, highly fragmented landscape that supported a pika metapopulation of many small local populations. And Menges (1988, 1990) reported a pioneering study on a plant metapopulation in which population turnover (extinctions and colonizations) was largely driven by successional changes in the habitat. But the empirical and theoretical studies were largely separated from each other. A few exceptions in this respect were studies by Hanski and Ranta (1983) on competing metapopulations of water fleas in rock pools and by Nachman (1988) on mite predator–prey metapopulations in a greenhouse. It may be added that spatial ecology in general was in its infancy, and landscape ecology was still being defined (Forman and Godron 1986, Turner 1989, Wiens et al. 1993).

The *Melitaea cinxia* study was conceived from the beginning as a test system for the prevailing metapopulation ideas and model predictions (Hanski 1999b). There is an interesting link to the long-term *Euphydryas editha* study in Stanford apart from Susan Harrison's work. Paul Ehrlich's visit to Helsinki in autumn 1990 influenced the decision by Hanski to select a checkerspot rather than some other species to start a large-scale metapopulation project. The first goal was to find out whether *M. cinxia* had a spatial population structure that would qualify it as a model system for classic metapopulation—in other words, whether it consisted of many local populations inhabiting small patches of habitat and connected by migration. It did (Hanski et al. 1994). Even before knowing that, the ambitious goal was to test the existence of alternative stable states in metapopulation dynamics, which was achieved in 1995 (Hanski et al. 1995b). But before going into any more details, let us take a closer look at the central organism.

Melitaea cinxia is widely distributed in the Palaearctic region from western continental Europe to the steppes in southern Siberia (Tolman and Lewington 1997) and up to the Arctic Circle in central Jakutia (Korshunov and Gorbunov 1995). In Europe, *M. cinxia* is relatively continuously distributed in the eastern and many southern regions, but toward the north the distribution becomes increasingly fragmented. There is also a disjunct occurrence of the species in the Atlas Mountains in North Africa (for a detailed map of the European distribution, see Hanski and Kuussaari 1995). Over the past decades, *M. cinxia* has declined substantially in many parts of central Western Europe due to human destruction of its habitats (e.g., Maes and Van Dyck 2001). A similar though not as severe decline has occurred in northern Europe, for example, in Finland, where *M. cinxia* is now restricted to the Åland Islands in the northern Baltic. The extinction of *M. cinxia* from mainland Finland was due to cessation of cattle and sheep grazing and overgrowth of meadows. However, even in the historical records the species only occurred sparsely in the extreme southwestern corner of Finland, most likely restricted by its host plants (described below), which occur at their northern range limits in southern Finland.

Melitaea cinxia occurs in a highly fragmented landscape in our main study area in the Åland Islands. Suitable habitat is similarly highly fragmented elsewhere at the northern range boundary—for example, in parts of Estonia, Sweden, and Great Britain, as well as at higher altitudes on mountains, such as in southeastern France. The pattern of distribution at the southern range boundary in Spain and elsewhere is not well known, but it is likely to be fragmented as in the north. Between the range boundaries, open grassland suitable for *M. cinxia* still occurs in quite extensive areas—for instance, in western Estonia (the Saaremaa island), southeastern Sweden (the Öland island), southeastern France, and Tuva province in Russia (southeast of Lake Baikal). Some of the open grassland habitats suitable for *M. cinxia* are natural; others have been created and maintained to varying degrees by traditional agriculture and other forms of human land use. The extensive steppes in southern Siberia have grassland suitable for *M. cinxia* (with appropriate host plants), but the actual habitat for reproduction may be highly fragmented (on north-facing slopes where the host plants do not regularly dry out) and temporally unstable (due to extensive grass fires).

In this chapter, we first present a brief overview of the life history of *M. cinxia* and the structure of its habitat in the Åland Islands. The backbone of the *M. cinxia* project in Åland is an annual monitoring of all existing local populations and unoccupied habitat patches, which we describe in section 4.3. This work also involves the systematic description of the landscape structure using the parameters that have been deemed significant for the butterfly. Sections 4.4 and 4.5 examine local population dynamics and large-scale dynamics, respectively. Though space does not allow a comprehensive treatment of everything that has been done and that might be of importance, we nonetheless attempt to discuss systematically all the factors and processes known to significantly influence the population dynamics of *M. cinxia*. Section 4.6 focuses specifically on patterns and processes of classic metapopulation dynamics, which include the effects of habitat fragment size and connectivity on the processes of local extinction and establishment of new populations. In the final section we justify the conclusion that the *M. cinxia* metapopulation in the Åland Islands is indeed an excellent model system to study classic metapopulation dynamics and metapopulation biology more generally. The discussion is then expanded to cover some more general messages and implications of our work.

4.2 Life History and Habitat

Melitaea cinxia is mainly univoltine in Europe, though bivoltine populations occur in parts of southwestern Europe. In Åland, adults fly in June to early July. Females usually mate once in their lifetime (chapter 5), and they lay eggs on either the ribwort plantain (called narrow-leaved plantain in North America), *Plantago lanceolata,* or the spiked speedwell, *Veronica spicata* (figure 4.1). These two plants are closely related, and currently the two genera are considered to belong to the same family "clade Veronicaceae" by Olmstead et al. (2001) (though the valid name for the family appears to be Plantaginaceae [Judd et al. 1999, Wahlberg 2001c]). Several other species of *Plantago* and *Veronica* in Åland are never or extremely rarely used (see table 7.2). Eggs are laid in clusters of 100–200 eggs on the underside of leaves located close to the ground (2.5 cm above the ground on average; see plates IX and X). Larvae hatch in two to three weeks and forage gregariously. They spin a communal web on the plant, and they stay inside the web during the night and in bad weather. They also frequently molt in the web. Half-grown larvae overwinter in a compact winter nest (figure 2.5, plate IX), which they spin at the end of August. Larvae continue feeding in the spring as soon as the host plants commence growth in late March or early April. Larvae continue to feed gregariously until the final instar. Pupation takes place in May within tussocks or in litter. The pupal stage lasts two to three weeks. Further details of the life cycle can be found in Kuussaari (1998).

The two host plants grow on dry, rocky outcrops, meadows, pastures, and road verges (figure 4.2). *Plantago lanceolata* occurs in all parts of Åland, but *V. spicata* mainly occurs in the west, northwest, and north, and it is completely lacking from the eastern Åland Islands (figure 2.3; see also chapter 7). We have located about 4,000 habitat patches that contain one or both host plant species. Many of the patches are very small (average area 1500 m^2, median 300 m^2, maximum 10 ha, n = 3951; based on 1999 data). The habitat patches form clusters or habitat patch networks (figure 4.3), which are occupied by butterfly metapopulations

Figure 4.1. The larval host plants of *Melitaea cinxia*, *Plantago lanceolata* (left) and *Veronica spicata* (right). Drawings by Zdravko Kolev.

that are dynamically relatively independent of each other. A habitat patch network may also be completely unoccupied in a given year if the respective metapopulation has gone extinct. There is a substantial amount of movement of butterflies among patches within networks, but little movement between neighboring networks.

There are several key features of the landscape and the species that make *M. cinxia* in the Åland Islands such a convenient model system for the study of classic metapopulation dynamics. The habitat occurs in distinct patches that can be readily delimited from unsuitable areas. The habitat patches are mostly located close to cultivated areas and farms, which means that they are relatively easy to find and are also easy to access during field work. Furthermore, none of the patches is very large, rather, most are very small, resulting in small population sizes and therefore in high yearly turnover rate. The key feature of the butterfly that makes it such a convenient study species is the high detectability of all developmental stages except the final larval instar and the pupal stage. In particular, the winter nests and adult butterflies are easy to find, and hence presence/absence of breeding populations on meadows can be detected with high reliability by trained observers.

4.3 Monitoring Metapopulations in a Highly Fragmented Landscape

Description of Landscape Structure

We have assembled a detailed description of the landscape structure in Åland during 1998–2001. For each habitat patch, a surveyor first determined the boundaries of the patch by thoroughly walking through the patch and its close surroundings. Decisions about boundaries were based on simple rules. First, all open area within a meadow belongs to the same patch, but parts of large, open areas are separate patches if they are separated from each other by approximately 20 m of unsuitable habitat (e.g., cultivated field, tall grass) or by approximately 50 m of otherwise suitable habitat but which does not contain host plants. Small patches often have a high density of host plants, whereas the largest patches often include a large proportion of habitat with adult nectar plants but no larval host plants. Second, all unsuitable habitat at the edge of the patch is left out of the patch. We use this rule to avoid including sometimes vast areas (compared to patch sizes) of completely unsuitable tall grassland. Third, parts of a single pasture are different patches if the host plants are highly concentrated in those parts.

Figure 4.2. Many habitat patches (dry meadows) used by *Melitaea cinxia* in the Åland Islands are located close to houses (see also plate XII). Photo by Saskya van Nouhuys.

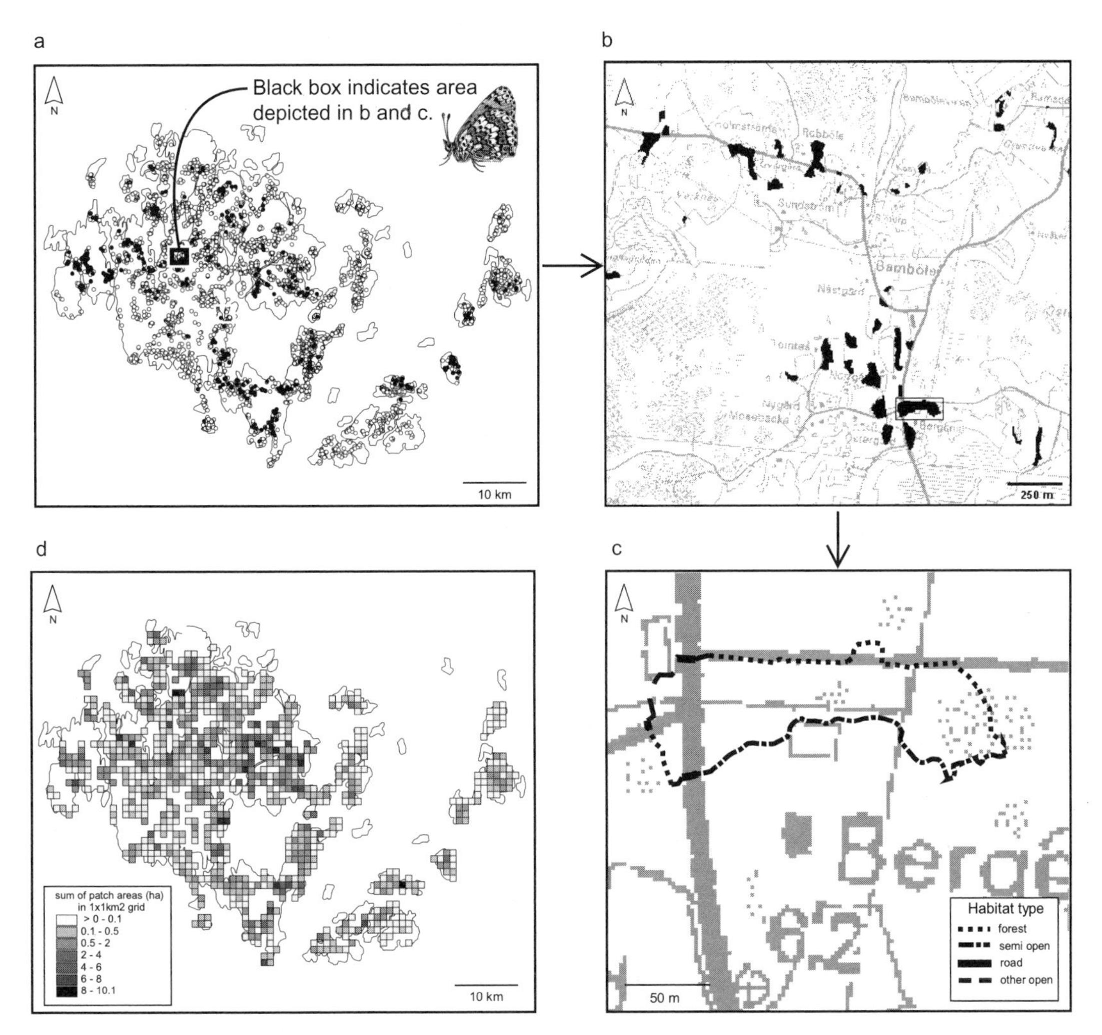

Figure 4.3. (a) Habitat patches that were occupied by *M. cinxia* (black dots) or were empty (open circles) in 2001. The box indicates the location of the area in (b). (b) An example of geographic positioning system (GPS)-delimited habitat patches. The box indicates the location of the area in (c). (c) An example of a GPS-delimited habitat patch with surrounding habitat types indicated. (d) The density of suitable habitat for *M. cinxia* in 1-km^2 squares.

We use this rule to include only those parts of large pasture areas that could potentially support some larval groups of *M. cinxia*. However, decisions cannot always be based only on such rules due to the heterogeneity in the field. For example, ecotones between habitat patches and surrounding habitats can be problematic when an open meadow gradually changes into semi-open woodland with small openings. Sometimes the surveyor must arbitrarily decide where to draw the boundary, as it could be anywhere within several or even tens of meters. Furthermore, some clearly separate landscape features such as rocky outcrops among cultivated fields define different patches even if they are located closer to each other than 20 m. The abundance of host plants also changes all the time, and some patches may not have more than single host individuals in a given year, whereas in the next year there may be several clusters of host plants. Such changes lead to the risk of omitting patches that are seemingly unsuitable in one year but suitable in the long run.

After a habitat patch was defined, the surveyor walked around the patch at its boundary and collected boundary coordinates with a submeter-accurate global positioning system. At the same time, codes of the surrounding habitat types were stored in the coordinate file (figure 4.3). These data were subsequently transferred into a geographic information system (GIS) where they could be laid over 1:20,000 basic topographic maps. The outcome of this exercise was the exact location, area, and perimeter of a patch and information on the habitat surrounding the patch. With GIS, maps are prepared for use in the field and various spatial analyses can be performed.

Annual Monitoring of Population Sizes

The occurrence of *M. cinxia* larval groups in each habitat patch is monitored in the autumn just after the larvae have spun the conspicuous winter nests (at the end of August). The monitoring of about 4000 habitat patches is conducted by 40 students working in pairs for 2 weeks. Each habitat patch is visited for a fixed period of time, the length of which is related to patch area. Additionally, if no larval webs are found in the given time, the patch is surveyed for same amount of time again to make the record of presence/absence as reliable as possible. The numbers of larval groups on each host plant species are counted. Of the approximately 15,000 larval groups recorded in Åland so far, less than 0.5% have been found on *Plantago* or *Veronica* species other than the actual host species. During the same visit to the habitat patch, several environmental variables are recorded, including grazing status, relative abundance and coverage of each host plant species, proportion of desiccated host plants, degree of overgrowth, and the abundance of nectar plants. Each of these variables affects the dynamics of *M. cinxia*.

Using various controls, we have estimated that around 50% of all larval groups are found. Unfortunately, even if extra time and effort is spent in finding out the true occupancy status of each habitat patch, we estimate that 10–15% of occupied patches are falsely recorded as empty. In the context of metapopulation modeling, false absences in large or isolated habitat patches may severely bias parameter estimates. To overcome this problem, Moilanen (2002) developed a Monte Carlo method that takes into account the probability of false absences while assessing extinction and recolonization rates. This method requires that the probability of false absences is estimated in an independent study, and Moilanen recommended that a check should be made on the sensitivity of other model parameters to small errors in the estimation of the probability of false absences.

We recheck the status of each local population found in the autumn the following spring. The spring survey provides information about the extent of winter mortality, and it also allows the numbers of cocoons of the specialist parasitoid *Cotesia melitaearum* to be counted (section 8.2).

4.4 Dynamics of Local Populations

The local populations of *M. cinxia* in Åland are mostly so small and prone to extinction that the bulk of the research has naturally become directed toward the processes of local extinction and establishment of new populations rather than toward classical life-table (Varley et al. 1973b) and *k*-factor analyses (Varley and Gradwell 1960, Manly 1977). Nonetheless, we start with the results of a life-table study that was conducted over 1.5 generations in 16 local populations in 4 different patch networks. We then consider both the endogenous and exogenous processes that are known to influence local population sizes of *M. cinxia* in Åland.

Life-Table Study

There is much spatial variation in survival both between populations within a metapopulation (e.g., Överö in table 4.1) and between metapopulations. For example, survival from the first larval instar to the third instar was less than 50%, on average, in the Västerkalmare metapopulation in summer 2000, but it was much higher in the Stålsby metapopulation (omitting patch 1405; table 4.1). The causes of larval mortality in the field are not well known. There are sporadic observations of entire larval groups disappearing because of heavy rainfall, predation, grazing by cattle, and various human activities. Dead larvae are quite often found in the field. In the spring 2000, larvae with slow development were collected from the field to observe their causes of mortality in detail. Most of them died during rearing and the most important mortality factor was parasitism by *Hyposoter horticola* (section 8.2). The causes of mortality also vary considerably among metapopulations. For example, mortality due to a

Table 4.1. Summary of a life-table study in four metapopulations.

Metapopulation and Patch No.	Survival								Causes of Mortality											
	Spring 2000				Summer 2000–Spring 2001				Spring 2000				Summer 2000				Spring 2001			
	4–5%	5–*A*%	*N* groups	*N* larvae	1–3%	1–6%	*N* groups	*N* larvae	H	F	O	U	E	F	O	U	D	F	O	U
Böle									12	—	9	44	67	1	7	30	17	—	—	24
573	78	4	4	490	—	—	—	—												
576	90	21	9	486	57	2	17	561												
1050	33	22	3	183	74	2	5	299												
1051	—	—	—	—	21	1	7	268												
Stålsby									13	3	—	14	—	51	1	14	8	10	1	19
1377	75	10	4	240	85	5	6	216												
1387	57	25	3	294	56	6	18	966												
1405	41	2	3	219	0	0	1	12												
1406	52	5	2	358	97	25	3	36												
Västerkalmare									26	—	1	14	38	2	13	18	25	4	2	9
875	19	21	4	52	50	15	8	556												
876	20	7	6	193	33	9	20	472												
877	87	14	5	194	44	13	16	467												
3555	24	9	2	85	62	12	3	135												
Överö									33	—	4	8	9	1	4	19	15	—	7	1
503	49	12	10	1302	56	6	16	1011												
934	57	20	2	30	0	0	4	57												
4812	91	18	1	11	42	0	1	36												
4816	91	27	1	22	—	—	—	—												

Survival percentages are calculated between the two larval instars or fifth larval instar and adults (*A*) as indicated. Numbers of adult butterflies observed are comparable only among habitat patches belonging to the same area, where all the patches were visited during the same days. Causes of mortality are pooled for metapopulations and reported for larvae with delayed development in the field (spring 2000) or for observations of dead larvae/eggs in the field (summer 2000 and spring 2001). H = *Hyposoter horticola*, F = fungus, O = other (e.g. human-caused disturbances, unsuccessful molting, drowning, predation), U = unknown, E = non-hatched eggs, D = dead larvae in the winter nest. Data from H. Paulomäki and K.-R. Valosaari.

fungal infection was frequent only in one metapopulation (Stålsby) in spring 2001 (apparently the generalist pathogen *Beauveria bassiana*; identified by Larry Huldén).

Competition

Resource Competition at the Larval Stage The role of resource competition in insect population dynamics has been debated for decades (for reviews, see Sinclair 1989, Hanski 1990, Turchin 1995, Hassell 1998). Ecologists now agree that though the primary causes of population oscillations in many insect populations have no bearing on population size or density, there is no such thing as density-independent regulation, and it would be surprising to have a population persisting for a long period of time without the vital rates ever being affected by population density at some spatial scale (Hanski et al. 1996a). In the case of *M. cinxia*, assessing the role of density dependence in population dynamics is complicated by gregarious larval behavior, as we describe below. Furthermore, density dependence may be inverse rather than direct—for instance, due to the difficulty of locating a mate and high emigration rate at low density (Kuussaari 1998). And the agents of density dependence are many. It is also worth pointing out here that metapopulation-level density dependence is commonplace: the rate of establishment of new populations is constantly limited by the availability of currently empty but suitable habitat patches within the colonization range of the butterfly.

Intraspecific resource competition takes place in high-density populations of *M. cinxia*, but the direct sign of it, food depletion, is rarely observed. However, intraspecific competition due to resource limitation may occur even in populations with low overall larval density because foraging by gregarious larvae is necessarily highly localized and the movement ability of larvae is limited (section 7.6). Availability of host plants varies greatly in time, up to by an order of magnitude from one year to another. Perhaps even more important than changes in host plant quantity are changes in host plant quality, which is most clearly manifested in drought years when the bulk of host plants may completely whither in the driest habitat patches and may cause 100% larval mortality (figure 4.4). Host plant senescence is the major cause of larval mortality in *E. editha bayensis* in California (chapter 3). Classical food depletion and resource competition are most frequent when a large proportion of larvae has reached the final instar, at which stage they gain the largest amount of body mass. At this stage, in late April, intraspecific competition seems to be relatively common, as is shown by the results of Kuussaari (described in more detail in chapter 7). In a study of 22 local populations in spring 1994, 12 out of 32 (38%) larval groups living where there was some scarcity of food showed signs of starvation, and all larvae died in 5 (42%) of these starved larval groups. The abundance of host plants near larval groups had a significant effect on both survival and larval growth.

The larvae of *M. cinxia* are not engaged in direct interspecific competition, as there are no other abundant leaf-feeding insect herbivores using *P. lanceolata* and *V. spicata* in Åland (there is naturally asymmetric competition with cattle and other vertebrate herbivores). The less common herbivores of *P. lanceolata* include unidentified aphids, thrips (*Haplothrips*), an agromyzid fly (apparently *Phytomyza plantaginis*), a specialist gracillariid moth (*Aspilapteryx tringipennella*), a specialist pyraliid moth (*Homoeosoma sinuellum*), several polyphagous moths (e.g., *Acronicta euphorbiae*), two weevils (*Gymnetron pascuorum* and *Trichosirocalus troglodytes*), and two fungi (the powdery mildew *Podosphaera plantaginis* and *Phomopsis subordinaria*, which infects the inflorescences; figure 4.5). *Phomopsis subordinaria* has only recently been recorded from Åland (A.-L. Laine pers. obs.), and its distribution and abundance are not yet known. *Veronica spicata* is also used by aphids, thrips, and polyphagous moths, as well as an oligophagous pterophorid moth (*Stenoptilia veronicae*). The only other checkerspot butterfly apart from *M. cinxia* that occurs in Åland, *Melitaea athalia,* feeds also on *P. lanceolata* and *V. spicata*, but only single larvae are found and, interestingly, often in *M. cinxia* larval groups. The main host plant of *M. athalia* in Åland is *Veronica chamaedrys* (Wahlberg 2000a).

Indirect competitive interactions may take place between *M. cinxia* and other species, but the importance of these interactions for population dynamics is not well known. One potential indirect competitor of *M. cinxia* is the weevil *G. pascuorum*, whose larvae feed on the seeds of *P. lanceolata* and thus reduce the seed set. Some natural populations of *P. lanceolata* are infected by the powdery mildew *P. plantaginis*. Field observations indicate that *M. cinxia* larval groups feeding on infected plants do poorly, and there are also several observations

Figure 4.4. Larval group of *M. cinxia* on a completely desiccated host plant. Photo by Saskya van Nouhuys.

of high mortality of larvae feeding on infected leaves (A.-L. Laine pers. obs.). In common garden and laboratory experiments, the survival and growth of larvae declined when feeding on mildew-infected diet. The decrease in larval fitness is most likely due to poor nutritional quality of infected leaves rather than to direct toxic effects of the mildew (A.-L. Laine in prep.). A powdery mildew infection can significantly affect the local dynamics of *M. cinxia*, as the prevalance of the mildew in local populations can be as high as 80% in September (A.-L. Laine unpubl. data) and as the gregarious larvae have a limited movement range (Kuussaari 1998).

Natural Enemies

Predators We have no observations of predation by vertebrates on any developmental stage of *M. cinxia*, even though the larval groups as well as the adult butterflies are highly conspicuous. The aposematic coloration of larvae seems to function well against visually searching vertebrate predators. Ants are often very numerous in the meadows, but only red ants (*Myrmica rubra*) have occasionally been detected to remove eggs. *Formica* ants attack adult butterflies wherever the opportunity arises, which may have surprising consequences for population dynamics (section 8.1). Adult butterflies are also occasionally caught in spider webs and by dragonflies. Of the predators (table 8.1), pentatomid bugs, lacewing larvae (Chrysopidae), and lady beetle larvae (Coccinellidae) may potentially have strong local effects on *M. cinxia* populations, as predation may explain the many instances in which egg batches or groups of small larvae have simply disappeared. However, we do not know how common such situations are, and we cannot assess the possible density-dependent effect of predators. Pupae of *M. cinxia* are cryptically colored and located, and we have no records of predation on them, though carabid beetles and shrews may well be important predators.

Parasitoids *Melitaea cinxia* has two primary larval parasitoids, which both have strong though dissimilar impacts on the dynamics of host populations. Chapter 8 describes in detail the interactions between the host and the parasitoids, and here we present a brief summary from the perspective of the host butterfly.

One of the two primary parasitoids, *Hyposoter horticola* (Ichneumonidae), is a large, solitary wasp with one generation per host generation. It has a

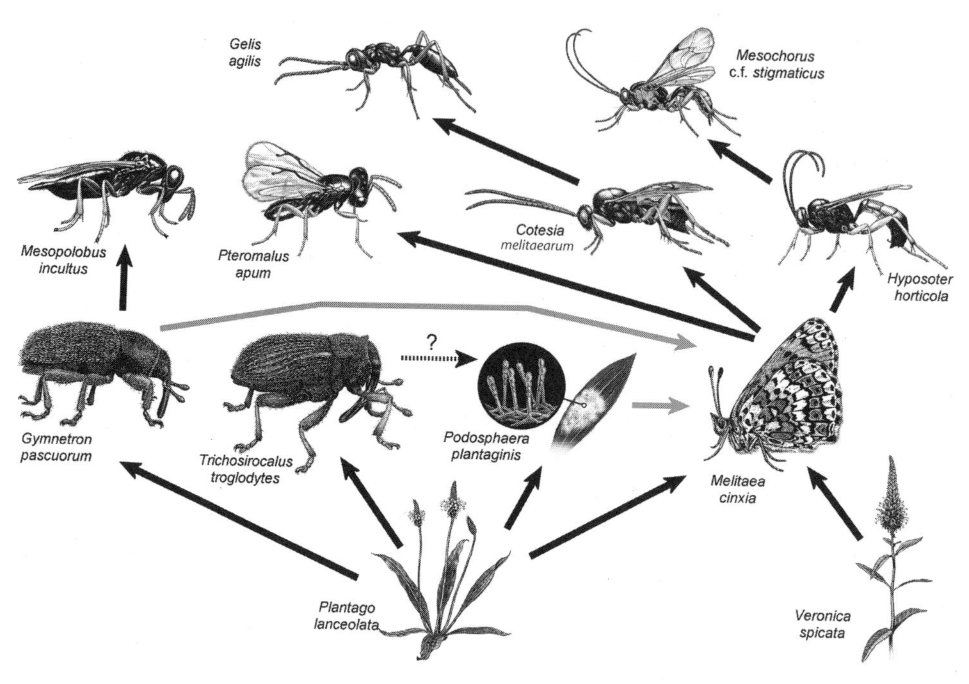

Figure 4.5. Interactions of *M. cinxia* and *P. lanceolata* with other organisms in the Åland Islands. Black arrows: direct interactions; gray arrows: indirect negative interactions; and dashed black arrow: suspected vector function. Drawing by Zdravko Kolev.

large dispersal range, and it is an exceptionally efficient parasitoid in terms of finding host individuals, even though most host individuals are susceptible for parasitism for an amazingly short period of time, less than 24 hours. If there were no other factors to limit parasitism (which are discussed in section 8.2), *H. horticola* would impose massive mortality on *M. cinxia*. As it is, some 30% of *M. cinxia* larvae die because of parasitism by *H. horticola*, and the level of parasitism is surprisingly stable both in space and in time, and it is apparently not influenced by host population density. Nonetheless, parasitism by *H. horticola* reduces local population sizes and thereby must make populations more vulnerable to various factors that increase extinction rate in small populations (section 4.6)

The story is completely different for the other primary parasitoid, *Cotesia melitaearum*, which is a gregarious braconid parasitoid with two or three generations per host generation and therefore a high intrinsic rate of population increase. *Cotesia melitaearum* has a shorter dispersal range than the host butterfly (van Nouhuys and Hanski 2002a; chapter 8), and hence its occurrence is largely restricted to those habitat patch networks where there is a large and well-connected host metapopulation. During our study, 5–19% of host populations have been annually occupied by this parasitoid. Even within particular networks, the incidence of *C. melitaearum* is higher in larger and better-connected host populations than in small and poorly connected populations. Parasitism is thus density dependent in the field (Lei and Hanski 1997), and under conditions that are most favorable for the parasitoid, parasitism has been observed to directly increase the risk of host extinction (Lei 1997). There are, however, two factors likely to increase the stability of the interaction between *M. cinxia* and *C. melitaearum*: density-dependent hyperparasitism by the ichneumonid *Gelis agilis* (Lei 1997, van Nouhuys and Tay 2001) and variable asynchrony in the phenology of *M. cinxia* and *C. melitaearum* (Lei and Camara 1999, S. van Nouhuys and G. Lei, in prep.). In brief, the coupling between *M. cinxia* and *C. melitaearum* is dynamically strong in parts of the Åland Islands during periods when the host population density is generally high, but in other areas the density of suitable habitat patches for the host is so low that the parasitoid is not able to maintain a viable metapopulation at all. The interested reader should consult chapter 8 for details, including the competitive interactions between the two primary parasitoids, interactions between the primary parasitoids and hyperparasitoids, and the influence of the host plant of the host butterfly on parasitism.

Interactions among and Movements of Adult Butterflies

Interactions among adults may influence population dynamics in many ways. Adult butterflies obviously interact in the course of mating (chapter 5), but local butterfly density may also influence the tendency of individuals to emigrate and immigrate. The rate of emigration from small habitat patches is sufficiently high to affect local and regional dynamics. Hanski et al. (2000) estimated that the probability of leaving a habitat patch of 1 ha in one day would be around 0.1 and even greater for smaller patches. In a previous study, about 15% of males and 30% of females were found within a week in some other habitat patch than the patch of first capture (Hanski et al. 1994). An interesting though still insufficiently studied difference appears to exist in emigration rate between the sexes: a fraction of males appears to stay put in their natal habitat patch, whereas all females that remain alive for a long time seem to eventually emigrate from small habitat patches (Kuussaari et al. 1996). This difference is probably related to the mate-searching strategy used by males, most of whom mainly perch while some also patrol (Wahlberg 2000b).

Migration distances are clearly restricted. The movement distances observed in mark–recapture studies have been generally <500 m, and the longest distance observed was 3.1 km (Hanski et al. 1994, Kuussaari et al. 1996). Daily migration distances, as inferred from a statistical model fitted to the mark–recapture data, were 0.18 km for males and 0.53 km for females (Wahlberg et al. 2002b). The long-term records on the actual colonization distances support the results of mark–recapture studies concerning lifetime movement distances: most colonizations have occurred within 2.3 km of the nearest source population and no colonization beyond 6.8 km has ever been recorded, though there have been ample opportunities for longer colonization distances in our studies (van Nouhuys and Hanski 2002a; figure 4.6). As we described above, the entire network of about 4000 habitat patches in Åland is divided into tens of habitat patch networks, within which the distances among nearby patches are usually hundreds of meters—that is, within the range that allows rapid recolonization after local extinction. If the butterfly metapopulation in an entire patch network goes extinct, recolonization is expected to take a longer time, depending on the occupancy status and population sizes in the neighboring patch network(s).

Migration has a significant inverse density-dependent component: emigration rate decreases with increasing butterfly density (Kuussaari et al. 1996). The increased emigration rate from low-density areas, as well as decreased mating success with decreasing population density (Kuussaari 1998), lead to reduced population growth rate in the smallest *M. cinxia* populations. This Allee effect is one of the factors that contributes to the extinction proneness of small populations (see section 4.6 for further discussion). We note in passing that inbreeding in small populations also reduces population size and further elevates the risk of extinction (Saccheri et al. 1998, Nieminen et al. 2001; section 10.5). Female *M. cinxia* tend to emigrate more frequently from habitat patches with no preferred host plant species than from patches containing the preferred host plant (Hanski et al. 2002; chapter 9) Additionally, the "match" between oviposition preference of migrating females and the host plant composition of currently empty but suitable habitat patches influences the rate of successful colonization (Hanski and Singer 2001; chapter 9).

Weather and Climatic Factors

Weather plays a key role in the dynamics of *M. cinxia* populations in Åland. The climatic conditions change unpredictably from one year to another, and there are vast differences in the prevalence, extent, and severity of weather phenomena that deviate from the average conditions. Different weather factors are most critical at different stages of development, but the amount of rainfall, the temperature, and/or the amount of direct sunlight are generally important.

The number of sunshine hours is a key factor in the spring after diapause. On sunny days the larvae can maintain a high body temperature, important for metabolism, in spite of low ambient temperatures, by basking in the sun in aggregates (plate IX). Measurements taken by Kuussaari (pers. comm.) showed that the difference between ambient temperature at 10 cm above ground and the body temperature of basking larval groups is 20°C on aver-

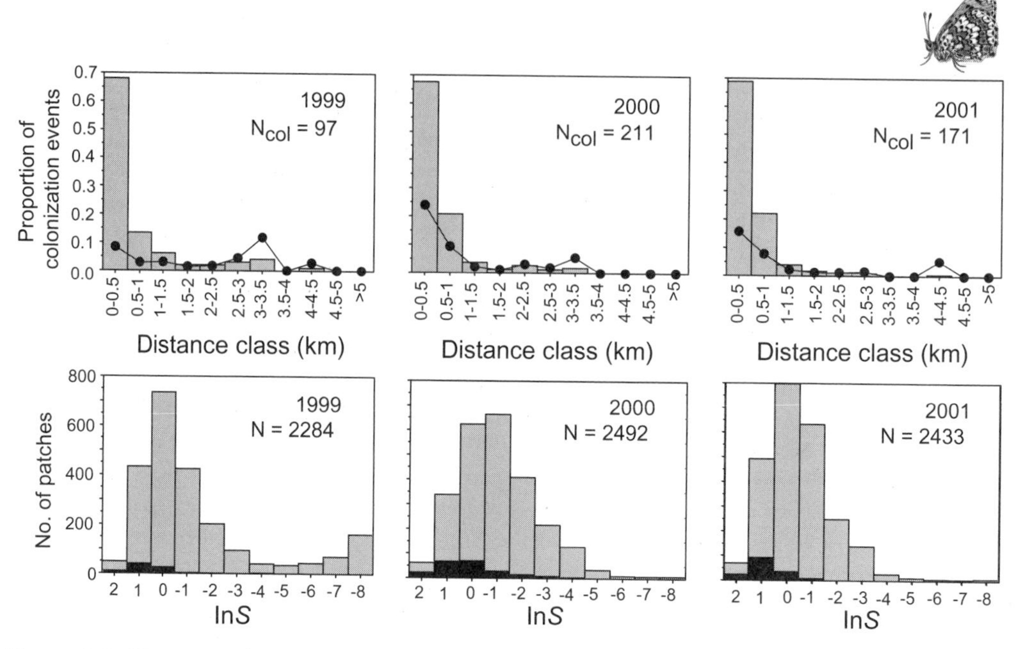

Figure 4.6. Upper panels: minimum colonization distances of *M. cinxia* in 1999–2001. Distance classes show the distance to the nearest population. Black dots show the proportion of colonized habitat patches out of empty patches at each distance class. Lower panels: numbers of habitat patches that either remained empty (gray bars) or became colonized (black bars) as a function of connectivity (decreases to the right).

age. The growth rate of the final instar larvae in late April and early May is particularly significant for those populations that have the primary parasitoid *Cotesia melitaearum*, which has the final generation developing in the last instar host larvae. There is substantial regional and year-to-year variation in the synchrony of development of the host and the parasitoid, and in some situations a large fraction or even all of the host larvae have managed to pupate before the adult wasps emerge to search for the final instar larvae to parasitize (chapter 8; S. van Nouhuys and C. Lei, unpubl. data).

During the pupal stage, the influence of the ambient temperature on the rate of development is probably the most significant factor. *Melitaea cinxia* pupates within tussocks of grass or herbs and in litter, where humidity is usually high enough for successful development. The longer the pupal period, the longer the time that the pupae are available for parasitoids and predators (Lei 1997).

In early summer when the adults are out, sunshine enhances mating, nectar feeding, oviposition, and migration. Windy weather, which is frequent in the archipelago, reduces butterfly activity considerably, even if the temperature is relatively high. In midsummer, irregular short-term droughts are frequent in Åland, and they appear to have become increasingly frequent over the past decade. During the 1990s, eight summers (June–August) had at least one month with less than half of the average rainfall as recorded for the past 116 years (more about this in sections 4.5 and 4.6). Moreover, even though the total precipitation is used to define dry and wet months, this is not the whole story. Some very dry months may still appear as close to the average, such as June 2001 (figure 4.7), when 36 mm out of the monthly total of 42 mm of rainfall occurred during one day. Such drought periods can considerably reduce the quality of host plants available for the newly hatched larvae. The newly hatched larvae desiccate easily and have basically no dispersal ability, which means that it is critical that the host plant individual on which they hatch is in good condition. If a dry period coincides with the hatching of larvae, widespread mortality ensues. Recall that host plant senescence is a key mortality factor in *E. editha bayensis* (chapter 3).

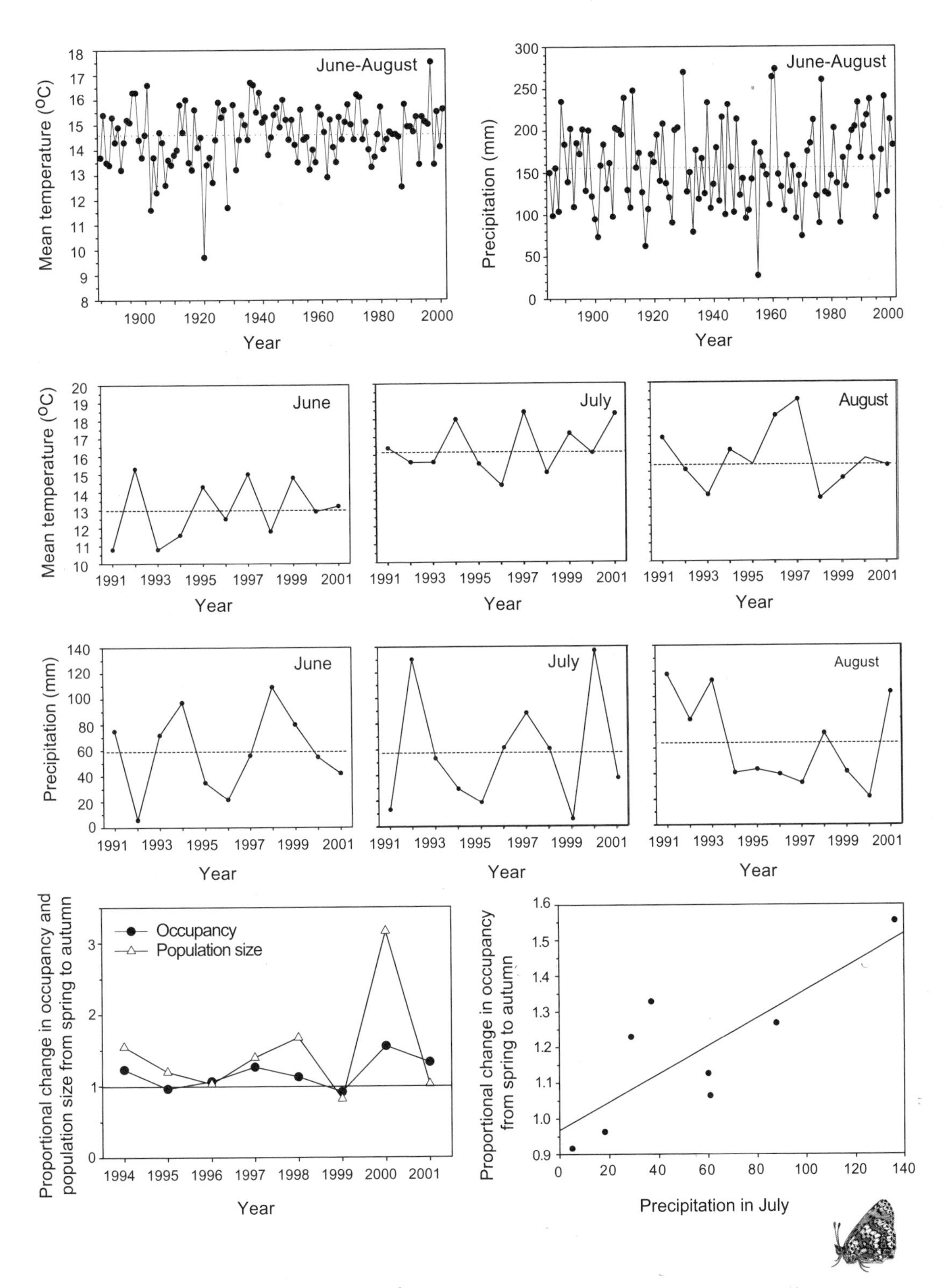

Figure 4.7. 116-year weather record from Åland (data for 1929 are missing). Top panels: Mean temperature and total precipitation in June to August. Dotted lines show the means for 1885–2000. Middle six panels: monthly weather records for 1991–2001; dotted lines show the mean values. Bottom panels: proportional changes from spring to autumn in habitat patch occupancy and total population size of *M. cinxia* in 1994–2001. Proportional change from spring to autumn in habitat patch occupancy in 1994–2001 plotted against total precipitation in July ($N = 8$, $r_S = .714$, $p = .048$; regression line is included for the purpose of illustration).

Older prediapause larvae are also susceptible to the desiccation of host plants. Only larvae that are already about to molt to the diapause stage may survive without food in early August (S. van Nouhuys and M. Singer, unpubl. data). In an experiment, all larvae survived for at least six days, but only the ones that were able to molt to diapause survived 11 days. Prediapause *M. cinxia* larvae do not try to find new hosts but just sit and wait on the desiccated plant. Generally, even a very rainy midsummer is more favorable for *M. cinxia* than is continuously warm and sunny weather. The latter would benefit the adult butterflies (as long as some nectar is still available), but the effect of plant condition on larvae is an even more important factor. A clear indication of the effect of rain is the significant positive correlation between the total precipitation in July and the proportional change in occupancy of habitat patches from spring to autumn (figure 4.7, lower right panel). In contrast, heavy rains related to thunderstorms can cause severe mortality of egg batches or newly hatched larvae by mechanically breaking up or even drowning larval groups (M. Nieminen, H. Paulomäki, and K.-R. Valosaari, pers. obs.). Even those larvae that survive a summer drought may have problems later, as the physiological condition in which they start overwintering is important for winter survival (possibly further depending on weather during the winter). Starved larvae are probably not capable of spinning a proper winter nest, which affects their survival through winter (as shown for poor-quality nests constructed by inbred larvae; Nieminen et al. 2001).

During winter, cold periods without snow cover may increase mortality, even though *M. cinxia* must be adapted to moderately low temperatures. Periods of mild weather after considerable snowfall followed by cold spells are more dangerous than low temperatures per se, because melted snow or ice forms small puddles that freeze over, killing the larvae (M. Nieminen, pers. obs.).

The infrequent years with average weather are apparently the most favorable ones for the survival of *M. cinxia* at every stage. The average season in Åland is as follows. The spring is relatively sunny, cool, and dry. The early summer is still dry, but temperatures rise (typically between 20° and 25°C) and rain showers become more frequent. The midsummer is warm and often half-cloudy, hot days (above 25°C) are uncommon, and showers are relatively frequent. Toward autumn the weather turns more cloudy and rainy, and the temperatures decrease, though nights remain relatively warm (often 10°–15°C). The winter is mild with occasional snow cover and long periods above freezing. The coldest periods are late in the winter, and in the spring nights can be very cold. In the eastern part of Åland the weather is more extreme on average: less rainfall and more sunshine hours in summer, generally windier weather, and cooler spring and milder autumn and winter due to the influence of the Baltic Sea. The often dry summers on the eastern islands are reflected by more violent population fluctuations of *M. cinxia* there than on the main Åland Islands.

Habitat Loss and Alteration

Melitaea cinxia lives in a highly fragmented landscape in Åland, where the habitat patches suitable for the butterfly are scattered throughout the landscape, though there is substantial variation in the regional density of these patches (figure 4.3). Fragmentation is partly due to the natural fine-scale topographic and soil-type structure in the Åland Islands, but there is also strong and historical human influence. The habitat patches least influenced by humans occur on rocky outcrops, where the vegetation is sparse due to limited soil and drought years that frequently kill or reduce the cover of most plant species. Grazing by cattle, sheep, and horses and hay cutting have kept large meadow and pasture areas open and potentially suitable for *M. cinxia* and its host plants. Construction and other small-scale disturbances create suitable but generally temporary habitat for *M. cinxia* on road verges, for example. Without traditional small-scale farming and related activities, there would be much less suitable habitat for *M. cinxia.* On mainland Finland and in many other areas in northern Europe this is a critical factor. Since the late 19th century, traditional farming has declined markedly and meadows and pastures have become scarcer. This process caused the extinction of *M. cinxia* from the Finnish mainland and southwestern archipelago apart from the Åland Islands (Kuussaari et al. 1995).

In the Åland Islands, variation in habitat quality generated by year-to-year variation in weather is superimposed on changing landscape structure operating at a slower time scale. The gradual decline in meadow area has affected the Åland Islands for tens of years. The current speed and severity of habitat decline appears to vary among different parts of Åland, being most extensive in the eastern part. At present, the suitable habitat for *M. cinxia* covers a total area of about 6 km², which is only about 0.6% of the land area of the main Åland Islands. Even though there are no exact records from Åland, where the situation is better than in mainland Finland, we may roughly estimate, based on the records from the mainland (Lappalainen 1998) and southwestern Finnish archipelago (L. Lindgren pers. obs.), that the current area of meadows and pastures is about 20% of their area only 50 years ago. Figure 4.8 gives a well-documented example covering a period of 20 years, with the predicted metapopulation response to habitat loss. Fortunately, in recent years grazing has again increased in some parts of Åland and overgrowth of meadows has stopped or the trend has even been reversed.

Summary

Factors and processes influencing the dynamics of local populations of *M. cinxia* are many, and most of them have well-studied consequences. Some of these factors are strongly spatially correlated, such as weather phenomena, as will be discussed in the next section, whereas others operate locally. The density-dependent processes that are known to influence local dynamics are described below:

- Food shortage is infrequent at the scale of entire populations but relatively frequent at the scale of larval groups.
- Parasitism by the primary parasitoid *Cotesia melitaearum* is only significant in those regions and years when there are large well-connected host populations.
- Inversely density-dependent emigration and immigration and difficulty of finding mates in low-density populations contribute to an Allee effect in local dynamics.
- Inbreeding depression in small populations increases the risk of population extinction (chapter 10).

4.5 Large-scale Dynamics

Spatial Scales from Larval Groups to the Åland Islands

The spatial and temporal changes in population sizes of *M. cinxia* in the Åland Islands can be examined at several scales. The smallest meaningful biological unit is a larval group, which, because of the limited mobility of larvae (mean distance moved by larval groups in the autumn 0.2 m, in the spring 0.5–1.7 m; see figure 7.7), corresponds to a small part of an individual habitat patch (meadow). Larval groups may be located in different places within the habitat patch in different years, but there are also plants and groups of plants that receive eggs more often than others, most likely because they are located in microhabitats preferred by egg-laying females. These microsites often have most of the larval groups in the habitat patch (figure 4.9), excepting drought years, when females are forced to lay eggs on plants growing in generally non-preferred microhabitats, as the driest parts of the habitat may have not a single fresh host plant. It becomes evident in such years how small-scale heterogeneity in habitat patch characteristics is important for population persistence (Kindvall 1996, chapter 3).

Local populations within habitat patches are ephemeral. Of the 1452 habitat patches known in 1993 and monitored ever since, 842 (58%) have been occupied at least once. But only 33 of them (3.9%) were continuously occupied throughout 1993–2001 (figure 4.10), and even in those patches population sizes showed substantial fluctuations. The next largest geographical units are patch networks, which are potentially occupied by metapopulations that are dynamically more or less independent of metapopulations in other patch networks (section 4.3). Of the networks, 25% have been occupied every year between 1993 and 2001 (figure 4.10), and the fraction of networks that are occupied has varied between 0.38 and 0.55. Patch networks are grouped within regions, which are delimited by clear geographical boundaries such as bodies of water and vegetational boundaries such as extensive forested areas. Figure 4.11 shows the changing occupancy patterns within two regions. Of the 17 regions, only 1 has not been occupied every year. This region (Vårdö) consists of several islands off the main Åland Island, of which only the one that is located closest to the other regions on the main Åland Island has ever been occupied. The reasons that this region has been unoccupied probably include low density of habitat patches and high extinction rate due to frequent dry summers (see above). Generally, as expected, the incidence of habitat patch occupancy is low in regions where the density of patches is low (Hanski et al. 1995a; section 4.6).

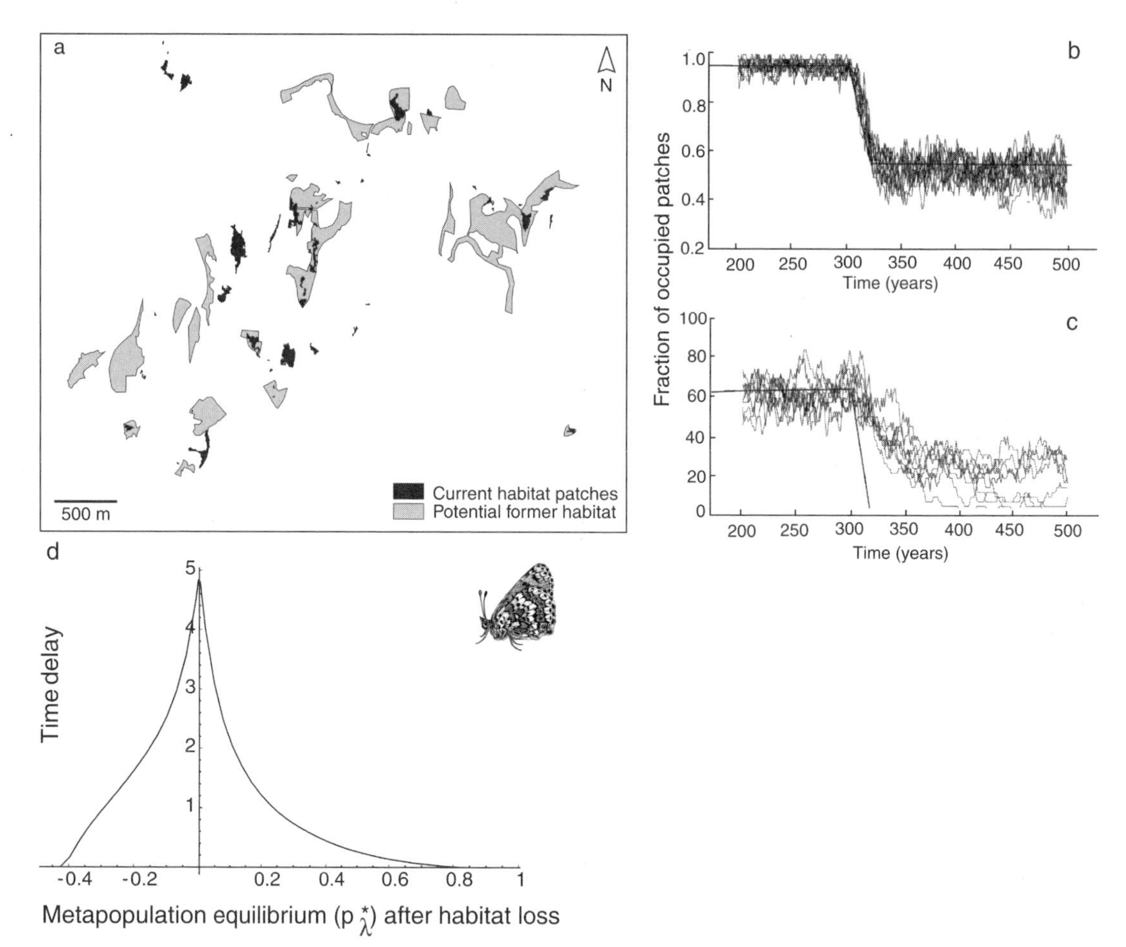

Figure 4.8. (a) Changing landscape structure in Åland. Gray areas indicate the partly overgrown areas that were suitable habitat for *M. cinxia* in 1973 but not 20 years later. Black areas were suitable in 1993. (b) Modeling results giving the metapopulation response to habitat loss in (a). Before and after the 20-year period when habitat was lost, the amount of habitat is assumed to stay constant. The thick line is the predicted equilibrium metapopulation size corresponding to the current structure of the landscape (amount of suitable habitat and its fragmentation). The thin lines show the model-predicted trajectories of metapopulation size before, during, and after the observed reduction in habitat area. (c) Similar results for a scenario of further loss of 50% of the area of each of the remaining patches in 1993. The equilibrium now moves to metapopulation extinction, but the actual predicted change in metapopulation size shows a long transient time (panels a–c from Hanski et al. 1996b). (d) The length of time delay in metapopulation response (vertical axis) to a change in landscape structure. The horizontal axis gives metapopulation size after the change in landscape structure. Note that the time delay is especially long when the metapopulation occurs close to the extinction threshold (as in panel c) (from Ovaskainen and Hanski 2002a).

Spatio-temporal Changes in Population Size

We have emphasized that all local populations of *M. cinxia* in Åland are more or less ephemeral: there are no large permanent "mainland" populations. The population units that may survive for a long time are metapopulations inhabiting patch networks. At this scale there is substantial, though by no means complete, synchrony in population changes, which to some extent must be due to migration. The asynchrony is crucial for stabilizing metapopulation dynamics and for increasing the time to metapopulation extinction (figures 4.12 and 4.13). The long-term records show that population trends are often similar in patch networks within large areas, often including parts of several regions (figure 4.12). In contrast, metapopulation sizes in

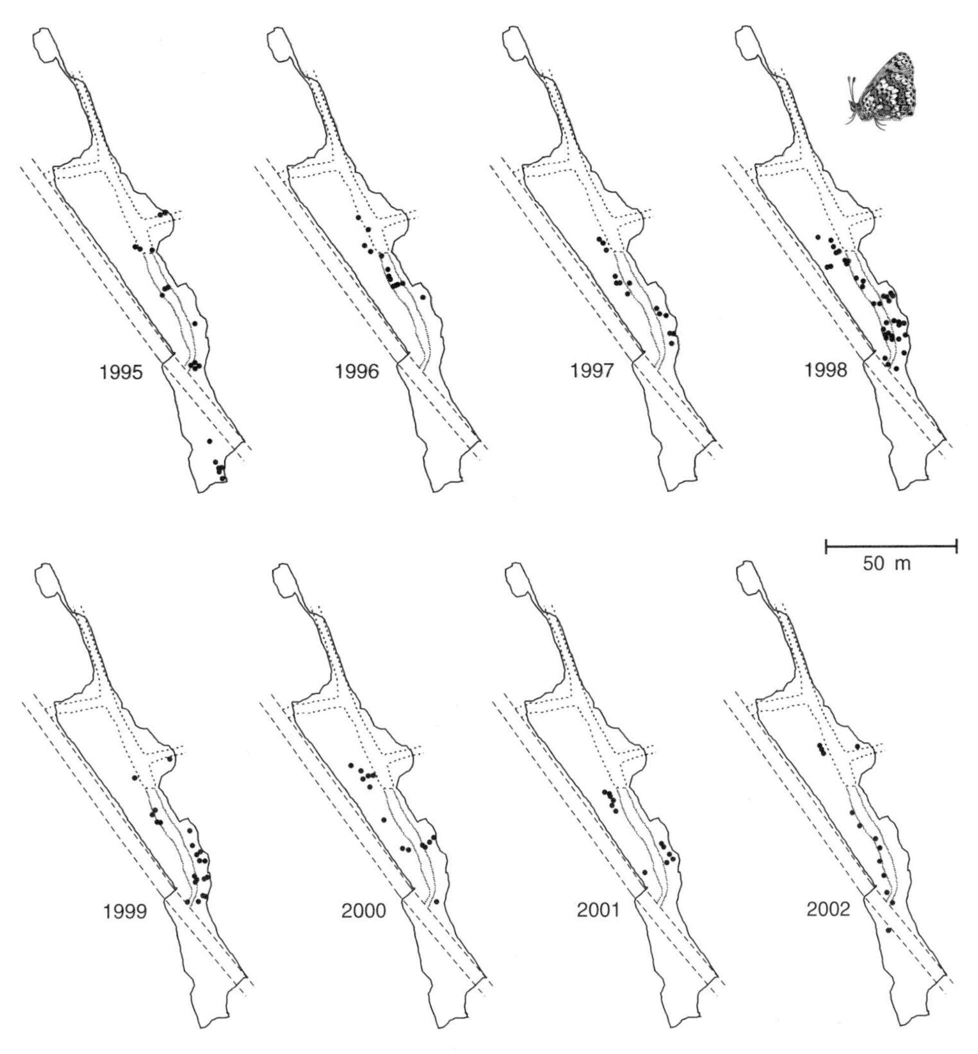

Figure 4.9. Maps of oviposition sites of *M. cinxia* within a habitat patch (continuous line) recorded during annual censuses in 1993–2002. The locations of larval groups (black dots) are based on hand-drawn sketches and therefore are approximate. The patch was unoccupied in 1993 and 1994. Both the northernmost and the southernmost part of the patch (the latter still had larval groups in 1995) have become severely overgrown. Long-dash lines: paved road; short-dash lines: gravel road; dotted lines: abandoned trail.

neighboring patch networks often change in opposite directions from one generation to another, demonstrating their dynamic independence and generating a mosaic of areas of increasing and decreasing density. The direction of population trends at the level of patch networks can be completely reversed in just one year from strong decline to substantial increase and vice versa (figure 4.13).

The largest spatial unit comprises the entire Åland Islands, which has a completely isolated "megapopulation" (population of metapopulations; Hanski 1999b) of *M. cinxia*. Some extreme years create large changes during one generation even at this scale, but more often the change is moderate or small. Survival over winter is one part of the life cycle during which the prevailing weather factors can have a great effect on abundance over large areas. This is illustrated by a positive correlation between the winter survival and the change in the numbers of larval nests over one year within regions. The weather conditions of the previous summer and autumn are also important because the condition in which

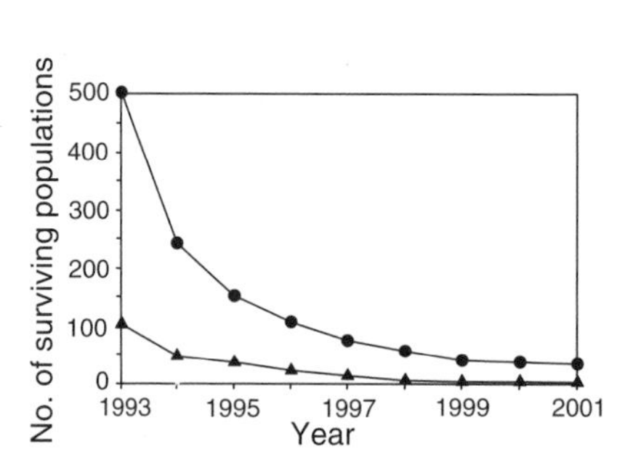

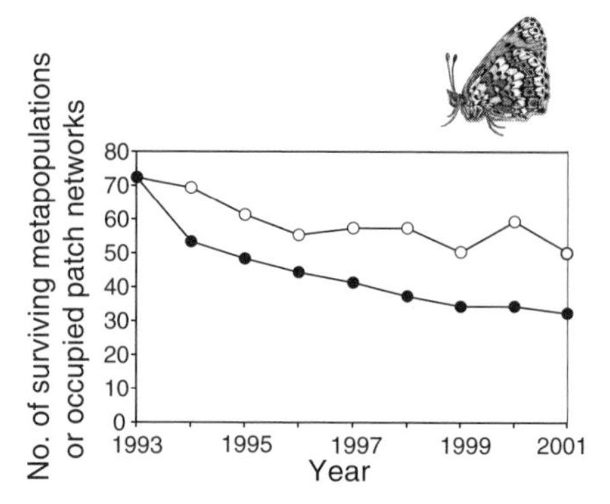

Figure 4.10. Decay plots of surviving populations and metapopulations of *M. cinxia* in Åland. Left panel: numbers of populations that have survived since 1993 (circles), and numbers of populations that have survived since 1993 and have had at least five larval groups in each year (triangles). Right panel: numbers of metapopulations that have had at least one population every year since 1993 (filled circles), and numbers of occupied patch networks (metapopulations) each year (open circles).

larvae enter diapause makes a big difference for overwinter survival. The survival of larval groups over winter has varied from 50% to 84%.

It is easy to demonstrate the kinds of large-scale changes in population size that have occurred. It is much more difficult to uncover the mechanisms driving particular changes, though the overall list of possible mechanisms can be compiled as we did above. The ideal would be to conduct an intensive life-table study in many populations scattered across the Åland Islands over many years. The data that are available cover 1.5 generations in 2000–01 in 16 populations belonging to 4 metapopulations. Spatial differences in survival can be large between metapopulations, as well as between habitat patches within metapopulations (table 4.1).

Some years ago one of us (I.H.) suspected that host–parasitoid dynamics involving the braconid *Cotesia melitaearum* might be responsible for the large-scale changes in host population sizes shown in figure 4.12. This seemed a credible hypothesis in the light of the general theory of spatially extended host–parasitoid interactions (Hassell et al. 1991, Nee et al. 1997, Hassell 2000a) and considering the fact that *C. melitaearum* increases the extinction rate of host populations (Lei 1997). However, since then it has become apparent that the conditions under which the coupling between *M. cinxia* and *C. melitaearum* is dynamically strong are restricted, and most of the time and in most places the parasitoid has no or very minor impact on host populations (chapter 8). Furthermore, as there is no clear year-to-year continuity in the spatial patterns (figure 4.12), such as one would expect if the patterns were driven by host–parasitoid interaction, this hypothesis is no longer considered viable.

At present, it appears evident that spatially correlated weather conditions are the primary cause of spatially correlated changes in population sizes. The reasons for the varying extent of spatial correlation and the direction of the change depend on the peculiarities of the weather in each year, and even short-lasting weather events can have profound regional effects. Some very large-scale weather phenomena might be expected to create a uniform population trend across the entire Åland Islands, though in practice such effects appear rare (figure 4.12). In the past 10 years, there has always been substantial asynchrony between the metapopulations in Åland, with the exception of the transition from 1998 to 1999 (figure 4.12). Such asynchrony naturally represents an important stabilizing factor at the largest spatial scale.

One counterargument to the idea that spatially correlated weather conditions create the large-scale patterns in the change in population sizes is that weather conditions might not show sufficient spatial variation at the scale of the Åland Islands, an area that is only 50 × 70 km. Data from a weather radar on the southwestern coast of Finland, which records the amount of daily rainfall with the spatial resolution of 1 km^2 (figure 4.14), can be used

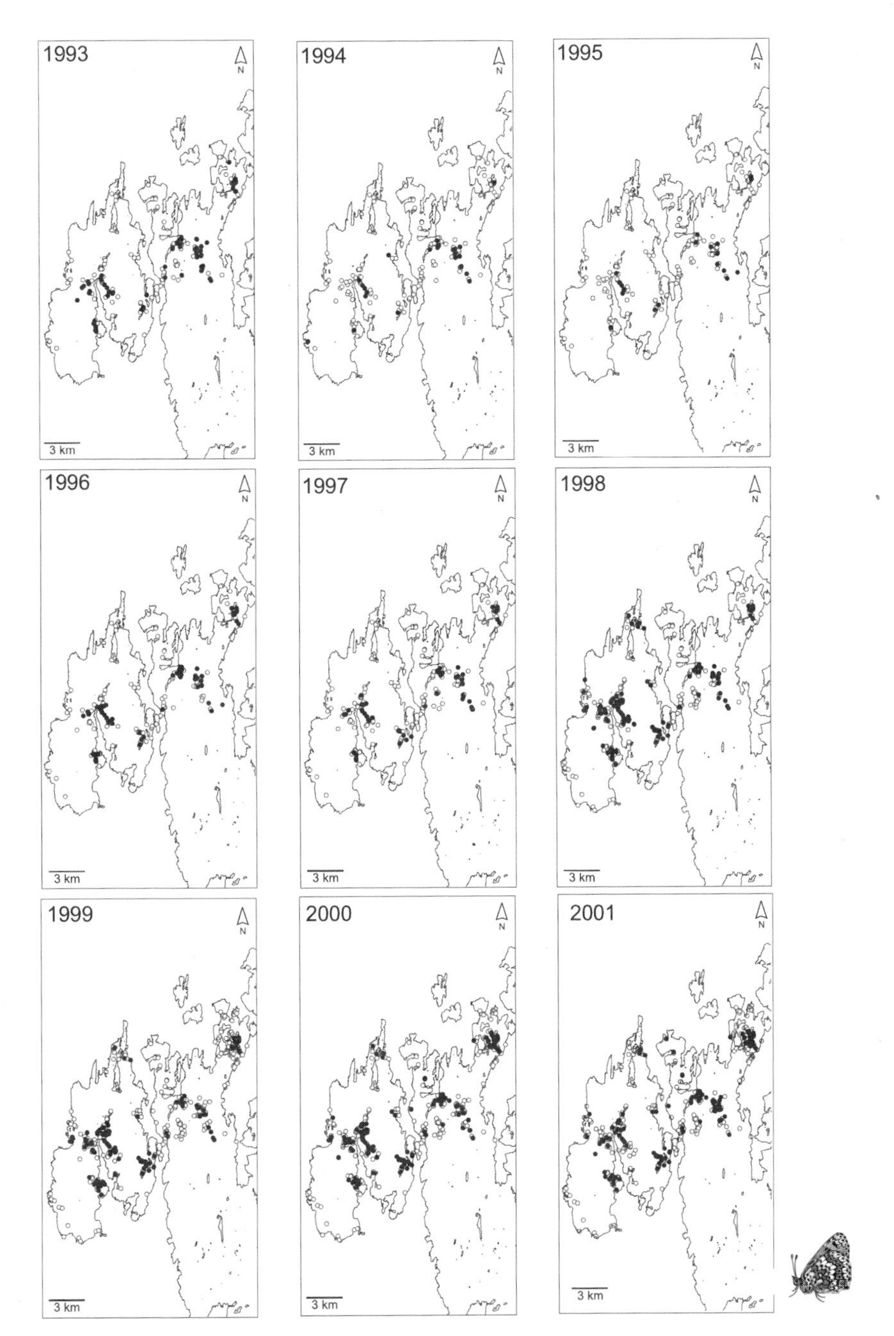

Figure 4.11. Temporal changes in the occurrence of *M. cinxia* in two regions in Åland (each side of the strait forms a region). Open circles: empty patches; large black dots: occupied patches; small dots: patches belonging to other regions (southeastern part of the area) or patches not known before 1998–99.

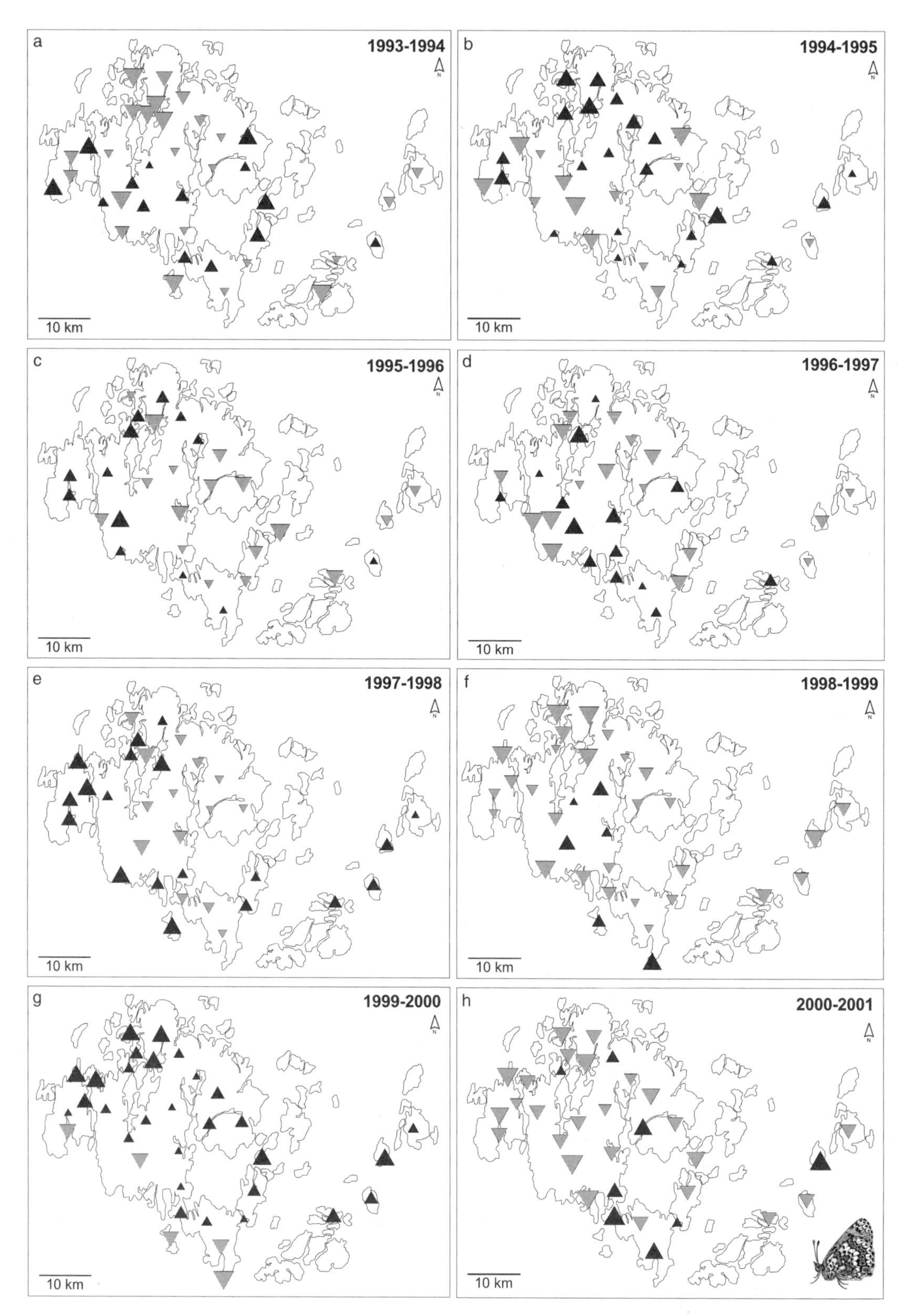

Figure 4.12. Population trends of *M. cinxia* in habitat patch networks in the years 1993–2001. Gray downward-pointing triangles indicate decrease in metapopulation size and black upward-pointing triangles indicate an increase. The size of the triangle shows the relative change in metapopulation size in the network.

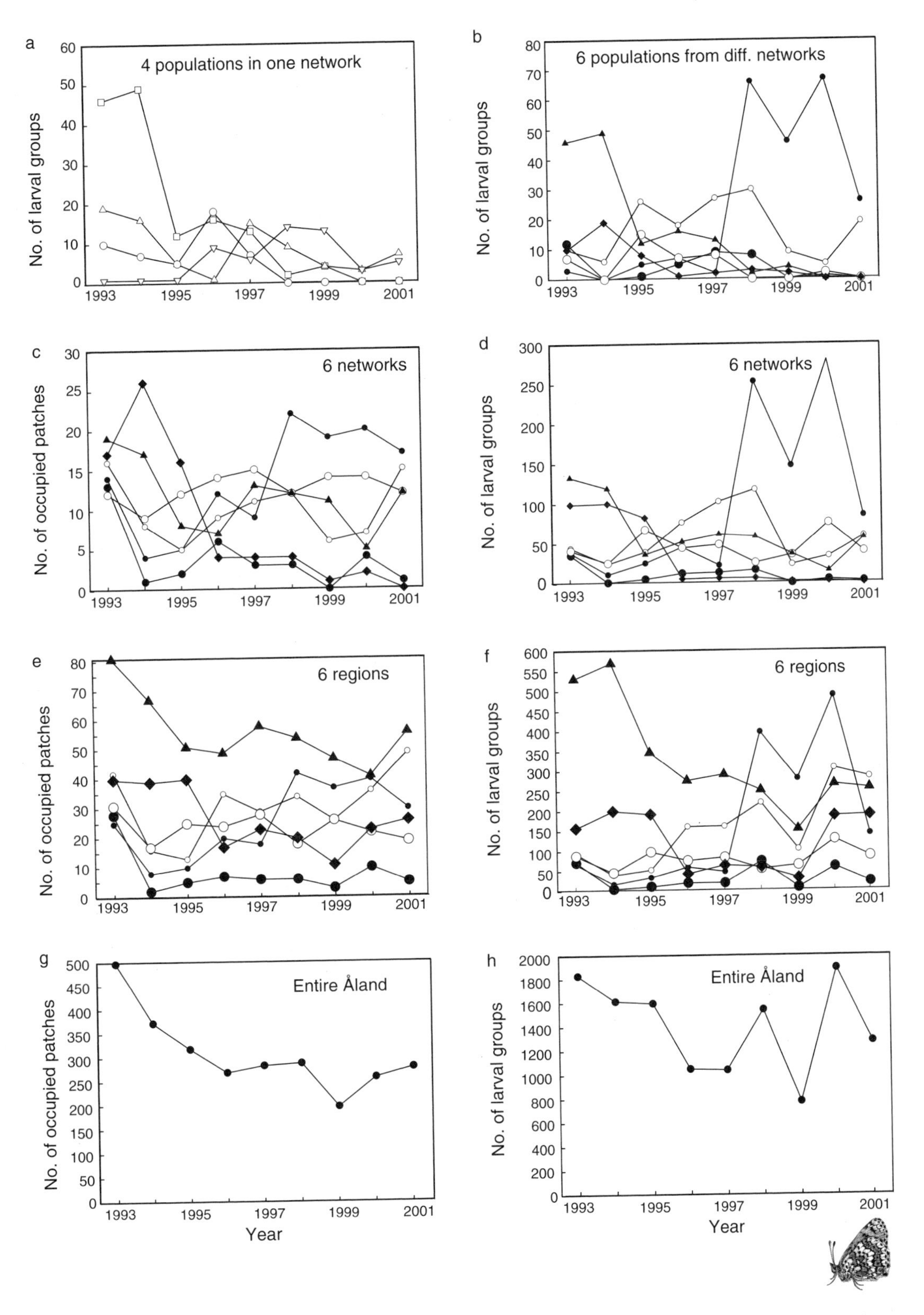

Figure 4.13. Temporal changes in habitat patch occupancy and population size of *M. cinxia* at several spatial scales. (a) Four populations within a metapopulation, (b) six populations in the same six metapopulations as in (c) and (d) (indicated by similar symbols), (c) and (d) six metapopulations in the same regions as in (e) and (f) (indicated by similar symbols), (e) and (f) six regions, and (g) and (h) the entire Åland.

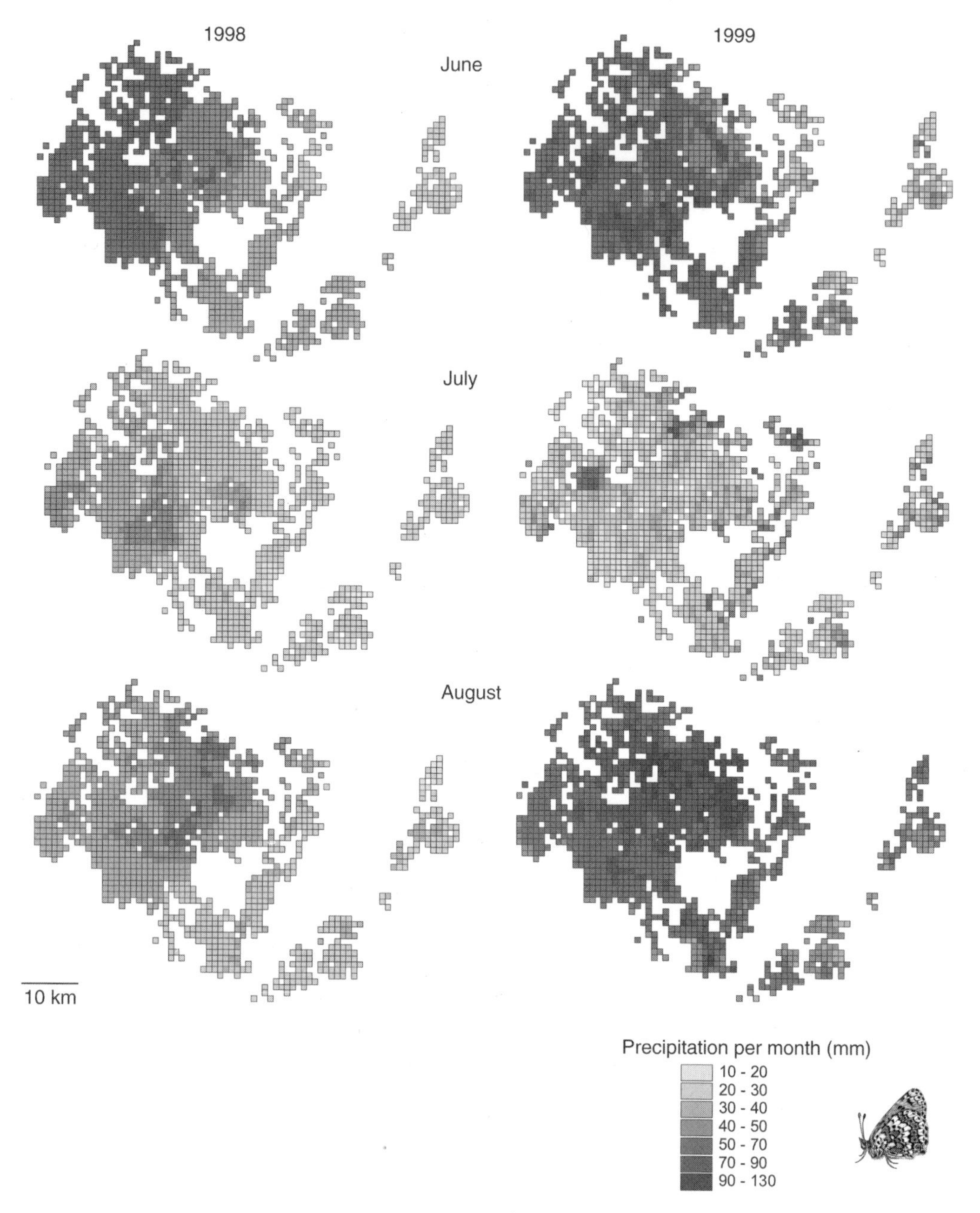

Figure 4.14. Precipitation within 1-km² squares in Åland in June–August in 1998 and 1999 (weather radar data, monthly sums per square).

to examine this question. In 1998, there was a trend of decreasing rainfall from west to east, especially in June and July, whereas in August there was more small-scale variation in rainfall. The pattern was quite different in 1999, when small-scale variation dominated the picture throughout the summer. These two examples of monthly rainfall patterns demonstrate that spatial variation in precipitation is considerable in the Åland Islands. The amount of variation is even greater in weekly and daily data.

Summary

Small local populations of *M. cinxia* can very rapidly move from one extreme (high density) to another (local extinction). Rapid changes appear also at the scales of patch networks and regions, though the probability of such a change becomes smaller with increasing spatial scale, and we have not yet recorded rapid changes at the scale of the entire Åland system. Various weather factors dominate among the causal agents of large-scale changes, whereas biotic factors primarily influence dynamics within habitat patches and patch networks.

4.6 Metapopulation Processes and Patterns

The *Melitaea cinxia* project has co-evolved with the development of the spatially realistic metapopulation theory (Hanski 2001, Ovaskainen and Hanski 2001) in the Metapopulation Research Group in Helsinki. The theory was initially conceived in the form of the incidence function model (Hanski 1994b, 1999b), which is an example of stochastic patch occupancy models (Verboom et al. 1991, Gyllenberg and Silvestrov 1994, Day and Possingham 1995). Much effort has been put into developing methods of parameter estimation that allow the model to be fitted to empirical data (Hanski 1999b, Moilanen 1999, 2000, 2002, O'Hara et al. 2002), and more recently a general mathematical theory has been constructed (Hanski and Ovaskainen 2000, Ovaskainen and Hanski 2001, 2002, 2003). (Box 12.1 presents a brief summary of the key elements of the theory.)

The focus of the spatially realistic metapopulation theory is on the influence of habitat patch area and connectivity on the processes of local extinction and colonization (patch area and connectivity may be corrected for spatial variation in habitat quality). In this section, we first summarize the mechanisms of extinction and colonization in the *M. cinxia* metapopulation in the Åland Islands. Though there is much more to local extinction than just the patch area effect, and there is more to colonization than the connectivity effect, patch area as a proxy for population size is nonetheless an important correlate of extinction risk, and connectivity is the primary determinant of colonization rate. The significance of the area and connectivity effects on population turnover is reflected in the instantaneous patterns of patch occupancy, which we review in the second part of this section.

Processes: Local Extinction and Colonization

Figure 4.10 clearly demonstrates that local populations of *M. cinxia* are at high risk of extinction: there are no large extinction-resistant populations that would guarantee the long-term persistence of the species in the Åland Islands. Instead, long-term persistence is maintained by classic metapopulation dynamics, and the species persists in a balance between extinctions and colonizations. Table 4.2 summarizes the yearly extinction and colonization rates in the years from 1993 to 2001. In this section, we present an overview of the actual causes and mechanisms of population extinction and establishment of new local populations.

Causes and Mechanisms of Population Extinction

Given the small size of most local populations of *M. cinxia* in Åland, one might expect that demographic stochasticity is all that is needed to explain the pattern of extinctions. In reality, this is not at all the case, and many other processes have been shown to make a contribution to population extinction (Hanski 1998a, 1999b, Hanski 2003). Nonetheless, it is true that an important correlate of extinction risk is small population size. Small populations are likely to go extinct for various reasons:

- Commonly there is just a single larval family, or a few larval families, in a local population in a given year. Because the survival of larvae within one family is largely correlated (Kuussaari 1998), demographic stochasticity is evidently a major cause of extinction in such small populations, even though there may be 100 or more larvae in one family. Only a small fraction of the larvae reach the adult stage,

Table 4.2. Summary of the observed extinction and colonization rates and the total numbers of occupied and empty habitat patches in the autumns of 1993–2000.

Year	No. of Extinctions	Proportion Extinct	No. of Colonizations	Proportion Colonized	No. of Populations	No. of Empty Patches
1993					483	749
1994	234	0.50	111	0.15	383	872
1995	155	0.41	106	0.13	356	926
1996	170	0.48	101	0.12	303	1026
1997	131	0.43	142	0.14	332	1146
1998[a]	131	0.39	143	0.13	430	2322
1999[a]	225	0.53	97	0.04	375	3507
2000[b]	135	0.37	230	0.09	473	2547
2001	152	0.32	172	0.07	496	3470

[a]A resurvey of Åland for previously unknown habitat patches was performed in 1998–1999.
[b]Most patches <100 m^2 and most patches on the small islands where *M. cinxia* has not been observed since 1991 were not censused in 2000.

where additional demographic stochasticity operates, including chance variation in the sex ratio (Wahlberg et al. 2002b).

- Allee effect reduces the growth rate of small *M. cinxia* populations. It has been found that a substantial fraction of females in small populations remains unmated (Kuussaari et al. 1998).
- Inbreeding depression further reduces the fitness of individuals in small populations and increases the risk of local extinction (Saccheri et al. 1998, Nieminen et al. 2001). The fitness components affected by inbreeding have been investigated by Nieminen et al. (2001) and Haikola et al. (2001).

Populations living in small habitat patches have a higher risk of extinction than populations living in larger patches. This relationship is partly due to correlation between patch size and population size and the fact that small populations have a high risk of extinction, but other factors play a role as well. First, small habitat patches tend to have less spatial heterogeneity in environmental conditions than large ones, and such heterogeneity tends to buffer populations against year-to-year variation in weather conditions (environmental stochasticity). Similar results have been well documented for *E. editha* (section 3.4).

Second, butterflies living in small areas frequently encounter the boundary of the habitat patch, which may be expected to increase emigration rate. Hanski et al. (2000) and Wahlberg et al. (2002b) showed for *M. cinxia* and related checkerspots in Finland that the emigration rate of butterflies from habitat patches of around 1 ha is around 10% per day, which is of the same magnitude as the daily mortality rate. In patches < 1 ha, which include most patches in the Åland system, emigration rate is even higher. One might expect that many of the smallest patches are below the critical minimum patch size for deterministic persistence (Skellam 1951, Okubo 1980). This may well be the case (see C. D. Thomas and Hanski 1997), but the dynamics of such small populations are so unstable that it is difficult to tease apart the different mechanisms of extinction in these populations. Finally, because emigration is inversely density dependent (Kuussaari et al. 1998), low-density populations suffer especially great losses due to emigration and receive little reinforcement in the form of immigration.

Population extinction is increased by unfavorable environmental conditions, which becomes clearly manifested in spatially correlated extinctions, generated by spatially correlated environmental stochasticity (regional stochasticity; Hanski 1991), as we have described in the previous section. Several other factors are also known to influence extinction risk:

- Habitat loss and degradation, including successional overgrowth of meadows and pastures that have been abandoned by farmers and habitat destruction caused by construction of roads, buildings, and so on, increase extinction risk.
- A small population of host plants in a patch increases extinction risk. Host plant density

fluctuates from one year to another due to variation in environmental conditions.

- Parasitism by the wasp *Cotesia melitaearum* increases extinction risk under conditions that are favorable for the parasitoid (section 4.4).
- Parasitism by the wasp *Hyposoter horticola* reduces population sizes by about 30%, leaving populations more vulnerable to extinction due to other processes.
- The immediate effect of grazing is negative due to trampling and feeding of plants with larvae or eggs, but the long-term effect is positive because grazing maintains early successional vegetation in which the host plants flourish. Other types of human disturbances operate similarly (e.g. road construction often creates new habitat but has a negative immediate impact on populations).

In summary, a large number of processes has been demonstrated to contribute to population extinction of *M. cinxia* in the Åland Islands. Our ability to demonstrate all these effects is enhanced by large sample sizes, but this is only part of the story, as most mechanisms have had a strong effect in at least some parts of the study area in some years. It is indeed another important lesson from these studies that the actual causes and mechanisms of extinction are variable in space and in time even in this single system. Considering the primary effects of the landscape structure on population extinction, it is apparent that high extinction risk is associated with small habitat patch size. Reduced habitat quality decreases expected population sizes further and thereby increases extinction risk and so does increased isolation, which reduces the potential of demographic and genetic rescue effects to counter local extinction.

Population Establishment Not surprisingly, the key factor determining the rate of establishment of new local populations in the currently empty but suitable habitat patches is connectivity to existing populations (figure 4.6). In this context, connectivity is a population dynamic measure, reflecting how isolated the focal patch is from other occupied patches, not a measure of landscape structure independent of the locations of existing populations (Moilanen and Hanski 2001). As described in box 12.1, connectivity increases with increasing number, increasing sizes, and decreasing distances to local populations that occur within the migration distance from the focal habitat patch. Though distance to the nearest local population from the empty patch that may become colonized is not an equally good measure of connectivity (Moilanen and Nieminen 2002), the distance to the nearest population is generally correlated with true connectivity, and its significance is easy to conceive. In the *M. cinxia* system, the average colonization distance to the nearest existing population has been 0.6 km, 95% of colonizations have occurred within 2.3 km from the nearest source, and the longest recorded colonization distance has been 6.8 km (van Nouhuys and Hanski 2002a). Given the limited migration range of *M. cinxia*, habitat patches that are located farther away than 3–4 km from the nearest populations have little chance of becoming colonized until some populations have been established in between. Van Nouhuys and Hanski (2002a) estimated that some 10% of the habitat patches in Åland have thereby been out of reach of the butterfly in any one year.

Apart from connectivity, other factors are known to influence the rate of successful colonization:

- Propagule size is expected to increase the per capita rate of successful colonization because butterflies are more likely to stay at sites where there are other butterflies (Kuussaari et al. 1996). Therefore, when several butterflies have entered an empty patch more or less simultaneously, colonization probability is increased.
- Size of the habitat patch increases immigration (Kuussaari et al. 1996, Hanski et al. 2000, Wahlberg et al. 2002b) and colonization rates (Moilanen and Nieminen 2002), apparently because larger patches are larger targets for migrating butterflies.
- Reduced amount of host plants and increasing fraction of desiccated host plants in a patch decrease colonization rate (Hanski 1999b).
- Grazing decreases colonization rate, apparently because of its negative immediate effects (Hanski 1999b).
- Great abundance of nectar plants increases immigration and thereby presumably also colonization (Kuussaari et al. 1996).

Recent studies have revealed that one life-history trait of migrating females, the host plant oviposition preference in relation to the host plant species composition of the habitat patch, has a strong and systematic influence on the rate of successful colonization. These findings are described in detail in chapter 9, and it suffices here to emphasize that the quality of habitat patches as perceived by migrating females is really a function of the phenotype of individuals rather than a characteristic of the species.

Patterns: Area and Connectivity of Habitat Patches Influence Distribution

Local Populations There is a positive effect of habitat patch area on occupancy as well as on population size (figure 4.15). Population size is expected to increase with patch area because the quantity of larval resources generally increases with area. But there is also much variability in butterfly density, especially in the larger patches, because local dynamics are only weakly density dependent and plant density varies greatly. The observed increase in the probability of patch occupancy with increasing area is obviously a consequence of populations in small patches having high extinction risk. The other key pattern is decreasing probability of occupancy with decreasing connectivity, which we can ascribe to reduced colonization rate with increasing isolation.

There is some discussion in the literature on butterfly conservation about the relative strengths of the effects of patch area and connectivity versus patch quality on occupancy (e.g., C. D. Thomas et al. 2001; chapter 13). We emphasize that the metapopulation theory does not assume that habitat patch quality makes no difference. On the contrary, it is self-evident that habitat quality is important; the theory assumes that suitable habitat can be delimited from unsuitable matrix in the description of the patch network. There may also be substantial variation in habitat quality among the patches that are generally suitable for occupancy. The simplest way of accounting for the effect of varying habitat quality is to weight patch areas by quality, with the aim of measuring effective patch areas, which would correlate better than true patch areas with expected population sizes.

Metapopulations The area and connectivity effects on habitat occupancy are expected to occur also at the patch network level. Hanski et al. (1995a) showed that the fraction of occupied patches in a network increases with average patch area and average level of connectivity in the network. A more effective analysis uses a single number, the metapopulation capacity of a fragmented landscape (box

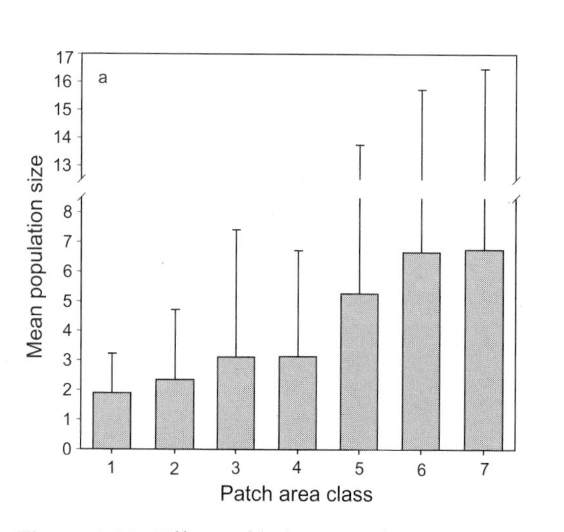

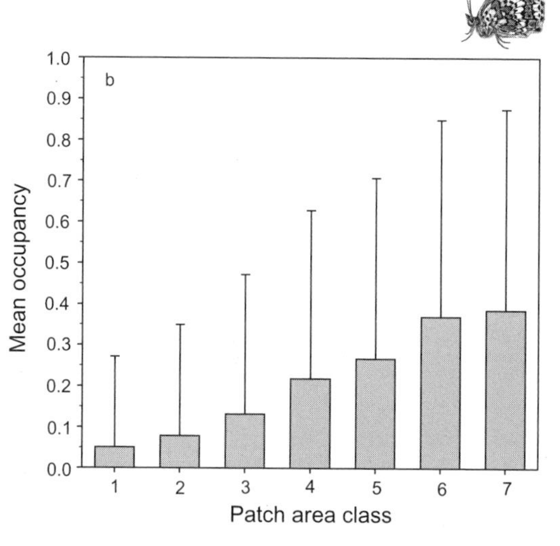

Figure 4.15. Effects of habitat patch area on population size (a) and occupancy (b) of *M. cinxia*. Patch area classes: 1 = 1–50 m^2, 2 = >50–100 m^2, 3 = >100–500 m^2, 4 = >500–1000 m^2, 5 = >1000–5000 m^2, 6 = >5000–10,000 m^2, and 7 = >10,000 m^2, The lines show standard deviations.

12.1), as a measure of landscape structure. This measure is derived from the spatially realistic metapopulation theory (Hanski and Ovaskainen 2000, Ovaskainen and Hanski 2001), and its value increases with the number, average size, and average connectivity of the habitat patches. Thus, the metapopulation capacity not only measures the amount of habitat in the landscape, but it also takes into account the spatial configuration of the habitat patches. The theory predicts a threshold value in the quality of the landscape, below which metapopulation is predicted to go extinct. Figure 12.4 illustrates the threshold condition for metapopulation persistence in *M. cinxia* in Åland: networks with metapopulation capacity lower than a critical value were unoccupied, whereas the level of occupancy increases, as theory predicts, with metapopulation capacity in the networks above the threshold value.

The metapopulation theory (Hanski 1985, Gyllenberg and Hanski 1992, Hanski and Gyllenberg 1993) predicts that if migration from large populations substantially influences the dynamics of small populations (the rescue effect), metapopulation dynamics may exhibit two alternative stable states, of which one corresponds to metapopulation extinction, and the other one to a positive metapopulation size. Immigration to small populations decreases extinction risk in *M. cinxia* (Hanski 1999b), and Hanski et al. (1995b) have reported patterns of habitat occupancy strongly suggesting the presence of alternative stable states.

Traditional metapopulation models assume independent local dynamics and hence independent extinction and recolonization rates. However, as we have seen, there is substantial spatial correlation in population turnover in *M. cinxia* (figures 4.12 and 4.13), primarily caused by the effect of spatially correlated weather effects on population dynamics (section 4.5). Ovaskainen (2002a) has made an important start in incorporating spatially correlated dynamics into metapopulation models. For one of the patch networks in Åland, he estimated the model parameters separately for seven year-to-year transitions in patch occupancy (spatial correlation in dynamics will manifest in temporal variation in parameter values). Figure 4.16 shows the model-predicted quasi-stationary distribution of metapopulation size both when the observed year-to-year variation in parameter values is taken into account and when this variation is ignored. Note that in the former case the model predicts a much greater variance in metapopulation size, and hence a much greater risk of metapopulation extinction, than when variation in parameter values is ignored.

Summary: Classic Metapopulation Structure and Dynamics

It is evident from the above description that the *M. cinxia* metapopulation in Åland represents an excellent model system of classic metapopulations. Hanski et al. (1995a) summarized the key characteristics of classic metapopulations as follows:

- The suitable habitat occurs as small discrete patches (section 4.2). The habitat patches support local breeding populations in *M. cinxia*, as 60–80% of butterflies spend their lifetime in their natal patch.
- All local populations have a high risk of extinction. Only 4.3% of populations that existed between 1993 and 2000 have survived throughout that period, and none of them has been large all the time (section 4.5).
- The patches are not too isolated to prevent recolonization. Only about 10% of all habitat patches in any one year are outside the recorded maximum colonization distance (section 4.4).
- Local and regional dynamics are asynchronous enough to make simultaneous extinction of all local populations unlikely (section 4.3).

4.7 General Lessons

Density Dependence and Long-Term Persistence

Study of the local dynamics of *M. cinxia* has shown the overriding importance of density-independent factors, especially various weather phenomena, in generating population oscillations, which often lead to population extinction. How should the observed dynamics be described in terms of the standard population dynamic theory?

First, it is clear that density dependence will occur whenever population size becomes very large. At high population density, food becomes depleted, and the parasitoid *C. melitaearum*, if present, may quickly reduce the host population size. But in-

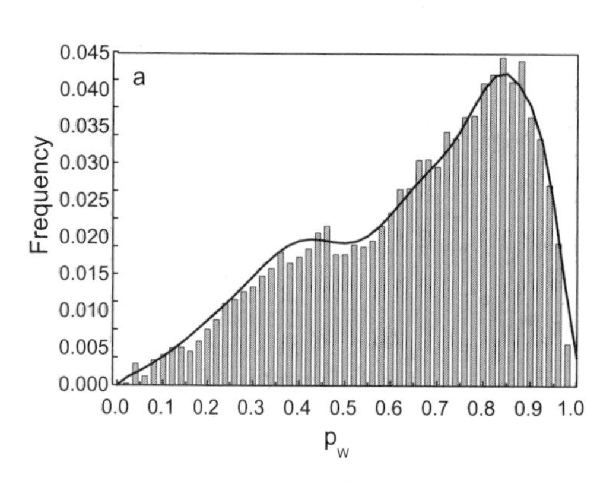

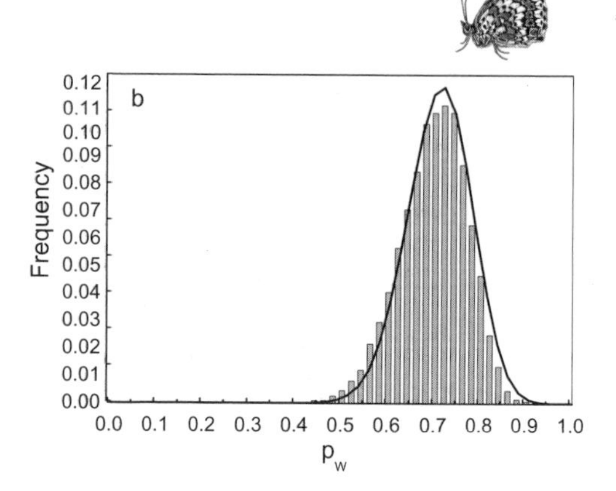

Figure 4.16. Model-predicted quasi-stationary distribution of metapopulation size of *M. cinxia* in one patch network in the Åland Islands. The bars are based on 10,000 simulated occupancy states of the metapopulation, and the line shows the analytically determined result based on Ovaskainen's (2002b) theory. In panel (a), the model was parameterized taking into account temporal variation in parameter values, but such variation was ignored in panel (b). Metapopulation size is measured by an appropriately weighted average of the patch occupancy states (p_w; from Ovaskainen 2002b).

stances of food depletion and *C. melitaearum*-caused population declines are infrequent. At the metapopulation level, there is obvious density dependence in colonization, as there is a finite set of habitat patches in any one network that are available for colonization.

Taking all these considerations into account, a plausible model for the dynamics of *M. cinxia* consists of a random walk of population size (largely driven by weather conditions) whenever the population is below the local "ceiling"; density dependence in the vicinity of the ceiling (population size cannot exceed the ceiling); and standard model of colonization, including distance-dependent migration. This model has been studied by Hanski et al. (1996a). They showed that the metapopulation may persist with a low frequency of local density dependence, but this happens only when the intrinsic rate of increase is low. On the other hand, if the intrinsic rate of increase is low, the rate of extinction is high, and the colonization rate may be insufficient to generate a positive balance between extinctions and colonizations. Only by selecting the model parameters carefully one may generate the combination of long-term metapopulation persistence and infrequent local density dependence, and even then the incidence of local density dependence is around 5% (Hanski et al. 1996a). It appears that *M. cinxia* in the Åland Islands is an example of such a species with infrequent local density dependence but long-term persistence at large spatial scales.

Impact of Global Changes

The major global changes that influence the dynamics of species worldwide include habitat alteration and loss, interactions with alien species, exploitation and persecution by humans, and climate change. *Melitaea cinxia* and its two host plants *Plantago lanceolata* and *Veronica spicata* have probably been at their northern range limit in the Åland Islands for hundreds of years and possibly for much longer. Traditional small-scale agriculture, which has created and maintained a large fraction of the habitat into which these species are confined, started in southwestern Finland about 3000 years ago. There are no obvious more recent alien species with which *M. cinxia* would interact either directly or indirectly. Habitat loss and alteration, primarily driven by changes in agriculture, have been less marked in the Åland Islands than in mainland Finland and in many other parts of northern Europe. Habitat alteration includes overgrowth of meadows, which is partly due to reduced grazing and abandonment of former pastures. The mode of grazing has changed from extensive to intensive, which means a smaller but more heavily grazed area. Another cause for overgrowth of meadows is the increased use of fertilizers, as well as nitrogen deposition, which was around 640 mg N/m^2/year in southern Finland in the first half of the 1990s (Wahlström et al. 1996).

Climate change is thought to have extended the northern range boundary of 34 species of nonmigratory butterflies in northern Europe during the past century (Parmesan et al. 1999). One might think that *M. cinxia* belongs to the species that would benefit from climate change, but this is unlikely to be the case for two reasons. First, in mainland Finland there is so little suitable habitat, and it is so fragmented, that the opportunities for range expansion are limited or nonexistent. Warren et al. (2001) have recently demonstrated for British butterflies that while half of the mobile habitat generalist species have expanded their northern range boundary over the past 30 years, 89% of habitat specialist species, including *M. cinxia*, have declined because of habitat loss and fragmentation. Second, if the recent high incidence of dry and warm summers is due to climate change, the net impact on *M. cinxia* is negative because dry periods in summer cause host plants in the naturally dry, rocky outcrops to whither, and thereby dry periods lead to massive larval mortality. The lesson here is that a particular global change (climate change ameliorating environmental conditions for butterflies in general) may strongly interact with species-specific factors (natural habitat of *M. cinxia* is susceptible to drought) and other global changes (habitat loss) to reverse the expected trend.

5

Checkerspot Reproductive Biology

CAROL L. BOGGS AND MARKO NIEMINEN

5.1 Introduction

Reproductive behavior, physiology, and ecology affect the number of offspring that an individual butterfly contributes to the next generation; therefore, these factors play an integral mechanistic role in influencing population and evolutionary dynamics. Critical questions include, how do males and females find each other and select mates? Why have particular mating systems evolved? What are the nutrient sources for offspring production and how much should females or males invest in each offspring? What determines temporal patterns of egg production? How do all these tie together to influence lifetime reproductive success? And how are these decisions influenced by the environment and by selection?

Untangling the answers to these general questions for insects and placing those answers in an appropriate evolutionary and ecological context requires a solid understanding of the reproductive biology of a set of model insect species. Ideally, the biology of these species should span a range of mating behaviors and numbers, feeding habits, ovarian dynamics and parental care, and age-specific reproductive and survival values to facilitate comparative studies (Boggs 1986). Due to contrasts with other butterflies, checkerspots have proven useful as one model group of species within the set used to address these questions. For *Euphydryas editha bayensis* in particular, the number of matings by females is generally low, the adult diet is nutritionally incomplete, and females emerge with a fixed number of oocytes in the ovaries, a relatively large percentage of which are chorionated and ready to be laid (Labine 1964, 1968, Boggs 1997b). Contrasting species include some *Heliconius*, whose females may mate many times, eat a nutritionally complete adult diet, exhibit continual oogenesis throughout life, and emerge with no eggs chorionated (Gilbert 1973, Dunlap-Pianka et al. 1977, Boggs 1990). Equally important, *E. editha* was also the subject of early studies on male and female reproductive success as outlined below, laying a framework for understanding these aspects of reproductive biology in insects.

In this chapter, we synthesize the contributions of checkerspot studies to our understanding of sex ratios, mating strategies, reproductive allocation, and lifetime reproductive success. Along the way, we show how reproductive biology connects to the broader issues of population and evolutionary dynamics as well as conservation and highlight key unanswered questions that could be addressed with the checkerspot system.

5.2 Sex Ratios

Sex ratios in a butterfly population have important implications for the evolution of mate location strategies and the operation of sexual selection, as well as for the effective population size (N_e) and contributions of each sex to N_e (see chapter 10). Determining actual sex ratios can be as difficult as it is important. In considering actual sex ratios in checkerspots, we start with the analysis of Ehrlich et al. (1984), who distinguished among a population's observed sex ratio (or sex ratio of captured adults), the realized sex ratio (or actual sex ratio of adults), and the operational sex ratio (or the sex ratio of adults available to mate; e.g., Odendaal et al. 1985). These distinctions were applied to a remarkably complete mark–release–recapture data set for *E. editha bayensis*, collected in 1981 from population H at Jasper Ridge (figure 1.2). The work was specifically aimed at elucidating factors that contribute to the realized sex ratios as well as examining the implications of those contributing factors for N_e.

Observed and Realized Sex Ratios

In all sex ratio estimates, we must begin with an observed sex ratio. For many butterfly studies, that ratio is biased by differences in catchability between the two sexes (e.g., Tabashnik 1980, Boggs 1987b). In the case of the 1981 *E. editha* data set, essentially all males and 90% of the females were captured at least once, resulting in little difference between the observed and realized sex ratios. Both sex ratios were male-biased, with an observed and estimated realized sex ratio of 100:51 males:females or 1.95.

The adult male bias in 1981 must result from factors operating after egg formation, given that a sample of 63 *E. editha* adults collected as postdiapause larvae and reared in the laboratory resulted in an adult sex ratio that did not significantly deviate from 1:1. Potential contributors include differential preadult mortality, adult mortality rates, and adult emigration rates (Ehrlich et al. 1984). Differential preadult mortality is due at least in part to the longer postdiapause (but not pupal) development times of females in this species. Rough calculations of likely mortality rates due to differences in emergence dates and estimated postdiapause daily mortality suggested that this excess mortality experienced by females would likely shift the sex ratio from 1:1 to 100:69, or 1.45 (Ehrlich et al. 1984). This is still less than the estimated realized sex ratio of 1.95 for that year.

Thus, factors affecting adults must also contribute to the realized sex ratio bias. Indeed, recapture data for 1981 indicated that females were more likely to emigrate and may have had a higher adult mortality rate than males, which would affect the realized sex ratio. There are a variety of mechanisms that could produce sex differences in mortality or dispersal. One is simply a difference between the sexes in their propensity to emigrate in response to density, as seen in *E. anicia* (Odendaal et al. 1989; see also chapter 9). Others include differential bird predation, which resulted in more female deaths in *E. chalcedona* (Bowers et al. 1985), and hazards encountered during mate searching, which resulted in male deaths in *E. editha* in the Sierra Nevada mountains in California (Moore 1987).

Similar factors may affect the realized sex ratio of other checkerspots. With respect to adult migration, *Melitaea cinxia* females have a higher emigration rate than males (e.g., Hanski et al. 1994, Kuussaari et al. 1996, Wahlberg et al. 2002b; chapter 4), which would favor male-biased realized sex ratios. Wahlberg et al. (2002b) did a broader analysis of movement patterns in Finnish checkerspots, including two additional *Melitaea* species and two *Euphydryas* species. They found no difference in emigration propensity between the sexes in any of these four species. Nonetheless, examining the mark–release–recapture data set for *M. diamina* in more detail, Hanski et al. (2000) and Wahlberg et al. (2002b) found that the local sex ratio affected patch-specific movement rate of males, which was exceptionally high in a strongly male-biased population, suggesting a role for adult migration under some circumstances. With respect to differential adult mortality, which may influence the realized sex ratio, females of *Melitaea cinxia*, *M. athalia*, *Euphydryas aurinia*, and *E. maturna* have higher death rate than males, whereas *M. diamina* showed no difference between sexes in the average life span (Wahlberg et al. 2002b).

Variation in Realized Sex Ratios

Less detailed mark–release–recapture data for *Euphydryas editha* at Jasper Ridge in years other than 1981 have provided an extensive data set illustrating the tremendous variation in realized sex ratios across years and populations (Hellmann et al. 2003). Using only data from years with reliable population estimates, the sex ratios were significantly different from 1:1 in 9 of 14 years in area C and in 11 of 17 years in area H. Six of those years were female-biased in area C, while all other significant deviations were male-biased. Sex ratios differed between the two populations, which highlights the largely independent population dynamics in these populations (figure 3.6). The average sex ratio in area C was half of that in area H (1.08 vs. 2.02), and the variance in sex ratio in area C was small compared to that of area H (0.13 vs. 1.12). Finally, sex ratio and population size were significantly correlated (larger populations had a greater male bias) in area C but not in area H, although there was no correlation between sex ratio and male population growth rate in either population. This could mean that males of this species are prone to leave small populations, as is the case for females in *M. cinxia* (see also chapter 9). The implications of this temporal and spatial variation remain unexplored, but must result in significant variation in the contribution of each sex to N_e and in the relative strength of selection or drift for each sex. This degree of variation also suggests that adult sex ratio in this species is not closely controlled by constant selection pressures, which is consistent with predictions of classical theory for species without parental care of offspring (Fisher 1930, G. C. Williams 1975).

Operational Sex Ratio

The operational sex ratio depends on both the realized sex ratio and on the fraction of individuals available to mate (e.g., Ahnesjö et al. 2001). As outlined below (section 5.3), female *Euphydryas editha bayensis* usually mate only once, although occasionally second matings are observed. In this case, virgin females form the primary pool of females in the operational sex ratio, whereas all males, virgin or not, who are not currently mating may be included in the pool of males. In general, the operational sex ratio is strongly male-biased under these circumstances. *Euphydryas editha*'s operational sex ratio in the Jasper Ridge area H population thus fluctuated through the season in 1981, as the first female did not eclose until 6 days after the first male was caught.

Ehrlich et al. (1984) calculated that nearly 12% of the males did not live long enough to see a female. This suggests that N_e for males (or, more exactly, the male component of N_e) may be small compared to the actual number of males flying because even in butterfly species without sexual differences in adult emergence time, roughly half of the males never mate (Boggs 1979, for *Heliconius* and *Dryas*; Wahlberg 1995, for *Melitaea*; K. Oberhauser pers. comm, for *Danaus*). Subtracting an additional 6% (half of 12%) would reduce male N_e to around 44% of the male population, due to mating alone. Given a realized sex ratio of 1.95 in 1981 for *E. editha*, this suggests that the sex ratio of individuals actually mating (as opposed to available to mate) may have been closer to 86:100, or approximately 0.85, which is female rather than male biased.

5.3 Mating Strategies

Sex ratios, and particularly the operational sex ratio, interact strongly with individuals' mating strategies because the number of individuals of each sex available to mate determines mating opportunity. Mating strategies include mate acquisition behavior, mate choice, numbers of matings, male reproductive investment, and sperm precedence. Numbers of matings interact with the operational sex ratio to influence protandry, or sex-biased differences in the timing of adult emergence. Many components of mating strategies may vary over time and space, both within and among individuals or populations. The strategy adopted by a given individual may be conditional on environmental factors, including strategies adopted by other individuals. For example, the satyrine *Lasiommata megera* switched between perching and patrolling behavior for mate searching on hilltops (Wickman 1988), and the nymphalid *Chlosyne californica* switched between territorial defense on hilltops and patrolling down valley, depending on time of day (Alcock 1994). As a result, not only the mean strategy, but also its variants, are important for understanding the evolutionary dynamics of mating biology.

Below, we first dissect the mating strategies of checkerspot butterflies, then integrate across strategy components, highlighting the contributions that work on this group of butterflies has made to our general understanding of such strategies.

Mate Acquisition

Butterflies as a whole exhibit a diverse array of mate acquisition strategies (summarized in Rutowski 1991). Historically, these have been classified as "perching" and "patrolling" (e.g., Scott 1974b). In the former case, the male butterfly perches on the ground, on a tree, or on a similar structure. As females (or any object that could possibly be a female) fly by, males dart out from the perch to investigate. Patrolling involves male flight over an area, searching for receptive females. However, Rutowski (1991) points out that this dichotomy is more accurately broken down to include site tenacity, site defense, and behavior while at the site, as perching in particular may be associated with different levels of site tenacity and defense.

Checkerspots exhibit a range of search behaviors, even within one species. *Euphydryas editha bayensis* males generally patrol for mates within a small area, *E. editha* males at Almont in Colorado also perch (Ehrlich and Wheye 1986), and *E. chalcedona* males in Arizona primarily perch, but with low site tenacity and defense (Rutowski 1991). Male *M. cinxia* in Finland both patrol to look for females and perch to wait for females, with the majority of search effort devoted to perching (Wahlberg 2000a). In Britain, the mixed strategy in *M. cinxia* is tilted in the opposite direction: males primarily patrol to look for females (Bourn and Warren 1997).

Factors influencing the specific strategy adopted by a given male of a given species include the operational sex ratio as well as the spatial distribution of receptive females. Distribution of receptive females may be influenced both by the distribution of larval and adult resources and nonresource landmarks and by the defensibility of those sites, resulting in differences in site tenacity and defense by males (Emlen and Oring 1977, Bradbury 1985, Rutowski 1991).

For *Euphydryas editha* and *Melitaea cinxia,* and presumably to a lesser extent *E. chalcedona*, the operational sex ratio is generally biased toward males due to the small number of matings by each female (section 5.2). In this situation, females are hypothetically able to choose among males, which are looking for virgin females. In general, males likely locate mates using visual cues. For example, Moore (1987) found that males were fatally attracted to conspecifics of either sex trapped in spider webs.

For *E. chalcedona* in Arizona, which feeds on bushy *Keckiella antirrhinoides* that can support large numbers of larvae, pupation sites and thus eclosing females are distributed patchily near the host plants (Rutowski 1988). Males in this population use the larval host plants as encounter sites, perching and flying to detect receptive females. Site tenacity is low, as is site defense, which is consistent with the occurrence of many plants in each population (Rutowski 1991).

The situation for *E. editha* is more complex. Although eggs are laid in clusters, presumably giving some spatial structure to the locations of eclosing females, in general larvae disperse during mid-development, both to diapause and because one host plant individual often cannot support development of the entire egg cluster (chapter 7). Thus, although males might be expected to limit their search to areas with host plants, male search locations should also reflect any heterogeneity in female eclosion sites or in initial female movement patterns.

However, if such heterogeneity is limited and virgin females are widely distributed across the landscape, a system may evolve in which males and receptive females congregate for mating at visible landmarks such as hilltops that may not contain adult or larval resources (Shields 1967, Scott 1968, Thornhill and Alcock 1983, Alcock 1987, Wickman 1988, Daily et al. 1991; but see Alcock 1994). This behavior is particularly likely when adult population density is low. In contrast to vertebrate lek systems, landmark or hilltop mating systems in butterflies generally do not include long-term territorial stability, with territory holders in the center being more attractive to females (but see Knapton 1985, Daily et al. 1991).

Shields (1967) asserted that *E. editha* exhibits hilltopping behavior at Dictionary Hill, San Diego, California. However, his evidence was sketchy, and all three hilltop females that he dissected had mated once, giving no indication that virgins congregate at hilltops to find mates. Additionally, subsequent studies of *E. editha* at Jasper Ridge area C and at Rabbit Meadow indicate that the butterflies may not exhibit a classic hilltopping strategy (Baughman et al. 1988b, 1990b, Singer and Thomas 1992), perhaps due to their relatively sedentary nature in these locations, which have sufficient adult nectar and larval hostplants (cf. Gilbert and Singer 1973, Singer and Thomas 1992).

The question of whether a hilltopping strategy has been adopted remains open for a population at Almont in Colorado (figure 5.1; Ehrlich and Wheye 1984, 1986). Here, sex ratios were male biased on ridgelines in comparison to lower slopes (figure 5.2). For a thorough demonstration of the hilltopping strategy, one would need to track the movements of virgin and newly mated females, as well as of males congregating for mating. Although this was not done at Almont, mark–release–recapture data indicated that recapture rates (no. recaptured/no. marked) were significantly lower for females on the ridgelines than on the slopes, suggesting lower residence times on the ridgelines, as expected for a hilltopping strategy. Further support for this

Figure 5.1. (a) Aerial photo showing road and benches at Almont Summit, Colorado. White lines show part of transects used in research. Transect point 1 is on the road, 33 is on the first bench, and second bench is at the farthest left point on the transect. Third bench is 300 m south of point 33. (b) Photo of road taken from between points 7 and 1 in photo (a), looking toward point 1, elevation about 3000 m. Male *Euphydryas editha* concentrate on the bare areas of the road on the right; first bench is visible on the slope dropping down to the left. Photos by Paul R. Ehrlich.

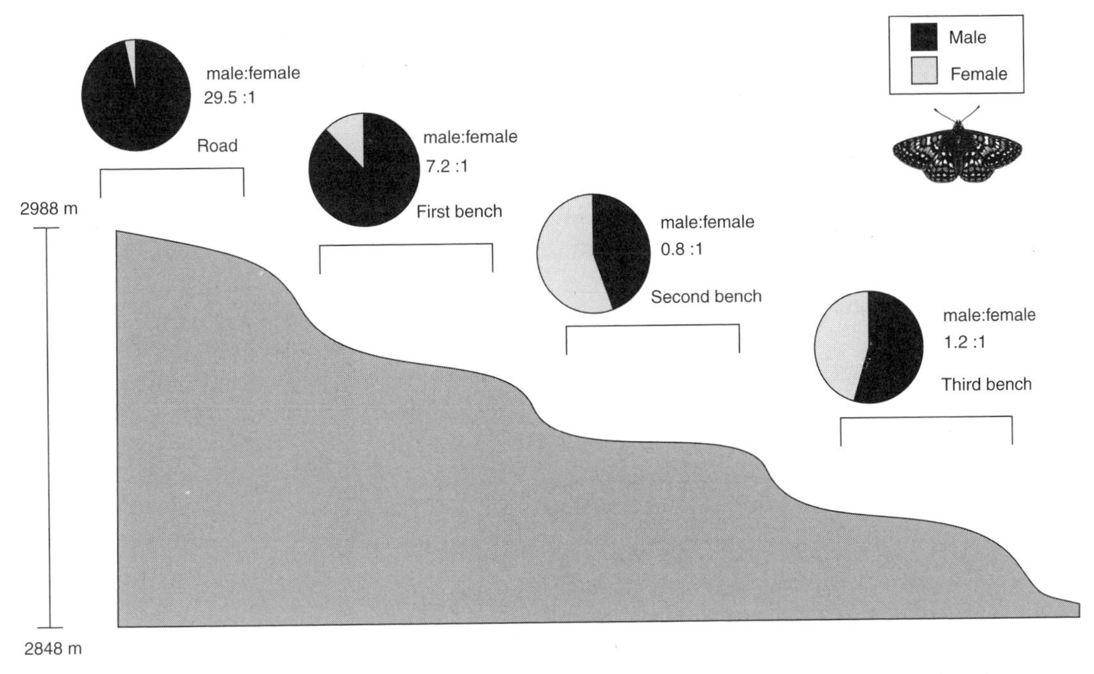

Figure 5.2. *Euphydryas editha* sex ratios as a function of topography in the Almont, Colorado, population in 1984. Data from Ehrlich and Wheye (1986).

idea comes from the observation that in years of low density the populations retracted to ridgelines, with proportionately more females found at higher locations than previously (Ehrlich and Wheye 1988). In 1984, however, a greater proportion of marked males moved from the ridgeline to the slopes than vice versa (8.5% vs. 3.1%, Goldstein's x^* = 3.82, one-tailed $P << .005$), suggesting that the lower residence on ridgelines may not be sex specific (see also Singer and Thomas 1992).

Whether these skewed sex ratios and patterns of population structure reflect larval survival and emergence patterns, adult habitat preferences (Odendaal et al. 1989), the results of random male searches for receptive females (Odendaal et al. 1988) and harassment of mated females, or a true hilltopping strategy with males and virgin females moving to hilltops remains to be explored. Any final explanation must account both for the relatively low movement rates observed and the skewed sex ratios. Nonetheless, the implications for management efforts are clear: both slopes and hilltops are important to the maintenance of the populations (chapter 13).

Regardless of whether *Euphydryas* follows an evolved hilltopping strategy for mate location, the marked difference in sex ratio across topography within the Almont population raises the question of whether male mating success varies with location. Paul Ehrlich and Darryl Wheye tested this with an ingeniously novel experimental technique. They dusted genitalia of field-caught male *E. editha* with fluorescent powdered pigments (plate VII.4) and then looked for the presence of pigments on subsequently captured mated females (Wheye and Ehrlich 1985). Presence or color of pigment had no effect on a male's ability to mate, which was tested in *E. chalcedona* under laboratory conditions (Fleishman et al. 1993). Ehrlich and Wheye (1986) concluded that males on ridgelines had less mating success than did those on lower slopes, after applying a correction factor to account for differences in the proportion of males dyed in each area. The same conclusion arises if one assumes that migration on the order of 3–8% is sufficient to consider the two areas as containing two loosely connected populations, one on the ridgeline and one on the slope (no. females carrying dye from the site/no. males dyed: ridgeline, 29/340; slope 88/709; Goldstein's x^* = 1.87, one-tailed $P = .03$). Thus, males in the population subarea with a higher sex ratio had lower mating success, assuming that the fraction of multiple matings by males did not differ by location.

Only if one assumes that the Almont *E. editha* population shows sufficient within-population movement to be considered one population can one conclude that there is no differential male mating success between males on ridgeline as opposed to the slope (Ehrlich and Wheye 1986, Singer and Thomas 1992). Given that the observed migration rates are within the realm of those seen among populations in metapopulations, the assumption of independence of the ridgeline and slope populations at Almont seems reasonable.

This case has implications for population genetic structure and local differences in contribution to effective population size (see chapter 10). Differences in distribution of males and females across a landscape may affect mating success and hence effective population size and genetic structure. In the case of Almont and Jasper Ridge, the distribution differences fluctuate with population size. The impacts of this variation could strongly influence the evolution of mating systems and other traits within the population, but remain to be modeled.

Mate Choice

Once a potential mate has been located, mate choice may come into play. Although essentially nothing is known about mate choice in checkerspots, butterflies in general exhibit female mate choice based on wing coloration, courtship persistence, pheromone titer, or indirectly through ability of males to fly at low temperatures (e.g., Rutowski 1979, 1982b, 1984, 1985, Watt et al. 1986, Wiernasz 1989, Sappington and Taylor 1990, Brunton and Majerus 1995, Kaitala and Wiklund 1995). Male choice among female mates based on age, body size, or wing coloration has also been documented (e.g., Rutowski 1982a, Cook et al. 1994, Wiernasz 1995, Ellers and Boggs 2003). Discrimination based on relatedness has not been reported in butterflies. In species with metapopulation structures, such as those observed in many checkerspots and in *M. cinxia* in particular, small population size leads to a high rate of inbreeding if avoidance of mating with close kin has not evolved. In *M. cinxia*, substantial inbreeding depression follows from close-kin matings (Saccheri et al. 1998, Haikola et al. 2001, Nieminen et al. 2001). Nonetheless, Haikola et al. (2004) recently demonstrated that *M. cinxia* females do not discriminate against close kin (brothers), indicating that inbreeding avoidance has not evolved.

Numbers of Matings

In both *Euphydryas editha* and *Melitaea cinxia*, females mate once (Figure 5.3), or occasionally twice, but males may mate multiple times with multiple females, a mating system referred to as polygyny. In samples of 23 and 22 field-collected female *E. editha bayensis,* 41% and 27% had mated twice (Labine 1964, Ehrlich and Ehrlich 1978). Labine (1964) reported that the first and second spermatophores appeared to be roughly the same age in 7 of the 9 multiply mated females in the sample of 23 females. In samples of 167 and 121 wild-caught mated female *M. cinxia*, only 8% and 6.5% had mated twice (Kuussaari et al. 1998), while in a sample of 28 caged males, 36% mated more than once (Wahlberg 1995). In contrast, *E. chalcedona* females may mate as many

Figure 5.3. *Euphydryas editha* mating at Almont Summit, 1984. Photo by Paul R. Ehrlich.

as three times in the field in Arizona (Rutowski and Gilchrist 1987).

Euphydryas editha have at least two mechanisms that contribute to a polygynous mating system by preventing female remating (Labine 1964), one physical and one neurological/behavioral. As a physical barrier to remating, the spermatophore deposited by males in the female bursa copulatrix in *E. editha* has a long neck that plugs the female genital tract and often extrudes from the genital opening, thus sealing it (Labine 1964; figure 5.4). Once this plug hardens, further mating is difficult, if not impossible, as long as the plug remains in place. However, Labine (1964) noted that a few females mated twice. In some cases, a second mating can occur before the plug has hardened; alternatively, the plug may eventually erode, which allows females to remate. Such plugs are fairly widespread in butterflies (Ehrlich and Ehrlich 1978) and include elaborate external sphragi produced by *Parnassius* (Scott 1986) and several Australasian genera (Orr 1988, 2002, Orr and Rutowski 1991). Indeed, ephemeral sperm plugs have subsequently been found in snakes (Devine 1975, 1977, Ross and Crews 1977) and mammals (Hartung and Dewsbury 1978, Voss 1979), including *Homo sapiens* (Martan and Shepard 1976, Dewsbury 1984, Baker and Bellis 1995)—another example of discoveries made in the checkerspot system that have much more general significance.

The second mating inhibition mechanism in *E. editha* is neurological and behavioral. Mated females exhibit active mate rejection behavior, with or without intact plugs. Labine (1964) hypothesized that this behavior was stimulated by neural perception of bursal distension by the spermatophore. Surgically cutting the bursal nerves along with removal of the plug resulted in females that remated. Labine's work was the first demonstration of such a neural mechanism controlling mate refusal behavior in insects. Sugawara (1979, 1981) later demonstrated this phenomenon decisively in the pierid *Pieris rapae*, and Obara (1982) found evidence that mate refusal behavior may be maintained in the longer term by a hormonal factor. More complex, related mechanisms have since been found in several moths (e.g., Giebultowicz et al. 1991). It is also noteworthy that unlike *E. editha*, Finnish *M. cinxia*

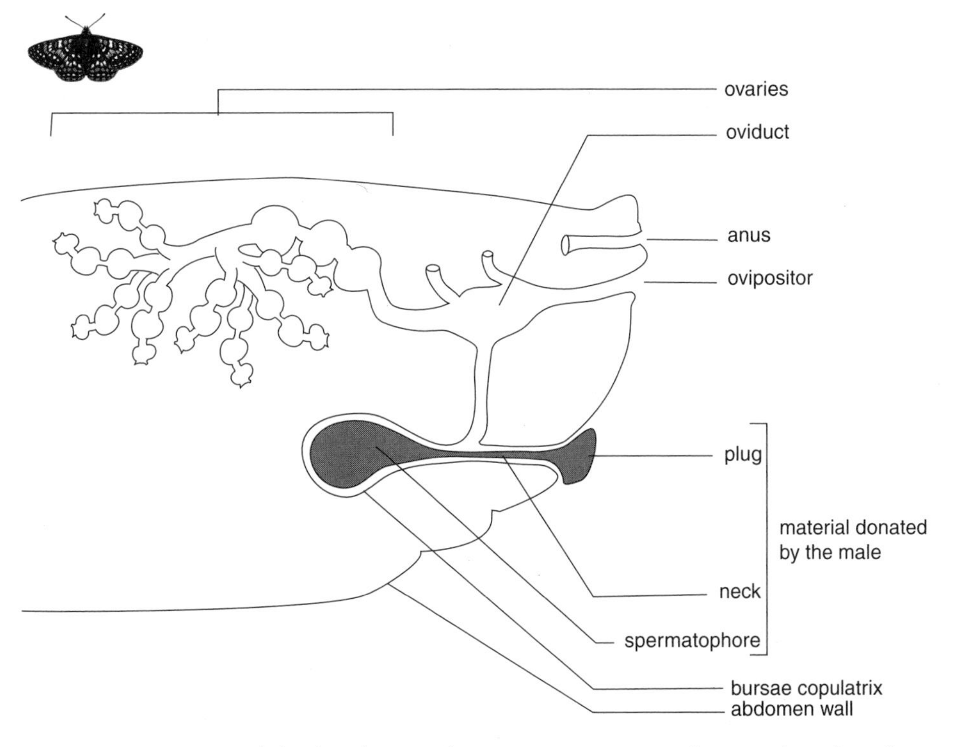

Figure 5.4. Structure of the female reproductive tract, spermatophore, and mating plug in *Euphydryas editha*, after Labine (1966b).

show no obvious evidence of a mating plug (Wahlberg 1995). Because the frequency of remating is low (<10%), there must be mechanisms other than the mating plug that reduce the rate of remating.

In spite of its apparent greater number of matings, *E. chalcedona* has the same strategy as *E. editha* for preventing female remating. Females are plugged, and mating induces mate rejection behavior. In controlled field experiments, Dickinson and Rutowski (1989) showed that both mating plugs and active mate rejection behavior influence female mating probability. Virgin females with transplanted plugs were less likely to be mated than were normal virgin females. While males courted mated females longer than virgin females, they had lower mating success due to female behavior. Further, the difference in courtship persistence indicates that the plug itself does not signal the female's mating status to the male. It is unknown whether there are differences between the two *Euphydryas* species in rates of degradation of plugs or in variance in plug size that contribute to the apparent differences in female mating frequency. Rutowski et al. (1989) showed that the area covered by the plug was not correlated with female age in *E. chalcedona* and found no evidence for plug loss.

An alternative explanation for second matings in the presence of plugs is that those matings may only occur if males abrade the plug or remove it during a longer courtship than occurs with virgins (M. Singer, pers. comm.). In such cases, plugs would both increase the first male's security of paternity and increase the control of females over second matings, since females would have to remain still for plugs to be abraded or removed. Further, plugs may also increase the risk of injury to females on a second mating, affecting female propensity to remate.

Other mechanisms inhibiting female remating occur in other butterflies. In particular, female remating is inhibited by pheromones that are either passed from the male to the female at mating or whose production is stimulated by mating (e.g., the nymphalids *Heliconius charitonius*, *H. erato*, *H. sara* [Gilbert 1976], and the pierid *Pieris napi* [Andersson et al. 2000]). Such pheromones have not been looked for in checkerspots, but their presence would suggest particularly strong selection for redundant systems to prevent female remating. This could either result from a long history of intersexual conflict over mating numbers or from strong selection on males to guard their genetic investment.

Protandry

As described above, a strong factor determining the operational and realized sex ratios in *Euphydryas editha* is few matings by individual females, which may select for emergence of adult males before adult females. This emergence pattern is called protandry. For *Melitaea cinxia*, and likely other species as well, protandry is primarily due to differences in postdiapause larval development time, rather than in pupal development time. In a laboratory experiment on *M. cinxia,* postdiapause larval development time averaged 28 days (SD = 6.3, n = 70) for males and 32 days (SD = 5.4, n = 52) for females, whereas pupal development times were 18.9 days (SD = 3.1, n = 70) for males and 19.2 days (SD = 3.0, n = 52) for females (S. Haikola, unpubl. data). The four-day difference in postdiapause larval development time can easily explain the observed protandry of two to three days observed in the field (Wahlberg 1995).

The evolution of protandry in polygynous insects has long been ascribed to the selection pressures on males to emerge and mature as adults before females, so as to have a good chance of encountering virgin (receptive) females (e.g., Wiklund and Fagerström 1977; see also Morbey and Ydenberg 2001), and checkerspots have played a crucial role in testing this phenomenon.

Iwasa et al. (1983) developed a game theoretical model to test the hypothesis that both the shape of the male emergence curve and the date of the peak of male emergence are selected to maximize probability of male encounter with receptive females. In other words, the emergence curves result in equal probabilities of male mating success across all male emergence dates. Their model included both preadult and adult mortality. The resulting predicted emergence curve was tested against *E. editha* emergence data for Jasper Ridge area H in 1981 (Iwasa et al. 1983), as well as the more topographically homogeneous area C using data for several years (Baughman et al. 1988a). Although the model was generally supported, several specific predictions were not. Most important, the model predicted a truncation in the male emergence curve, which was never seen in the data, and the peak of male emergence often lagged a day or two behind the predicted peak (figure 5.5). As suggested by Baughman et al. (1988a), these differences may be due to recombination affecting the variance in the male emergence curve, even if it is under selection (see also Bulmer 1983); constraints due to selection on female emer-

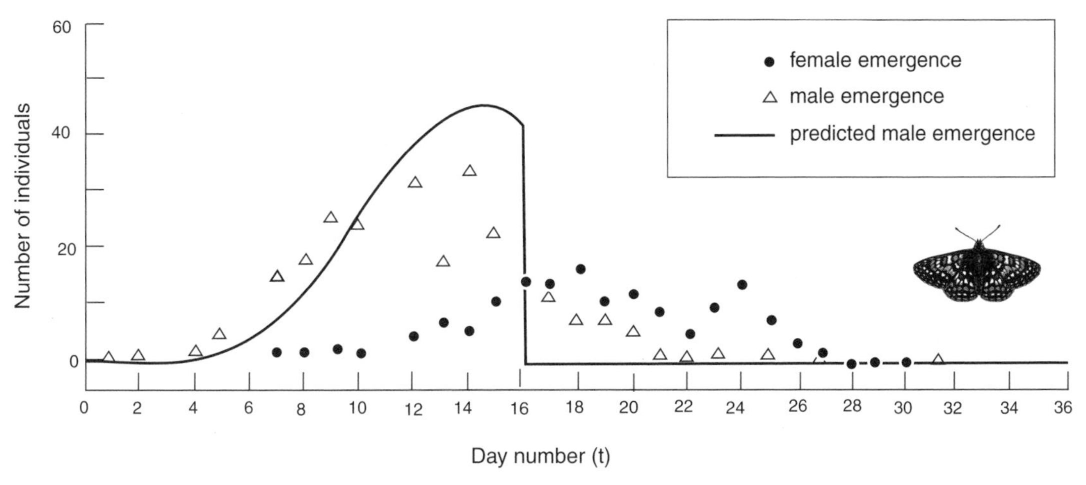

Figure 5.5. *Euphydryas editha bayensis* emergence curves, predicted for males using the model of Iwasa et al. (1983), and observed for both sexes.

gence time; or small microhabitat differences affecting larval developmental rates. Emergence time can also be strongly affected by weather on the immediate day, as high humidity favors emergence and results in more time for drying wings. It is interesting, however, that the underlying hypothesis of equal mating success by males emerging at different times in the season does hold true for *E. editha* in Jasper Ridge area H (Baughman 1991), suggesting that the model's basic assumption of selection is at least in part true. Laboratory experiments are needed to fully address the relative roles of environmental variation and selection in determining the male emergence curve.

Male Reproductive Investment

Both of the known mechanisms responsible for inhibition of female remating depend on male transfer of large amounts of accessory gland material and on the toughness, or resistance to absorption or abrasion, of such materials once they have hardened. Data from *Euphydryas* are consistent with this. As would be expected if there were little or slow absorption of the spermatophore, variation in spermatophore size has no significant effect on female reproductive output (Jones et al. 1986), and ^{14}C or ^{3}H radiotracers derived from male-donated nutrients appear only in low quantities in eggs (Boggs 1997a).

The relative amount of accessory gland material transferred in *Euphydryas* is in the middle of the range for butterflies. This is as might be predicted based on the limited role of the *Euphydryas* male accessory gland products as paternal investment, but given their likely importance in mate guarding (Boggs 1981). Specifically, Jones et al. (1986) estimated that male *E. editha bayensis* transfers 3.4 mg, or 3.5% of the male's body mass, whereas male *E. chalcedona* from California transfers 2.5 mg, or 1.9% of the male's body mass. These estimates were derived by first calculating the difference in male mass before and after mating, then correcting that difference by changes in the mass of control males held in similar conditions but not mated. The estimated values are likely to be underestimates, as control males may have been more active than males that mated.

In contrast, Rutowski and Gilchrist (1987) weighed bursal material dissected from newly mated, field-caught female *E. chalcedona* from Arizona, obtaining a value of 4.6±0.3 mg, or 7±1% of the male's wet body mass. This is larger than the 2.5 mg/1.9% of Jones et al. (1986), yet smaller than the uncorrected value of 10.7 mg mass loss by Jones et al.'s mated males. Furthermore, Rutowski and Gilchrist (1987) did not include the plug material, but Jones et al. (1986) did. Finally, male mass recorded in California was about twice that recorded in Arizona, which could either be due to population differences or because the butterflies in California were well fed and kept in captivity.

Taking all this into account and given that the uncorrected mass transferred for *E. editha* in the

Jones et al. study was 8 mg, it is reasonable to estimate that *E. chalcedona* male contributions (including the plug) are around 7 mg, representing 7–10% of the male's body mass, while those of *E. editha* are around 6 mg, representing as much as 10–12% of the male's body mass. Reported mean values for relative male reproductive investment in butterflies range from 0.5 to 36.6 mg for total ejaculate mass, representing 1.4% to 15.5% of the male's wet body mass (Rutowski et al. 1983, Svärd and Wiklund 1989, Vahed 1998). Thus the *Euphydryas* male investment is in the middle of the range.

These data highlight the possibility that male investment varies from population to population, as is known for female age-specific fecundity schedules in *E. editha* in California (Moore and Singer unpubl. ms.). Although studies in other insects have documented effects of changes in relative reproductive investment between the sexes in response to adult food availability within populations (e.g., Boggs 1990, Fox 1993, Simmons 1994; see also Boggs 1995), comparatively little attention has been paid to environmental or genetic factors influencing differences among populations in relative investment levels of either sex (but see Savalli et al. 2000). In the case of these *Euphydryas* populations, it is unknown whether their environments differ in the quality or quantity of available larval or adult food or whether other environmental factors (or genetic drift) have shaped the observed differences in male size and reproductive investment.

Sperm Precedence

Mechanisms preventing female remating have a flip side: they can help guarantee male paternity. If the first male's sperm do not fertilize all offspring produced after a subsequent mating, it is in his interest to prevent female remating. In *E. editha*, Labine's (1966a) pioneering work on sperm precedence indicated that the last male to mate fathered subsequent offspring. This is not true in all butterflies (e.g., Drummond 1984). For instance, Bissoondath and Wiklund (1997) found that the biggest male to mate fathered most of the offspring, regardless of mating order, in *Pieris napi*. Nonetheless, sperm precedence has strong implications for sexual conflict of interest with regard to remating and for the expectations of investment by males in paternal versus mating effort. In short, with complete sperm precedence, a male should be under intense selection to invest in mating effort, aimed at preventing female remating, while females may be selected to thwart the male's intentions and remate, in particular if genetic diversity among her offspring is advantageous to the female. This again is consistent with patterns seen in *Euphydryas*: complete sperm precedence and investment by males in remating prevention.

The combination of sperm precedence and mechanisms preventing remating has other general implications for *E. editha* evolutionary dynamics. Individual movements are not equivalent to gene flow (chapter 10). Females moving to a new location carry the genetic representation of their mates from the original location, while males moving to new locations do not automatically leave a genetic legacy at those new sites. Although sperm precedence patterns are unknown in *M. cinxia,* these considerations are consistent with recent findings that females, but not males, from newly established *M. cinxia* populations are especially dispersive, most probably because the new populations are established by more dispersive females than the average in the metapopulation (Hanski et al. 2002).

Factors shaping the evolution of particular combinations of remating inhibition mechanisms, sperm precedence, and consequent individual number of matings have long been of interest to biologists (see Parker 1970). In the case of butterflies and insects with similar reproductive biology, the particular inhibition mechanisms of a species may depend on adult life span and on the battering that plugs suffer over time (given that plugs do eventually erode), historical constraints associated with pheromonal and behavioral remating inhibitors, as well as unknown factors associated with the physiological and physical mechanisms underlying sperm precedence.

Evolution of Mating Systems

The mating system described above for checkerspots is one approximating polygyny, with some variation in female mating numbers; intermediate levels of male investment in reproductive effort, which likely varies across populations; complete sperm precedence, at least in *E. editha bayensis*; adult sex ratios that may vary considerably over time and space; and male mate-location strategies that are likely driven by the distribution of eclosing females, in at least some populations. This combination is consistent from an evolutionary perspective and constitutes one combination among an array of mating systems observed in butterflies and other insects.

Proposed evolutionary explanations for the observed numbers of matings include increased genetic diversity of mates in cases of high numbers of matings among females (see, Jennions and Petrie 2000) and avoidance of time costs associated with mating for low numbers of matings among females. For the number of matings to be a character under selection, it must have a genetic basis, which has recently been shown to be the case in the pierid *Pieris napi* (Wedell et al. 2002). No detailed analysis of the number of matings as a strategy has yet been done in checkerspots, but we would expect the observed patterns to result from trade-offs between female time costs associated with mating and male harassment and genetic benefits of multiple mating in small, inbred populations. Additionally, because females apparently derive little nutritional benefit from mating, and toxins passed to females by males at mating have not been reported in butterflies (Fowler and Partridge 1989, Chapman et al. 1995 for *Drosophila*), nutritional factors are not likely to play a major role in sexual selection affecting the evolution of number of female matings. The size of the male investment, in contrast, could make nutrients a factor in determining male mating patterns.

The degree to which checkerspot' mating systems are a product of historical constraints versus innovation could be examined with a proper phylogenetic analysis; pioneering analysis of variation in mating system components has been done for other butterflies (e.g., Wiklund and Forsberg 1991). The degree to which particular mating systems tend to be associated with other aspects of the butterfly's life history, allocation patterns, or environment remains to be explored. Given our knowledge of checkerspots, they will serve as an excellent model system for one end of the observed variation in mating systems in butterflies (chapter 12).

5.4 Reproductive Allocation

A female insect's reproductive strategy includes not only her mating habits, but also her age-specific fecundity schedule and oviposition habits. The latter is discussed in chapter 6. The age-specific fecundity schedule is the result of female allocation of nutrients from larval, adult, and male sources to reproduction and other activities, combined with the characteristic ovarian dynamics of the species. Below we first examine age-specific fecundity patterns of checkerspots and place them in the context of those of other butterflies and insects; then we explore the sources of nutrients used in reproduction and the constraints these nutrients place on the butterfly's life history; and finally we consider the relative role of larval- and adult-derived nutrients for reproduction in the context of the female's reproductive strategy and the effect of that on the female's ability to buffer environmental variation.

Age-specific Reproductive Patterns

Labine (1968) first described the age-specific fecundity pattern for *Euphydryas editha bayensis*, placing it in the context of similar patterns found in other butterflies (figure 5.6). Labine found that females from the Woodside population near Jasper Ridge emerged with 17.8% of the egg complement in the ovaries mature and ready to be fertilized and laid. The percentage for females from the Morgan Hill population of the same subspecies was remarkably similar, with 17.4% mature (Boggs 1997b). This fraction of potential lifetime fecundity represents the number of eggs laid in about the first two days of oviposition (Labine 1968, Boggs 1997b).

Emergence with a complement of mature eggs has several important implications. First, females can and do begin laying eggs soon after eclosion and mating, which is advantageous in a system with a phenological race between host plant senescence and prediapause larval development (chapter 3). Second, those eggs laid in the first two or so days are made completely from larval-derived nutrients, reducing reliance on adult resources. Third, because of high reliance on larval-derived nutrients in early-laid eggs, newly emerged females are relatively heavy, with high wing loading, making flight expensive and difficult, which also suggests that newly emerged females may be unlikely to seek landmarks for mating purposes if males are abundant enough to be able to find the females. Finally, at least some eggs are likely to be laid in the female's natal habitat.

The age-specific fecundity pattern shows some variation among ecotypes of *E. editha*. Females in the Sierra Nevada General's Highway population lay proportionately more eggs later in life. In this population females are also lighter at eclosion than are *E. editha bayensis* females (Moore and Singer unpubl. ms.), suggesting that fewer eggs may be mature at adult eclosion. Additionally, mean time to first reproduction is longer for females at Rabbit Meadow (close to General's Highway) than for *E. editha bayensis* (Singer 1986). This suite of traits

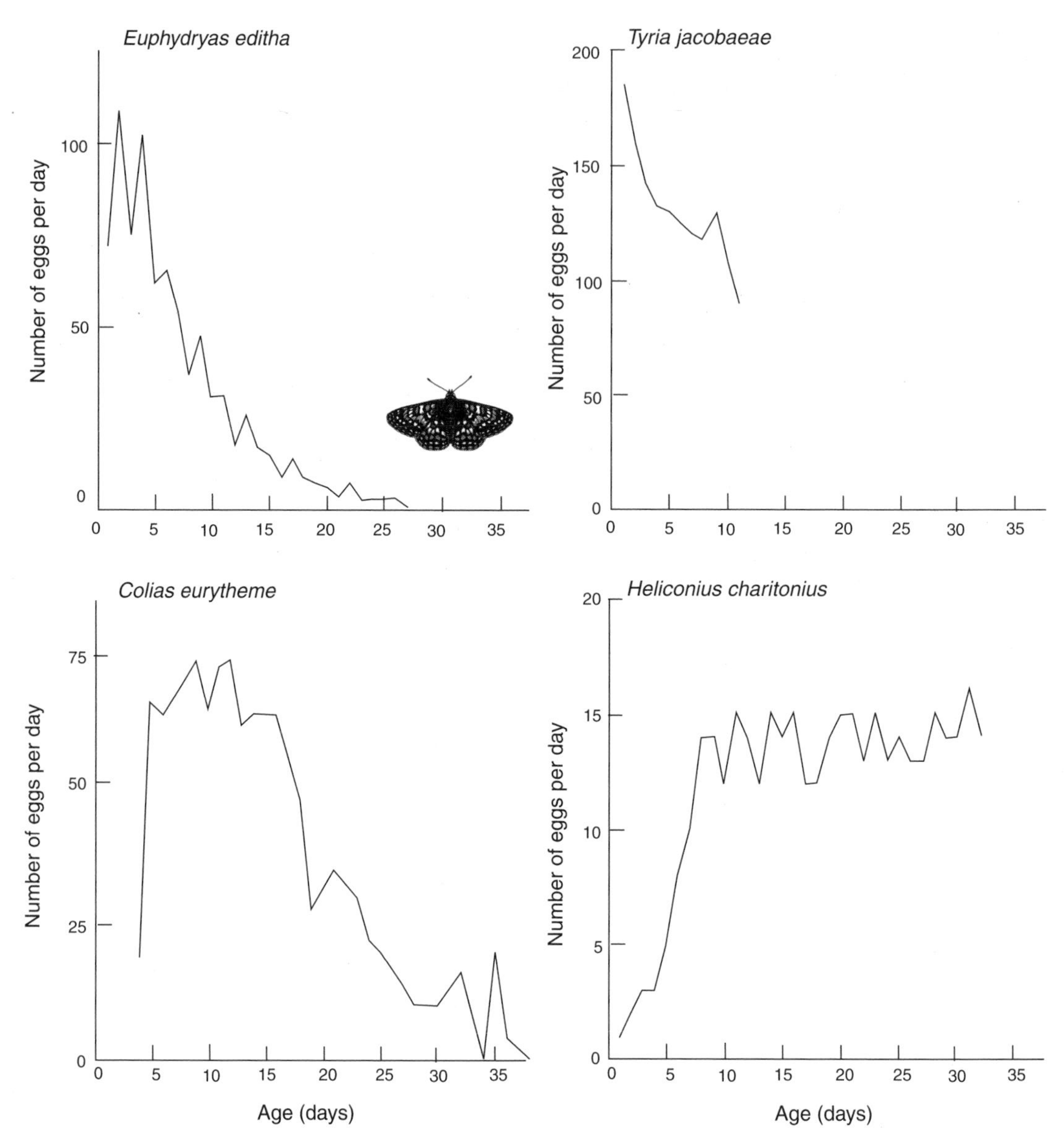

Figure 5.6. Age-specific fecundity curves for 53 *Euphydryas editha bayensis* females, from Labine (1968). To obtain these results, last-instar larvae were collected in the field and reared to adulthood. Females were fed 30% honey-water once a day and maintained in pint containers at 25°C. Data on *Tyria jacobaeae* (adapted from Richards and Myers 1980), *Colias eurytheme* (adapted from Stern and Smith 1960), and *Heliconius charitonius* (from Boggs 1979) are shown for comparison.

is consistent with lower selection pressure for rapid reproduction in the Sierra populations. Boughton (1999) found that eggs laid late in the flight season in traditional Sierra habitats did not suffer a survival cost, unlike that seen for *E. editha bayensis* (Singer 1971b, 1972, Cushman et al. 1994, Hellmann 2002c; chapter 3). Concomitantly, adult resources may be more important in egg production, and female dispersal patterns may differ between the San Francisco Bay area and the Sierra.

In spite of these differences among subspecies, the general age-specific fecundity pattern seen for *E. e. bayensis* is also characteristic of other checkerspots, including *E. chalcedona* (O'Brien et al. 2004) and *M. cinxia* (Kuussaari 1998). This pattern lies at one end of the spectrum seen for butterflies. Among species, as females eclose with less mature eggs and as the adult diet becomes more nutritionally complete (meaning that it can play a greater role in egg production), the initiation of oviposition is delayed, and the plateau of age-specific fecundity is prolonged (Boggs 1986). This suggests that age-specific fecundity is constrained by or co-adapted with age-specific feeding and nutrient allocation patterns (see Boggs 1992).

Like other checkerspot species, female *E. editha bayensis* lay eggs in clusters (chapter 11). Labine (1968) reported an average cluster size of 45 eggs, with a range of 21–75, for females observed ovipositing in the field at Woodside. Similarly, females from Morgan Hill maintained in flight cages in a greenhouse with potted larval host plant and adult nectar plants had a mean cluster size of 42 eggs, with a range of 1–191 (unpubl. data for females in Boggs 1997b). Cluster size decreases significantly with cluster number and hence with age (Labine 1968, C. L. Boggs, unpubl. data). Multiple clusters may be laid on one day, with a maximum of four observed for captive Morgan Hill females (C. L. Boggs, unpubl. data).

Melitaea cinxia egg clusters are larger than those of *E. editha*, with 150–160 eggs on average and up to 7 egg clusters laid over the lifetime (Wahlberg 1995, Kuussaari 1998). As with *E. editha*, the number of eggs per cluster decreases with later clusters (Wahlberg 1995). The interval between oviposition bouts is usually two days, although occasionally egg laying occurs on sequential days. There have been no observations of more than one egg cluster laid in one day (M. Kuussaari and S. van Nouhuys, pers. obs.).

Both the total number of eggs that a female *E. editha bayensis* can lay over her lifetime and the number of mature eggs at eclosion are correlated with female mass at eclosion, but not with wing length (Murphy et al. 1983, Boggs 1997b). This is consistent with the large role of larval nutrients in manufacturing early eggs in particular. These relationships also suggest that wing loading of young females is not subject to heavy selection, as might be the case were early female dispersal important.

In many Lepidoptera, individual egg mass decreases in old females, as does cluster size for those species that lay eggs in clusters or daily egg number for other species. In *E. editha*, the rate of egg laying declines with age (e.g., Labine 1968, Boggs 1997b, Moore and Singer unpubl. ms.), although there is evidence that the relative role of cluster number and cluster size in this fecundity decline may vary among populations (Moore and Singer unpubl. ms.). Whether individual egg mass declines and the physiological and evolutionary factors underlying any such pattern have been debated for *E. editha*. Based on two published studies and unpublished data for females used in Boggs (1997b), individual egg mass is highly conserved at about 0.23–0.25 mg at least for the first 360 or so eggs laid (Murphy et al. 1983, Moore and Singer 1987), but the egg mass likely declines toward the end of the physiological life. Specifically, Murphy et al. (1983) tested females from Woodside and Edgewood in the laboratory using various dietary treatments containing amino acids, sugar, and/or water. Their data showed a trend across all diets for eggs in later clusters to be lighter than the first 360 or so eggs, but they did not test this trend for significance. Moore and Singer (1987) kept females from both Sierra Nevada and Morgan Hill populations under a variety of laboratory and field conditions and fed them a variety of adult diets. They argued that their results showed no linear decline of egg mass with age, although their statistical approach would not detect nonlinear effects. In a separate analysis, they did find that individual egg mass from either the first or penultimate clusters was significantly heavier than that from the last egg cluster, but there was no difference between the first and penultimate clusters. In exploring the relationship of egg mass and adult diet, Murphy et al.'s (1983) analysis indicated that females fed diets without amino acids had significantly lighter late eggs than females fed diets with amino acids. Moore and Singer (1987) argued that Murphy et al.'s statistical analysis was flawed and that their own results showed no effect of diet type on egg mass, but, again, Moore and Singer's analysis tested only for linear effects with age. Thus, egg mass appears to decline at the end of female life, but resolution of the role of adult diet components in maintaining egg mass still awaits proper analysis, ideally using a variant on the statistical approach in Boggs (1997a, 1997b) for cases with repeated measures but unequal sample sizes within individuals.

At least two sets of hypotheses have been advanced to explain the generally observed decline in egg mass and/or investment with female age in Lepidoptera and other organisms. Both sets of hypotheses assume that offspring fitness is a function of egg mass. Indeed, for *E. editha bayensis*, lighter eggs have a lower hatch rate than heavier eggs (Murphy et al. 1983).

One hypothesis is adaptive and assumes that selection has shaped the relationships among egg size, number, and survival along with adult survival to optimize female fitness. One such model (Begon and Parker 1986) assumes that adult-derived resources do not significantly influence egg production and thus predicts that egg size or number should decrease with age, as investment in offspring should show an early peak. Until recently, it has been assumed that this is the case for nectar-feeding butterflies. However, recent work shows that nectar-derived resources can be critical for egg production (see below).

A second hypothesis argues that physiological constraints operate on egg mass (e.g., Telfer and Rutberg 1960, Campbell 1962, L. J. Richards and Myers 1980, Wiklund and Karlsson 1984, Boggs 1986, 1997b), suggesting that adaptation is constrained by nutrient availability and the details of the holometabolous lifestyle. This set of hypotheses assumes that some particular nutrient is limiting to reproduction and is available only from larval feeding. If the process of incorporating these nutrients into eggs depends on the concentration of the nutrient in the maternal stores, then as the nutrients are depleted, smaller amounts will be deposited in each egg. These small amounts will then be matched with small amounts of other nutrients (which may actually be available in surplus), resulting in small eggs. The assumption underlying this hypothesis is supported by an analysis of reproductive effort for *E. editha bayensis*, which indicates that a constant fraction of stored nutrients is allocated to reproduction and other metabolic uses at each age (Boggs 1997b).

Sources of Nutrients Used in Reproduction

The above discussion regarding individual egg mass highlights the interaction of reproduction with foraging and the importance of understanding the sources and types of nutrients used in reproduction. Checkerspots in general, like many temperate-zone butterflies, feed on plants as larvae and drink nectar as adults and may also sip moisture from mud ("puddling"). They may also obtain fluids from dung or carrion. Plants eaten by larvae provide carbohydrates and the full spectrum of essential and nonessential amino acids needed by adult butterflies, but the plants are low in sodium. Nectar from flowers frequently visited by butterflies contains a mix of sugars plus some amino acids and traces of a variety of other chemicals (for a review, see Boggs 1987a); the amount of amino acids present, in particular, may vary with soil fertility (Gardener and Gillman 2001). Puddling is believed to be stimulated by and to provide sodium to feeding adults (Arms et al. 1974), although evidence is accumulating that proteins, amino acids, or other nitrogenous compounds may be thus obtained by species in some families, including nymphalids (Beck et al. 1999, Boggs and Dau unpubl. ms., Walling and Boggs unpubl. ms.). Generally, young males puddle, and male virility can be increased by puddling (Lederhouse et al. 1990). In at least some species, males pass sodium to females during mating (Pivnick and McNeil 1987). Older females of species with a low average number of matings will feed at puddles, presumably after exhausting the nutrients derived from the male (Sculley and Boggs 1996). Finally, male and female nutrient budgets are linked to some degree in Lepidoptera, as males donate nutrients to females at mating through the spermatophore or accessory fluids, which the female uses to varying extents for egg manufacture or general maintenance (e.g., Boggs and Gilbert 1979, Boggs 1990, 1995).

Euphydryas editha bayensis feed on flowers and exhibit pronounced preferences for specific nectar sources at Jasper Ridge (Murphy 1984). *Lomatium macrocarpum* was visited in greater proportion than its availability early in the season in both areas H and C, while the same was true for *Lasthenia chrysostoma* in area H and *Layia platyglossa* in area C later in the season. These preferences were as consistent as those described as "majoring" in bees, suggesting that butterflies could be effective pollinators for these plants. However, the amount of pollen vectored by *E. editha* for *Lomatium* was very small; the butterflies carried much more *Lasthenia* pollen. The data thus indicate that the relative importance of the plant and the butterfly to each other is not symmetrical.

Consistent with this observed shift in feeding preference through the season, McNeely and Singer (2001) showed that *E. editha* from the Sierra Nevada could learn alighting preferences for flowers and could learn to reduce handling time when searching for nectar within a flower. This differs from the lack of learning exhibited by the same species for oviposition hosts and is consistent with the hypothesis that learning is involved in behaviors that are both repeated and have a low fitness cost if not carried out properly (McNeely and Singer 2001).

The distribution of nectar plants can influence the adult movement patterns (Gilbert and Singer 1973) and the spatial distribution of egg clusters. Working with *Euphydryas chalcedona* in the Sierra Nevada, Murphy (1983) showed that females became more concentrated near nectar sources as the season progressed. As a result, egg clusters were most common on host plants directly adjacent to nectar. The finding was duplicated for *E. chalcedona* on Jasper Ridge (Murphy et al. 1984). The distribution of nectar plants also affects the movement patterns of *M. cinxia*, as increasing flower abundance decreased emigration rate and increased immigration rate of males (Kuussaari et al. 1996). Such influences on population dynamics can be important in checkerspots and other species living in highly fragmented landscapes.

Puddling can also influence movement patterns of adult *Euphydryas editha*, although it has been rarely documented in this genus. In 1975 at Pioneer Resort, Colorado (Ehrlich and White 1980), and in 1990 at Morgan Hill in California (Launer et al. 1993), large numbers of individuals commuted from slopes and ridgelines to a seep or creek on the valley floor to sip moisture from puddles. At the Pioneer Resort location, adult nectar plants were abundant, while the Morgan Hill site was suffering from a drought in 1990. Similar movements at Morgan Hill have not been observed since in consistent surveys over 10 years.

Both males and females puddled at Pioneer Resort and Morgan Hill. In fact, the sex ratio at the Morgan Hill puddling site was more female-biased than the sex ratio at the ridgeline early in the season, in contrast to male-biased sex ratios seen at puddles for many species (Launer et al. 1993). Both males and females at the puddling site tended to be older than the population as a whole early in the season and as old as the population as a whole late in the season. Puddling by older females is consistent with females receiving a smaller number of matings, absorbing relatively little of the male-donated nutrients, and needing to replenish puddle-derived nutrients that might otherwise come from males (Sculley and Boggs 1996). However, why puddling should be observed so infrequently remains a mystery.

These sex-specific puddling patterns highlight another characteristic of the *Euphydryas* nutrient budget: male and female budgets are not closely tied together through male nutrient donations at mating. Because females do not utilize much of the male-derived nutrients in reproduction (and presumably maintenance), this means that there is relatively little opportunity for male-derived nutrients to buffer female nutrient budgets against variation in nutrients available in the environment. In other words, in bad years, females cannot readily expect to "make up" for unavailable nutrients through use of male-derived nutrients. This relative independence of nutrient budgets affects not only female capacity to buffer against variation in food availability, but it also means that the intensity of sexual selection on male nutrient donations at mating is not likely to fluctuate much across years in response to food availability.

Relative Role of Adult/Larval Nutrient to Reproduction

With the above-described knowledge about sources of nutrients used in reproduction as well as about age-specific reproductive patterns and ovarian dynamics, we can now examine the relative role of adult- and larval-derived nutrients in reproduction for female *Euphydryas editha bayensis*. Given that about 17.5% of eggs are already mature at adult eclosion and that the adult diet of nectar and possibly puddling is relatively incomplete, the prediction is that larval nutrients should be of great importance to early eggs, decaying with time, but not to zero. Data from various sources support this expectation.

Murphy et al. (1983) examined the percentage of abdominal fat body remaining in females at various times throughout life. Their sample sizes were small, yielding low power for statistical tests. Nonetheless, two trends emerged: relatively more fat body remained after about 360 eggs had been laid in groups of butterflies that were fed a more complete adult diet of sugars plus amino acids, but relatively less fat body remained in individuals in the same groups at death. As Murphy et al. (1983) speculated, these trends are consistent with a delay in depletion of larval reserves early in butterflies fed the more complete diet, yet consistent with a fuller use of those reserves by the end of life given the availability of amino acids from the adult diet. This speculation is supported by the fact that the amino acids fed to the adults were proline, serine, alanine, arginine and lysine; arginine and lysine are essential amino acids for insects, which cannot be made by adult butterflies and must otherwise come from larval feeding on plants. Thus, their presence in the adult diet could affect the timing of breakdown of storage proteins and use of other stored reserves from larval feeding (see also O'Brien et al. 2002). Stjernholm and Karlsson (2000) have since shown for *Pieris napi* that increased numbers of matings by females (and hence increased amount of male nutrient donations received) result in loss of thorax mass as well as more complete usage of abdominal fat body, consistent with the results for *E. editha*.

Radiotracer data provide a second line of evidence supporting the hypothesis that larval nutrients should be important to early-laid eggs, decaying with time, but not completely (Boggs 1997a). Late last instar larvae were fed ^{14}C-glucose, while adults were fed ^{3}H-glucose, for a double-label design. Previous experiments with *Speyeria mormonia* had shown that similar results are obtained if the labels were reversed between the two developmental stages. Eggs were then assayed for the presence of ^{14}C and ^{3}H. The experiment was repeated with amino acids instead of glucose. For *E. editha*, incorporation into eggs of label from larval-derived glucose showed a significant linear decline with age (figure 5.7), whereas the label from adult-derived glucose increased initially, leveled off, and then declined slightly. Incorporation of the label from amino acids showed a similar pattern to that of glucose, except that the label from larval-derived amino acids first increased slightly and then declined with age. This initial increase may be due to the relative timing of label administration relative to maturation of the first eggs laid. None of these changes in the amount of label incorporated into eggs was due to changes in egg mass with age, as inclusion of individual egg mass in the regressions did not change the relationship between amount of label and age.

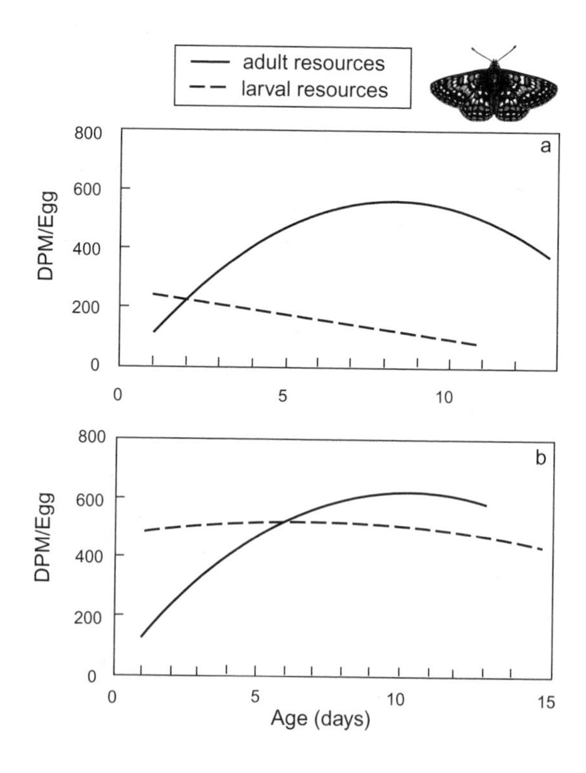

Figure 5.7. Incorporation into eggs of radiolabel from adult and larval resources by *Euphydryas editha,* using labeled glucose (a) and labeled amino acids (b). Data from Boggs (1997a).

Stable carbon isotope data provide a final line of evidence supporting the hypothesis that larval nutrients should be important to early-laid eggs, and decay with time. The experimental design exploits the fact that C3 and C4 plants differ in their isotopic ratios of ^{13}C to ^{12}C. Thus larvae can be fed on their normal C3 host plant, and fed either cane (C4) or beet (C3) sugars as an adult (O'Brien et al. 2000, 2002, 2004). Unlike studies with radiotracers, this design results in labeling the entire larval and adult nutrient pools, allowing quantitative analysis of allocation. Data for *E. chalcedona* supports the hypothesis (O'Brien et al. 2004), showing essentially no contribution from adult carbon sources in early eggs and an increase to a low-level plateau of adult carbon usage in later eggs (figure 5.8). Additionally, *E. chalcedona* incorporates less adult carbon into its eggs at equilibrium than do other nectar-feeding Lepidoptera examined, including the nymphalid *Speyeria mormonia*, the pierid *Colias eurytheme*, and the sphingid *Amphion floridensis*. Nonetheless, a large proportion of carbon incorporated into later eggs from the larval diet is in the form of essential amino acids, indicating that nonessential amino acids can be made from adult nectar via transamination of the glucose-derived backbones, although this proportion is lower than that seen in the other species.

The pattern of use of adult and larval nutrients in egg production should insulate early egg production from adult nutrient stress. Nonetheless, nutrient stress in early adult life may still impact subsequent fecundity to the extent that stored larval reserves are incapable of buffering those stresses. This is suggested by Murphy et al.'s (1983) data for adults fed only water versus sugars plus amino acids, and Boggs' (2003, unpubl. data) data on *S. mormonia* showing that early adult food restriction has larger effects on fecundity than does restriction later in life. Further, the reproductive allocation pattern observed for *E. editha bayensis* indicates that females will be particularly affected by food shortages that occur in the last larval instars. The timing of the life cycle relative to plant phenology makes it unlikely that such shortages will occur due to plant senescence (chapter 3), but they could occur due to host defoliation in small habitat patches. The differences in reproductive strategies between *E. editha* from the San Francisco Bay area and the Sierra Nevada suggest that the ecotypes also differ in their relative sensitivities to adult versus larval food resource stress.

Finally, the relatively strong dependence on larval resources for egg production seen in *Euphydryas* could be adaptive, through several different mechanisms. First, if survival probabilities differ in the larval and adult stage, the relative amount of time spent in each stage may be subject to selection, with dependence on larval resources for production of eggs being simply a result of more time spent in the larval stage. Alternatively, dependence on larval resources could be the result of strong selection for rapid onset of reproduction in the face of a short growing season. This hypothesis assumes that larval feeding supplies the full complement of nutrients needed for egg production and is more efficient for early rapid egg development than is the accumulation of baseline larval stores that would be supplemented with adult feeding to produce eggs. Appropriate models to explore these ideas are lacking.

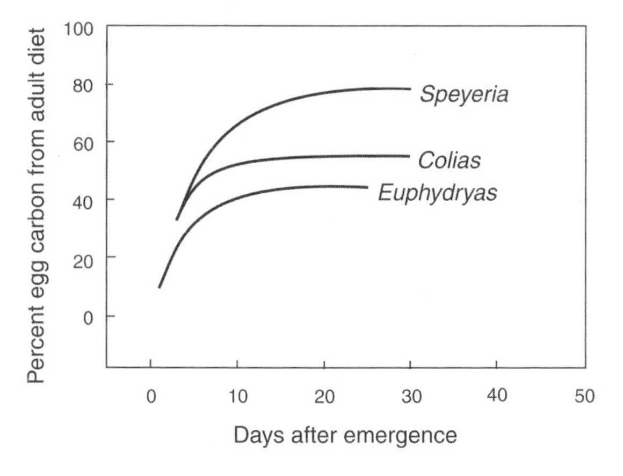

Figure 5.8. Proportion of adult-derived carbon incorporated into eggs through time for *Euphydryas chalcedona*. Data for *Speyeria mormonia* and *Colias eurytheme* are shown for comparison. From O'Brien et al. (2004).

5.5 Lifetime Reproductive Success

Lifetime reproductive success of a given female will depend on her age-specific fecundity and survival schedules, as well as the environment into which her offspring are placed. The scenario is more complex for males: not only are his own mating and survival probabilities important, but the success of his mate(s) is also critical.

Building on information available for females, Cushman et al. (1994) estimated lifetime success for female *Euphydryas editha bayensis* eclosing in different portions of the flight season at Morgan Hill with prediapause larval starvation as the only source of larval mortality. In this sense, the model was optimistic about female success. However, it was pessimistic in ignoring the possibility that females could oviposit on *Castilleja,* a rarer alternative host to *Plantago,* which senesces later in the season (chapter 3), and in ignoring the possibility that females might select host plants for oviposition that were more likely to be nonsenescent several weeks later. Nonetheless, for the 1992 flight season, the model estimated that only 8–21% of the eggs from females eclosing on the first day of the flight season would reach diapause, while the percentage dropped to 1–5% for females emerging on the seventh day of the flight season. The pattern is in agreement with studies by Singer (1972) showing that prediapause mortality is a major factor affecting population size in *E. editha bayensis.*

Cushman et al. (1994) used rough values available for mortality during diapause, postdiapause, and pupal development to estimate that only 11–33% of the flying females were able to replace themselves in the next generation, with the range resulting from assumptions about the length of the prediapause developmental period. This suggests that N_e for the female population was considerably lower than the number of female adults, which will further reduce N_e for males, given that male success is dependent on female success. Similarly, female N_e is much smaller than the number of adult female butterflies in *M. cinxia* as well. Survival from the first instar to post-diapause larvae varied from 0 to 25% (7% on average) in a life-table study (section 4.4), indicating that many females have no progeny in the next generation.

A critical remaining question concerns the role that the massive prediapause mortality plays in shaping female reproductive patterns in an evolutionary context. An underlying theme throughout the chapter is that prediapause larval mortality indeed plays a crucial role in shaping female reproductive patterns in checkerspots, yet the critical experiments remain to be done. For example, what are the fitness costs and benefits of emerging slightly earlier at a smaller size and leaving fewer offspring that might have a better chance of survival? Does the pattern of allocation of larval nutrients to reproduction play a strong role in the observed results of any such trade-off? Does variation among years in timing of host plant senescence relative to the adult flight period affect adaptation to the mean timing (e.g., Kingsolver and Watt 1983)? And how does the pattern shift as the developmental constraints are altered, first in the Sierra *E. editha* populations, then in other checkerspots such as *E. gilletti* with multi-year development or *M. cinxia* with a great role of parasitoids in its population dynamics (chapters 4 and 8)?

5.6 Concluding Remarks

We have highlighted two themes throughout this chapter. One is the pervasiveness of within-species and often within-population variation in the reproductive biology of checkerspots. This includes numbers of matings by either sex, strategies of mate acquisition, male and female reproductive effort, and sex ratios. It is also likely that patterns of allocation to egg production or male accessory gland products will show such variation. Variability within populations was once thought to be unlikely for traits that are closely tied to fitness because one variant with the highest fitness was expected to be fixed. However, such thinking ignored some realities of evolutionary processes in the wild: variation in environmental conditions and selection pressures can result in the maintenance of variation for traits strongly tied to fitness, even ignoring potential constraints on adaptation that may preserve variants for a trait (see Price and Schluter 1991). Variability among populations likewise could result not only from genetic drift, but also from variation in local environments and selection pressures, leading to local adaptation of suites of reproductive traits. Both the expected structure of those suites of adaptive traits, as well as the implications for population and metapopulation dynamics of such variation, remain to be fully explored.

A second underlying theme in this chapter has been the extent to which checkerspot butterflies can buffer reproductive success against variation in the environment and the effect of environmental variation when buffering fails. For *Euphydryas editha bayensis*, environmental variation has strong effects on prediapause larval survival. It is as yet unclear how well- or ill-adapted female emergence and oviposition strategies are with respect to buffering their offspring against that environmental variation. Likewise, the strategy of "reproduce fast, die young" that female checkerspots employ reduces their susceptibility to variation in adult nectar supply, but at the same time this strategy increases their reliance on larval resources for reproduction, thereby increasing their exposure to risks from variation in the larval food supply. The relatively low usage by females of male-derived nutrients in reproduction means that the reproductive nutrient budgets of the two sexes are not well integrated. This reduces the possibility that male-derived nutrients are used to buffer female's resources for egg production when food shortages occur. At the same time, it also reduces the likelihood of reversals of sex roles and sexual selection pressures due to the relative importance of male and female investment to offspring production, as has been observed in some Orthoptera (Simmons 1992).

6

Measurement, Correlates, and Importance of Oviposition Preference in the Life of Checkerspots

MICHAEL C. SINGER

6.1 Introduction

For almost all butterflies, oviposition preference is the principal mechanism by which the insect–host plant relationship is established. It is this trait of the insect that interacts with spatial distributions, abundances, and acceptabilities of plants to generate patterns of insect–host association across the landscape. There are a few putative exceptions. *Parnassius apollo* in Sweden and *Melanargia galathea* in Britain are described as scattering eggs without regard to host distributions (Ford 1945, Wiklund 1982). Nonetheless, all aspects of the distribution and abundance of butterflies are influenced by host plant relations.

All butterflies lay their eggs one at a time. Most females leave an individual egg on a plant and move on before laying another. Checkerspot butterflies, in contrast, lay their eggs in clusters of several to many dozens, a strategy that has great importance in determining the fitness of individuals, the fate of populations, and our ability to study oviposition behavior.

Several authors have argued that butterflies that leave solitary eggs must find 20–50 oviposition sites per day and therefore are more time limited than egg limited, especially in cool, unpredictable climates (Courtney 1982). A time-limited female that locates a host, even a relatively poor quality host, should normally oviposit if there is some chance of offspring survival. She should then recommence her search without delay. She should not spend much time assessing host quality when she could otherwise be searching for a new host. This does not hold true for checkerspots. Their characteristic laying of eggs in large clusters of variable size (chapter 11) dramatically reduces the cost of prolonged search or host assessment. A highly discriminating female that delays her oviposition by rejecting most hosts can simply lay a larger cluster when she eventually does oviposit, though the delay may have its own separate fitness costs. The insect's fecundity suffers little or not at all from a decline in frequency of oviposition caused either by host scarcity or butterfly choosiness.

A checkerspot female thus normally has time to make sophisticated discriminations among hosts. Whether this discrimination is indeed sophisticated the reader may judge from this chapter, but it is clear that the decision-making process takes much longer than in most butterflies. For example, I have observed that the swallowtail *Iphiclides podalirius* usually deposits its single egg within half a second of first tasting the accepted host leaf, and spends even less time inspecting leaves that it rejects. On the other hand, *Melitaea cinxia*, *Euphydryas aurinia*, and *Euphydryas editha* may spend several minutes examining each of many host individuals before ovipositing. The average time for a single oviposition search by *E. editha* is on the order of 30 minutes (Mackay 1985). When a searching *M. cinxia* alights on a host, it typically tastes it, then rests for a while before walking to another part of the plant and tasting that part, then resting again. We should not, of course, be anthropomorphic, but these butterflies give the appearance of concentrating their every neuron on the task of host plant assessment. Accordingly, the first part of this chapter describes the behavioral mechanisms involved in checkerspot host preference, the advantages and disadvantages that these mechanisms generate for preference testing, and the development of the unorthodox preference-testing technique that is used by my research group. The second part of this chapter considers the results of oviposition deliberations by checkerspot butterflies and the relationship of this decision making to the rest of their physical and behavioral traits.

6.2 Definition of Terms

Some terms must be defined before discussing checkerspot oviposition preference. An "encounter" occurs between plant and insect when the insect arrives at a distance from which it could perceive stimuli emanating from the plant. For example, an insect may encounter a plant visually, perceive it, approach or alight upon it, and then encounter it chemically. "Acceptance" is a positive response made by an insect to a plant that has been encountered. For example, an insect may accept visual stimuli by turning toward a plant and alighting on it. The insect may then accept (or reject) contact chemical stimuli by feeding or ovipositing (or not).

Regarding insect traits, "motivation" is a general tendency to feed or oviposit, without reference to any particular host. A motivated insect is sensitive to stimuli that may lead it to feed or oviposit (Singer et al. 1992b). "Perceptual ability" is the set of likelihoods of perceiving a specified set of plants that are encountered. "Preference" is the set of likelihoods of accepting a specified set of resources that are perceived (Singer 1986, 2000). In practice, preference would normally be measured as the set of likelihoods of accepting resources that are encountered. "Specificity" has been defined (Singer 1986) and used by some butterfly biologists (e.g., Courtney et al. 1989, Thompson 1998) to mean the strength of preference, regardless of its direction. This usage has not penetrated the general plant–insect interaction literature, where "specificity" is sometimes synonymous with "host affiliation" or with "insect diet" and sometimes has a wider meaning incorporating both preference and performance (Van Klinken and Edwards 2002). In this chapter I abandon my previous usage of "specificity" (Singer 1986) and use "strength of preference."

Regarding plant traits, "apparency" is the likelihood that a plant will be perceived by some specified insect or set of insects (Feeny 1976, Singer 1986), and "acceptability" is the likelihood that a plant will be accepted after being perceived by some specified insect or set of insects (Singer 1986, 2000).

These definitions allow us to dissect the causes of observed events in the field. For example, if a set of insects all prefer host Y over host X, we may nonetheless observe some of them feeding on X in the field. Some individuals may do this because their strength of preference is low, while others may do so because their motivation is high (see section 6.7; Singer et al. 1992b).

The definition given here renders "preference" a useful trait in evolutionary biology because it is a trait of the insect whose variation can be measured among individuals and populations. However, preference is often defined by ecologists as the proportion of a particular resource in the diet as a function of the availability of that resource in the habitat (Hassell and Southwood 1978, Crawley 1984). Preference is an emergent trait of the plant–insect interaction rather than a trait of insect or plant (Singer and Parmesan 1993, 2000). A partial solution to the difficulty posed by different meanings of "preference," which I adopt here, is to use the term "electivity" (Ivlev 1961) for the ecological parameter and reserve "preference" for the behavioral parameter (Kuussaari et al. 2000, Singer 2000).

With the definition of "preference" that I use here, experiments in which an insect is offered a single plant (no-choice tests) should not strictly be called tests of "preference," although a series of such tests might merit such an appellation. An insect cannot have a "preference" for a single resource. We might better describe the insect as having an "affinity" for such a resource. A preference test should examine the relative likelihood that the insect will accept the different members of a specified set of plants that it encounters (recall that acceptance constitutes a positive response to an encounter). A typical butterfly encounters a plant by perceiving visual or chemical stimuli while in flight and may accept it by alighting (Feeny et al. 1989). It then assesses chemical and tactile stimuli with contact receptors and may accept the plant by ovipositing.

We might therefore consider pre-alighting and post-alighting preferences as potentially separate (Rausher et al. 1981). Below, the sequence of events in checkerspot oviposition is described before considering how their pre- and post-alighting preferences might best be assessed.

6.3 The Need for Behavioral Tests in the Study of Insect Oviposition

The most frequent practical application of oviposition tests to herbivorous insects is to assess the insects' potential, when used as agents of biological control, to attack both intended and unintended targets (Zwölfer and Harris 1971, McEvoy 1996, McFadyen 1998). The principal recent issue has been the possibility that insects introduced to control an exotic weed may attack endangered native plants (Simberloff and Stiling 1996, Louda et al. 1997, Van Klinken and Edwards 2002). This possibility is best evaluated by no-choice tests rather than by choice tests. Suppose that a weevil introduced to attack an exotic weed will also attack an endangered native plant when given a no-choice test. It does not matter whether choice tests may indicate consistent preference for the target over the nontarget plant. The reason is that insects normally encounter hosts sequentially rather than simultaneously and that what may appear to the experimenter to be a choice situation may, from the insects' perspective, be a series of no-choice situations. Even if this is not the case, some individual insects will wander away from the hosts on which they developed, encounter habitats that do not contain those hosts, and experiment with hosts that they would not have attacked if they had stayed at home (C. D. Thomas et al. 1987, Singer et al. 1992b). Situations are bound to occur where the biological control insect is exposed to the endangered native plant and not to the introduced weed. The important question then becomes, not which plant is preferred, but whether the endangered native is acceptable to the insect in a no-choice situation.

Choice tests also have their uses, both to ecologists and evolutionists, as tools to investigate the mechanistic causes of spatial and temporal patterns of plant–insect association. This investigation is the principal theme of a separate review of checkerspot oviposition (Singer 2003). Suppose we observe that insects select different host species in spatially separated habitats with similar host composition. We can investigate the mechanisms responsible for this pattern by asking whether the host species vary in acceptability among sites and/or whether the insects vary in their preference for one host species versus the other. Choice tests have been used to show that both insect and plant variation contribute to observed spatial pattern (Singer and Parmesan 1993). Suppose we observe that, over time, the proportion of eggs laid by an insect population on host A increases and the proportion laid on B decreases. We can investigate the mechanism by asking whether the preference of the insects for A versus B has changed and/or whether the relative acceptability of the plants has changed. Again, choice tests are appropriate for both questions, though only the first one has been investigated (Singer et al. 1993, Singer and Thomas 1996).

6.4 Sequence of Events in Host Search: Responses to Visual, Chemical, and Physical Stimuli

In general, butterflies are attracted to alight on plants by mixtures of visual and olfactory stimuli (Rausher 1978, Feeny et al. 1989). In *Euphydryas editha* at Rabbit Meadow, Sequoia National Forest, California, alighting seemed to be primarily or entirely in response to visual stimuli, as evidenced by strong relationships between alighting bias and plant visual traits. Parmesan et al. (1995) calculated alighting bias toward or away from each plant species by recording the proportion of alights on the focal plant species by each butterfly, taking a weighted mean of these proportions across all butterflies, and dividing it by the proportion of ground cover that the focal plant represented in line transects. These alighting biases were apparently fixed and not at all dependent on adaptive learning. Host finding was no more efficient by experienced insects than by naive butterflies performing their very first host searches (Parmesan et al. 1995). Table 6.1 shows alighting biases for three plant species at Rabbit Meadow. Two of the plants are hosts and the third, *Chaenactis*, is not. The strongest observed alighting bias was toward the nonhost *Chaenactis* (Asteraceae), which visually resembled one of the hosts (figure 6.1; Parmesan et al. 1995). Although the butterflies frequently alighted on *Chaenactis*, they never oviposited on these plants because after alighting they tasted the plants with their atrophied foretarsi, ascertained that they were

Table 6.1. Representation of hosts and nonhosts in the vegetation and in alights of searching female *E. editha* at Rabbit Meadow.

	Proportion Vegetation Area	Proportion Naive Alights	Proportion Experienced Alights	Alighting Bias
Pedicularis (traditional host)	0.08 (0.03)[b]	0.44 (0.07)[a]	0.45 (0.14)[a]	5.5
Collinsia (novel host), plot 1	0.14 (0.02)[b]	0.10 (0.02)[ab]	0.06 (0.03)[ab]	0.5
Collinsia (novel host), plot 2	0.14 (0.02)[ab]	0.16 (0.04)[a]	0.12 (0.05)[b]	1.1
Chaenactis (nonhost)	0.01 (0.004)[b]	0.16 (0.05)[a]	0.21 (0.06)[a]	19.0

From Parmesan et al. (1995).

Standard errors are in parentheses. Values with different superscript letters along rows are significantly different (P <.05).

not hosts, and flew away. The high alighting biases toward both *Pedicularis* and *Chaenactis* and the bias away from *Collinsia* suggest that alighting biases were similar for plants of similar visual appearance regardless of their status as hosts or nonhosts. This suggestion was confirmed by multiple regression analysis in which plant shape and leaf shape explained a high proportion of variation in alighting bias among all plant species present in the habitat, again, regardless of their host/nonhost status (Parmesan 1995). In contrast to other butterflies (Feeny et al. 1989), checkerspots seem not to use olfaction at close range in their decisions to alight. It is possible that olfaction assists them in choosing a general area in which to search, but this is not known (see Odendaal et al. 1989).

After tasting a plant and finding it chemically acceptable, the checkerspot curls its abdomen under a leaf, extrudes its ovipositor, and probes the lower surface of the leaf, often sweeping the ovipositor from side to side across the area that would receive eggs if they were laid. This probing is clearly a response to chemistry because it can be stimulated by placing the insect on a dampened filter paper on which an ethanol wash of host leaf surface has been evaporated. There is apparently no chemical sense on the ovipositor: eggs are readily laid on nonhost or even nonplant material; all that is necessary is

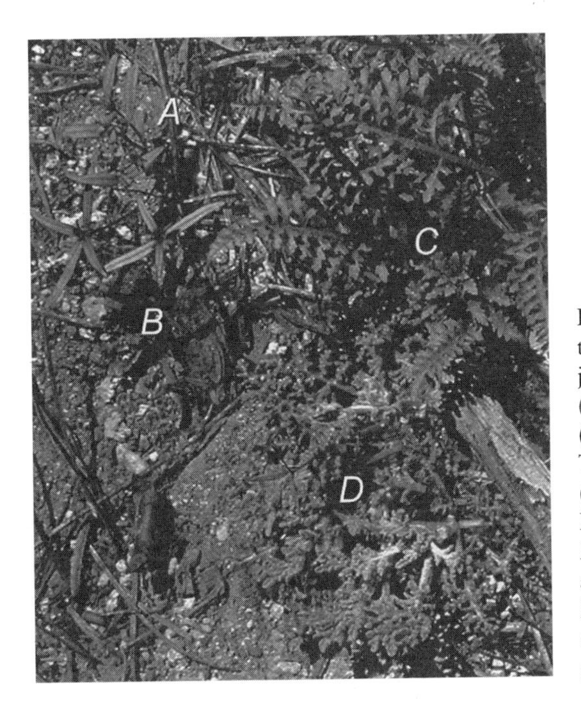

Figure 6.1. Photograph showing two hosts and two nonhosts of *E. editha* growing naturally juxtaposed at Rabbit Meadow. The hosts are (A) *Collinsia torreyi* (Antirrhinaceae) and (C) *Pedicularis semibarbata* (Orobanchaceae). The nonhosts are (B) *Gayophytum diffusum* (Onagraceae) and (D) *Chaenactis* sp. (Asteraceae). Butterfly alighting was strongly biased toward *Pedicularis* and *Chaenactis* and slightly biased away from *Collinsia* and *Gayophytum*. To humans, an individual *Pedicularis* is more apparent than an individual *Collinsia*. Photograph by Rob Plowes.

that the tarsi of the insect contact the host; the ovipositor does not need to do so. Once the plant has been chemically accepted, oviposition seems to depend principally on physical features of the site such as the size, shape, and orientation of the leaf and the extent to which it yields when pressed by the ovipositor. A role for olfactory stimuli is also implied from observations that females dip their antennae toward plants that they are tasting, but this role has not been investigated.

To summarize what is known of the sequence of events in checkerspot oviposition: a positive response to visual stimuli is to alight and taste a plant; a positive response to chemical stimuli is to curl the abdomen and press the ovipositor against a leaf; and a positive response to tactile stimuli is to oviposit. The following sections discuss the roles of plant and insect traits in the decisions that insects make to alight and then to oviposit.

Pre-alighting Butterfly Preference and Host Apparency

Apparency was first characterized by Feeny (1976) as a visual or chemical property of a plant that influenced its susceptibility to being found by herbivores. Singer (1986b) defined it as the likelihood that a plant will be perceived by some specified insect or set of insects. The property of the insect that interacts with plant apparency is the insect's perceptual ability, defined above as the set of likelihoods of perceiving a particular specified set of plants that are encountered. The concept of apparency has fallen into disuse. This may be because apparency is in practice hard to measure and to separate from perceptual ability. We usually cannot tell whether an insect that passes over a plant without stopping fails to perceive the plant or perceives it and decides against alighting on it. Whether the plant is perceived depends on an interaction between insect perceptual ability and plant apparency. Whether the insect alights on a plant that it has perceived depends on an interaction between the insects' pre-alighting host preference and plant acceptability—in the case of checkerspots, plant visual acceptability.

With present knowledge and techniques, these factors may be difficult or impossible to tease apart in practice. However, defining them in principle is necessary to provide the impetus to understand the mechanisms. In one case, our group has made progress in identifying pre-alighting preference rather than the apparency/perceptual ability relationship as a cause of observed patterns of alighting. *Euphydryas editha* at Rabbit Meadow had added to their diet a novel host, *Collinsia torreyi* (figures 6.1 and 6.2), just a few (< 20) generations before our study. The butterflies found this novel host very inefficiently: in the habitat patches where it was the principal host, the proportion of alights upon it was lower that the proportion the butterflies would have achieved by alighting on vegetation at random (table 6.1; Mackay 1985, Parmesan et al. 1995).

The inefficiency of finding *Collinsia* could have two causes. The first is an evolutionary lag in the insects' pre-alighting preference, such that *Collinsia* was perceived but not preferred for alighting (even though many insects accepted it readily on contact). The second is a failure of the insects to perceive *Collinsia* by virtue of the interaction between its apparency and the insects' perceptual abilities. The first explanation was suggested by Mackay (1985). At that time *Collinsia* had been used by the Rabbit Meadow butterflies for < 20 generations. It seemed likely that there had been insufficient time for the insects to evolve adaptive behavioral responses to perceived stimuli emanating from this novel host. Parmesan (1991) investigated this by comparing host-finding efficiencies at two sites where *E. editha* had independently added novel hosts. The first site was Rabbit Meadow where the traditional host was a perennial rosette, *Pedicularis*, averaging about 150 cm^2 in area, while the novel host was a small, erect annual *Collinsia*, averaging about 8 cm^2 (figure 6.1). At the second site, Schneider's Meadow, the situation was reversed: the novel host was a perennial rosette, *Plantago*, averaging about 80 cm^2, while the traditional host was *Collinsia* (*C. parviflora*), averaging about 5 cm^2 (figure 6.2).

In strong contrast to the Rabbit Meadow insects, Schneider's Meadow butterflies found *Collinsia* efficiently (Parmesan 1991). Host-finding ability evolves sufficiently rapidly that it can vary strongly among conspecific populations. Parmesan argued that behavioral preferences are likely to be more evolutionarily labile than perceptual abilities, so that insects in conspecific populations should be less likely to differ in what is perceived than in their behavioral responses to what is perceived. Parmesan therefore attributed the inefficiency of finding *Collinsia* at Rabbit Meadow to an evolutionary lag in the response to natural selection of pre-alighting preference, rather than to an evolutionary constraint associated with the failure of the insects to perceive *Collinsia* at all.

Figure 6.2. Typical shapes of the novel and traditional hosts of *Euphydryas editha* at Rabbit Meadow and at Schneider's Meadow. (A) *Collinsia torreyi* (Rabbit Meadow, novel); (B) *Plantago lanceolata* (Schneider's Meadow, novel); (C) *Pedicularis semibarbata* (Rabbit Meadow, traditional); and (D) *Collinsia parviflora* (Schneider's Meadow, traditional).

Post-alighting Oviposition Preference and Host Acceptability

In the same manner that perceptual ability can be viewed as an insect property that interacts with plant apparency, preference can be viewed as an insect property that interacts with plant acceptability (Singer 1986, 2000). The simultaneous variation of both preference and acceptability creates difficulties for experimental design and interpretation (section 6.12, Singer 2000, 2003). Preference tests usually report either observations of host acceptance and rejection, or observations of the distribution of feeding or oviposition. They are therefore equally sensitive to variation in both preference and acceptability. If such a test holds the plants constant and varies the insects, we might regard it as a test of variation in preference. If it holds the insects constant and varies the plants, we might regard it as a test of variation in acceptability. Often neither of these designs can be achieved, and both plants and insects may vary among experimental replicates. When this happens, entomologists tend to assume that the plants are really all the same and the insects are variable in preference, while botanists tend to assume the converse. This should tell us that neither assumption is valid!

Below I describe the forms taken by different preference tests, including the testing techniques that our group has developed specifically for checkerspot butterflies. Bear in mind that none of these experiments is strictly a test of preference. Depending on the variables involved, they are tests of preference, acceptability, or some combination of these traits.

6.5 Standard Method for Choice Tests: Design and Applicability to Checkerspots

The most common form of preference testing used in butterflies is to place the butterfly in a cage with several test plants and allow oviposition to occur for a day. At the end of the day the eggs on each plant are counted. The positions of the plants are then rotated to control for position effects, and the experiment is repeated the following day (Thompson 1993, Bossart and Scriber 1995, Wehling and Thompson 1997).

This standard method has been used with success in species that lay single eggs. It has also been applied to compare two populations of a checkerspot, *Euphydryas phaeton* (Bowers et al. 1992). Insects from a population of *E. phaeton* that had switched to a novel host, the exotic *Plantago lanceolata*, were significantly more likely to oviposit on this species than butterflies from a population that had retained its traditional diet. Mazel (1986) used a similar method to show that *Euphydryas aurinia* from populations that were monophagous on *Cephalaria*, *Lonicera,* and *Gentiana* all tended to prefer *Succisa* over their own hosts. These examples applying the standard method were successful with checkerspots because they tackled questions about population-level mean preferences in which each individual insect provided a single data point. The sample sizes are therefore the numbers of insects independently acquired from their populations. In theory the upper limit to such a sample is the number of females in a population, and if this is large, the power of the test can be high, and comparisons of preference among populations can be robust.

Problems in application of the standard method to checkerspots arise when comparing individuals as well as groups. Each captive *M. cinxia* or *E. aurinia* is likely to lay three to eight egg clusters in its lifetime. If each egg cluster were a data point, as in the standard method, the power to compare individuals is so low that statistically significant differences among individuals could only be detected if they were extreme—for example, if some insects were extremely unlikely to accept host A and others extremely unlikely to accept host B. Suppose that a particular individual would, if it had an infinite life span and were equally exposed to two different plants, lay one-sixth of its eggs on plant type A and five-sixths on B, while a second individual had the opposite preference. Suppose that we obtain six egg clusters from each individual, distributed among the hosts exactly as expected. One butterfly lays five egg clusters on plant A and one on B; the other butterfly does the reverse. By Fisher's Exact test, the ratio 5:1 differs from 1:5 with $P = .08$, failing to reach statistical significance. The substantial difference in preference would, in practice, remain undetected by the standard method because of the limit to sample size set by the lifetime production of egg clusters.

The example above shows that it is impractical to use the standard method to ask whether two individual checkerspots have different preferences. Suppose that we were to use this method to test many individuals from the same population, a population in which preference is continuously variable. There would, of course, be some tendency for the distribution of eggs laid in these standard tests to be correlated with the true preferences of the insects, but the correlation would be low because of the effects of chance on the distribution of eggs laid by each individual. Therefore, if we want to correlate individual preference with other traits such as fecundity, age, or offspring performance, the test again suffers from low power stemming from low numbers of egg clusters laid by each individual.

A separate difficulty attending use of the standard method with checkerspots is that the insects do not duplicate natural flight behavior in cages, even in quite large cages. This means that during the time when a caged butterfly would naturally be searching for hosts if it were in the field, it is likely instead to sit on the walls of its cage. Therefore, by the time the butterfly does move sufficiently to encounter plants, the time period during which it might have discriminated among different hosts has usually passed, and it is likely to accept the first host that it finds (see below).

6.6 Preference-Testing Checkerspots: Sequential Choice Test

Melitaeologists need a technique to test post-alighting preference that generates more data from each individual than the number of egg clusters it lays. The sequential choice test (Singer 1982) overcomes the problem of low sample size by staging a series of encounters between a butterfly and the test plants and recording the outcome of each encounter, while preventing the insects from actually ovipositing. An insect that is not allowed to oviposit will continue to show acceptance and rejection of plants that it encounters, thereby providing more information than could have been obtained from a single oviposition.

This test takes advantage of the manipulability of checkerspots (chapter 9). A female checkerspot placed gently on a host, either in the field or in the greenhouse, appears to behave as though she had naturally alighted on that host. But is this appearance correct? We have tried to test this in several ways. First, Rausher et al. (1981) followed individual *E. editha* naturally alighting on *Pedicularis* plants in the field. Each plant had been numbered with a flag, so that the sequence of landings on individual plants could be noted with the aid of a tape recorder. When a butterfly accepted a plant by curling its abdomen fully, probing under a leaf with the ovipositor, and remaining still for three seconds, it was manually removed from the plant before the first egg had been laid. It was then put in a cage for five minutes and replaced, in randomly chosen order, on the *Pedicularis* that it had accepted and on the last *Pedicularis* that it had rejected. We found a significant positive association between the responses shown before the insect had been captured and after it had been subjected to staged encounters with the same plants. Most manipulated butterflies accepted the plant they had accepted when at liberty and rejected the plant they had rejected when at liberty (Rausher et al. 1981).

Our second test of the relevance of manipulated trials to actual host use was done by testing the preferences of insects captured that were naturally ovipositing on different species at the same site (Singer 1983, Singer et al. 1993). There was a strong association between the tested preferences and the observed ovipositions (table 6.2). Again, this shows that a test using manipulated butterflies measures a factor connected with the observed variation of host use.

How to Perform a Sequential Choice Test

We have established that manipulated tests are pertinent to events in the field. What are the actual procedures involved in the testing? Each insect is offered a series of staged encounters at, say, 15-minute intervals. Each trial lasts a maximum of three minutes. Acceptance is judged from pressing the extruded ovipositor against the plant for a count of three seconds. Rejection is the absence of this behavior during the entire three-minute period. An insect that accepts is not allowed to oviposit but is manually removed from the plant before the first egg has been laid.

The test is based on the observation that, as time passes, the probability that a particular plant would be accepted, if it were encountered, jumps from 0 to almost 1 very rapidly, in the space of just a few minutes (Singer 1982). That probability then remains close to 1, at least during the principal hours when oviposition is likely (noon to 4 PM) until oviposition occurs. Suppose that an insect is offered the same plant over and over again, in repeated staged encounters, and is prevented from actual oviposition as described above. There is a rejection phase when the plant is consistently rejected, followed by a acceptance phase when the plant is accepted about 95% of the time (Singer 1982). Now suppose that the same insect is offered staged encounters with two plants,

Table 6.2. Tested preferences of butterflies captured ovipositing on *Collinsia* and *Pedicularis*.

	Preferring *Collinsia*	No Preference	Preferring *Pedicularis*
Ovipositing on *Collinsia*	12	8	2
Ovipositing on *Pedicularis*	0	4	26

Ignoring the "no-preference" category, the contrast is highly significant (1.6×10^{-7}) by Fisher's Exact test. Data from Singer (1983).

X and Y, in alternation. If we indicate a rejection by R and an acceptance by A, we may observe one of three types of sequence, shown below. Each acceptance or rejection in these sequences is the result of an entire three-minute trial, with no account taken of the time to acceptance within any such trial:

1. RX; RY; RX; RY; RX; AY; RX; AY; RX; AY; AX; AY; AX; AY; AX
2. RX; RY; AX; RY; AX; RY; AX; RY; AX; RY; AX; RY; AX; RY; AX; AY
3. RX; RY; RX; RY; RX; RY; RX; RY; AX; AY; AX; AY; AX; AY; AX; AY

In case 1, we would say that Y is preferred because X is rejected in encounters that follow acceptances of Y. In case 2, X is preferred, and in case 3, no preference is detected.

To the extent that the behavior of manipulated butterflies really represents what they would do if they were at liberty, then the result shown in case 1 above estimates the length of time a butterfly would search in the motivational state where encounter with Y but not X would result in oviposition, before reaching the motivation at which either X or Y would be accepted, whichever was the next plant to be encountered. This length of time is called the "discrimination phase" (figure 6.3). It is a measure of the strength of preference for Y over X. The discrimination phase in case 1 above is shorter than that in case 2, so the preference for Y over X shown in case 1 is weaker than the preference for X over Y in case 2.

Because the insect cannot be offered continuous exposure to both plants, the length of the discrimination phase cannot be measured precisely. Its minimum length is the time difference between the first acceptance of the preferred host and the last rejection of the second-ranked host. In practice, this minimum length is the value that has been used, partly because the maximum cannot be estimated for insects that never accept the second-ranked host. Use of the minimum value gives us the freedom, if we so choose, to use data from butterflies that escape or die before they have accepted all the hosts in the test series.

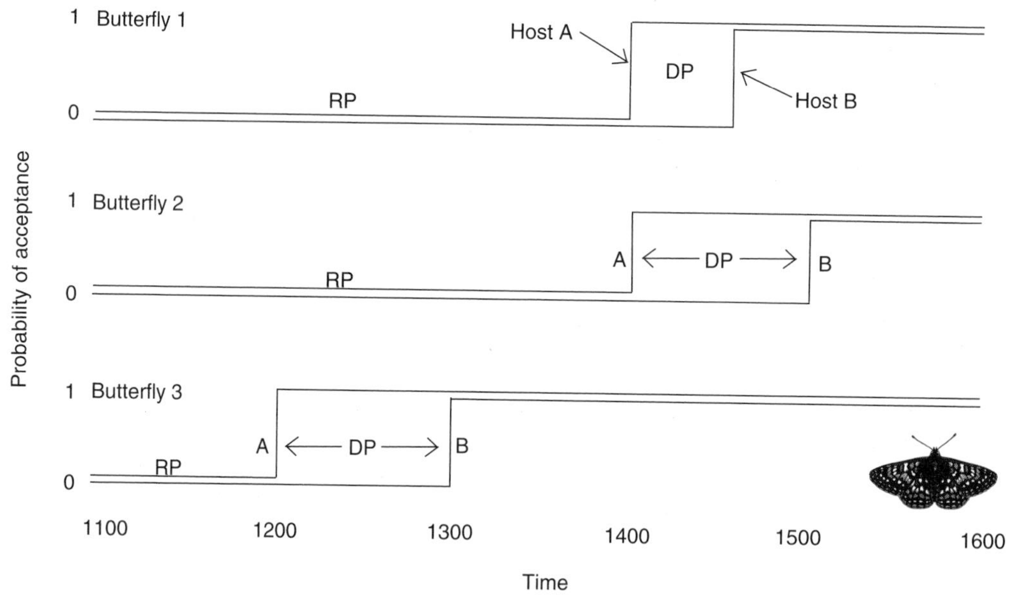

Figure 6.3. Stylized depiction of the responses of female butterflies to two plants when repeated encounters are staged and oviposition is not allowed. In this model the probability of acceptance of each plant is initially zero because the insect is not motivated to oviposit. When the insect becomes motivated the probability of accepting the preferred host jumps instantaneously from zero to 1, remaining at 1 until after oviposition has occurred. The same happens for the second-ranked host at a later time. The text discusses how closely this model corresponds to reality. Records are shown for three butterflies, each of which prefers plant A over plant B. Rejection phases indicated by RP; discrimination phases by DP.

Note that it is impossible to estimate the length of a discrimination phase unless one begins the test sequence before any of the test plants are acceptable. If the first staged encounter with just one of the test plants results in acceptance, then the discrimination phase has begun. If this happens, it may be possible to obtain a rank order of preference, but estimating the discrimination phase length would require allowing the insect to oviposit and recommencing the test. Note also that the test, once begun, should not be interrupted, though periods of cloudy weather may cause unavoidable interruptions. An insect that rejects both test plants before such an interruption may switch directly to accepting them both when testing recommences. In such a case the opportunity to discover which plant would have been accepted first has been lost. The insect should be allowed to oviposit and its test restarted.

Expressing the Results of the Sequential Choice Test

The original description of the sequential choice test (Singer 1982) suggested that the test allowed preference to be described in two ways. The rank order of preference is the order in which the plants become acceptable, while the strength of preference (then described as "specificity") is estimated from the length of the discrimination phase. This distinction has been adopted by some authors who have found it useful in order to argue that rank order is more highly conserved in evolution than strength of preference (Courtney et al. 1989, Thompson 1993). In checkerspots both aspects of preference can vary simultaneously, giving rise to a bell-curve of preferences (figure 6.4). In this figure strength of preference is depicted as the distance along the abscissa from the "no preference" point, and rank order is opposite on either side of this point. The minimum length of discrimination phase, again on the abscissa, is determined using only time differences during the period (11:30 AM to 4:30 PM) when oviposition is likely. Therefore five hours in Figure 6.4 is equivalent to 1 day, 10 hours to 2 days, and so on.

Figure 6.4 shows that the range of plants that would be accepted, if they were encountered, expands at different rates in different individual butterflies. In more recent work at the site depicted in figure 6.4, Rabbit Meadow, California, we have often capitalized on the extreme nature of this variation to stream-

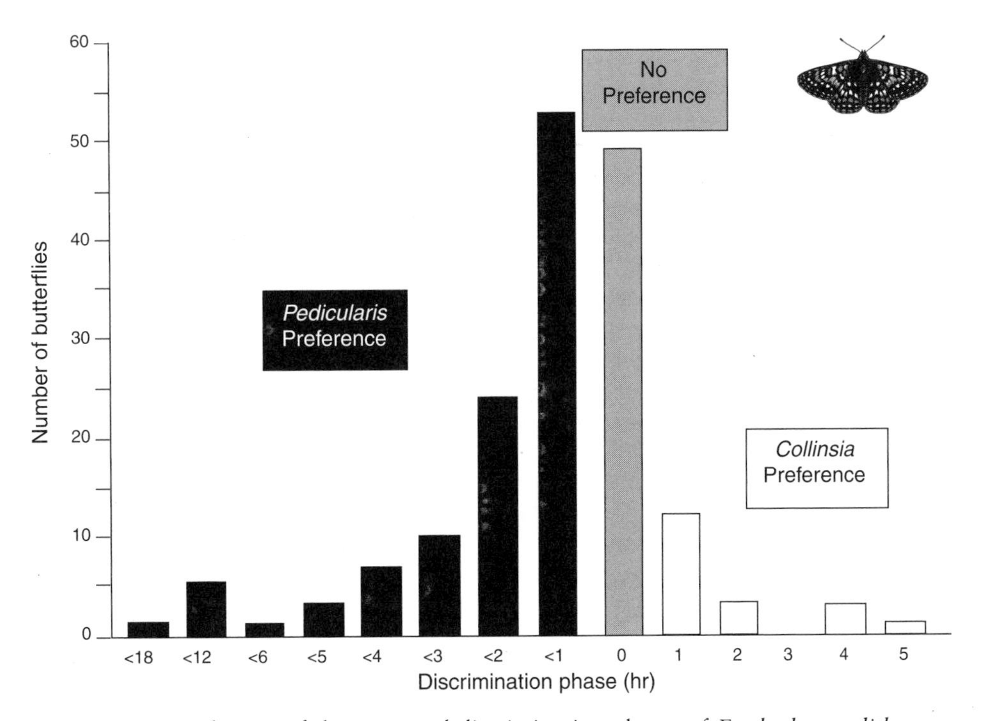

Figure 6.4. Distribution of the measured discrimination phases of *Euphydryas editha* at Rabbit Meadow in 1981. The data were pooled from two adjacent habitat patches, one where *Collinsia* was used and one where *Pedicularis* was used (from Singer 1983).

line our preference tests and place the insects into one of five preference categories with minimum effort. We begin the test of each insect in an intensive manner, staging encounters with *C* and *P* (*Collinsia* and *Pedicularis*, respectively) at frequent intervals, say 10–15 minutes. As soon as one plant is first accepted, we wait five minutes and test the other plant. At this stage we can place the insect into one of three categories: *P*-preferring, *C*-preferring, or no preference. This usually occurs between noon and 2:30 PM on the day that the insect is captured. Most butterflies, as figure 6.4 shows, are classed as *P*-preferring. As soon as a butterfly has been classed as *P*-preferring, it is set aside and tested on *C* between 3:30 and 4:00 PM on the same day. If it rejects *C*, it is set aside and retested on *C* between 3:30 and 4:00 PM the following day. This relatively simple procedure allows us to capture fresh butterflies and initiate testing every day, because intensive testing is confined to the first day, and only one test (on *C*) is required for those insects that still need testing on day 2. The overall two-day test places insects into the following five categories:

C: Preferring *Collinsia*
N: No preference
*P*0: Preferring *Pedicularis*, but accepting *Collinsia* later on the same day as the first acceptance of *Pedicularis*
*P*1: Preferring *Pedicularis*, but first accepting *Collinsia* on the day after the first acceptance of *Pedicularis*
*P*2: Preferring *Pedicularis*, and not accepting *Collinsia* even at the end of the day after the first acceptance of *Pedicularis*.

We have frequently used tests of this form to compare butterflies from different habitat patches or nearby populations (Singer and Thomas 1996). The tests can be done either with freshly gathered plants or with the test plants undisturbed, still rooted in the ground and covered with small net tents to prevent insect escape. When we are comparing butterflies from different sites, we take care to synchronize their capture, lest any influence of nutritional level or other effects of length of captivity might cause bias.

Controversial Nature of the Sequential Choice Test

It should be clear from the examples given here that the sequential choice test mitigates the problem of low statistical power that would stem from treating egg clusters as data points. It does so by estimating a discrimination phase before each oviposition. Such an estimate, even in the categorical form described above, contains more information than the presence or absence of a single egg cluster on plant X or plant Y. This technique has enabled us to compare preferences of individual checkerspots and associate them with other traits of those individuals, despite the paucity of actual oviposition events. However, any preference test carries baggage in the form of assumptions of varying testability, and ours is no exception. In fact, these assumptions have seemed highly implausible to many workers in insect preference, and, as a result, the sequential choice testing technique has remained controversial for more than 20 years. The following section identifies some of our assumptions and discusses the extent to which they have been tested.

Testing Assumptions of the Sequential Choice Test

Assumption 1 There is a precise time at which an insect switches from a rejection phase, during which a particular plant would be consistently rejected if encountered, to an acceptance phase, during which that plant would be consistently accepted. The timing of the switch from rejection to acceptance differs in responses to different plant categories. If this assumption were not true, the discrimination phase would not be real. The truth of this assumption can be tested by determining the frequency of rejection of plants that have been previously accepted, when no oviposition has intervened. For plants that were moderately or highly acceptable to the insects, the frequency of such rejections was typically 5% or less, but in one case a plant that was first accepted several days after the highest-ranked host was never consistently accepted (table 6.3; Singer 1982). For most hosts, a plant that had been accepted was accepted again with about 95% probability, provided that no oviposition had occurred, that there was no adverse change in the weather, and that it was not near the end of the day.

Assumption 2 When a test covers more than one day, we assume that the motivational state of the butterfly at the beginning of the second day's test is the same as its motivational level at the end of the first day's test. This assumption is not easy to test, and we have not explicitly tested it, but the

Table 6.3 Frequency of rejection of previously accepted plants.

Population	Natural Host	Tested Host	No. of Opportunities for Rejection	No. of Unexpected Rejections
Rabbit Meadow	*Pedicularis*	*Pedicularis*	50	1
	+ *Collinsia*	*Collinsia*	50	4
	+ *Castilleja*	*Castilleja*	50	2
Woodside	*Plantago*	*Plantago*	77	0
		Collinsia	43	3
Yolla Bolly	*Collinsia*	*Collinsia*	17	0
		Plantago	28	0
Indian Flat	*Collinsia*	*Collinsia*	40	2
		Castilleja	16	9

The Rabbit Meadow data are previously unpublished; the rest are from Singer (1982).

behavior of *E. editha* is consistent with the assumption in the following manner. When we commence testing at some time between 11:30 AM and noon, we usually observe that the range of plants that are accepted resembles the range that had been accepted at 4 PM the previous afternoon. It is unusual for additional plants to be accepted at this time or for plants to be rejected that had been accepted the day before. We tentatively conclude that the passage of time between 4:30 PM and 11:30 AM has little effect, and time during this period is not included in the calculation of the discrimination phases.

Assumption 2 thus appears to be true for *E. editha*. It is less appropriate for *M. cinxia*, which will more frequently reject a plant that was accepted the day before. The insects sometimes pass through a long discrimination phase on day 1 of testing and a much shorter one at the beginning of day 2. In theory, this violation of the assumption could render the sequential testing technique inapplicable to *M. cinxia*. However, discrimination phases of *M. cinxia* for its two principal hosts in the Åland Islands (*Veronica spicata* and *Plantago lanceolata*) are so short that most tests can be completed within a single afternoon (Kuussaari et al. 2000), and at this time scale the important assumption is assumption 1, which holds quite well within each day of testing.

Assumption 3 We assume that an encounter with plant A at time *t* has no effect on the insect's responses to either plant A or plant B at some subsequent time. This assumption is obviously violated when a butterfly has just accepted a host and is transferred quickly to another one. There is a clear carryover effect, making the second host much more likely to be accepted than if the butterfly were allowed to rest in a cage with no hosts for a few minutes. Therefore, when we observe an acceptance, we allow the insect at least five minutes of rest before testing another plant. Apart from this effect, experiments consistently fail to show any effect of manipulated experience on host acceptance. Two such experiments are described below.

In the first experiment, we collected wild females each morning at Rabbit Meadow and split them into two groups. One group was offered 9–10 repeated encounters with *Collinsia*, the other with *Pedicularis*. In the afternoon of the same day, all butterflies were offered the same test plant species. This test plant was sometimes *C* and sometimes *P*, on alternate days. Thus, each day's experiment tested whether butterflies with different recent experience of host encounter differed in their responses to a single test plant. No such effects were found (table 6.4; C. D. Thomas and Singer 1987).

In the second experiment, we collected teneral females (as mating pairs, so with no prior host encounter) at Rabbit Meadow and offered some of them alternating encounters with *P* and *C*, while others were offered only *C* or only *P*. We then offered each insect a single test with either *C* or *P* on the afternoon of the second day of its adult life. Again, no effects were detected (table 6.5; Singer 1986).

Assumption 4 Handling the butterflies does not affect their responses. I have already referred to work showing that butterflies captured that were naturally ovipositing or accepting different plants subsequently differed in tested preference in the expected directions (Rausher et al. 1981, Singer et al. 1993; table 6.2). Handling the butterflies does

Table 6.4. Acceptance or rejection of *Pedicularis* (P) and *Collinsia* (C) in single trials after a series of staged encounters.

	Accept *C*	Reject *C*	Accept *P*	Reject *P*
Prior encounters with C	28	19	30	25
Prior encounters with *P*	30	19	26	27
	$p > .9$		$p = .7$	

p values are from Fisher's Exact test.

not eliminate their differences in preference, but this does not mean that handling has no effect. In fact, handling does have a clear effect: it increases the likelihood of oviposition. A butterfly can be encouraged to oviposit by being picked up and quickly replaced on the test plant. Perhaps picking the insects up and replacing them makes them respond as though they were encountering plants more frequently, and they may be sensitive to plant density. Our method of dealing with this violation of assumption 4 is that, whenever an insect appears to be rejecting a plant, she is picked up and replaced three times during each three-minute staged encounter, before the result of the test is recorded as rejection. By this means we attempt to encourage oviposition equally in all test subjects.

It is clear from this account that the assumptions of the technique are to some extent violated and that subjectivity cannot be totally eliminated from these sequential choice trials. We cannot be sure of the exact relationship between the test results and the behavior of the butterflies in the field. However, we have performed several experiments blind. An example is the heritability experiment described in section 6.8. Results such as those, in which offspring tested blind nonetheless significantly resembled their parents in preference, give us confidence that our sequential choice testing technique is adequate for most purposes.

In counterpoint to the problems associated with violation of assumptions made in preference testing, the great advantage of checkerspots for preference work is that they are so manipulable and will duplicate their natural behavior when placed on a host, either in the greenhouse or in the field. This enables us to circumvent the problem of lack of natural flight in captivity by staging encounters with plants and observing the results. We can do so without allowing actual oviposition because the insects are so slow and deliberate that it is possible to tell that the decision to oviposit has been made and to remove the butterfly from the plant without allowing it to lay a single egg. Finally, checkerspot manipulability also allows us to ascertain the consequences of oviposition on different hosts under natural conditions. A butterfly held captive for one or two days can be placed experimentally in the field and caused to oviposit on a plant chosen by the experimenter. It is even possible to watch the oviposition, arrest it after a particular number of eggs has been laid (provided this is not too many to count), and transfer the butterfly to another plant, where it will peacefully continue to oviposit. The fates of the manipulated eggs can then be followed in the field (Singer 1984, Boughton 1999).

6.7 Distinguishing between Preference and Motivation

Starting with Dethier's (1959) paper on "mistakes" made by ovipositing butterflies and continuing to

Table 6.5. Host acceptance (*P* = *Pedicularis*; C = *Collinsia*) on the second day of life.

	Accept *C*	Reject *C*		Accept *P*	Reject *P*
Prior encounters with C	57	50	with P	33	15
Prior encounters with both hosts	49	31		39	8
	$p = .3$			$p = .15$	

p values are from Fisher's Exact test.

the present day, there has been discussion in the literature about the frequency with which insects oviposit on hosts that are suboptimal, hosts that are not preferred, or even on nonhosts that are toxic (Chew 1977, Chew and Robbins 1984, Singer 1984, Feldman and Haber 1998). These discussions have often involved questions about the roles played by unusual oviposition events in evolution of diet (e.g., Singer et al. 1971, C. D. Thomas et al. 1987). Perhaps such events are preludes to host shifts. Whether this is true depends on the behavioral mechanisms that cause unusual ovipositions and on the likelihood that insects performing such unusual acts do so because of their possession of particular heritable preferences.

In this context our ability to test preferences of freshly captured checkerspots in the field has enabled us to investigate the behavioral mechanisms that underlie observations of natural oviposition on low-ranked hosts. Why might a checkerspot be found ovipositing on a plant other than its preferred host, like the two *P*-preferring insects found ovipositing on C shown in table 6.2? There are two possibilities. First, the insect may have been searching for a long time without finding its preferred host. Second, the butterfly's discrimination phases are short, and it does not search for long before it accepts a second- or third-ranked host. In the first case we could describe the butterfly as highly motivated to oviposit. In the second, we could say that its preference is weak or its specificity is low. Why make this distinction? The evolutionary consequences are different in the two cases. Differences among individuals in motivation caused by differences in length of search are not likely to be heritable, but differences in length of discrimination phase could be heritable and, indeed, are likely to be (Singer et al. 1988, 1992b; see below).

To clarify the distinction between preference and motivation, I have depicted in figure 6.3 stylized records for three butterflies, two of which (#2 and #3) differ in motivation but not in preference, and two of which (#1 and #2) differ in strength of preference but not in motivation. The figure indicates that, at 1:50 PM butterfly #3 would accept plant A if that plant were encountered, but butterfly #2 would reject it. This would be ascribed to the difference in motivation. At 2:50 PM butterfly #2 would reject plant B, while butterfly #1 would accept it. This would be ascribed to their difference in strength of preference.

We performed an experiment to illustrate how and under which circumstances both preference and motivation may vary simultaneously in the field. We compared Rabbit Meadow butterflies from two large (>2 ha) habitat patches by initiating preference tests immediately after the butterflies were captured (this is the same site shown in figure 6.4, but not the same experiment). In the cleared patch the butterflies used a novel host, *Collinsia*, but mostly preferred their traditional host, *Pedicularis*, which was not present in the clearing. In the unlogged patch the insects remained on their traditional host. The insects from the cleared patch (that preferred *Pedicularis*) waited longer after their capture before they would accept *Pedicularis*, while insects from the unlogged patch (that preferred *Collinsia*) mostly accepted *Pedicularis* on their very first staged encounter or soon after (Singer et al. 1992b). It may seem unexpected that the *Collinsia*-feeding insects accepted *Pedicularis* more readily than the *Pedicularis*-feeding insects. However, once we incorporate the idea of motivation, it becomes logical that insects in the patch that lacked their preferred host should have a higher mean level of motivation and accept that preferred host sooner after capture.

The same experiment that revealed interpatch variation of motivation also documented a difference in preference between the two patches in the expected direction (Singer et al. 1992b). Insects from the *Pedicularis*-feeding patch had longer discrimination phases and thus stronger preferences for *Pedicularis* over *Collinsia*. The principal source of this difference was that the insects assorted themselves between the patches according to their preference genotypes (C. D. Thomas and Singer 1987, Singer and Thomas 1996). The somewhat paradoxical conclusion of this experiment was that, in the comparison of insects from the patches where C and *P* were used, the insects from the C-use patch were at the same time more likely to accept *P* rapidly and less strongly prefer of *P* over C. The first effect was interpreted as a difference in motivation and the second as a difference in preference.

These experiments and others (Singer and Thomas 1996, Thomas and Singer 1998) show that preference is systematically variable among habitat patches at Rabbit Meadow. It is also highly variable within patches (Singer 1983; figure 6.4). Variation of preference clearly affects diet (table 6.2), and diet clearly affects fitness (Singer 1984, Moore 1989a, Singer et al. 1994). Therefore, if preference were heritable, diet evolution should occur, driven by evolution of preference. But how heritable is preference? The next section addresses this question.

6.8 Heritability of Preference

We captured *Euphydryas editha* at Schneider's Meadow (Carson City, Nevada), and tested them in the field for preference for *Collinsia parviflora* versus *Plantago lanceolata,* using undisturbed test plants still rooted in the ground. We then raised the offspring entirely on *Collinsia* and applied to them a series of blind tests, in which the person doing the preference test had no knowledge of the identity of the insect (other than a coded number) and hence no expectation of its preference. Data from sibs were combined into family mean preferences and regressed on the maternal preference. On the assumption that males and females make equal contributions to inheritance of preference, we obtained an estimate of heritability of 0.89, with 95% confidence limits of 0.32 and 1.4 (figure 6.5). The assumption of equal male and female contribution is indirectly supported by the results of reciprocal interpopulation crosses between an *E. editha* population where insects strongly preferred *Collinsia* over *Plantago*, and one where *Plantago* was weakly preferred over *Collinsia*. Regardless of the direction of the cross, the F_1 females showed weak preferences for *Collinsia*, preferences lying outside the range we found in either of the parental populations (Singer et al. 1991).

These experiments indicate that preference variation in *E. editha* was genetically based both within and among populations. In *Melitaea cinxia,* interpopulation variation was likewise genetically based (Kuussaari et al. 2000) and was maintained through two laboratory generations in which all insects were raised on the same host species.

6.9 Relationship of Preference to Fecundity

The chapter title alludes to the "importance of preference in the life of the checkerspot." Thus far I have discussed only one type of importance—that preference affects diet. The further relevance of oviposition preference in herbivorous insects depends on its correlates with other traits of the female, especially with fecundity. It is widely believed by workers with insect oviposition preference that preference is driven by "egg load." An insect that feels increasing pressure from unlaid eggs ("egg presure") will be increasingly motivated to oviposit. Differences among individuals in fecundity should generate differences in strength of oviposition preference. Some authors have also argued that evolution of fecundity should tend to cause evolution of diet and that evolution of preference should cause evolution of fecundity (Courtney et al. 1989, Courtney and Hard 1990). Thus, preference may be intimately and intricately connected with life history.

These beliefs about the relationship between egg load and preference stem from more than anthropomorphic empathy with egg pressure. They are based on modeling (Mangel 1989) and on several independent findings, including one with *E. editha* (Moore 1989a), that insects captured naturally ovipositing on low-ranked hosts had higher mean egg loads than those captured ovipositing on high-ranked hosts (references in Agnew and Singer 2000). This observed correlation between host acceptance and egg load has been taken to imply a cause–effect

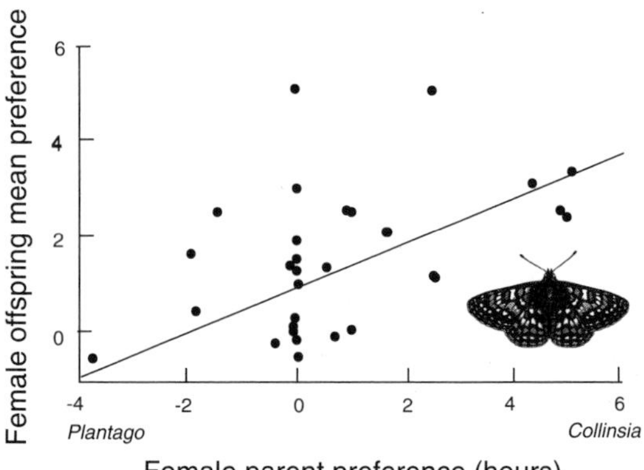

Figure 6.5. Regression of mean offspring preference on maternal oviposition preference for Schneider's Meadow butterflies, with the parents tested in the field and all offspring raised on *Collinsia* (Singer et al. 1988).

relationship between egg load and preference: variation in egg load drives variation in preference. However, acceptance is not the same trait as preference, and correlations do not necessarily indicate cause and effect. Below I describe experiments undertaken with *E. editha* to help understand what is happening in the field.

Consider the possible reasons a checkerspot may oviposit on a low-ranked host. In section 6.7 I argued that this could be either because the insect has been searching for a long time (high motivation), and/or because it has a short discrimination phase (low specificity or weak preference). To the extent that the first cause occurs, we would expect that such insects would indeed have high egg load. While the butterfly searches without finding a host, its egg load increases, as does the range of hosts it would accept if they were to be encountered. Because both motivation and egg load increase with time, we would expect them to be positively correlated in the field, whether or not there were any causal relationship between them.

Although egg load and host acceptance both increase with time during a search, they may do so independently of each other and in response to different physiological changes. Therefore, we cannot deduce a causal relationship between high fecundity and acceptance of low-ranked hosts from field studies in which butterflies accepting such hosts had higher current egg loads. How can we tell if such a causal relationship exists? Agnew and Singer (2000) were able to test this using the extensive variation of discrimination phase length in the Rabbit Meadow population of *E. editha* (figure 6.4). We performed two separate experiments. First, butterflies were captured and their discrimination phases measured. Immediately at the end of the discrimination phase, each individual was allowed to oviposit on its preferred host, *Pedicularis*. There was a strong positive association between the length of the discrimination phase and the size of the egg cluster (Agnew and Singer 2000; figure 6.6). Dissections showed that each butterfly laid all its mature eggs at each oviposition, regardless of host acceptability (cf. Pilson and Rausher 1988). Therefore, we can reject the hypothesis that the second-ranked host is first accepted at a particular egg load.

A second set of experiments also measured discrimination phase but allowed oviposition on *Pedicularis* a fixed length of time (48 hours) after this preferred host had been first accepted. In this case, there was no significant relationship between egg cluster size and discrimination phase (figure 6.7). If the insects with short phases had been those whose eggs were maturing fastest, as most workers would expect, then we should have obtained a negative association between discrimination phase length and fecundity in the second experiment and no relationship in the first. In fact, we obtained a positive relationship in the first experiment and none in the second. Collectively, these two experiments strongly indicate that preference and fecundity are not causally connected in Rabbit Meadow *E. editha* (Agnew and Singer 2000). Both egg load and host acceptance increase with time, and they are phenotypically correlated as a result. However, the individuals whose eggs mature fastest are not the individuals with the fastest increase in their range of accepted hosts. This lack of causal connection between fecundity and preference is useful in a practical sense to Melitaeologists because it allows us to model the evolution of diet breadth without in-

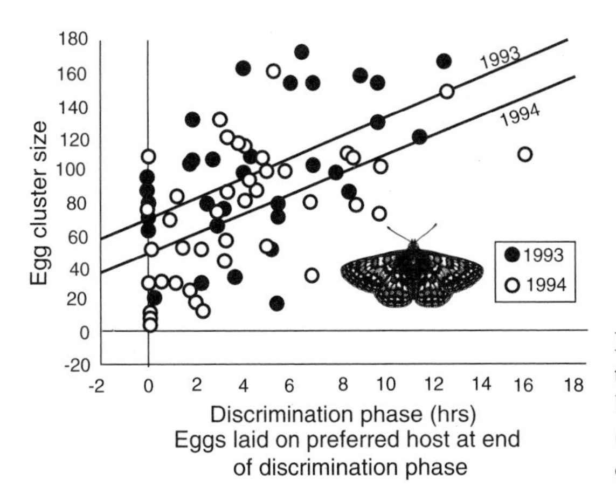

Figure 6.6. Regression of egg cluster size against the length of the discrimination phase when butterflies laid on their preferred host (*Pedicularis*) immediately at the end of the discrimination phase (Agnew and Singer 2000).

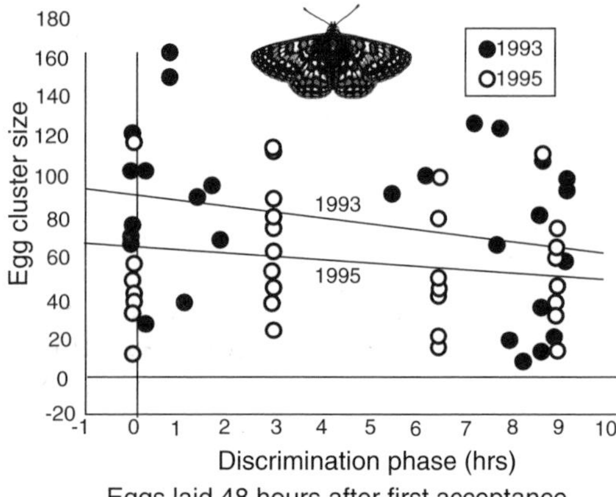

Figure 6.7. Regression of egg cluster size against the length of the discrimination phase when butterflies laid on their preferred host 48 hours after it was first accepted. In this case timing of oviposition did not depend on the length of the discrimination phase (Agnew and Singer 2000).

corporating constraints from correlated effects on fecundity.

6.10 Relationship of Maternal Preference to Offspring Performance

Discussion of relationships between preference and performance typically confounds several different questions. Three of the most important ones are:

1. Is preference correlated with performance among populations? In other words, is preference variation among populations associated with performance variation in the same set of populations?
2. Is preference correlated with performance within populations? In other words, do individual mothers with particular preferences produce offspring with particular performances?
3. Is host choice adaptive at the population level? Within a population, is the host type that is generally preferred for oviposition the same host type that supports highest offspring survival? More generally, to what extent is the rank order of plants in the butterflies' preference hierarchy concordant with the rank order of the same plants in their ability to support larval growth and survival (see Wiklund 1975)?

Answers to question 3 are often described as "preference–performance correlations." We have argued (Singer et al. 1994) that this is an incorrect use of the term "correlation," which should describe a relationship between two variables. In question 3, "preference" is not treated as a variable at all, the only parameter used is the mean population preference. Answers to questions 1 and 2 are more properly described as preference–performance correlations. We have investigated all three of these questions with *E. editha* and two of them with *M. cinxia*. Up to now our conclusion is that *E. editha* shows all three possible types of preference–performance association, while *M. cinxia* shows neither of those investigated (questions 1 and 3).

With respect to the first question, variation in preference among populations is associated with parallel variation in offspring performance in *E. editha* (Rausher 1982) but not in *M. cinxia*. Interpopulation variation of preference is pronounced in *M. cinxia* (Kuussaari et al. 2000), but it is not associated with performance, despite the fact that performance is also variable (van Nouhuys et al. 2003). The results from *M. cinxia* resemble those of another nymphaline, *Junonia coenia* (Camara 1997c).

With respect to the second question, adult *E. editha* with particular preferences did produce offspring with particular performances. We have demonstrated this phenomenon twice, once using variation in preference for different host species, and again using variation in preference for particular plant categories within a single host population. The first experiment was done at Schneider's Meadow, where we measured preferences of adults in the field and then placed offspring of each butterfly on one of the two host species, again in the field. We regressed on maternal preference two separate measures of larval performance: growth rates measured on *Plantago* and on *Collinsia*. The two regressions differed significantly (Singer et al. 1988).

The second experiment was done at Rabbit Meadow using individual *Pedicularis* plants. Ng (1988) found that approximately half the butterflies discriminated among *Pedicularis* that were similar in size, appearance, phenology, and exposure. The other butterflies did not discriminate among the test plants under field conditions. All of the discriminators made the same choices: if one butterfly preferred plant #52 over plant #28, others tested on the same plant pair would either do likewise or fail to discriminate. None would prefer #28 over #52. Ng was able to use this feature of the system to classify butterflies as discriminators or nondiscriminators and plants as accepted or rejected by discriminating butterflies. Accepted plants retained their status from year to year and were in practice more likely to receive eggs that plants rejected in the preference trials (Ng 1988).

From each family of larvae, a group of 20 neonates was placed on a rejected plant and 20 on an accepted plant. Relative survival on the two plant classes differed significantly between offspring of discriminators and nondiscriminators (figure 6.8; Ng 1988). The slopes of the reaction norms differ significantly between the data shown in the two panels of the figure.

From the perspective of the butterflies, the preference–performance correlations at Schneider's Meadow (for *Plantago* and *Collinsia*) may have the same physiological basis as those at Rabbit Meadow (for accepted and rejected *Pedicularis*). However, from the evolutionary perspective the two correlations have different importance. The first is relevant to evolution of the number of host species in the insects' diet. For example, if a novel host becomes available, it is more likely to be incorporated into the diet if the females most likely to accept it for oviposition are the same females that produce offspring most likely to survive on it. The second relationship is pertinent to coevolution of the butterfly population with the *Pedicularis* population. Whether it is also pertinent to evolution of the number of host species in the diet depends on the nature and mechanistic basis of associations between preferences expressed within and among those host species (section 6.11)

The third question in the list of types of preference–performance association was, is host choice adaptive at the population level? Is the host type that is generally preferred for oviposition the same host type that supports highest offspring survival? Once again the answers from *M. cinxia* and *E. editha* seem completely different. The extensive interpopulation variation of preference in *M. cinxia* is associated with two factors: first, the relative abundance of the two hosts in the region of each focal patch, and second, the relative isolation of the focal patch from *M. cinxia* using the same and the other host species (Kuussaari et al. 2000). This is what would be expected if there were both strong natural selection on oviposition preference generated by relative abundance of the hosts and strong interaction of individuals among the patches that is mediated by preference-associated dispersal (C. D. Thomas and Singer 1987, Hanski and Singer 2001). There is a strong case for this explanation, both by elimination of competing expla-

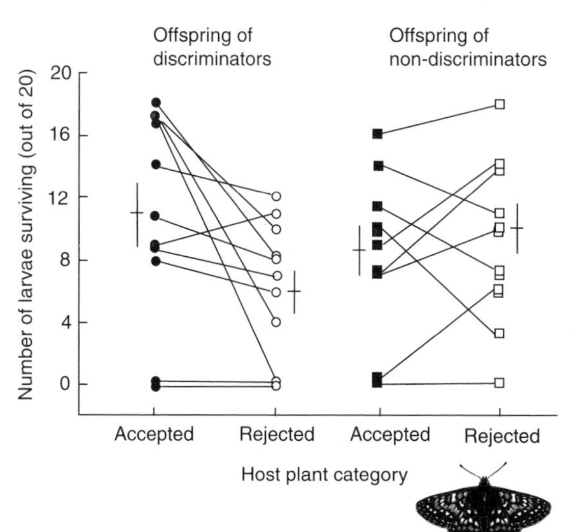

Figure 6.8. Relative survival on accepted and rejected *Pedicularis* plants of the offspring of discriminating and undiscriminating butterflies. Classification of plants as accepted and rejected had been made by the discriminating butterflies before the experiment. Each line joins points depicting offspring of the same individual mother; the data are the numbers of larvae surviving in each larval group. Initial group size was 20 larvae (Ng 1988).

nations and by analogy with the colonization effect, a parallel effect of the same type of patch isolation on colonization rates of empty patches (Hanski and Singer 2001; chapter 9). However, the selective mechanism(s) by which host abundance may influence local evolution of post-alighting preference are not fully understood. There are no detectable local adaptations in larval performance that could either drive the local adaptation of preference or be driven by it (van Nouhuys et al. 2003). Instead, the evolution of local adaptation of preference may stem from positive associations between pre-alighting and post-alighting preferences that could cause insects that prefer the locally abundant plant to find hosts efficiently (Schöps and Hanski 2001).

Euphydryas editha shows a complex pattern of geographical variation in both diet and preference (Singer 1971b, White and Singer 1974, Holdren and Ehrlich 1982, 1984, Singer et al. 1994, C. D. Thomas and Singer 1998, Pratt et al. 2001). Diets of the populations that are discussed in detail in this chapter are shown in table 6.6, and the general pattern of geographical variation of diet is illustrated in figure 12.3. At most sites we can classify as oviposition hosts those plant species that naturally receive eggs and classify as potential hosts those species that are used elsewhere but are present and not used in the target habitat. Singer et al. (1994) and C. D. Thomas and Singer (1998) performed a series of experiments in which oviposition preference ranks were tested for actual and potential hosts at a set of nine sites. Neonate larvae from the local habitat were then placed on all tested plant species at their natal site. No larvae were transplanted among sites in these experiments.

Within each of the nine populations, host preferences were strongly adaptive: the rank order of potential host species in terms of their acceptability to ovipositing adults was identical to their rank order in supporting larval survival (Singer et al. 1994, C. D. Thomas and Singer 1998). Because we used only local plants and insects in these experiments, we cannot know the basis of the differences among sites in the identity of the plant species that was most accepted and that supported the highest larval survival. This basis could be interpopulation variation in plants and/or in insects. Other experiments involving reciprocal 'transplants' clearly implicate genetic variation in both plants and insects (Singer and Parmesan 1993).

The nine populations described above were not undergoing rapid diet evolution. At two additional sites where anthropogenic changes had triggered rapid diet evolution, early work showed that most insects preferred their traditional host, but offspring survival was higher on the novel host. There was strong selection favoring preference for the novel host. Both sites later achieved the apparently more general concordance between preference and performance, at one site (Schneider's Meadow) because rapid evolution of preference rendered the novel host more preferred than the traditional host, and at the other site (Rabbit Meadow) because the direction of natural selection on preference was reversed: the less preferred novel host became less suitable than the traditional host (Singer et al. 1994, C. D. Thomas and Singer 1998). Thus, in this set of experiments and observations, diet variation among *E. editha* populations was generally adaptive. Observed maladaptations resulted from recent anthropogenic disturbance and were ephemeral.

The expectation from work on other insects is that maladaptation of host preference should be frequent because of evolutionary constraints on

Table 6.6. Past and present diets of *E. editha* populations discussed in the text.

Population Name	Past Diet	Diet Changes
Yucca Point	*Collinisa tinctoria*	None
Tamarack Ridge	> 95% *Collinsia torreyi*	None
Colony Meadow	*Pedicularis semibarbata*	None
Rabbit Meadow	~95% *Pedicularis semibarbata*, ~5% *Castilleja disticha* before 1970	*Collinsia torreyi* and *Mimulus whitneyi* added to diet ca. 1976; *Mimulus* deleted ca. 1988; *Collinsia* deleted ca. 1998
Schneider's Meadow	82% *Collinsia parviflora*, 12% *Plantago lanceolata*, 6% *Penstemon rydbergii* in 1982	96% *P. lanceolata*, 4% *C. parviflora* in 2002

preference rank (Courtney et al. 1989, Thompson 1993, Wehling and Thompson 1997), manifested as lack of relevant genetic variability for selection to act (Karowe 1990). This expectation is not supported by any of the results for *E. editha*. However, maladaptation does exist in *E. editha* preference and has two causes. The first is an evolutionary lag behind rapid anthropogenically driven fluctuations of natural selection, as discussed above, and the second is asymmetrical gene flow among habitats with different selection regimes (Singer and Thomas 1996). Our work on *Euphydryas aurinia* at first suggested the type of maladaptation expected by Courtney et al. and by Thompson, stemming from evolutionary constraints of preference rank. However, detailed field experiments revealed that the appearance of maladaptation was misleading (see section 6.12).

6.11 Role of Oviposition Preference in Evolutionary Host Shifts

Observed Evolution of Preference during Host Shifts

Checkerspots do not provide widely cited examples of shifts in host plant use, perhaps because the principal aim of so many workers on host shifts has been to study speciation, and checkerspots do not usually speciate in response to host shifts (C. D. Thomas and Singer 1998). Yet checkerspots provide two of the few examples in which ongoing host shifts have been observed and their mechanisms studied, and the only examples in which rates of change of insect traits have been measured during these ongoing host shifts (Singer et al. 1993b, Singer and Thomas 1996). We have also been able to compare host preferences of insects in populations that have recently undergone host shifts with preferences in populations that have not done so (see Tabashnik 1983, C. D. Thomas and Singer 1987, Singer et al. 1992a, Camara 1997c).

Figure 6.9 tracks changes in preference that occurred during the host shift from *Collinsia* to *Plantago* at Schneider's Meadow, a shift that has now gone almost to completion: in 2002 we found 96% of larval groups on the novel host (M. Singer and R. Plowes, unpubl. data). Because the data in figure 6.9 were obtained with contemporary plants and insects in each year, we cannot say from these data alone whether the changes were in insect preference and/or in plant acceptability. However, we were able to partially address this question. The plant pots used for the heritability experiment in 1983 (section 6.8) had been retained with their soil in the laboratory until 1990. Therefore, when the diet change between 1983 and 1990 became evident in the field (Singer et al. 1993), we watered these plant pots and were able to obtain offspring of the test plants used in the heritability experiment. We preference-tested the lab-raised offspring of field-caught 1990 butterflies on these plants. The family mean preferences among these offspring of butterflies captured in 1990 differed significantly from the previous family mean preferences obtained in the heritability experiment, the preferences of lab-raised offspring of butterflies captured in 1983 (Singer et al. 1993). This demonstrates that genetic evolution of preference was involved in the host shift.

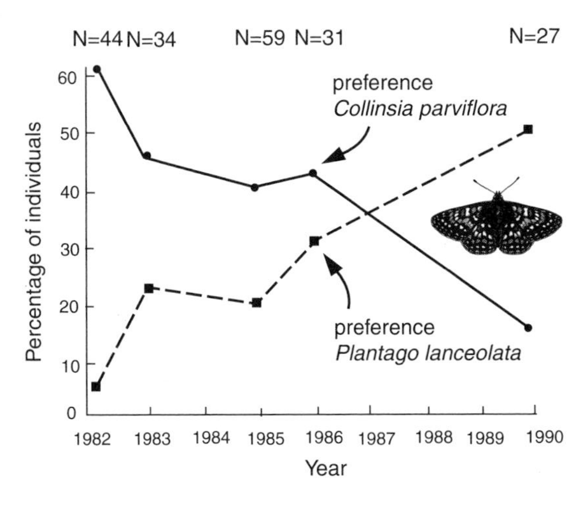

Figure 6.9. Long-term changes in oviposition preference at Schneider's Meadow. Points on the graph indicate proportion of insects preferring the traditional host *Collinsia parviflora* or the novel host *Plantago lanceolata*. The two graphs are not exact mirror images because females with no preference are not shown. Tests were done with contemporary plants; hence the trends could represent changes in plants or in insects or in both. Laboratory experiments implicated genetic changes in butterfly preference (Singer et al. 1993).

We were also able to document strong and significant temporal trends in oviposition preference associated with the independent host shift from *Pedicularis* to *Collinsia* at Rabbit Meadow (Singer and Thomas 1996). In this case our evidence that evolutionary changes of insect preference had occurred came from side-by-side preference comparisons between butterflies from the changed populations and those from a nearby (12 km) unchanged population, Colony Meadow, in Sequoia National Park. That park population represented the putative ancestral condition of Rabbit Meadow butterflies in the 1960s, before the logging that had triggered their bout of diet evolution (Singer et al. 1992a). After a midsummer frost had extirpated the Rabbit Meadow butterflies on their novel host and left the metapopulation once again restricted to its traditional host (see chapter 9), the affinity of the remaining butterflies for the novel host that they no longer used nonetheless remained much higher than that of butterflies in the park. Singer and Thomas (1996) argued that this difference reflected an evolutionary legacy among insects in patches of the traditional host of gene flow from patches where the novel host had been used. Below I give examples showing how the processes involved in the host shift can be understood by examining interactions of the insects with their novel and traditional hosts. The relevance of this to the chapter topic is that evolution of oviposition preference turns out to play a major role in diet evolution, both as a mechanism by which diet evolves and as a constraint on the direction of evolution.

One way to understand why the diets of herbivorous insects typically remain narrow is to ask what obstacles to diet expansion are faced by an insect population. Our group was able to answer this question by examining the interactions of butterflies with their novel and traditional hosts at Rabbit Meadow just a few generations (<20) after diet expansion had occurred. This set of studies revealed a set of behavioral barriers to diet expansion that had been temporarily overcome in this case but that would strongly inhibit incorporation of the novel host into the insects' diet.

Anatomy of Host Shifts Revealed from Interactions with Novel and Traditional Hosts

In sections 6.4 and 6.7 I used the recent addition of *Collinsia* to the diet of *E. editha* at Rabbit Meadow to discuss the role of visual stimuli in host finding and the relationship between preference and motivation. This diet expansion involved colonization of *Collinsia torreyi* after it had been rendered suitable for the insects by the U.S. Forest Service. Logging removed the butterflies' traditional host, *Pedicularis semibarbata*, from habitat patches that were logged and simultaneously extended the life span of *Collinsia* by soil fertilization. The butterflies colonized *Collinsia* in the cleared patches while retaining their traditional diet in the undisturbed habitats. This diet expansion is discussed in chapter 9 with respect to the movements of butterflies among cleared and undisturbed habitat patches. Here I detail a set of experiments and observations indicating that the insects behaved toward the novel host, *Collinsia*, as though it were the traditional host, *Pedicularis*, and that these behavioral patterns were maladaptive for insects that did use the novel host. The following paragraphs discuss the sources of maladaptation to *Collinsia*, including those that have been discussed previously.

Host finding was efficient for the traditional host and worse than random for the novel host (section 6.4; Mackay 1985, Parmesan et al. 1995). We could say that the insects retained their pre-alighting preferences for the traditional host, with consequent inefficiency of search for those individuals that used the novel host because they lived in habitat patches from which the traditional host had been extirpated.

Most, but not all, of the butterflies also retained their post-alighting preferences for *Pedicularis* (Singer 1983, Singer and Thomas 1996). After finding *Collinsia* inefficiently, most insects rejected it, even though it would have supported higher survival of their larvae than *Pedicularis* (Moore 1989a, Singer and Thomas 1996, C. D. Thomas et al. 1996, Boughton 1999). Even some insects that developed naturally on *Collinsia* from eggs laid on it in the field found it totally unacceptable as an oviposition host. Insects that did accept *Collinsia* preferred senescent over blooming plants, thereby incurring offspring mortality that could have been avoided by preferring younger individual plants (Boughton 1999).

Butterflies ovipositing on the rosette-shaped *Pedicularis* (figures 6.1 and 6.2) were positively geotropic: on encountering a chemically acceptable stimulus, the butterflies typically dropped to the ground and searched for a low-lying leaf under which to place their eggs. Likewise, butterflies that alighted on the tip of a *Collinsia* and accepted the taste stimuli that it provided usually responded by dropping to the ground, curling the abdomen and probing with the ovipositor to find a low-lying leaf. However, *Collinsia* is a small, erect annual (figures 6.1 and 6.2), not a perennial rosette, and it has no low-lying leaves large enough to receive an egg cluster. So, the butterflies could be seen walking through the vegetation with their ovipositors extended in front of them, searching for a substrate that provided physically acceptable stimuli for oviposition. This often led them to oviposit on nonvegetation—for example, under fallen twigs—another maladaptive trait that can lead to egg mortality in hot weather.

Because Rabbit Meadow *E. editha* laid all their mature eggs at each oviposition, butterflies that underwent prolonged search because of failure to find their preferred host eventually laid large egg clusters. By this mechanism, mean egg cluster size was higher on *Collinsia* than on *Pedicularis* (Singer 1986, Moore 1989a). This was another maladaptive behavior. Unpublished experiments in which larval group size was manipulated indicated that, because of predators, the optimum group size for larval survival was much lower on *Collinsia* than on *Pedicularis* (details in chapter 7). Indeed, cluster size in populations adapted to *Collinsia* was much smaller than in populations adapted to *Pedicularis* (see below).

Despite high larval survival on *Collinsia*, butterflies tended to emigrate from patches of this host and accumulate in patches of *Pedicularis*, increasing competition and causing heavy larval mortality on the traditional host (Boughton 1999, 2000; see chapter 9). This tendency to emigrate prevented the colonization of small *Collinsia* patches and reduced butterfly density in moderate-sized patches, despite high suitability of the host plants in these patches (C. D. Thomas and Singer 1998).

These various maladaptations to the novel host, *Collinsia*, composed a powerful barrier to its colonization by the butterflies at Rabbit Meadow. The barrier was overcome for a while in the 1980s, when butterflies developed on *Collinsia* at high density (>1 adult/m^2), despite their lack of adaptation to this host (Singer 1983, Singer and Thomas 1996, C. D. Thomas et al. 1996).

The constraints to evolutionary diet expansion imposed by these maladaptations to *Collinsia* at Rabbit Meadow are not permanent. We can tell this by examining the interactions of *Euphydryas editha* with the same species of *Collinsia* in a large metapopulation at Tamarack Ridge 50–80 km north of Rabbit Meadow, where *C. torreyi* is the traditional host. *Pedicularis* does occur at Tamarack, where it is used exclusively by *Euphydryas chalcedona*. But most *E. editha* adults from Tamarack refused to oviposit on *Pedicularis*, and larvae would not even attempt to feed on it, even if we offered them *Pedicularis* from Rabbit Meadow, where it is the traditional host of *E. editha*. We conclude from these extremely negative responses to *Pedicularis* that it is not the traditional host at Tamarack. By elimination, this role at Tamarack is filled by *Collinsia*.

How do Tamarack *E. editha* respond to *Collinsia* where it is the traditional host? They find it more efficiently than do Rabbit Meadow butterflies, and they prefer blooming individuals over senescent ones. They lay small clusters of eggs (mean cluster size = 5.5, compared to 52 on *Collinsia* at Rabbit Meadow) at the top of the plant and are able to use small host patches, presumably because they do not preferentially emigrate from them. (Some of the evidence summarized here is culled from unpublished observations of M.C. Singer, D. Boughton, C. Parmesan and C. D. Thomas; the rest can be found in Singer [1994, 2003], Parmesan et al. [1995], C. D. Thomas and Singer [1998], and Boughton [1999].)

A phylogeographic study using mtDNA found the *E. editha* metapopulation at Tamarack to be closely related to the population at Rabbit Meadow (Radtkey and Singer 1995). We therefore conclude that the suite of maladaptations to *Collinsia* observed at Rabbit Meadow did not constitute a long-term constraint to the use of this species, to which the Tamarack butterflies were better adapted in all respects. However, despite the possibility for the Rabbit Meadow butterflies to have overcome what we can now perceive as short-term constraints to colonization of *Collinsia*, and despite the high densities that the insects achieved on this host for more than 10 years, they have failed to accomplish this host shift. In 2002 there were many thousands of butterflies at Rabbit Meadow, but we found no evidence of oviposition on *Collinsia*. At the same time we performed a field experiment in which we staged encounters at Rabbit Meadow between the two hosts and butterflies from both origins (Tamarack and Rabbit Meadow). We then followed the fates of egg clusters that these butterflies laid. By far the highest fitness was achieved by Tamarack butterflies ovipositing on *Collinsia*. Their success was dramatically higher than that of Rabbit Meadow butterflies on *Pedicularis*, which in turn exceeded the fitness of Rabbit Meadow butterflies on *Collinsia*

(M. C. Singer, R. Plowes, and M. Butcher, unpubl. data). The mechanisms whereby Tamarack butterflies were able to outperform Rabbit Meadow butterflies on Rabbit Meadow *Collinsia* in the field lay in two of the maladaptations of the Rabbit Meadow butterflies listed above. Rabbit Meadow butterflies were often unable to oviposit when they tried to because they could not locate physically acceptable substrates, and when they did oviposit they preferred phenologically more advanced host individuals. Additionally, about half of the Rabbit Meadow butterflies simply rejected *Collinsia*; this effect was not included in the fitness calculations.

Our conclusion, then, is that maladaptations to *Collinsia* have prevented the Rabbit Meadow butterflies from climbing an adaptive peak (*Collinsia* use) that, at least in 2002, was higher than the peak (*Pedicularis* use) on which they currently rest. This raises the question of how the original colonization of *Collinsia* by Rabbit Meadow butterflies was achieved in the 1970s. This occurred at a time when the influence of logging on *Collinsia* was recent and the phenological heterogeneity of this host (to which the butterflies respond so maladaptively) may have been less. Old plant samples and field censuses exist that could be used to investigate this question.

Role of Associations among Preferences in Host Shifts

In section 6.9 I summarized evidence from *E. editha* challenging the general view that strength of preference is driven by egg load or fecundity. A corollary of the view that we challenged would be that preferences for different hosts should be associated. This is because if egg load drives preference, then an increase in egg load should reduce strength of preference for all hosts, not just for some. This would cause associations among preferences that could be important drivers of host shifts.

Our behavioral evidence against the egg load hypothesis would not lead us to expect the associations among preferences predicted by that model. However, we did find such an association at Rabbit Meadow. At this site, insects that were nondiscriminators among individual *Pedicularis* plants were also less likely to discriminate between *Pedicularis* and *Collinsia*. There was a highly significant association between discrimination within the traditional host species and acceptance of the novel host. If this association among preferences were fixed, it could affect the evolutionary expansion of the diet. As acceptance of the novel host increased, adaptive discrimination among individuals of the traditional host would be lost (see Janz and Nylin 1997). In this case, because of the preference–performance correlation (section 6.10), it is not clear that such a loss would be opposed by selection. However, we were able to show that the relationship among preferences was labile and rapidly evolving, and it did not constrain the observed evolutionary diet expansion. We compared *E. editha* from the Rabbit Meadow population (that had added *Collinsia* to their diet within the past 20 generations) with butterflies from the population 12 km away at Colony Meadow in Sequoia National Park, where neither logging nor the consequent diet evolution had occurred.

If the association among preferences constituted an evolutionary constraint, we would expect that, as acceptance of *Collinsia* increased, the proportion of butterflies discriminating within *Pedicularis* would decrease. This did not happen. Comparison of butterflies between Rabbit Meadow and the putative ancestral condition at Colony Meadow showed a highly significant difference in acceptance of *Collinsia* but no trend at all in discrimination within *Pedicularis*. The reason for this was that the association between the two measures of preference was evolutionarily labile and significantly different between the sites. Two traits had evolved significantly over a period of less than 20 generations: a simple trait, acceptance of *Collinsia* and a more complex one, the relationship between *Collinsia* acceptance and discrimination within *Pedicularis* (Singer et al. 1992a). In this case the insects could have their cake and eat it, too: they added a novel host to the diet without diminishing adaptive discrimination of the traditional host among individual butterflies.

This result contrasts sharply with evidence from a comparison among nettle-feeding butterfly species (Janz and Nylin 1997) that shows an expected trade-off that *E. editha* seems not to have suffered, at least in this case. Generalist species made poorer discriminations than specialists among individual nettle plants of dissimilar quality. It is interesting that evidence for the trade-off between diet breadth (number of host species) and adaptive discrimination within host species appears in the interspecific comparison done by Janz and Nylin, but not in this observed bout of rapid evolution at the intraspecific level.

6.12 Novel Axes of Variation Revealed by Preference Testing of Checkerspots Reveal General Problems in Experimental Design

Herbivorous insects in a population feeding on more than one host species are presented with variation both among conspecific plant individuals and among plant species (Dolinger et al. 1973, Lewis 1982, C. D. Thomas 1987, Singer et al. 1992a, Janz and Nylin 1997, Mopper 1998). We might expect differences within plant species to be less important to the insects than differences among species. However, a recent survey (Strauss and Karban 1998) indicates that effects of within- and among-species plant variation on insect fitness were on the same order of magnitude, within the range of host species that an insect uses. It is therefore unsurprising that checkerspots make substantial discriminations within as well as among host species. This process generates complexity because discrimination within species is not nested within discrimination among species, as one might reasonably expect. Preference-testing of insects on conspecific and heterospecific plants has revealed novel axes of variation, the existence of which threatens many standard and apparently sensible experimental designs in plant-insect interactions. Three examples are discussed below.

Individual *Melitaea cinxia* varied in the relative importance they assigned to variation within and among host species (Singer and Lee 2000). We applied sequential choice preference tests to lab-raised butterflies from two populations of Finnish *M. cinxia*, one of which fed naturally on *Plantago lanceolata* (*P*), the other on *Veronica spicata* (*V*). All butterflies were tested on the same six individual plants, three of each species. We found repeatable differences among individual insects from the same site in their tendency to rank plants as species or as individuals. Some insects preferred all *P* over all *V* or vice versa. We designated these as "pure" species ranks. The pure ranks all favored the host species used by the population of origin of the butterflies. Other insects, in both populations, produced mixed ranks, preferring some individual *V* over some *P*, and some individual *P* over some *V*. The differences among individuals in production of pure or mixed ranks were highly repeatable on second testing of each butterfly. We retested 21 insects, of which 19 were classified in the same category both times. Thus, there were real differences among the butterflies in the ways in which they responded to variation within and among plant species.

How might a "pure-rank" butterfly differ from a "mixed-rank" butterfly? The pure-rank butterfly could either discriminate more between the plant species and/or less within them. Thus, a butterfly can rank all members of one species above all members of another by either discriminating strongly between species and/or by discriminating weakly within them. This observation has a cautionary message for some experimental designs that estimate variation in either insect preference or plant resistance. Consider an experiment in which a set of butterflies from the population of *M. cinxia* where all pure-rank butterflies prefer *V* are tested on just two plants, one *Plantago* and one *Veronica*, in an attempt to assay variation in their preference for the two plant species. Assume that the difference between insects producing pure and mixed ranks stems from differences in within-species discrimination. If these insects are offered a low-ranked plant of their favored host species, *V*, and a high-ranked *P* plant, then the mixed-rank insects would prefer *P* and the pure-rank insects would prefer *V*. The variation in discrimination within species would falsely appear as variation in discrimination among species. Depending on the nature of the relationships between discrimination within and among species, this could be an important general but overlooked problem in experimental design.

The message from these experiments with *M. cinxia* seems to be that preference tests should not be carried out with just one plant chosen arbitrarily to represent its species. If we want to know which of two plant populations (or species) is generally preferred by butterflies from a particular population, we should randomly sample insects and plants independently from their populations and then test each insect on a different pair of plants. This procedure was used by Singer and Parmesan (1993) and by Hanski and Singer (2001) to determine whether the rank order preference of insects from different populations for two plant species depended on the origin of the insects, the origin of the plants, or both. So are all our problems resolved when we sample plants and insects at random to represent their populations? Apparently not. The examples below show why.

When *Euphydryas aurinia* butterflies and their hosts were sampled randomly, we obtained the odd result that insects from populations feeding on *Gentiana, Lonicera,* and *Cephalaria* all preferred *Succisa pratensis*, which they never encountered in the field, over their own hosts (Mazel 1986, Singer et al. 2002). If the sequential choice test reasonably represents reality, as has been argued here, then these butterflies waste hours or even days searching for a nonexistent plant species before they will accept their own host species. This constitutes an apparent maladaptation that could result from any of the usual causes of maladaptation: gene flow, evolutionary constraints on preference hierarchies, and so forth. However, the appearance of maladaptation is false. It is an artifact of forcing plants and insects to interact in random combinations when they do not do so in nature. This effect disappeared when the host populations were sampled differently, using naturally accepted plants (Singer et al. 2002). This result casts doubt on all experiments that test whether host choice is adaptive by manipulating insects to feed on randomly chosen members of different host species. Alas, this category includes many of our own experiments (Singer et al. 1994).

Euphydryas editha at Rabbit Meadow were offered *Pedicularis* plants in sequential choice trials. Newly hatched larvae were then placed on the plants to assess whether plants that were generally preferred supported higher offspring survival. They did not (Ng 1988). However, this apparently simple result, that discrimination is not adaptive, disguises the complexity that was shown in figure 6.8. as a preference–performance correlation. If we put the question in the form, are the plants that are most acceptable in a general sense also the most suitable in some general sense, we get a very misleading result.

The existence of preference–performance correlations carries yet another message for experimental design. Consider the case in which an experimenter tests butterflies to ascertain their general rank order of preference for plants in their habitat and then places larvae in the field to determine whether the rank order of plants in the preference hierarchy corresponds to the rank order of larval survival. Suppose that preference for plant X is rare, but that the few butterflies that prefer plant X produce larvae that are very well adapted to it and that survive better on it than any other larvae on any other plant. In the experiment, plant X will falsely seem unsuitable. Because preference for it is rare, that plant will receive almost exclusively larvae that are offspring of butterflies that prefer other plants.

The solution here might seem to be to place larvae only on plants that their mothers prefer. But mothers do not always do what they prefer; sometimes even the majority may fail to find their preferred hosts (Singer et al. 1989). And whether or not a mother succeeds in finding and using her preferred host may be related to her general health, and so to the health and performance of her offspring. Taken together, these three experiments reveal the existence of unsuspected axes of variation in the plant–insect interaction. Each new axis that is discovered exposes hitherto unsuspected assumptions in common experimental designs and renders correct designs, targeted to specific questions, more and more difficult to achieve.

A recent workshop in Finland incorporated a section titled "Beyond Checkerspots." If we try for a moment to think about plant–insect interactions beyond checkerspots, even beyond butterflies, we find that several recently exposed phenomena in plant–insect interactions make a plethora of difficulties for experimental design. It is now clear that touching a plant while measuring it, or even just standing next to it in the field, affects its subsequent resistance to herbivory and that these effects are species specific. Insect eggs induce plant resistance even before they hatch (Agrawal 2000), and at the same time they are imbibing plant odors through their micropyles and inducing their own digestive enzymes.

These effects create considerable difficulties for the design of experiments that manipulate plants, and we should not allow butterflies and their hosts to be left out in the cold by this trend. The preceding section showed that checkerspots can play a unique role in revealing unexpected phenomena that expose unstated (and unthought of) assumptions in common experimental designs. We may appear to be rendering experimental design more complex, difficult, and obscure, and, in fact, we are. However, these phenomena are real and need to be taken into account. Who knows how many more of them lie in wait for us? There is still work to do, for those with the optimism to undertake it.

6.13 Conclusions

Development of techniques for testing insect oviposition preference should be informed by the specific limitations and behavioral repertoires of the species studied. The sequential choice test was developed for checkerspots in the light of the paucity of oviposition events and the ease with which the butterflies can be manipulated. The problem was overcome by using as data the results of encounters between insects and plants, rather than the number of egg clusters laid. The resulting data are, at least at present, more amenable than other types of data to incorporation into models that predict distribution of insect eggs across landscapes where hosts of different quality vary in both distribution and abundance.

Preference can be measured as a continuous variable, so correlations between preference and traits such as fecundity and offspring performance can be used to test theories about the evolution of insect diet and diet breadth. When insects discriminate simultaneously both within and among host species, as checkerspots clearly do, the application of oviposition preference tests reveals hitherto unsuspected axes of variation in the plant–insect interaction. The consequences for experimental design of the existence of these variables are awful to contemplate.

Despite the procedural problems we have encountered, oviposition preference testing of checkerspot butterflies has resulted in improved sophistication in our understanding of the behavioral mechanisms that generate patterns of association between plants and their herbivorous insect parasites. Our general conclusion does not support the views of researchers who have considered preference to be a simple variable severely limited in evolutionary dimensionality and tied to evolution of life-history traits such as fecundity. Oviposition preference of checkerspots is a multidimensional trait capable of complex responses to natural selection on diet. It is both an important driver of the evolution of host affiliation and an important constraint upon it. It helps to drive host shifts when insect populations respond to natural selection on diet by evolution of preference. It constrains evolution when maladaptive discrimination among individual plants of a highly suitable host species obstructs evolutionary incorporation of that species into the diet of an insect population (section 6.11).

Jaenike, in a 1990 review, discussed the preference–performance correlation of Rabbit Meadow *Euphydryas editha* shown in figure 6.8 (Ng 1998). Jaenike suggested that the insects varied in their state of health or nutrition and that healthy insects would have faster maturing eggs and produce healthier offspring. The faster egg maturation should increase egg load and thereby reduce discrimination among hosts. The increased health of offspring might have disproportionate effects on performance on the different host categories. Therefore, the preference–performance correlation would result from direct effects of variable nutrition on oviposition preference and on offspring survival. I hope that section 6.9 convinces readers that the rate of egg maturation does not affect oviposition preference in *E. editha* and that Jaenike's explanation is unlikely. However, I have a more general point to make. Jaenike's argument stems from a tendency to view variation of preference as a mechanistic consequence of variation in other traits of the organism. One of the main messages of this chapter is that effects of other traits on preference are less frequent than generally expected. For example, the association of preference with migration among patches (C. D. Thomas and Singer 1987) was thought by a series of anonymous reviewers to reflect an effect of migration on preference rather than vice versa: host encounter followed by effects of encounter on preference. Now that we have more evidence, we can argue with conviction (Hanski and Singer 2001; chapter 9) that these associations really do stem from effects of preference on migration, not vice versa. In general, when thinking about the underlying causes of a correlation between preference and trait X we should give equal weight to the possibility that preference variation drives trait X as to the possibility that trait X is driving preference. In checkerspots preference seems to be a complex, potentially independent trait whose evolution is not detectably constrained by selection acting on life history or other traits. Preference is a trait in its own right and should be studied as such.

7

Larval Biology of Checkerspots

MIKKO KUUSSAARI, SASKYA VAN NOUHUYS,
JESSICA J. HELLMANN, AND MICHAEL C. SINGER

7.1 Introduction

Butterflies have achieved prominent status in population and metapopulation biology. Accordingly, butterfly biologists should distinguish between events that influence internal dynamics of populations and those that influence interactions among populations. Interactions among populations are driven principally by decisions made by adult insects and by the consequences of those decisions, though highly mobile parasitoids can also be important (van Nouhuys and Hanski 2002b; chapter 8). Within-population dynamics could in theory be driven principally by variation in fecundity of adults or in mortality of any of the four life-history stages: eggs, larvae, pupae, and adults. In practice, as we show here, patterns of larval mortality are often of paramount importance in population dynamics. Negative effects on larvae are often related to spatial and temporal variation in the quality and availability of host plants. The numbers and proportions of individuals attacked by generalist and specialist predators and parasitoids vary greatly. When survival is exceptionally high, butterfly populations may exhibit explosive growth, whereas high larval mortality can lead to population extinctions. To understand the dynamics of a butterfly population, one needs to know the causes of variation in larval survival (Singer 1972, Dempster 1983; chapters 3, 4, and 8).

Because of their limited mobility, small butterfly larvae need to find the right host plant in the appropriate environmental conditions near the spot where they hatch from the egg. As the larvae grow, their capacity to move increases, but they are still limited to host plants in the area where they hatched. The presence of the larval host plant alone is not enough. For successful larval development, the host plant needs to grow under appropriate environmental conditions, which often means, in the case of checkerspots, a warm, dry microclimate. To successfully conserve a butterfly population, one needs to know which factors facilitate larval growth and survival (Singer 1972, J. A. Thomas 1984, 1991, 1995a, Ehrlich and Murphy 1987a, New et al. 1995).

As an introduction to checkerspot larval biology, we first describe three particular features of checkerspot caterpillars: group living, obligatory diapause, and unpalatability and aposematic coloration. We then present an overview of larval development in our two focal species, *Euphydryas editha* and *Melitaea cinxia*. In the rest of the chapter we focus on the following aspects of the checkerspot larval biology, emphasizing the comparison between *E. editha* and *M. cinxia*: geographic patterns in host plant use, variation in larval performance among and within host plant species, chemical defence of host plants, sources of larval mortality, the advantages of gregarious larval behavior and the role of group size in larval behavior and survival. Finally, we describe spatial and temporal variation in larval survival in *E. editha* and *M. cinxia* and discuss the consequences of such variation for population dynamics.

7.2 Three Particulars in Checkerspot Larval Biology

Egg Clusters and Gregarious Larval Behavior

Checkerspots lay their eggs in clusters (chapter 5; table 11.1), which make them somewhat exceptional among butterflies, as 90–95% of lepidopteran species lay their eggs singly (Stamp 1980, Hebert 1983). There is evidence that species that lay their eggs in clusters have greater population fluctuations and more outbreaks than species that lay their eggs singly (Nothnagle and Schultz 1987). Tendency for great fluctuations in numbers may also lead to increased vulnerability to local extinction, as suggested by high extinction rates of *M. cinxia* in the Åland Islands (Hanski et al. 1995b, Hanski 1999b; chapter 4) and of *E. editha* across its range (Parmesan 1996).

How long larvae remain in groups varies much among species (chapter 11). All checkerspot species live gregariously at least during the first one or two larval instars (Wahlberg 2000b), but it is not rare for the larvae to remain gregarious for several instars. In some species, such as *M. cinxia* (J. A. Thomas and Simcox 1982, Hanski et al. 1995a) and *Euphydryas aurinia* (Porter 1981, Warren 1994, Lewis and Hurford 1997), the conspicuousness of larval groups allows relatively reliable censuses during the larval stage (chapter 4). Because larval groups move slowly and only short distances from one host plant to another, tracking larval survival is possible in the field. Gregarious larvae spin webs on the host plants on which they live, and these webs can be fairly easy to find, even when larvae are small and would otherwise be cryptic. In species such as *Euphydryas phaeton* (Stamp 1982a) and *M. cinxia* in Finland (Kuussaari 1998), larvae diapause as groups and tend to remain gregarious until the last molt before pupating. Low mobility, web-building, and conspicuousness of larval groups also facilitate the study of some of the parasitoids that attack larvae and are important agents of mortality in many checkerspot populations (Ford and Ford 1930, Porter 1981, Lei and Hanski 1997; chapter 8).

Larval Diapause

A second key feature of checkerspot larval biology outside the tropics is the ability of larvae to diapause through extreme heat and cold and to facultatively reenter diapause several times, thereby extending their life spans to two or more years when conditions are adverse (Bowers 1978, Singer and Ehrlich 1979). In hot, dry climates, larvae typically enter diapause in late spring, remain in diapause during summer and early winter, break diapause some time between mid-winter and early spring, and then rapidly produce a single generation of adults. Larvae that break diapause in unusually adverse conditions (in years of drought or low host density) can reenter diapause after a small amount of feeding; they then attempt to complete their development in the following year. The nature of diapause is variable among populations as well as among species. For example, diapause is obligate in laboratory rearings of *M. cinxia* originating from low elevation sites in Finland and Andalucia (Spain) and from high elevation sites (1800 and 2000 m) in the French Alps. These sites close to the species' latitudinal and elevational limits have a single butterfly generation per year. In contrast, *M. cinxia* populations in the center of the species' range (low elevations in southern France, including the French Alps) have a variable number of generations per year and show facultative diapause in laboratory rearings (M. Singer, pers. obs.).

Unpalatability and Aposematic Coloration

Checkerspot larvae tend to be brightly colored (e.g., black with bright white, yellow, or orange stripes or spots; plate XI), which makes them highly conspicuous to visually searching predators. Birds and some invertebrate predators tend to avoid attacking the larvae. The fact that the majority of checkerspot butterflies use only host plants containing iridoid glycosides (Wahlberg 2001b; chapter 11) suggests that checkerspots use these compounds as a defense against predators. Detailed studies have shown that checkerspot larvae sequester iridoid glycosides and related iridoids produced by their host plants and use them for their own defense (Bowers 1988, 1991, Camara 1997b, Suomi et al. 2001). In both laboratory (Dyer and Bowers 1996, Theodoratus and Bowers 1999) and field experiments (Camara

1997b), generalist arthropods (insects and spiders) have been shown to be deterred by sequestered iridoid glycosides. Unpalatability of checkerspot larvae for birds has been shown using cage experiments. There is, however, much variation in the extent of unpalatability both among the species and populations using different host plants (Bowers 1980, 1990, 1991).

7.3 Overview of Larval Development

Euphydryas editha in California and *Melitaea cinxia* in the Åland Islands represent typical checkerspot butterflies, but they differ from each other in the details of their larval biology. The two species are contrasted in table 7.1.

Euphydryas editha

Euphydryas editha eggs are typically laid in batches of about 40 eggs, but the range is a few eggs to a couple hundred (Labine 1968, Singer et al. 1994, C. Boggs and M. Singer, unpubl. data). The eggs hatch synchronously after about two weeks, and the larvae often spin a web immediately upon hatching. The larvae live in the web for only several days in some populations of *E. editha bayensis*, but up to the entire prediapause development in many other populations (Moore 1989a, Hellmann 2002c).

Checkerspot larvae living in habitats where host plants tend to be ephemeral must be able to move from their natal plant to individuals of the same or other host species. For example, in *Euphydryas editha bayensis*, larvae search for food fairly widely if host senescence occurs. In this ecotype, prediapause development takes approximately three to five weeks and varies as a function of the abiotic conditions and host plant quality. Larvae diapause singly without a web on the ground, in plant litter, or under rocks. In this and other coastal ecotypes of *E. editha*, diapausing larvae may become aggregated simply because they share favorable overwintering sites. There can be important variation in body size among individuals at diapause because larvae vary in the timing and stage at which they enter diapause (third versus fourth instar; Singer 1971a).

Larval diapause is broken by winter rains on the coast and by snowmelt at higher elevations. Larvae feed for several weeks before finding pupation sites on the ground, under rocks or litter, or in pine cones, at the sixth or later instar (Singer 1971a, Singer et al. 1994). Postdiapause larvae forage singly and can move greater distances than prediapause larvae. The pace of growth and development of postdiapause larvae is strongly influenced by microclimate (figure 7.1) because larvae behaviorally thermoregulate by basking. To raise their body temperature to levels for optimal growth (30–35°C; Porter 1982; chapter 3), postdiapause larvae move among microclimates and cross slopes of different topographic exposure (Weiss et al. 1987, 1993). Working on *E. editha bayensis*, Weiss et al. (1987) found that postdiapause larvae disperse from a release site as much as 10 m per day, presumably in search of suitable foraging conditions. Dispersal involves a short-term trade-off against growth, but is compensated by the benefits of locating an area where body

Table 7.1. Comparison of larval life history in *Euphydryas editha* and *Melitaea cinxia*.

Trait	*Euphydryas editha*	*Melitaea cinxia*
Egg batch size	5–90	150–200
Web building and gregarious behavior	During first to second instar	From first until penultimate instar
Time of larval dispersion	In first and second instar	In last instar
Diapause	Solitarily in third or fourth instar	As a group in fourth or fifth instar
Flexibility of diapause	Reentering diapause possible	Reentering diapause not possible
Winter nest	No	Yes
Basking	Solitary basking in spring	Group basking in spring
Aposematism	Yes	Yes
Unpalatability	Yes	Probably

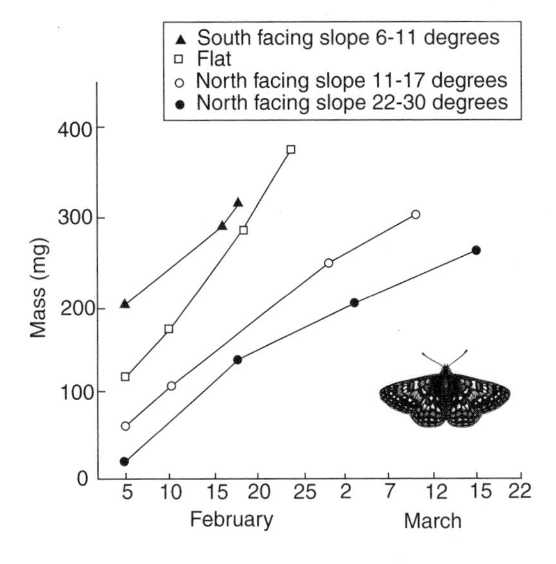

Figure 7.1. Mean postdiapause *Euphydryas editha* larval mass in the field on different slope exposures (Weiss et al. 1987).

temperature can be elevated to the optimal range for foraging and development.

Melitaea cinxia

In the Åland Islands, *M. cinxia* flies in June and lays large batches of eggs (typically 150–200) underneath the leaves of its host plants, *Plantago lanceolata* and *Veronica spicata* (Kuussaari et al. 1995, Kuussaari 1998; Plate IX). Larvae hatch after two to four weeks depending on the temperature. As soon as they have hatched in July, larvae spin a communal web on the host plant. Because of their restricted mobility, small larvae depend on the plant on which their mother laid the eggs. They feed gregariously in their web and grow slowly during the rest of the summer. The larvae prepare for diapause by spinning a dense winter nest (figure 2.5, plate IX), within which they molt for the last time before diapause and remain as a compact group over the winter. At this molt the color of the larvae changes from pale brown to black, and the head capsule becomes bright orange, giving the larvae the aposematic appearance that they will have after diapause and the deep black color that will assist in thermoregulation in the weak northern spring sunshine (Plates IX and XI). During early September, the larval groups are easy to find because the newly constructed winter nests are bright white and conspicuous on the green host plants.

The larvae become active during sunny days as soon as the snow has melted in late March or early April. The black larvae remain in tight aggregates and spend much time basking in the sun and spinning new webs as they move from one plant to another. The movements of the larval groups are often easy to follow based on the webs and defoliated host plants that they leave behind. During cold and cloudy weather the larvae tend to remain within the webs, where they are able to maintain a higher body temperature than outside (S. van Nouhuys, unpubl. data). The larvae also often stay in the webs overnight. During sunny weather the larvae usually feed and bask just outside the web. Postdiapause larvae molt into the penultimate instar inside a web, often synchronously as large clusters. As development proceeds, less time is spent basking and spinning webs, and the larvae tend to split into smaller subgroups (see section 7.6). In the last instar, larvae are much more mobile, and they tend to bask singly or in groups of only two or three larvae. Depending on the abundance of host plants, they may remain gregarious within an area of a few square meters, or they may become solitary and disperse long distances of >10 m. Larvae pupate within the vegetation close to the ground in early May.

7.4 Larval Host Plant Use and Performance

Checkerspot butterfly eggs are laid on plants belonging to 16 different families distributed throughout the Palaearctic, Nearctic, and Neotropics (chapter 2). All but two of the families are in the single subclass Asteridae, and members of all but two of these families produce iridoid glycosides as plant secondary compounds (Jensen et al. 1975, Higgins 1981, Tolman and Lewington 1997, Olmstead et al. 2000, Wahlberg 2001b). A checkerspot species may feed on many plant species in several families throughout its range, but individual populations are typically restricted to only one or to a couple of host genera, or to a single host species (Wahlberg 2001b; chapter 6). The plant on which an egg batch is laid depends on the phylogenetic history of the butterfly species, the evolutionary history of the particular population, the preference of the ovipositing female, and the array of available host plants (Singer 1984, Singer et al. 1992a; chapters 6 and 9). Once the eggs hatch, survival depends on the suitability of the host for larval development, environmental conditions (such as drought), natural enemies, and the tendency of the larvae to move among plants. Many of these factors can and usually do differ among host plant species, as well as among individuals of the same plant species.

While checkerspot larvae are all constrained somewhat by their mother's oviposition choice, the range of host plant use by larvae is usually less restricted than adult host plant range for oviposition. There is, however, great variation in the ability of larvae to move among plants (Warren 1987a, Tolman and Lewington 1997, Kuussaari 1998, Hellmann 2002c); therefore some species are more dependent on the host plant choice of ovipositing females than others. *Euphydryas editha* is an example of a checkerspot that can adjust to differences in host phenology by moving between host plant species within a generation. *Euphydryas phaeton* lays eggs on *Chelone glabra*, but postdiapause larvae are mobile and feed on a wide range of plant species (Bowers 1980). *Melitaea cinxia* larvae, on the other hand, move relatively little, and though they may be able to eat several plant species, they are restricted by adult oviposition preference. We use *E. editha* and *M. cinxia* to illustrate the pattern of host use within and among host plant species and then discuss the fitness consequences of the plants on which they feed. Because one of the characteristics that makes checkerspots a distinct group is their use of host plants that produce iridoids, we also summarize what is known about the ecological consequences of feeding on iridoid-producing plants.

Pattern of Host Plant Use in Euphydryas editha

Adults of *Euphydryas editha* oviposit on the leaves and flowers of hosts in nine genera: *Antirrhinum*, *Collinsia*, *Cordylanthus*, *Castilleja*, *Mimulus*, *Pedicularis*, *Penstemon*, *Plantago*, and *Veronica* (White and Singer 1974, Ehrlich and Murphy 1981b, Radtkey and Singer 1995, G. Pratt pers. comm.). Females often select just a subset of the plants available locally on which oviposition and larval development is possible. Populations with the same potential host resources often differ in the frequency with which they choose those resources and the order of their preference for them (chapter 6). For example, two populations on the eastern slopes of the Sierra Nevada, separated by only 150 km, are found in habitats where both *Collinsia parviflora* and *Penstemon rydbergii* are present (Singer and Parmesan 1993, Singer 1994). One population laid predominantly on *P. rydbergii*, while the other used *C. parviflora*, though in both habitats *C. parviflora* was the more abundant host. This divergence is explained both by genetic differences in female preference among the populations and genetic differences in the resistance of plants to oviposition among the sites (i.e., differences in plant chemistry as detected by ovipositing females).

If there is variation among populations in host use and host preference, an obvious question is whether females make the best choices for their offspring. Even in cases where larvae disperse from the natal host, we can ask whether females choose a suitable foraging arena for their young. To answer these questions requires knowledge of both oviposition choice and larval performance. We have explained above that oviposition (and oviposition choice) varies among populations; it turns out that *E. editha* larvae also vary in their ability to survive on the same host species across populations (Rausher 1982). In other words, knowing whether females make the right choices (i.e., whether they are adapted to their environment) is a local problem. Singer et al. (1994) studied the concordance of oviposition choice and larval performance in a series of *E. editha* populations, where larvae were assumed not to disperse from the natal host within 10 days of hatching. In eight populations where diet had been constant over a period of several years, individual females shared the same preference ranking of potential host plant species present in the habitat. The preference ranking was also adaptive. The rank order of plant species in the preference hierarchy of the adults, therefore, was the same as the rank order of those plants in their ability to support survival of experimentally placed neonate larvae. In contrast, two populations in which diet was observed to be undergoing rapid change (after anthropogenic disturbance) showed variation among females in the rank order of host preference, and the plant species preferred by most females was not the plant species that supported highest larval survival. This situation was temporary during the 1980s, and in both of these rapidly evolving populations, preference ranks were adaptive by the early 1990s. In one case this was achieved by rapid evolution of preference, such that the less preferred host became the more preferred (Singer et al. 1993). In the second case, the direction of natural selection on diet was reversed, and the host that had been less preferred but was more suitable became less suitable and remained less preferred (Singer et al. 1994, Singer and Thomas 1996). These results suggest that female *E. editha* tend to choose well for their larvae except in circumstances where anthropogenic influences have dramatically changed the relative suitabilities or availabilities of potential host plant species.

Depending on the degree to which larvae disperse from the natal host plant, the relationship between female oviposition choice and larval diet can be tightly or loosely coupled. In populations where larval groups stay together for most or all of the growing season and disperse little, larval diet and female oviposition choice are identical. As discussed above, however, larvae in some populations, particularly in coastal areas in California where plants are ephemeral, leave their natal host plant and explore neighboring areas for alternative plant resources. In these cases, larval diet and female oviposition choice can be distinctly different. *Euphydryas editha bayensis*, for example, often use multiple host species during larval development (Hellmann 2002c). Under conditions that accelerate host senescence, larvae survive best on *Castilleja* because they remain edible longer than the more common host species, *Plantago erecta* (Singer 1972, Hellmann 2002c; chapter 3). By moving, larvae can achieve the survivorship benefit regardless of the plant their mother chose as long as *Castilleja* are available in nearby areas (Hellmann 2002c). However, the likelihood that a larva will encounter a host species other than the one on which it was laid ultimately is determined by the neigborhood (not the individual plant) in which the female lays her eggs.

Pattern of Host Plant Use in Melitaea cinxia

Melitaea cinxia lays eggs on plants in the genera *Plantago* and *Veronica*, both of which are currently considered to be in the family Plantaginaceae (Judd et al. 1999, Olmstead et al. 2001). *Melitaea cinxia* has been recorded from six host species, with the most widely used one being *Plantago lanceolata* (table 7.2). *Melitaea cinxia* larvae are sporadically found feeding on related plant species (Tolman and Lewington 1997, Kuussaari 1998, Wahlberg 2001b), and, where host plants are scarce, postdiapause larvae have been occasionally observed feeding on unrelated species such as *Trifolium repens* and *Lotus corniculatus*. It is unlikely, however, that they could develop on these plants (Kuussaari et al. 1995, Kuussaari 1998).

Some populations appear to use all host species available for oviposition, while other populations exclude what appear to be suitable and abundant host plants, or at least do not use all available species in proportion to their abundances (Kuussaari et al. 2000). For example, on the island of Saaremaa in western Estonia, *M. cinxia* almost exclusively use *Veronica spicata*, even though *P. lanceolata* is equally or even more abundant. In Åland, *M. cinxia* uses *P. lanceolata* in the east where it alone is available, uses both *V. spicata* and *P. lanceolata* in the central parts of the main Åland Island, where both plants are present, but uses *V. spicata* with disproportionally greater frequency in the west where both plant species are present (Kuussaari et al. 2000; figure 7.2).

Table 7.2. Host plants used by *Melitaea cinxia* in Europe.

Host Plant	Predominance of Use[a]	Reference[b]
Plantago alpina	Primary	4
Plantago lanceolata	Primary and secondary	1, 6, 7
P. major	Occasional	1, 3
P. maritima	Occasional and primary	1, 3
P. media	Occasional	1, 3
Veronica[c] *incana*	Primary	2
V. longifolia	Occasional	1
V. serpyllifolia	Occasional	1
V. spicata	Primary and secondary	1
V. teucrium	Primary	5, 6
V. officinalis	Occasional	8
V. chamaedrys	Occasional	8
Centaurea sp.	Unknown	6

[a]The relative use of a particular plant where it is known.
[b]References: 1, Kuussaari (1998); 2, Wahlberg et al. (2001); 3, S. van Nouhuys (pers. obs.); 4, M. C. Singer (pers. obs.); 5, Weidemann (1988); 6, Tolman and Lewington (1997); 7, Higgins and Riley (1983); 8, M. Kuussaari and M. Nieminen (pers. obs.).
[c]The genus *Veronica* has been in the family Scrophulariaceae, but now it is considered Plantaginaceae (Olmstead et al. 2001).

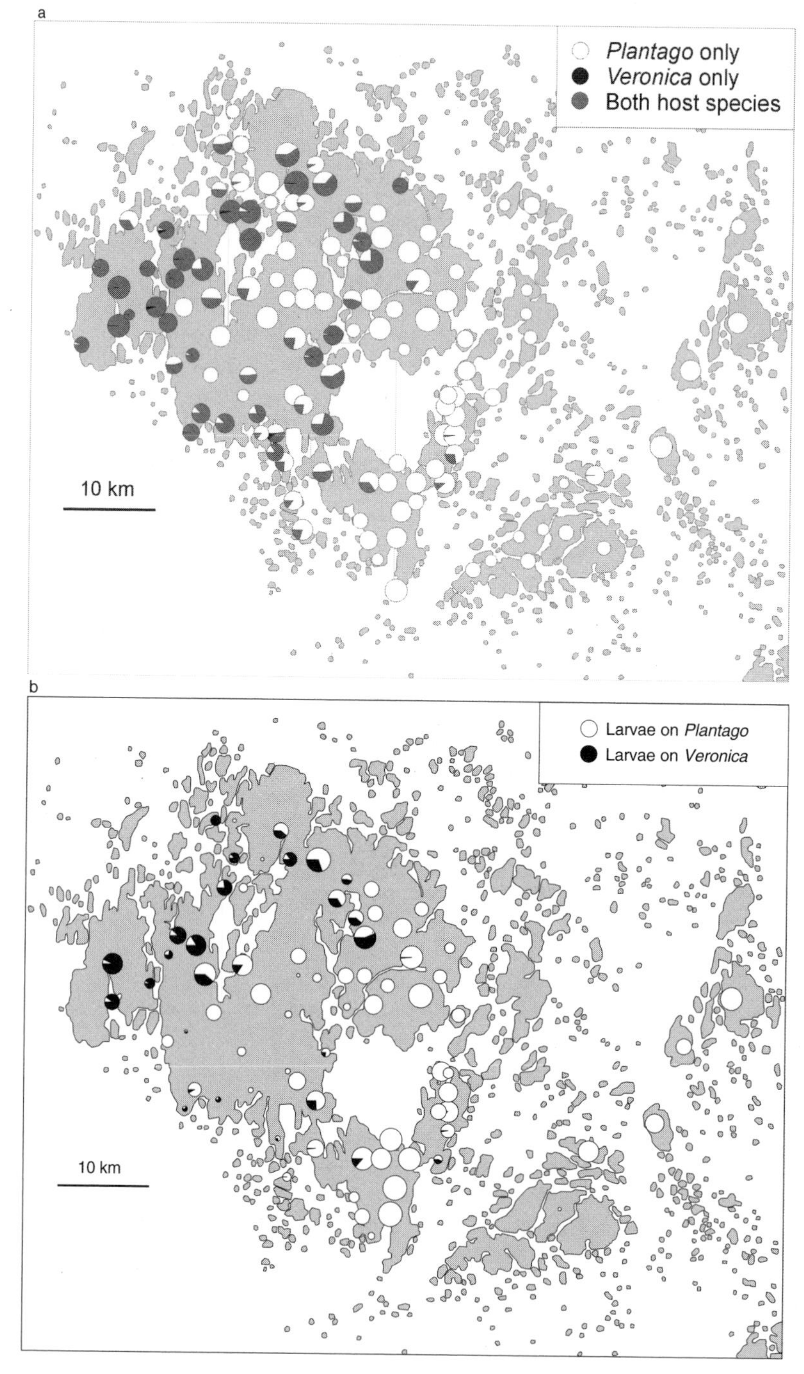

Figure 7.2. Map of *Melitaea cinxia* host plant use in the Åland Islands (modified from Kuussaari et al. 2000) (a) occurrence of the two host plant species, (b) observations of larval occurrence, (c) proportionality of larval host use.

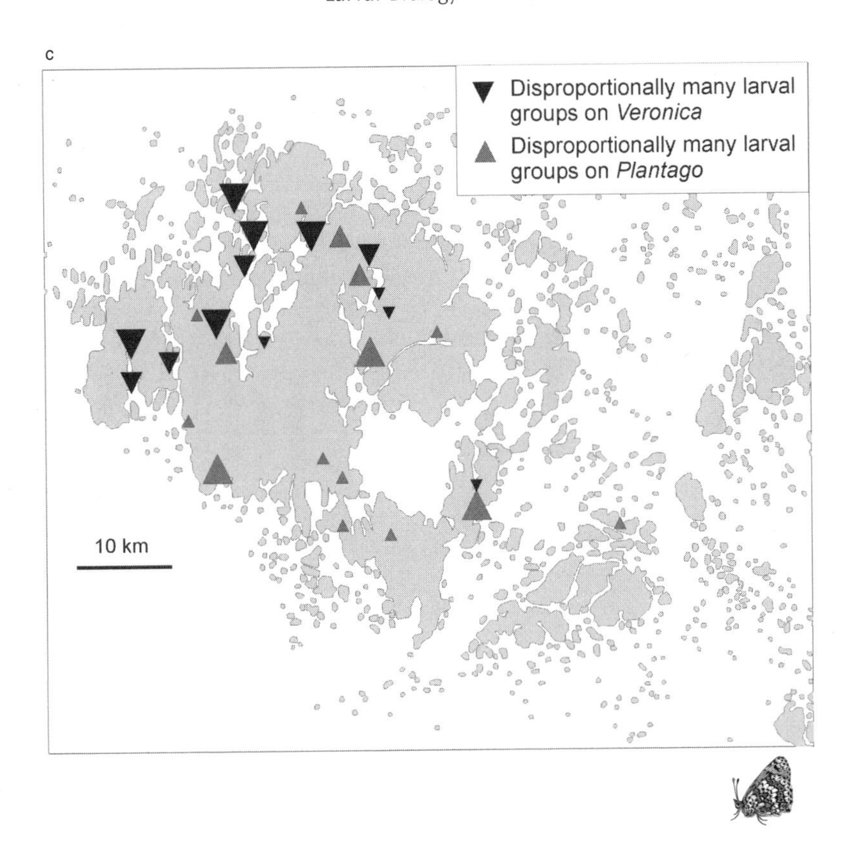

It is clear that variation in host plant use in Åland is a consequence of both plant distribution and of variation in the oviposition preference of adult butterflies (Kuussaari et al. 2000, Hanski and Singer 2001; chapter 6). In contrast to *E. editha* (Ng 1988, Singer et al. 1988), there is no straightforward link between adult oviposition preference and the performance of *M. cinxia* larvae in the Åland Islands. There is a genetically determined southeast to northwest gradient in host plant preference in adult butterflies (Kuussaari et al. 2000; chapter 6), but there appears to be no corresponding variation in host plant suitability for larval development or local adaptation by larvae (van Nouhuys et al. 2003).

Records of survey data covering all of the populations in Åland over six years (1994 to 1999) show that in some years larval survival (number of larvae per group surviving until the spring) when feeding on *P. lanceolata* is high, whereas in other years survival on *V. spicata* is high. In spite of the large sample size (about 300 populations in each year), survival on the two host plants did not show an overall statistical difference (van Nouhuys et al. 2003; figure 7.3). This is the case even though parasitism by *Cotesia melitaearum* is more common among *M. cinxia* feeding on *V. spicata* than on *P. lanceolata* (van Nouhuys and Hanski 1999; chapter 8). In laboratory experiments the performance of larvae appears to depend on how the test plants are chosen, but there is some evidence that larvae grow larger and faster on *V. spicata*. In one laboratory experiment, van Nouhuys et al. (2003) compared the performance of larvae from two contrasting habitat patch types. Both patches contained a high density of *V. spicata* and *P. lanceolata*, but one was in an area where adult butterflies prefer *V. spicata* (site ID 21), and the other one was in an area where butterflies preferred *P. lanceolata* (site ID 1075). The progeny of butterflies from each population were compared on randomly selected suitable-looking host plants of each species from both populations. Larvae grew larger (ANOVA, $P = .001$) and groups were more likely to survive (logistic regression, $P = .02$) on *V. spicata* than on *P. lanceolata*, regardless of their origin. Larvae feeding on *P. lanceolata* from the habitat patch in which females laid eggs on *V. spicata* performed as well as larvae feeding on *P. lanceolata* from the

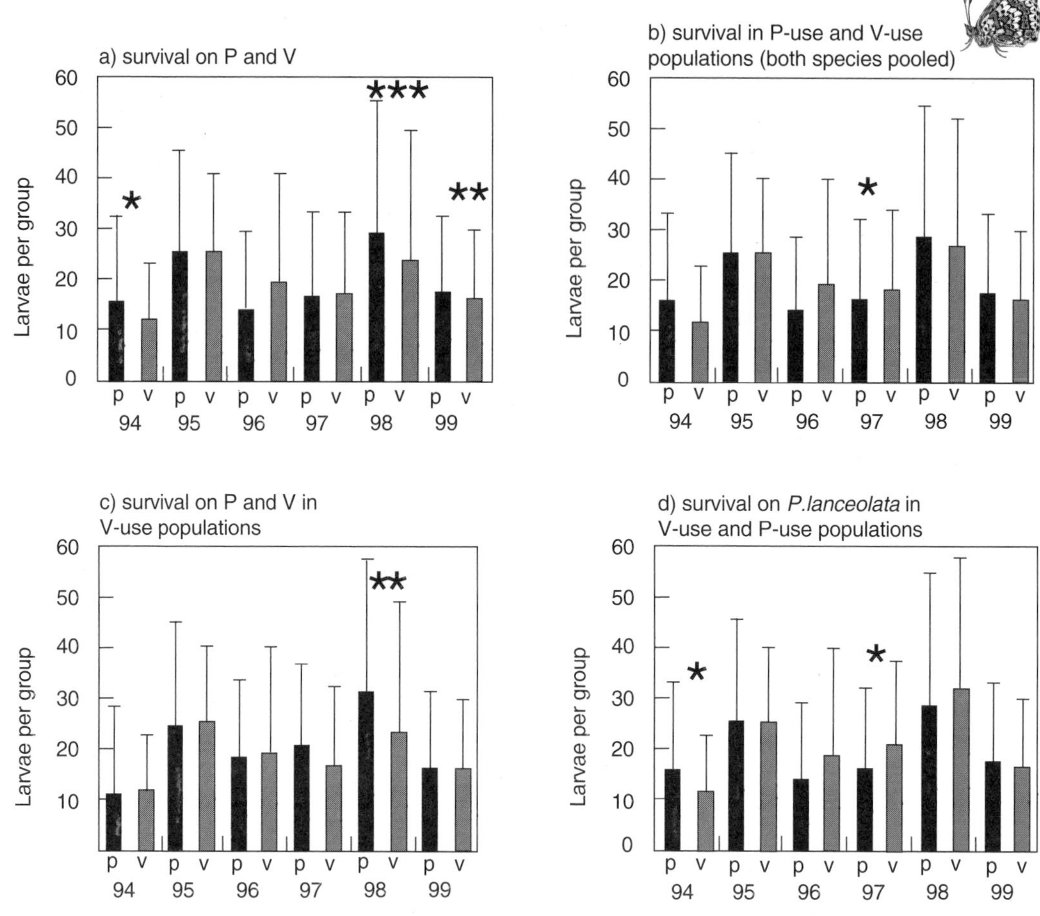

Figure 7.3. Annual variation in larval survival on *Plantago* (P) and *Veronica* (V) in the Åland Islands (van Nouhuys et al. 2003). The mean number (±1 SD) of *Melitaea cinxia* larvae surviving per nest from spring 1994 to spring 1999 (a) on each plant species (n = 1923); (b) in each host habitat patch type (both species pooled) (n = 1923); (c) on each plant in mixed-use habitat patches (n = 578); and (d) on *P. lanceolata* in *Plantago*-use and mixed-use habitat patches (n = 1545). Within-year differences between groups were tested as planned contrasts using analysis of variance. $^{*}p < .05$, $^{**}p < .01$, $^{***}p < .0001$.

habitat patch in which it was used by ovipositing butterflies. These results (and similar corroborating experiments) indicate that the suitability of individual host plants does not vary spatially, nor are larvae locally adapted to the locally preferred host plant.

Melitaea cinxia larvae move among plants less than do *E. editha* larvae (see section 7.6). In Åland, most larvae only have access to a single plant species both because *V. spicata* is available in a small fraction of the habitat patches and because the plants grow in single-species clumps. But, in the uncommon mixed-habitat patches, larval groups may benefit by successfully moving to the alternate host species as *E. editha* do. This is because, although there is no great difference in host phenology and drought tolerance between the two host plant species, one or the other may senesce earlier, depending on spring and summer rain. For example, in 1995 larvae were placed on both host plants in one habitat patch before summer drought. Only those on *V. spicata* and those that were able to move to *V. spicata* survived (Kuussaari 1998).

Checkerspot Larvae and the Chemical Defense of Their Host Plants

Checkerspot butterflies are almost exclusively associated with plants that produce iridoids, mostly iridoid glycosides and seco-iridoids, which are noxious and thought to be produced by plants as defense against herbivory (Kooiman 1972, Bowers 1983b, 1988, Jensen 1991, Seigler 1998, Wahlberg 2001b). Iridoid glycosides deter many generalist herbivores (Bernays and DeLuca 1981, Puttick and Bowers 1988). Generalist herbivores that do feed on iridoid-producing plants or on an artificial diet containing iridoids do not often sequester the compounds (Bowers and Puttick 1986, Bernays 1988, Bernays and Cornelius 1988) but develop slowly or just to a small size (Bowers and Puttick 1988, Puttick and Bowers 1988; but see Stamp and Bowers 1994). For example, Puttick and Bowers (1988) found that survival and weight of the southern army worm, *Spodoptera eridania*, fed with an artificial diet without iridoid glycosides were greater than when their diet included the iridoid glycosides catalpol, loganin, or aucubin.

The situation is likely to be different in herbivores specialized in feeding on iridoid-producing plants, though there have been few experimental tests of the direct effects of iridoid glycosides on the larval growth in specialists. Camara (1997a) found no effect of iridoids on the weight of *Junonia coenia*. Harvey and van Nouhuys (unpubl. data) found that *Melitaea cinxia* larvae grew faster and in some cases became larger pupae when feeding on *P. lanceolata* with a high iridoid concentration than on *P. lanceolata* with a low iridoid glycoside concentration. Adult *M. cinxia* appear to oviposit more frequently on *P. lanceolata* containing higher than average concentrations of the iridoid glycoside aucubin in natural populations in Åland (Nieminen et al. 2003). However, the host plant *V. spicata*, which contains lower levels of the iridoids aucubin and catalpol than *P. lanceolata*, is used when available (Kuussaari et al. 2000) and is perhaps a more suitable host for larval development (van Nouhuys et al. 2003, M. Saastamoinen, unpubl. data). *Veronica spicata* does have other iridoids (Suomi et al., 2002) that could be converted to catapol and then sequestered.

For the most part, checkerspot larvae feeding on iridoid-producing plants are able to sequester iridoids and become distasteful or noxious themselves (Bowers 1980, 1981, 1990, Bowers and Puttick 1986, Stermitz et al. 1986, 1994, Franke et al. 1987, Gardner and Stermitz 1988, Belofsky et al. 1989, L'Empereur and Stermitz 1990a, Suomi et al. 2001). Additionally, iridoid glycosides function as oviposition stimulants for some specialist butterflies (Pereyra and Bowers 1988) and as feeding stimulants for at least some checkerspot butterflies (Bowers 1983b). It is generally thought that specialist herbivores benefit from the defensive chemistry of their hosts through reduced competition and inhibition of generalist natural enemies. However, some herbivores have primary natural enemies with narrow host ranges that are not hindered by the plant defensive chemicals sequestered by their hosts. It may even be that these specialist enemies avoid competition with generalist natural enemies, and perhaps direct predation, by using chemically defended hosts. Experimental studies on iridoid-producing plants, their herbivores, and the natural enemies of those herbivores, generally support these ideas (Bernays and DeLuca 1981, Bernays 1988, Puttick and Bowers 1988, Dyer 1995, Dyer and Bowers 1996, Camara 1997b, Theodoratus and Bowers 1999, Stamp 2001, Nishida 2002).

Most generalist predators appear to avoid larvae that ingest iridoids. Theodoratus and Bowers (1999) tested this by measuring the predation of iridoid-sequestering *Junonia coenia* larvae feeding on two different species of *Plantago* by lycosid spiders. They found in both laboratory and field experiments that the spiders ate more caterpillars feeding on *P. major*, which had a lower concentration of iridoid glycosides than on *P. lanceolata*. Similarly, *J. coenia* fed with diets of high iridoid glycoside concentration are rejected by at least three ant species (Dyer and Bowers 1996, Camara 1997b), predatory wasps, and stink bugs (Stamp 1992). Birds are similarly deterred by iridoid-containing larvae in cage experiments (Bowers 1980, 1991, Bowers and Farley 1990).

Specialist natural enemies of checkerspots cannot avoid iridoids, and, unlike generalist natural enemies, they must not be deterred by these compounds. The only known specialist enemies of checkerspot are internal larval parasitoids (chapter 8). These parasitoids must have mechanisms to detoxify plant defensive compounds. The effect of sequestered plant chemical defenses on parasitoids has been little studied. The plant secondary compound tomatine is detrimental to parasitoid wasps (Campbell and Duffy 1979), as is nicotine. Thorpe and Barbosa (1986) found that nicotine ingested by the moth larva *Manduca sexta* caused some mortality of immature *Cotesia* parasitoids, but it did not affect the overall size of surviving parasitoids. Barbosa et al. (1986) found that the negative effects of nicotine were greater for the generalist parasitoid *Hyposoter annulipes* than for the specialist *Cotesia congregata* (it is important to note that these two parasitoids use their host insect resources differently). In a laboratory study, Harvey and van Nouhuys (unpubl. data) found that the specialist parasitoid *Cotesia melitaearum* developed equally well in *M. cinxia* larvae feeding on high- and low-iridoid *P. lanceolata*. The parasitoid may be so well adapted to using a toxic host that there is no measurable cost, or the cost may be apparent only under some conditions (Nieminen et al. 2003; chapter 8).

If a specialist herbivore, which sequesters iridoids, and a generalist herbivore that does not concentrate iridoids, are presented to generalist predators, the specialist herbivore should be at an advantage (for reviews, see Bowers 1990, Camara 1997b). However, this pattern may not manifest if generalist herbivores have other means of defense. For example, Stamp and Bowers (1992) found that the specialist caterpillar *Junonia coenia* was more often killed by predatory stink bugs than a generalist caterpillar, *Pyrrharctia isabella*, while feeding on *P. lanceolata* because the generalist behaved more cryptically than the specialist.

The natural enemy community of checkerspots is a good model for the study of host specialization because so much is known about the ecology of the host butterflies and the host plant chemistry. Further comparative studies can be designed using host plant species and populations that vary naturally in iridoid concentrations and types and are used differently by herbivores (Kooiman 1972, Bowers and Puttick 1986, Bowers et al. 1992, Bowers and Stamp 1993, Camara 1997b, Darrow and Bowers 1999, Stamp and Bowers 2000b, Nieminen et al. 2003). Researchers can test for the effects of iridoids (or other potential chemical defenses) on insect communities by manipulating iridoid production through artificial selection in the laboratory (Marak et al. 2000) and by changing the iridoid content of artificial diet (Puttick and Bowers 1988, Lei and Camara 1999).

7.5 Larval Survival

Adult checkerspot butterflies produce 100–1500 eggs in their lifetime (chapter 5). The majority of the offspring perish before reaching adulthood, and the larval phase is often the riskiest stage for a holometabolous insect (Zalucki et al. 2002). Individuals that survive the foraging period typically have a high probability of surviving to reproduce. Because larval survival is often the key determinant of population size and distribution, understanding the sources and variation in larval mortality is essential to understanding and predicting butterfly population dynamics. We now turn to the factors that influence whether a larva survives to pupation. Because *E. editha* and *M. cinxia* feed on similar plant species in similar types of environments, they obviously share several mortality risks.

Sources of Larval Mortality

Larval mortality in checkerspots and in many other species generally stems from one of two sources: predation (including parasitism) and starvation (Haukioja 1993). Other factors such as larval desiccation, pathogens, cannibalism, and consumption by competitive herbivores also can cause larval death and are important in some lepidopteran populations (Gilbert and Singer 1975, Stamp and Casey 1993, Dwyer et al. 2000, Zalucki et al. 2002).

Predation and Parasitism

Potential larval predators include spiders, insect predators (e.g., stink bugs), parasitoids, and vertebrate predators (especially birds; see table 8.1). Vertebrate predation is often common during lepidopteran outbreaks (Crawford and Jennings 1989, Elkinton et al. 1996) and on species with normally high population density (Bowers et al. 1985, Alonso-Mejia et al. 1998), but it has not been frequently observed in checkerspots. Group feeding by larvae (see section 7.6) is an effective strategy against some predators, as is sequestration of defensive chemical compounds as discussed in section 7.4.

Laboratory experiments can assess whether a predator will consume a larva, and field monitoring of predator abundance indicates the potential for larval predation. But it is challenging to quantify larval mortality due to particular predators and parasitoids in the field. Luckily, parasitism is one source of mortality that is relatively easy to assess because some parasitoid species leave behind a signature of their presence. For this reason, and because parasitism is often heavy on checkerspots, the most thorough work on the predation of checkerspot larvae has focused on parasitism (chapter 8).

Natural enemies of *Euphydryas editha* larvae include spiders, predatory insects, and parasitoids (tables 8.1 and 8.2). The specialist hymenopteran parasitoids *Cotesia koebeli* and *Benjaminia fuscipenni* and the dipteran parasitoid *Siphosturmia melitaeae,* as well as several generalist parasitoids, attack *E. editha* (White 1973, 1986; table 8.3). However, there is no evidence that predation or parasitism explains a significant portion of larval mortality in this species (chapters 4 and 8). This may be because other forms of mortality, such as starvation, prevail, because defensive adaptations are relatively successful, or because predator populations are low where *E. editha* are found. Perhaps population fluctuations due to host plant availability make *E. editha* an unreliable host for specialist parasitoids.

True predation of *M. cinxia* larvae does occur, but it does not appear to account for a large part of larval mortality. Parasitism, in contrast, causes substantial larval death (tables 4.1, 8.1, and 8.3). Parasitism by the wasp *Cotesia melitaearum*, a checkerspot specialist, can be heavy in tightly clustered and high-density *M. cinxia* populations but not in regions where local populations are sparse and small. Thus *Cotesia melitaearum* is entirely lacking from a large fraction of *M. cinxia* populations (Lei and Hanski 1997, van Nouhuys and Hanski 2002a, 2002b). Parasitism by another specialist parasitoid, *Hyposoter horticola*, is less variable and typically causes about 30% mortality of postdiapause larvae (van Nouhuys and Hanski 2002b).

The prevalence of parasitism appears to be one difference between *E. editha* and *M. cinxia*. Further work is called for on parasitoid communities associated with *E. editha* to determine whether less-studied populations experience significant parasitism and under which conditions this might occur. For example, Moore (1989a) found up to 66% parasitism by the parasitoid *Cotesia koebeli* in one Sierra Nevada population of *E. editha*.

Starvation

A shortage of food can occur where host plant density is low or larval density is high or when host plants wither before caterpillars have finished development. All these factors appear to play a role in the mortality of checkerspot larvae, which feed on more or less ephemeral or patchily distributed host plants. The degree to which larvae are affected by food shortage can be mediated by larval mobility. Relatively immobile larvae are unable to locate new host individuals when their natal host plant becomes unsuitable or is consumed. Mobile larvae, in contrast, are able to leave an unsuitable host individual and locate another one (Dethier 1959; section 7.6). Whether larvae suffer higher rates of mortality while dispersing than if they had remained in place depends on the distribution of potential food plants and the vulnerability of larvae to predation en route.

Starvation is an important source of larval mortality in *E. editha* (Singer 1972, White and Singer 1974). Coastal populations of *E. editha* forage on host plants that undergo annual senescence before the onset of summer drought (chapter 3). In years when senescence occurs before most larvae have grown enough to enter summer diapause, larval mortality rates are high, in excess of 90% (Singer 1972, Hellmann 2002c). Hence, the relationship between timing of host senescence and egg hatching is important, and variation among years in this relationship leads to variation in the number of larvae that survive to diapause and ultimately to adulthood. This single factor explains a large amount of the population fluctuations in one well-studied population at Jasper Ridge, California, and similar considerations likely apply to other coastal populations (McLaughlin et al. 2002a; chapter 3). In montane populations of *E. editha*, larval mortality from starvation is also an important factor. In a metapopulation at Rabbit Meadow, 2350 m above sea level, there is often competition among larvae and subsequent starvation when large numbers of larvae defoliate an individual host (C. D. Thomas et al. 1996, Boughton 1999a). Frost events in montane areas also cause massive larval starvation when hosts are killed (Singer and Thomas 1996).

Because different plant species often senesce at different times, host use plays a role in determining rates of larval mortality in *E. editha*. In the San Francisco Bay Area, for example, larvae feed on two host species: one that senesces relatively early, *Plantago erecta*, and another one that senesces up to two weeks later, *Castilleja* (Weiss et al. 1988; chapter 3). The montane metapopulation at Rabbit Meadow also uses two hosts that differ in the same manner; *Collinsia torreyi* causes larval mortality due to early senescence, and *Pedicularis semibarbata* does not. Thus, larval mortality in any single habitat and year depends both on weather (because it influences host senescence) and on the fraction of larvae that forage on each of the two hosts (figure 7.4). In years when the longer lasting host persists long enough to sustain *E. e. bayensis* larvae while the other one does not, habitats with high abundance of *Castilleja* are likely to have greater larval survivorship than habitats where that host is sparse (Hellmann 2002c). Studies also suggest that population extinction rates vary with host plant use in *E. editha*; populations that forage on *Pedicularis* have a greater tendency to persist (M. Singer and C. Parmesan, unpubl. data).

Melitaea cinxia larvae also suffer from starvation. Large-scale starvation of prediapause larval groups is caused by occasional late summer droughts, which result in temporal withering or complete drying out of host plants over large areas in the Åland Islands (chapter 4). Although droughts affect entire landscapes rather than single habitat patches, typically there is variation in the severity of drought among different parts of Åland (figure

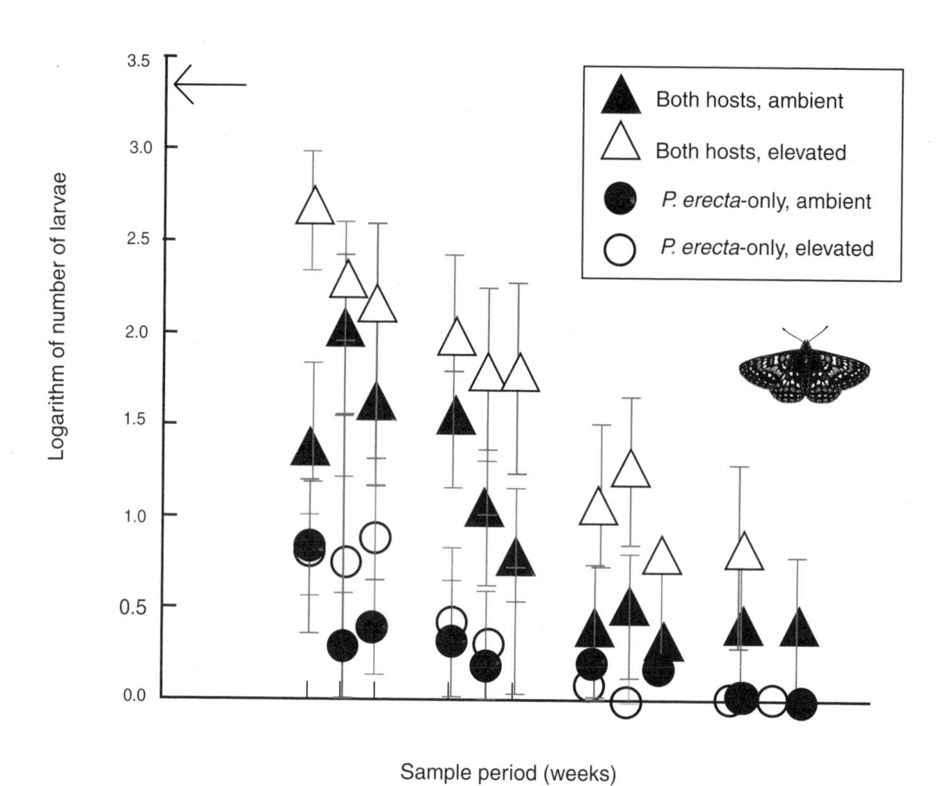

Figure 7.4. Host plant-dependent survival in *Euphydryas editha* (Hellmann 2002a). Prediapause larvae were grown on plots of plants containing either one (*Plantago erecta*) or two (*Plantago erecta* and *Castilleja exserta*) host plants under two temperature treatments in the greenhouse. Host plant availability treatments reflect distributional differences of the two hosts in the field. Temperature treatments represent different microclimatic conditions as would occur in nature across slopes of different exposure or across years as determined by weather. The number of larvae in the treatment with both host plants over time was significantly higher than in the treatment with only the common host (area under the abundance curve). Further, the survival of larvae is higher when both hosts are present under elevated temperatures than under ambient temperature. This result suggests that access to the less common but long-lasting host, *Castilleja*, confers a significant survival advantage to larvae foraging on senescing host plant material, and this advantage is greatest when temperature is high.

4.14). On one hand, drought may cause a cluster of local extinctions when all the larval groups in several habitat patches run out of food (Hanski 1999b). On the other hand, there is variation in habitat quality among the habitat patches, and during droughts larval survival may be relatively high in patches with soil that retains moisture well. Even within local populations, a fraction of larvae survive because microhabitat and soil type can vary at a small spatial scale.

Melitaea cinxia larvae can starve when egg batches are laid on plants growing in areas of low host-plant density. The larvae run out of food after defoliating the single or few host-plants in the close neighborhood and are not sufficiently mobile to find more food. In spring 1994, postdiapause larval survival was followed closely in all the larval groups that survived the previous winter in 20 local populations of *M. cinxia* in different parts of Åland. Nearly half (47%) of the larval groups were located in areas of such low host-plant density that shortage of food was likely to decrease larval growth (table 7.3; M. Kuussaari, unpubl. data). Clear signs of starvation (i.e., an entire larval group starved, some larvae starved or were found feeding on nonhost plants such as *Trifolium repens*) were observed in 38% of the larval groups in low host-plant density areas, whereas no signs of starvation were observed in areas of higher host plant density (table 7.3). Both postdiapause larval survival and growth rate were positively correlated with host plant density ($P < .01$).

Comparison with Other Life Stages

We have suggested that processes critical to the population ecology of checkerspot butterflies occur during the larval stage. In particular, variation in larval mortality due to weather, host plant use, predation, and parasitism causes fluctuations in population size. There are also important risks that limit survival of eggs, pupae, and adults and thus are important contrasts to the risks discussed above for larvae. Eggs can be preyed upon or fail to hatch. Eggs of *M. cinxia* are eaten by ants and by ladybird and lacewing larvae (table 8.1). The fraction of eggs consumed is a little-studied issue in both *E. editha* (Moore 1989a) and *M. cinxia* (chapter 4), but the fitness consequences of such predation may be high because entire larval groups are lost. Additionally, eggs can fail to hatch due to inbreeding depression (Haikola et al. 2001; chapter 10) or unsuitable abiotic conditions, such as heavy showers (chapter 4).

The extent of pupal mortality is even less well known, but it is clear that the sources and magnitude of this mortality factor vary greatly among species and populations. For example, the mortality of *E. editha* pupae placed in the field varied from 53% to 89% among habitat types and years (White 1986). Predation, cold weather, and parasitoids (table 8.3) are the likely causes of pupal mortality. A fraction (13%) of *M. cinxia* pupae placed in the field in one year was lost to parasitism (Lei et al. 1997).

Extreme weather events including frost and high winds have been known to kill adult *E. editha* butterflies, and even entire populations (Singer and Thomas 1996, C. D. Thomas et al. 1996, Hellmann 2002b). Females may be at particular risk of predation while ovipositing (Moore 1987); experimental manipulations in the San Francisco area showed that *E. editha* populations tend not to suffer heavy losses of adults (Ehrlich et al. 1975). In 1964, every captured butterfly was removed from the populations (area C: $n = 123$; area H: $n = 1353$), but abundance in 1965 was not obviously affected (Hellmann et al. 2003). There have been practically no observations of adult predation on *M. cinxia* apart from infrequent predation by dragonflies (I. Hanski pers. comm.), but extreme weather undoubtedly decreases adult longevity and limits reproductive success.

Table 7.3. Host plant density and the occurrence of postdiapause starvation in 20 local populations of *Melitaea cinxia* in the spring 1994 (M. Kuussaari, unpubl. data).

Host Plant Density[a]	No. of Larval Groups	Entire Group Starved	Some Larvae Starved	Feeding on Nonhost Plants	Some Signs of Starvation	%
High	14	0	0	0	0	0
Moderate	22	0	0	0	0	0
Low	32	5	3	4	12	39

[a]*Plantago lanceolata* density: low < 20, moderate = 20–35; high > 35 *Plantago* rosettes/m^2.

7.6 Role of Group Size in Larval Behavior and Survival

Advantages of Cluster Laying and Gregarious Behavior

We have already emphasized that laying eggs in clusters and gregarious larval behavior, at least during the first larval instar, are common traits among checkerspot butterflies (Stamp 1980, Wahlberg and Zimmermann 2000). Some advantages of these traits have been studied in checkerspots, including *Euphydryas aurinia* (Porter 1981, 1982), *E. editha* (M. Singer, unpubl. data), *E. phaeton* (Stamp 1981a, 1982a), *M. cinxia* (Lei and Hanski 1997, Kuussaari 1998, van Nouhuys and Hanski 1999), *Chlosyne lacinia* (Clark and Faeth 1997a, 1998), and *C. janais* (Denno and Benrey 1997). Increased larval growth rate and survival with increasing group size, especially during the first larval instars, have been demonstrated in several species of checkerspots (Clark and Faeth 1997a, Denno and Benrey 1997, Kuussaari 1998), as well as of other Lepidoptera (Lawrence 1990, Fitzgerald 1993).

Most hypotheses proposed to explain the evolution of egg clustering and gregarious larval behavior in Lepidoptera focus on the benefits of gregarious behavior for larval survival and growth rate (Clark and Faeth 1997a, Denno and Benrey 1997). These hypotheses are usually based on either increased foraging efficiency or enhanced defense against natural enemies with increasing group size, mechanisms that are not mutually exclusive. Other potential explanations for egg clustering include avoidance of egg desiccation (Stamp 1980, Clark and Faeth 1997a) and increased fecundity when females are time limited as opposed to egg limited (Courtney 1984, Parker and Courtney 1984). As noted by Bryant et al. (2000), there are so many potential causes and consequences of group living that it may be common for several different factors to affect the costs and benefits of gregariousness in a particular case.

The benefits of increasing group size vary during butterfly development from egg to pupae. Large clusters of eggs may avoid desiccation better than smaller clusters or single eggs, as was experimentally shown in the laboratory for *Chlosyne lacinia* (Clark and Faeth 1998). Decreasing group size increased egg desiccation also in a field experiment on *M. cinxia* (Kuussaari 1998). In the latter experiment eggs were placed under the leaves of host plants in small mesh baskets that excluded egg predators. Invertebrate predators like lacewing (Chrysopidae) and lady beetle (Coccinellidae) larvae have been observed to cause substantial egg mortality in *M. cinxia* populations locally, but it is not known whether the rate of egg predation is associated with egg cluster size. Parasitism and predation of *E. phaeton* eggs in one study was not affected by cluster size nor by the number of clusters per plant (Stamp 1981b). Similarly, parasitism of *M. cinxia* by *Hyposoter horticola*, which acts as an egg parasitoid because it lays eggs in host larvae that have not yet hatched, also appears to be unrelated to egg cluster size (van Nouhuys and Ehrnsten 2004).

Facilitation of feeding due to larval aggregation is likely to be most pronounced during the first larval instar, when the small larvae establish their first feeding site (Shiga 1976, Fitzgerald 1993, Clark and Faeth 1997a). At this stage, the physical plant defenses, such as trichomes, are most difficult to overcome (Young and Moffett 1979, Zalucki et al. 2002). Another critical ability likely to improve with increasing group size is building of the feeding web. Web building is best developed in species that have large or medium-sized egg clusters, whereas it can be nonexistent in species with small egg clusters. Constructing substantial webs may be advantageous when larval groups are large because the per capita energetic cost of building the web may be less for larvae in large than in small groups. This aspect of larval biology has not been studied.

Web building also varies significantly within species living under different conditions. For example, *M. cinxia* in low-elevation southern France and the Russian steppe appear to overwinter in small groups with little or no webbing (S. van Nouhuys and M. Singer, pers. obs.), while *M. cinxia* in Åland and the high Alps diapause as family groups in dense webs. Both weather and natural enemies are likely to be important selective forces. The defensive function of webs has not been well studied, but generalist predators are most likely hindered by a dense web, as are at least some parasitoids. Tachinid flies attempting to oviposit in *E. maturna* postdiapause larvae can only do so through holes in the web (Wahlberg 1998); however, specialist parasitoids of *M. cinxia* and *E. phaeton* readily move within the web to parasitize early instar larvae (Stamp 1982b, S. van Nouhuys pers. obs.). Webs actually stimulate the search behavior of *C. melitaearum*, which parasitizes *M. cinxia* (S. van Nouhuys unpubl. data).

Group living may have a substantial effect on growth rate, especially after winter diapause, when black larvae may bask in tight clusters and increase their body temperature close to developmental optimum during cold but sunny spring days (Porter 1982, Casey et al. 1988, Stamp and Bowers 1990, Casey 1993, Kuussaari 1998, Bryant et al. 2000). Rapid development is critical in populations in areas with a short growing season. Additionally, increased growth rate due to gregarious behavior may help the larvae escape specialist parasitoids, as the larvae may manage to pupate before their parasitoids emerge (Porter 1983). The gregarious basking of *M. cinxia* larvae during warm springs allows them to increase their development rate so that the majority of them pupate before the spring generation of the parasitoid *C. melitaearum* become adults. The parasitoid cocoons, which are white, immobile, and often in the shade, cannot control their own development rate. Consequently, during cool spring seasons the host and the parasitoid are developmentally better synchronized, and a larger fraction of larvae are available for parasitism, increasing parasitoid population size substantially (S. van Nouhuys and G. C. Lei unpubl. data).

Larval aggregations may avoid predation and parasitism more successfully than solitary larvae by various kinds of active defenses, such as head jerking (Stamp 1982b, 1984) and regurgitation (Stamp 1984, Peterson et al. 1987). Head-jerking behavior can knock attacking parasitoids off the larvae, an effect that is enhanced when many larvae jerk their heads simultaneously (Stamp 1982b).

Although active defenses may be enhanced in large groups, large groups may also attract more predators than small groups. In *M. cinxia*, parasitism by *Cotesia melitaearum* increases with increasing group size (Lei and Hanski 1997, van Nouhuys and Hanski 1999). Increased resource competition among larvae is another potential disadvantage of large group size. Gregarious larvae are more often distasteful and conspicuously colored than solitary larvae (Stamp 1980, Guilford 1988), which has led to the suggestion that gregarious larvae avoid predation, especially by birds, through aposematism (Bowers 1980, 1981, 1993, Stamp 1980). The strength of the warning signals of aposematic species may be amplified by group size (Bowers 1993, Fitzgerald 1993).

Because species putatively most closely related to the checkerspots also sometimes lay their eggs in clusters (Kallimini and Nymphalini: Harvey 1991; chapter 2), egg clustering may well be ancestral in checkerspots. The interesting question, therefore, is not why checkerspots lay their eggs in clusters, but rather why there are different degrees of larval gregariousness within the checkerspots. In the following sections we take a closer look at variation in group size and its effect on larval survival, as well as gregarious behavior and larval mobility, mostly based on data on larval behavior and development in the highly gregarious *M. cinxia* and in the less gregarious *E. editha* larvae.

Variation in Group Size

In *M. cinxia* in the Åland Islands the average egg cluster size is about 170 eggs, but it can vary from < 50 to > 350 eggs (figure 7.5). In the field, small egg batches may result from various disturbances to ovipositing females. For instance, an attack by ants (Wahlberg 1995; chapter 8) or other predators such as dragonflies or a sudden change in weather (M. Kuussaari pers. obs.) may cause the female to terminate oviposition. Especially large larval groups may result from fusion of groups when more than one egg batch has been laid on the same host plant. This happens even in relatively low-density populations of *M. cinxia* because some host plant individuals tend to be especially attractive to egg-laying females (Singer and Lee 2000) and because ovipositing butterflies are attracted to conspecific eggs (M. Singer and L. Ramakrishan unpubl. data). Multiple egg batches per plant is demonstrated by a survey of naturally occurring egg batches on 4295 randomly selected *P. lanceolata* plants in 5 *M. cinxia* populations. Of the 39 plants that had egg batches, 34 plants had 1, 4 plants had 2, and 1 plant had 3 batches of eggs. Thus, multiple egg batches on one plant were observed more often than predicted by chance (M. Kuussaari unpubl. data).

Nonrandom accumulation of egg batches on some host plant individuals is also well known in other checkerspot species, such as *E. aurinia* (Porter 1981) and *E. phaeton* (Stamp 1981a). In a Sierra Nevada population of *E. editha*, Moore (1989a) found up to 19 clusters of eggs on a single *Pedicularis semibarbata* individual. Most of the host plants receiving eggs received more than one egg cluster during the butterfly flight season. Studies by Rausher et al. (1981) suggest that eggs are contagiously distributed on *Pedicularis* in *E. editha* because females are more likely to alight and lay eggs on large, isolated plants.

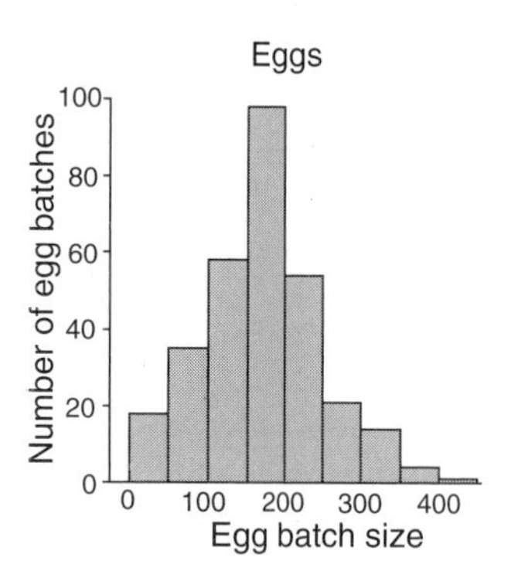

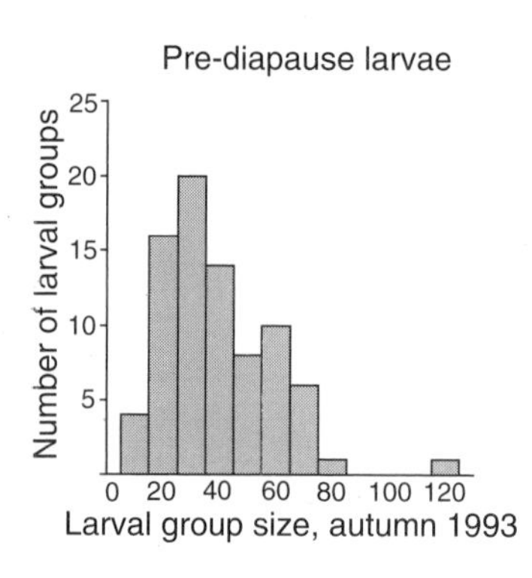

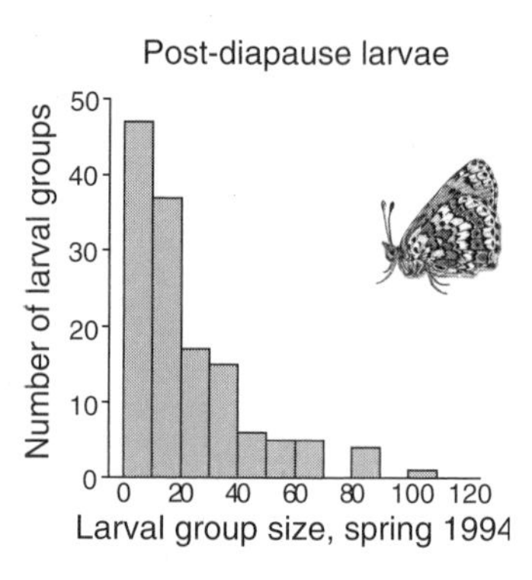

Figure 7.5. Distribution of *Melitaea cinxia* group size at the egg stage and before and after winter diapause (modified from Kuussaari 1998).

When the number of larvae per plant is high, the advantage of being on a populated plant must decline (Rausher et al. 1981, C. D. Thomas et al. 1996). In populations where hosts senesce, competition among larvae on the same plant may be limited because host plant quality, not host plant depletion, determines the need for dispersal (Singer and Ehrlich 1979). In such populations, larval competition may occur as the number of nonsenescent hosts declines, but this competition is a function of the number of larvae in a foraging area and not the number of groups on an individual plant. In all checkerspots, optimal group size may differ on different host plant species or in different foraging environments as a function of plant size, timing of senescence (Moore 1989a), and the availability of alternative host plants.

Group Size and Larval Survival

Variation in group size among populations of *Euphydryas editha* stems from genetic variation in egg cluster size and is associated with the species of host plant used (M. Singer, unpubl. data). For example, populations on *Pedicularis semibarbata* have mean egg cluster sizes of 50–90, those on *Collinsia tinctoria* 20–30, and those on *Collinsia torreyi* only 5–7. At Rabbit Meadow in the western slopes of the Sierra Nevada mountains, where *Collinsia torreyi* has been recently colonized by *E. editha* from *Pedicularis semibarbata*, the mean cluster sizes were larger on *Collinsia* than on *Pedicularis* (52 and 48 eggs on average, respectively; Moore 1989a). A similar difference was observed among captive insects split into two groups, with each group offered only one host species (Singer 1986a). In an experiment, newly hatched larvae were placed in groups of 1, 5, 20, and 50 on *Pedicularis* and *Collinsia* at Rabbit Meadow over 4 different years. In one year group size had little effect, but in the other years larval survival was highest in groups of 5 on *Collinsia* and in groups of 20–50 on *Pedicularis*. Single larvae had close to zero survival on *Pedicularis*, but 60% survival on *Collinsia*. Natural selection on group size, therefore, was different on the two hosts in a direction that would explain the evolved differences between geographically separate populations specialized on these hosts. In the final year of the study, a tanglefoot (predator exclusion) treatment was added to the group size experiment on *Pedicularis* (it was not feasible to exclude predators from *Collinsia*). The effect of group size on survival totally disappeared; hence, it appears that there was an interaction among fitness, group size, and host use that was predator mediated.

A striking feature of *M. cinxia* is the consistently positive effect of group size on survival throughout development from egg stage to the last caterpillar instar, which has been demonstrated by experimentally varying group size (figure 7.6). Figure 7.5 shows natural group size variation in *M. cinxia* just before and immediately after winter diapause. Groups with < 25 larvae have only a small chance of surviving over the winter. The likely reason is that small groups are unable to build a high-quality winter nest, which is necessary for successful overwintering (Nieminen et al. 2001). The quality of winter nests of experimentally inbred *M. cinxia* larvae was scored significantly lower (with thin silk and holes in the nest) than the quality of winter nests in control groups (dense silk, no holes; Nieminen et al. 2001). Consequently, winter mortality of the inbred groups was higher than mortality in the con-

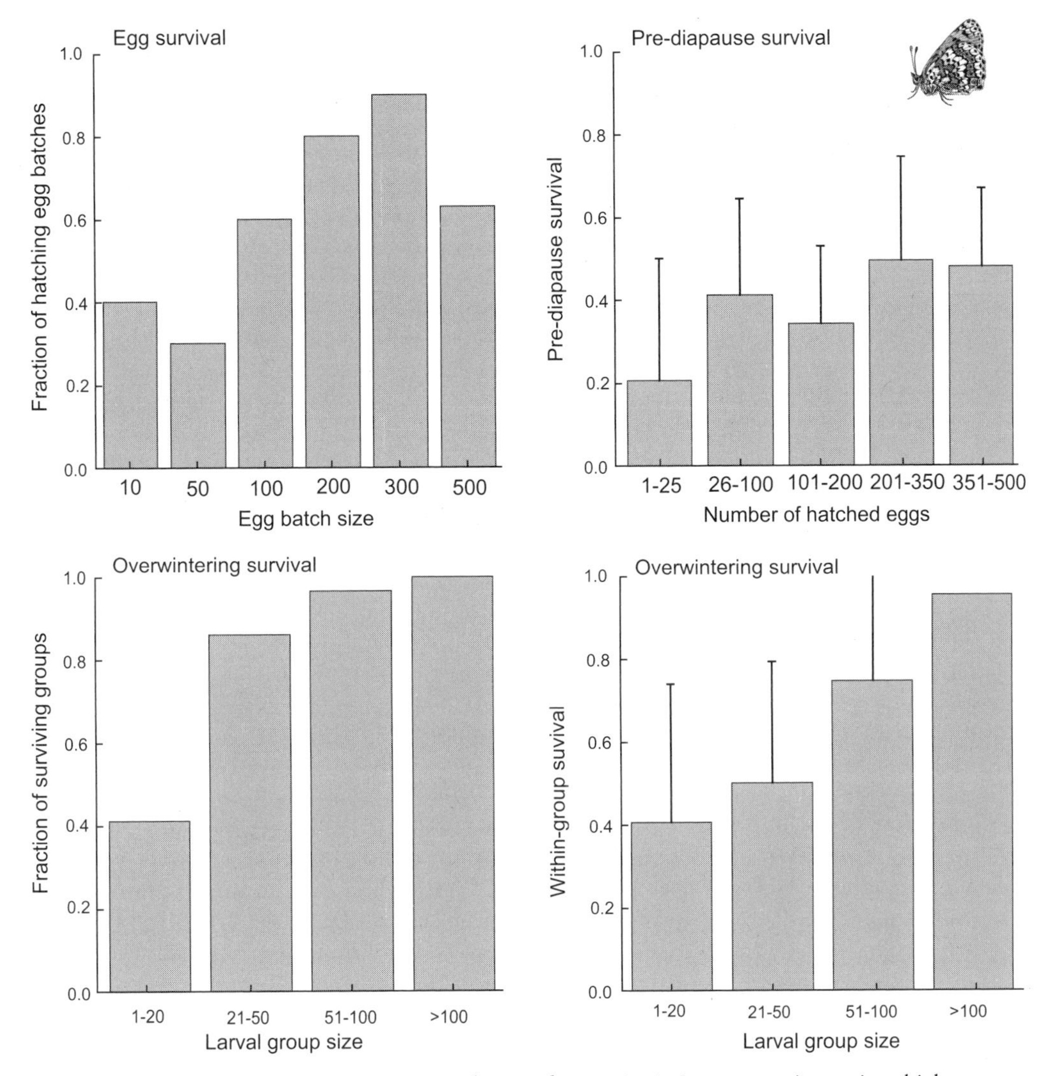

Figure 7.6. Effect of group size on survival in *Melitaea cinxia* in an experiment in which group size was manipulated (modified from Kuussaari 1998).

trol groups (Nieminen et al. 2001). Results from larval survival studies suggest that during the prediapause larval development of *M. cinxia*, the benefits of large group size are greater than the costs, even though the risk of parasitism (Lei and Hanski 1997) and the quantity of host plant tissue needed for development (Kuussaari 1998) increase with increasing group size. Because larval groups typically face a high level of prediapause mortality, large initial group size is a good strategy for preparing a high-quality winter nest necessary for successful overwintering. In this context it may not be accidental that attraction of females to conspecific eggs was documented in Finnish *M. cinxia* that spin winter nests, but not in *M. cinxia* from southern France that do not spin winter nests. The response to conspecific eggs differed significantly between *M. cinxia* from the two regions (M. Singer and L. Ramakrishan unpubl. data).

After diapause, large group size could be expected to have negative effects by increasing resource competition among larvae with growing food demands, but in practice, negative effects have seldom been observed. On the contrary, field studies conducted before the dispersion of larvae in the final instar suggest that both growth and survival increase with group size during postdiapause larval development in *M. cinxia* (Kuussaari 1998). This could be partly because larvae overwintering in large groups probably started out the spring with more energy reserves than larvae who had overwintered in small groups. Growth rate in large groups was also enhanced by group basking. The temperature in the aggregations of black larvae tended to be about 20°C above ambient temperature (30.6°C on average in larval groups in sunny weather) during early spring (Kuussaari 1998). When host plant availability is limited, larval groups tend to split into smaller subgroups and to move longer distances, thereby decreasing resource competition among siblings. Because a large portion of the plant consumption by larvae (>80%) occurs during the last instar when larval survival is difficult to monitor, it is not known to what extent the larvae suffer mortality caused by large group size at this stage.

Larval Dispersion and Movement Distances

After egg hatching, larval group size gradually declines because of larval mortality, group splitting, and larval dispersal. There is much variation both within and among checkerspot species in the timing of larval dispersal. Among the five species of checkerspots in southern Finland, *Melitaea athalia* larvae disperse soon after their first or occasionally after their second molt in July or August (Warren 1987a, Wahlberg 1997b, S. van Nouhuys pers. obs.), whereas the other four species remain in groups until diapause. *Euphydryas maturna* and *Melitaea diamina* disperse immediately after diapause (Wahlberg 1997a, 1998), but larvae of *E. aurinia* and *M. cinxia* remain in conspicuous groups and exhibit a similar basking behavior during cool spring days. *Euphydryas aurinia* larvae disperse after their first molt in the spring, but *M. cinxia* larvae tend to remain together for still another larval instar before dispersing in the final instar just before pupation. The reasons for these differences in the level of gregarious behavior among closely related species remain an interesting question for further research.

Variation among species in the benefits of increasing group size may be associated with the timing of larval dispersal. In a comparative study by N. Wahlberg (unpubl. data), the growth of the larvae of *Melitaea deione*, a species in which larvae disperse soon after hatching, was not affected by group size. In contrast, the larvae of *Melitaea parthenoides* grew faster in larger groups and remained gregarious until diapause in the field. It is interesting to note that *M. athalia* larvae, which disperse early, are cryptic and solitary and, unlike *M. cinxia*, respond to disturbance by dropping from the plant (S. van Nouhuys pers. obs.).

Prediapause larvae of *E. editha* tend to disperse individually. The distances that individual larvae are able to disperse are likely to increase with age, and the cost of dispersal probably decreases with age. Although groups do not appear to migrate as units, new groups can be formed in *E. editha* late in the season as the majority of hosts become unsuitable (senesce). In *E. editha bayensis*, groups can be found on long-lasting *Castilleja* individuals, although eggs were not laid on these plants; such groups probably represent a mixture of larvae from different egg batches (J. Hellmann pers. obs.). These late-forming groups have been occasionally observed to form loose webs. Competition among siblings and nonsiblings and the appearance of larval groups may also occur as the number of suitable host plants declines due to host depletion (C. D. Thomas et al. 1996).

After diapause, *E. editha* larvae often disperse great distances relative to their size, up to 10 m/day. Postdiapause larvae in coastal populations eat as they move, feeding almost exclusively on *P. erecta* (because *Castilleja* has not yet germinated) and consuming entire plants as they disperse. Weiss et al. (1987, 1988) studied the movement and growth of postdiapause larvae of *E. editha bayensis* and found that the position of larvae in the landscape significantly affected the accumulation of body weight. As mentioned before, postdiapause larvae bask in the sun to elevate their body temperature and to increase their rate of growth. Whether larvae are found on cool, north-facing or warm, south-facing slopes significantly affects the amount of sunlight they receive and hence the rate at which they grow (chapter 3). Weiss and colleagues found that larvae in low-insolation environments grew to pupation two weeks slower than larvae in high-insolation environments. Given temporal constraints on the feeding time of prediapause larvae (section 7.5 and chapter 3), it should be advantageous for prediapause larvae to move to sites where they can grow fast.

The movement capacity of the first two instars of *M. cinxia* is extremely limited. At this stage the larval group just expands its web and moves only on the host plant individual that it occupies. The mobility of the larvae increases as they grow but remains limited to a maximum of 1–2 m before winter diapause (Kuussaari 1998). The distances moved by prediapause larval groups get longer with increasing group size, but most groups move < 0.5 m before diapause (figure 7.7). In the spring, the movement distances increase with group size and developmental stage (figure 7.7) and decrease with increasing host plant abundance. Before the ultimate larval instar, the average distances moved even by large larval groups tend to be less than 1 m. However, large groups in areas of low host plant availability may move up to 6 m in search of food. Finally, last instar caterpillars are substantially more mobile, and although entire groups rarely move more than a few meters as a group, single larvae have been observed at distances more than 10 m from any known larval group. The maximum distance that a final instar *M. cinxia* caterpillar has been observed to move is 16 m, but no studies have been conducted specifically to quantify movement abilities of full-grown larvae.

Unlike *E. editha bayensis*, *M. cinxia* larvae are quite systematic in their foraging. When feeding on *Plantago lanceolata*, the primary host plant in the Åland Islands, the larvae usually do not leave a plant until it is completely defoliated. After defoliation of one host plant, they move as a group to a neighboring plant. If drought causes host desiccation, *M. cinxia* larvae stay in their web and wait for rain. In contrast, when prediapause *E. editha* larvae are faced with a senescent host, they keep searching until they starve or find a new host if they are not large enough to enter diapause. This difference reflects the difference between a north temperate and a Mediterranean climate and between a perennial and an annual host. An *E. editha* larva feeding on an annual host at the end of California spring cannot wait for rain to reinvigorate its food supply. Its only chance is to quickly find a patch where hosts are still edible.

In the spring, an average postdiapause larval group of *M. cinxia* consumes about 90 *Plantago* individuals (rosettes), and the largest larval groups defoliate up to 400 plants (M. Kuussaari unpubl. data). When movements of larval groups were intensively monitored in low-density *M. cinxia* populations in spring 1994, it was possible to measure the defoliated areas (area within which >90% of host plants were consumed) after the larvae had pupated. Typically, all host plants were completely devoured within distinct areas, while in the surroundings there were practically no signs of larval feeding. The average area defoliated was 3.5 m^2, and the maximum was 15 m^2 defoliated by the largest groups of larvae.

7.7 Spatial and Temporal Variation in Survival and Population Dynamics

The suitability and availability of food for checkerspot larvae is often closely tied to climate and weather

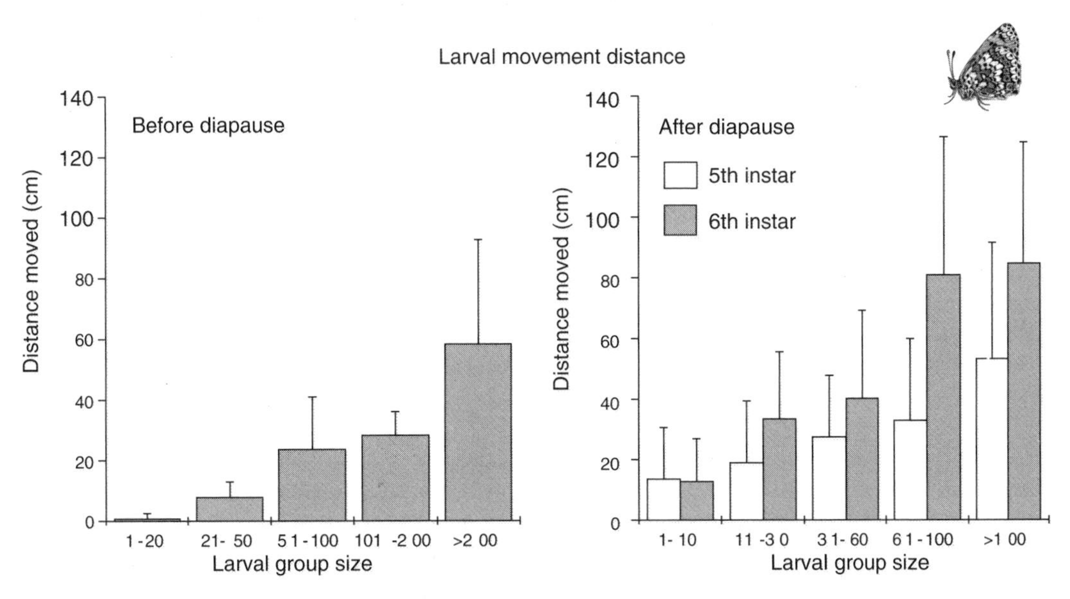

Figure 7.7. Pre- and postdiapause larval movement distances in *Melitaea cinxia* as a function of group size (modified from Kuussaari 1998).

and may vary widely within and among habitats and among years. Habitat alteration due to grazing, logging, and other human-caused environmental disturbances also influences the quality of foraging habitat for larvae over space and time and contributes to variation in population size among sites and over time (Singer et al. 1993, Hanski et al. 1995b, Singer and Thomas 1996, Weiss 1999). Also, populations of natural enemies are not constant over space and time (Lei and Hanski 1997, van Nouhuys and Tay 2001). Spatial and temporal variation in all these factors creates a mosaic of risks to larvae.

The primary source of variation in larval mortality within a habitat patch is variation in the weather that a patch experiences across years. A secondary source of variation is a factor that mediates the effect of climate: topography (important to *E. editha*) or within-patch heterogeneity (important to *M. cinxia*). In the case of coastal *E. editha* populations, for example, years with exceptionally low or high seasonal rainfall lead to population declines, presumably by affecting the phenology of prediapause larvae and their host plants (Ehrlich et al. 1980, Dobkin et al. 1987, McLaughlin et al. 2002a). Such extremes do not affect larvae equally, however. The distribution of postdiapause larvae in the year after an extreme year suggests that larvae on cool slopes fare better than larvae on warm slopes. Cool slopes offer a climatic refuge, presumably by maintaining suitable phenological overlap between developing larvae and senescing host plants (Singer 1972, Weiss et al. 1988, 1993; chapter 3).

Annual variation in weather also affects the survivorship of *M. cinxia* larvae. Summer drought decreases group size at diapause, and survival over the winter decreases with increasing severity of winter weather. Both summer drought and winter weather have an influence on the average group size in the following spring. Large groups survive to diapause and survive over the winter better than small groups. In addition to group size, survival appears to depend on the host plant distribution, soil water retention, and small-scale topography. There is no large-scale topographic variation (hills or mountains) in Åland, but small-scale topography, such as small slopes, rocks, and ant hills influence the effect of drought on plants, as does soil quality and host plant distribution. Larvae on a withering plant a short distance from suitable alternative host plants have a greater chance of persisting through drought conditions than larvae on isolated withered host plants or those surrounded by other withered plants.

Sources of larval mortality also vary among populations. Drought events, for example, are typically widespread (Ehrlich et al. 1980), but the intensity of drought varies across space so that conditions are more severe at some sites than at others. Differences in habitat quality among sites lead to systematic differences in larval mortality. For example, sites may differ in the total abundance or mixture of larval host plants, affecting the ability of larvae to move among hosts, the amount of time that larvae are able to forage before diapause (coastal *E. editha*), or the total number of larvae that reach adulthood (*M. cinxia* in Åland). For *M. cinxia*, mortality rates are significantly lower at sites lacking the parasitoid *C. melitaearum* than at sites where parasitoids are present at high density (Lei and Hanski 1997), and this parasitoid is more successful when it attacks larval groups on one of the two host plant species (*Veronica spicata*; van Nouhuys and Hanski 1999). However, *C. melitaearum* is not present in most populations, and although in some years larval survival is higher in habitat patches dominated by one host plant species or the other, on average *M. cinxia* survive equally well on both host plants (van Nouhuys et al. 2003; figure 7.3).

Habitat management influences larval survival. Although cattle grazing is the primary factor maintaining meadows open and suitable for *M. cinxia* in the long term, its short-term effects on larval survival and population persistence are negative. Larval groups on meadows occupied by grazing mammals are frequently trampled or eaten by sheep, cattle, and horses (M. Kuussaari and M. Nieminen pers. obs.). As a consequence, active grazing increases the risk of local population extinction and decreases the probability of colonization of currently empty habitat patches (Hanski et al. 1995b). Cattle grazing has similarly both positive and negative effects for *E. editha*. Near coastal populations it tends to decrease cover of invasive plants (Weiss 1999; chapter 12), but grazing presumably also causes some mortality due to incidental trampling. The initial introduction of cattle, of course, was ultimately responsible for the expansion of invasive plants in the first place in North America. In some locations, *E. editha* have even shifted their host use in response to habitat changes caused by grazing (Singer et al. 1993a). If cattle are removed, or if land management for cattle production is stopped, these populations could risk extinction from food shortage.

Density-dependent Larval Survival and Population Dynamics

We have shown that variable weather and topography cause variation in larval survival and that weather largely drives population fluctuations in checkerspot butterflies. This is not to say, however, that larval density would never influence checkerspot larval survival and population dynamics. Negative density dependence has been shown to affect local dynamics of *M. cinxia* (Hanski 1999b), and food limitation for large numbers of larvae has been observed in some populations of *E. editha* (White 1974, Boughton 1999). As we discussed in section 7.6, at the level of individual larval groups, "density" (number of larvae per group) has a generally positive effect, with the caveat that there is local adaptation in group size related to the size (and possibly other qualities) of the host plant primarily used in the population.

Increasing density is likely to increase mortality by two primary mechanisms, first by increasing competition for food and second by increasing predation or parasitism due to aggregation, or increase in population sizes of natural enemies. Both mechanisms are known to occur in *M. cinxia* populations in the Åland Islands. As shown by the results in table 7.3, food shortage causes starvation of *M. cinxia* larvae even in low-density populations of the butterfly, and higher larval density inevitably increases competition for food. The specialist parasitoid *Cotesia melitaearum* rarely persists in low-density *M. cinxia* populations and rarely significantly influences the population dynamics of the butterfly. However, in large and tightly clustered *M. cinxia* populations, *C. melitaearum* populations have been large, and the proportion of larvae parasitized has been high, causing both steep declines in local abundance and even local extinctions of the host butterfly (Lei and Hanski 1997, van Nouhuys and Hanski 2002b).

The role of larval density for *M. cinxia* population dynamics was studied experimentally in 22 local populations in Åland during two generations in 1993–94. In half of the populations local larval density was increased by relocating larval groups. In the remaining populations the larvae were also relocated but retaining the original local density (M. Kuussaari unpubl. data). The population sizes were compared in the next generation. As predicted, population growth rate was lower in the populations in which local density was experimentally increased than in the control populations, probably as a direct consequence of increased competition for food.

The results of the above experiment demonstrate that defoliation of host plants by *M. cinxia* can affect population dynamics. However, such extreme situations have been rare during the past 10 years of intensive research in Åland. The only case of a complete large-scale defoliation of practically all host plants in a habitat patch was observed in spring 1994, when the largest local population of *M. cinxia* in Åland was comprised of more than 100 larval groups. The habitat patch was a dry, sandy meadow of 2500 m^2 with abundant *P. lanceolata* covering more than half the area. A large proportion of the host plants in that meadow became defoliated two or more times during the spring, and the larvae had to wait for plant regrowth before being able to complete their development and pupate. Even though substantial larval mortality due to starvation probably occurred, a large proportion of the larvae were eventually able to pupate because of the relatively quick regrowth of *P. lanceolata* after defoliation. Consequently, a similar density of larval groups was observed in the same population in the next larval generation in autumn 1994. The population crashed, however, before the next adult generation, probably because of the very high parasitism rate by *C. melitaearum* in the spring 1995 (G. Lei pers. comm.).

Models of the well-studied *E. editha bayensis* populations at Jasper Ridge during their long decline to extinction (chapter 3) suggest a small role for density-dependent factors (McLaughlin et al. 2002a). Strong density dependence is unlikely to occur in *E. editha* except when populations are at exceptionally high levels, though the role of parasitoids in *E. editha* deserves further study.

7.8 Conclusion: The Checkerspot Larva as a Gambler

Checkerspots spend most of their lives as larvae. We began by remarking that negative events happen to these larvae and that the nature and frequency of these events are often key factors in butterfly population dynamics. However, we should not give the impression that larvae are powerless in the face of this onslaught. At several points in their lives they make active decisions that have dramatic consequences for the survival of themselves or their offspring. Checkerspot larvae are wonderful gamblers, and, because the fecundity of adults is so high, they play for high stakes. Certainly *M. cinxia* does. Larvae in Finland have an obligatory diapause, but they can decide whether to enter diapause at the beginning of the fourth or the fifth instar. This decision must be made before the end of the third instar. Once made, it is irrevocable, but, because these gregarious larvae do not investigate their surroundings, the decision must be made with little knowledge about food availability. We have watched a group of more than 100 larvae starve en masse because they chose to feed rather than diapause in fifth instar only to find, once they started to forage, that their host was totally defoliated with no others in reach. *Euphydryas editha bayensis* larvae likewise are gamblers. A female larva growing in the cool San Francisco Bay springtime has the option of pupating early and becoming an adult with relatively low fecundity and high likelihood of offspring survival. Alternatively, it could remain longer in the larval stage and become a highly fecund adult with high probability of offspring mortality from host senescence. Surprisingly, *E. editha bayensis* larvae take the second option (chapter 3), thereby driving the subspecies to the limits of its ecological tolerance and generating the highly stochastic population dynamics that are its trademark. These two examples illustrate how larvae are not mere passive feeding machines crawling around, eating their food, and waiting for events to happen to them. They assimilate and use information to make complex decisions that influence not only the events that happen to them but those that descend upon their offspring. However, the availability of information on which to base these decisions is often poor.

8

Natural Enemies of Checkerspots

SASKYA VAN NOUHUYS AND ILKKA HANSKI

8.1 Introduction

Butterfly eggs, larvae, pupae, and adults are attacked by an array of natural enemies. Many characteristics of butterflies, such as larval crypsis, distastefulness or toxicity, gregariousness, web building, host plant use, and adult wing patterns, are thought to have evolved as defenses against natural enemies (Gilbert and Singer 1975, Bowers 1981, Stamp 1981a, Dempster 1984, Bowers et al. 1985, Sillén-Tullberg 1988, Vulinec 1990, Montllor and Bernays 1993, Weseloh 1993, Ohasaki and Sato 1999, Hunter 2000). But, paradoxically, while population biologists widely agree that natural enemies have shaped the evolution of butterflies, the extent to which the observed abundances, population dynamics, and other ecological processes in butterflies are now influenced by natural enemies is more controversial. What is known from life-table analyses and experiments is that generalist arthropod and vertebrate predators often cause significant and typically density-independent mortality in butterflies and moths (Dempster 1983, Bellows et al. 1992). Disease can also cause massive mortality at high population densities, particularly in moths with outbreak dynamics (Anderson and May 1981, Dwyer 1994, Dwyer et al. 2000, Kamata 2000). In a review covering all insects, Hawkins (1994) found that parasitoids generally cause 20–60% mortality in their hosts, with the maximum level of mortality related to the host feeding niche. Insects feeding externally on leaves, such as butterfly larvae, tend to suffer lower parasitism than leaf miners, but higher parasitism than borers and root feeders. Parasitoid-caused mortality in butterflies is at least occasionally density-dependent and may thereby contribute to host population regulation (Ford and Ford 1930, O. W. Richards 1940, Dempster 1983). In general, the question of whether natural enemies often regulate their insect host populations and keep them at low density remains much debated (Price 1984, Walde and Murdock 1988, Strong 1989, Turchin 1995, Bernstein 2000, Hassell 2000a).

Studies of the natural enemies of checkerspots are necessary for a comprehensive understanding of the ecology and population biology of these butterflies, but such research also contributes to our understanding of the ecology and evolution of host–parasite and predator–prey interactions in general. Parasitoid–host interactions are among the most common interspecific interactions. They also include some of the most specialized relationships and pose challenging research questions about the ecology and evolution of interactions at all levels, from molecular (Henter 1995, Mikio et al. 2001) to population (Cappuccino and Price 1995, Godfray 2000, Hochberg and Ives 2000) to community (Hawkins 1994, Hawkins and Sheehan 1994) and to ecosystem (Marino and Landis 1996, Landis and Menalled 1998, Letourneau 1998).

Informative empirical study of the population biology and dynamics of natural enemies must take place in systems for which the host ecology is well known. This is especially true for parasitoids, as their biology and dynamics are intimately dependent on the dynamics of their hosts. Short-term field and laboratory experiments have shown that the strength of interaction in predator–prey and host–parasitoid systems depends on many factors, including the relative dispersal ability of the species, host density, direct competition, trophic complexity, and host resistance or parasitoid virulence (Morrison and Strong 1980, Rosenheim et al. 1993, Godfray 1994, 2000, Myers and Rothman 1995, Müller and Godfray 1998, Amarasekare 2000, Doak 2000, Hawkins 2000, Hochberg and Ives 2000, Hufbauer 2002, Stefanescu et al. 2003). How the results of such experiments play out over long times and in large areas is unknown. Large-scale empirical work on the population dynamics of parasitoids has been mostly conducted in the context of biological control. In these studies, changes in the sizes of parasitoid populations are measured with reference to the size and persistence of pest insect populations in managed ecosystems, and the two parameters of the greatest interest are the rate of spread of the parasitoid and the degree of host suppression (Clausen et al. 1965, Murdock et al. 1984, 1989, Roland 1994, Landis and Menalled 1998, Bernstein 2000, Hastings 2000). Studies in managed ecosystems are important for pest control and as tools to identify key components of species interactions, but they do not necessarily address the processes involved in long-term stable interactions that occur in natural, unmanaged systems (Hawkins et al. 1999).

Luckily, there are also a few studies of natural systems in which the population dynamics of parasitoids have been addressed along with the dynamics of their hosts: the winter moth *Cyzenis albicans* in the United Kingdom and in the United States, where it is a pest (Hassell 1969, Roland 1994), the forest tent caterpillar *Malacosoma disstria* (Roland and Taylor 1995), the gypsy moth *Lymantria dispar* (Elkinton and Liebhold 1990, Ferguson et al. 1994, Elkinton et al. 1996), the cinnabar moth *Tyria jacobaeae* (van der Meijden and van der Veen-van Wijk 1997), the Tussock moth *Orgyia vetusta* (Harrison 1997), the knapweed gall fly *Urophaga jaceana* (Varley 1947), the goldenrod gall maker (Abrahamson and Weis 1997), and willow gall-making sawflies (Price 1992, Roininen et al. 1996). In all these cases parasitoids appear to play an important role in host population dynamics under some ecological conditions but not under other conditions. Their role depends primarily on habitat structure and heterogeneity, phenology, and weather, but also on the dispersal behavior of the species involved and on their interactions with other species.

Euphydryas editha and *Melitaea cinxia* and their natural enemies provide excellent opportunities for the study of parasitoid biology and population dynamics because of the solid knowledge base provided by long-term studies of the host populations and metapopulations (chapters 3 and 4). Many of the parasitoids attacking checkerspots are specialists, and some of them even have what appear to be specialist hyperparasitoids. An example is the three-species chain *Melitaea cinxia* (host), *Hyposoter horticola* (primary parasitoid), and *Mesochorus* sp. cf. *stigmaticus* (hyperparasitoid) in the Åland Islands in Finland. The theory of host–parasitoid dynamics (reviewed by Holt 1997, Bernstein 2000, Hassell 2000b) is largely concerned with specialist parasitoids; hence well-studied systems like this example provide much needed empirical data to relate to the current theory.

In this chapter we first outline the natural history and general population ecology of checkerspot natural enemies. It is evident from this review that we know much more about parasitoids than about any other type of natural enemies, which probably reflects the generally dominant role of parasitoids among the natural enemies of checkerspot butterflies. We then discuss multitrophic interactions among the parasitoids, the host butterflies and their food plants, as well as interactions among the natural enemies themselves. Next we address the spatial population dynamics of the interacting species. The final section is devoted to some areas of potential future research.

8.2 Natural Enemy Communities Associated with Checkerspots

The natural enemies of checkerspots are united by the biology and behavior of their hosts. Larvae of most checkerspots are conspicuous, gregarious, and web building at least for a part of their lives, and most of them feed on plants that are chemically defended by seco-iridoid or iridoid glycoside secondary compounds (chapters 7 and 11). Generalist predators and parasitoids are to a smaller or greater extent deterred by the toxic chemicals sequestered by or associated with checkerspots (chapter 7;

Bowers 1980, Camara 1997b), and some natural enemies may be hindered by the silken webs and gregarious behavior of the host larvae (Morris 1972a, 1972b, 1976, Stamp 1981a, Moore 1989a). Nonetheless, and in spite of these defenses, larval mortality caused by generalist predators does occur and is occasionally severe (table 8.1 and chapter 7). For example, in the Åland Islands and elsewhere, eggs and larvae of *M. cinxia* are eaten by pentatomid bugs and coccinellid and chrysopid larvae (S.van Nouhuys pers. obs.; figure 8.1), and pupae are parasitized by generalist parasitoid wasps (Lei et al. 1997). Numerous *E. editha bayensis* have been seen with beak-shaped portions of their wings missing, or detached wings have been found on the ground (Bowers et al. 1985). In contrast, we have observed no avian predation of *M. cinxia*, but at least during one year in one region there was significant predation of adults by dragonflies (I. Hanski pers. obs.).

Even if mortality directly caused by generalist predators is limited, the behavioral responses of the prey to these predators may lead to significant population dynamic consequences. A good example is ant predation on *M. cinxia*. Ants do not touch the larvae, apparently because they are distasteful, but the ant species *Myrmica rubra* does attack eggs and adults (table 8.1). Adults can escape, but in the case of ovipositing females, harassment by ants can nonetheless be very detrimental. If ants disturb a female checkerspot during the one hour that it takes to lay a full clutch of eggs, she may be forced to leave early, after having laid only a part of the usual clutch of about 150 eggs. Because larval survival strongly depends on group size, larvae from a small egg cluster have only a limited chance of successful diapause (chapter 7). Thus, although ants are unlikely to catch the female, the disturbance that they cause may drastically reduce the survival of her offspring. It remains to be determined whether abundant ant populations might limit the abundance of local butterfly populations and even increase their risk of extinction. Another species of ant has a positive indirect effect on *M. cinxia*. In this case, larval host plants growing on ant mounds make especially good host plants because of warm microclimate and because the plants do not usually desiccate during the summer. These plants are often used by *M. cinxia* in spite of their proximity to ants, which are frequently seen attacking other, primarily moth, larvae (S.van Nouhuys pers. obs.).

Natural enemies with narrow host ranges, particularly parasitoids, appear to have evolved mechanisms to take advantage of the chemically and physically defended host larvae (Stamp 1981a, Price 1984). Parasitoids with narrow host ranges are involved in the natural enemy communities associated with checkerspots (table 8.2). Unfortunately, knowledge about parasitoid host ranges and systematics in general is limited (Shaw 1994). The parasitoids of checkerspots are no exception, and thus no systematic analysis of the host records can be attempted here. There are only a few checkerspot species for which the ecology of the parasitoids has been studied and published in any detail. These are *Euphydryas editha* (White 1973, 1986, Moore 1989a, 1989b) and *E. phaeton* (Bowers 1980, Stamp 1981a, 1982b, 1984) in North America, *E. aurinia* in England (Ford and Ford 1930, Fraser 1954, Porter 1981, 1983), and

Figure 8.1. While most work on natural enemies of checkerspots has been done on parasitoids, other organisms may have significant impacts on checkerspot populations. For instance, moth larvae (left) attack egg masses and mirid bugs (right) eat hatchlings of *Euphydryas gillettii* in Colorado (see table 8.1). Photos by Paul R. Ehrlich.

Table 8.1. Predators of checkerspot butterflies.

Butterfly	Predator	Prey Stage	Scale of Impact	Reference[a]
Chlosyne lacinia	Fire ant *Solenopsis xyloni*	Larvae	50% mortality in one population	1
	Pentatomid bug *Stiretrus anchorago*	Larvae	Almost entire larval nests in one population	2
	Polistes wasp	Larvae	Almost entire larval nests in one population	2
	Jumping spider	Adult	Unknown	2
	Orb weaving spider	Adult	Unknown	2
Euphydryas aurinia	Bird (Cuckoo)	Larvae	Unknown, one population	3
	Pterostichus versicolor	Larvae	Unknown, one population	3
	Bug	Larvae	Unknown, one population	3
	Frogs and toads	Larvae	Unknown, one population	3
E. chalcedona	Bird	Adults	Significant in some populations	4, 5
E. editha	Stilt bug (Berytidae)	Eggs	Minor	6
	Unknown	Eggs	0–85% mortality in different populations	7
	Herbivore	Eggs, larvae	0–8% in one population, four years	6
	Unknown	Pupae	23–34% in one population, three years	8
	Bird	Adults	Unknown	19, 10
	Spider	Adults	Causes male bias in a population	11
E. gillettii	Myrid bug(Myridae)	Eggs, larvae	Contributes to 30–80% mortality in two populations	12
	Erythraid mites	Eggs, larvae	Contributes to 30–80% mortality in two populations	12
	Beetle larvae	Eggs, larvae	Contributes to 30–80% mortality in two populations	12
E. phaeton	Unknown	Eggs	7% of eggs from 70% of clusters, one population	13
Melitaea cinxia	Lady beetle (Coccinellidae)	Eggs, larvae	Unknown, probably high predation of eggs; entire clusters eaten	14
	Lacewing (Chrysopidae)	Eggs, larvae	Unknown, variable entire egg clusters eaten	14
	Pentatomid bug (Pentatomidae)	Larvae	Unknown	14
	Red ant (*Myrmica rubra*)	Eggs	Unknown fractions of egg clusters eaten	15
	Spider	Adults	Unknown	14
	Dragon fly	Adults	Unknown	16
Melitaea athalia	Small mammals	Pupae	31–41%, one population over three years	17
	Carabid beetle	Pupae	2–6%, one population over three years	17
	Staphylinid beetle	Pupae	0–2%, one population over three years	17

[a]References: 1, Clark (1997b); 2, Drummond et al. (1970); 3, Wilkinson (1907); 4, Brown and Ehrlich (1980); 5, Bowers et al. (1985); 6, Moore (1989a); 7, White (1973); 8, White (1986); 9, Ehrlich (1965); 10, Hendricks (1986); 11, Moore (1987); 12, E. H. Williams et al. (1984); 13, Stamp (1981b); 14, S. van Nouhuys (pers. obs.); 15, J. Ehrnsten (pers. comm.); 16, R. Setchfield (pers. comm.); 17, Warren (1987c).

Table 8.2. Parasitoids of checkerspot butterflies.

Parasitoid[a]	Host	Geographic Area	Reference[b]
Hymenoptera			
Braconidae			
Cotesia acuminata agg. (S)[c]	*Euphydryas maturna*	Europe	1, 2, 3
	Melitaea didyma	Europe	4
	Melitaea latonigena	Siberia	5
	Melitaea phoebe	Europe, Siberia	5, 4
	Melitaea athalia	Europe	4
	Melitaea scotosia	China	4
Cotesia bignellii (S)	*Euphydryas aurinia*	Europe	2, 7
Cotesia cynthia (S)	*Euphydryas cynthia*	Europe	6, 6
Cotesia euphydridis (S)	*Chlosyne harrissii*	North America-E	8
	Euphydryas phaeton	North America-E	8
Cotesia koebelei (S)	*Chlosyne leanira*	North America	8
	Chlosyne neumoegeni	North America	8
	Euphydryas chalcedonia	North America-W	8
	Euphydryas editha	North America-W	8, 9, 10
Cotesia melitaearum (S)	*Euphydryas aurinia*	Europe	7, 5, 11
	Euphydryas aurnia davidi	Siberia	5
	Euphydryas desfontainii	Europe	1
	Euphydryas maturna	Europe	2,4
	Melitaea cinxia	Europe, Asia	12, 6, 5, 13
	Melitaea diamina	Europe	4
	Melitaea didyma	Europe	4
	Melitaea parthenoides	Europe	4
	Melitaea trivia	Europe	4
	Melitaea athalia	Europe	2, 13, 14
	Melitaea deione	Europe	3, 4
Apanteles lunatus	*Chlosyne lacinia*	North America-W	15
Other Ichneumonoidae			
Apechthis compunctor (G, P)	*Euphydryas maturna*	Europe	1
Benjaminia euphydridis (S)	*Chlosyne harrissii*	North America-E	16
	Euphydryas phaeton	North America-E	16
Benjaminia fumigatos (S)	*Melitaea didyma*	Europe	4
Benjaminia fuscipennis (S)	*Euphydryas editha*	North America-W	9
	Euphydryas chalcedonia	North America-W	16
Benjamina sp. (S)	*Melitaea didymoides*	Siberia	5
Benjamina sp. (S)	*Euphydryas gillettii*	North America-W	6
Cratichneumon vinnulus (P)	*Chlosyne lacinia*	North America-W	15
Gelis sp. (G, P)	*Melitaea athalia*	Europe	14
Hyposoter horticola (S)	*Melitaea cinxia*	Europe	12, 13
Hyposoter sp.	*Melitaea cinxia*	Europe	13
	Melitaea phoebe	Siberia	5
Ichneumon cynthiae (S, P)	*Euphydryas cynthia*	Europe	17
Ichneumon balteatus (P)	*Melitaea cinxia*	Europe	17
Ichneumon cinxiae (S, P)	*Melitaea cinxia*	Europe	17
Ichneumon gracilicornis (G, P)	*Euphydryas aurinia*	Europe	2
	Euphydryas maturna	Europe	1, 2
	Melitaea cinxia	Europe	12
Ichneumon pulvinatus (P)	*Melitaea cinxia*	Europe	17
Probolus culpatorius (G, P)	*Euphydryas maturna*	Europe	1
Stenichneumon culpator (G, P)	*Melitaea athalia*	Europe	17
Unknown Ichneumonid	*Chlosyne harrissii*	North America-E	18
	Melitaea cinxia	Siberia	5
	Melitaea deione	Europe	4

(*continued*)

Table 8.2. (*continued*)

Parasitoid[a]	Host	Geographic Area	Reference[b]
Chalcidoidea *Coelopisthia caledonica* (G, P)	*Melitaea cinxia*	Europe	12
Pteromalus apum (G,P)	*Euphydryas aurinia*	Europe	19, 20
	Euphydryas maturna	Europe	19
	Melitaea cinxia	Europe	19, 20
	Melitaea didyma	Europe	19
Pteromalus archippi (P)	*Chlosyne lacinia*	North America-W	15
Pteromalus puparum (G, P)	*Euphydryas phaeton*	North America-E	21
	Melitaea cinxia	Europe	12
*Pteromalus vanessae (*P)	*Euphydryas gillettii*	North America-W	6
Pteromalus sp. (P)	*Melitaea athalia*	Europe	14
Spilochalcis phoenica (P)	*Chlosyne lacinia*	North America-W	15
Trichogramma fasciatium undescribed	*Chlosyne lacinia*	North America-W	15
Trichogrammatid (egg)	*Euphydryas phaeton*	North America-E	22
Unidentified Chalcid (P)	*Euphydryas editha*	North America-W	10
Diptera			
Tachinidae			
	Euphydryas phaeton	North America-E	21
Compsilura concinnata			
Erycia fatua (S)	*Euphydryas aurinia*	Europe	7, 23
	Euphydryas maturna	Europe, Siberia	2, 23
	Melitaea britomartis	Europe	23
	Melitaea cinxia	Europe	24
	Melitaea deione	Europe	24
	Melitaea parthenoides	Europe	3, 23
	Melitaea athalia	Europe	24
Erycia festinans (S)	*Melitaea cinxia*	Europe, Siberia	5, 25
	Melitaea didyma	Europe	23
Erycia fasciata (S)	*Melitaea didyma*	Europe	23
	Melitaea latonigena	Siberia	5
Erycia furibunda (S)	*Euphydryas aurinia*	Europe	2, 3, 6, 23, 25
	Euphydryas desfontainii	Europe	25
	Euphydryas maturna	Europe	23
Eupharoceradaripennis (P)	*Chlosyne lacinia*	North America-W	15
Siphosturmia melitaeae (P)	*Chlosyne lacinia*	North America-W	15
	Euphydryas chalcedonia	North America-W	27
	Euphydryas editha	North America-W	28
Unidentified Tachinid	*Euphydryas editha*	North America-W	10, 29, 30
Other Palearctic Tachinidae genera recorded from Melitaeini	*Exorista, Compsilura, Nemorilla, Phryxe, Huebneria, Palis, Ceromasia, Masicera, Thelymorpha*		23

[a]The parasitoids emerge from the larval stage of the host unless otherwise noted. Species marked with (P) emerge from the host pupa, species marked with (S) are thought to be a specialist of Melitaeini, and species marked with (G) are known to use hosts other than Melitaeine.

[b]References: 1, Eliasson; (1991), 2, Komonen (1997); 3, Komonen (1998); 4, M. Shaw (pers. comm.); 5, Wahlberg et al. (2001); 6, E. H. Williams et al. (1984); 7, Porter (1981); 8, Marsh (1979); 9, White (1973); 10, Moore (1989a); 11, Ford (1930); 12, Lei (1997); 13, S. van Nouhuys (pers. obs.); 14, Warren (1987c), 15, Drummond et al. (1970); 16, Carlson (1979); 17, Meier (1968); 18, Shapiro (1976); 19, Askew and Shaw (1997); 20, Shaw (2002); 21, Stamp (1984); 22, Stamp (1981a); 23, Tschorsnig (pers. comm.); 24, Tschorsnig (1994); 25, Ford et al. (2000); 26, Belshaw (1993); 27, Brown and Ehrlich (1980); 28, White (1986); 29, C. Boggs (pers. comm.); 30, Mattoni et al. (1997).

[c]This is probably at least two species, M. Shaw, personal communication

Melitaea cinxia in Finland (Lei and Hanski 1997, Lei et al. 1997, van Nouhuys and Hanski 2002b).

The general importance of natural enemies for the population dynamics of their checkerspot hosts is largely unknown. This is because the impact of natural enemies is extremely variable, natural enemies have been studied in only one or a few locations for each host species, and the roles of predators, especially, is difficult to measure. There is, however, some indication that natural enemies play a smaller role in the dynamics of the well-studied populations of *E. editha bayensis* and *E. chalcedona* (Ehrlich et al. 1975, Lincoln et al. 1982) than in *E. aurinia* (Porter 1981), *E. phaeton* (Bowers 1980, Stamp 1984), and *M. cinxia* (Lei et al. 1997, van Nouhuys and Hanski 2002b). Such differences might be explained by differences in the larval interactions with host plants, climate, and the composition of the parasitoid communities.

Natural Enemy Community Associated with Melitaea cinxia

The parasitoid complex associated with *Melitaea cinxia* in the Åland Islands consists of two primary larval parasitoids, two abundant hyperparasitoids (figure 8.2), and several generalist pupal parasitoids (Lei et al. 1997). The specialist parasitoid *Cotesia melitaearum* (Wilkinson) (Braconidae: Microgastrinae; plate X) is a gregarious braconid which, depending on weather, has two or three generations per host generation (year) in Åland. *Cotesia melitaearum* is a relatively sedentary wasp that occupies only well-connected host populations (Lei and Camara 1999, van Nouhuys and Hanski 2002a). Where present, *C. melitaearum* can have a great impact on host populations, even to the point of causing local extinctions (Lei 1997). However, *C. melitaearum* is only found in about 10% of host populations, and most frequently it persists as small, extinction-prone populations that have little impact on local host population dynamics (van Nouhuys and Hanski 1999, van Nouhuys and Tay 2001). Because *C. melitaearum* is a poor disperser, the spatial structure of the host metapopulation and therefore the respective habitat patch network largely determines the occurrence and role of *C. melitaearum* in host dynamics (van Nouhuys and Hanski 2002a).

The second primary parasitoid attacking *M. cinxia* is *Hyposoter horticola* (Gravenhorst) (Ichneumonidae: Campoplaginae; plate X), which is a solitary wasp with a single generation per host generation. *Hyposoter horticola* is restricted to a few Melitaeini, and *M. cinxia* appears to be its only host in Åland (Lei et al. 1997, van Nouhuys and Hanski 2002b). Exceptionally, though it is a larval parasitoid, *H. horticola* lays eggs in the host larvae while they are still inside the eggshell (plate X; van Nouhuys and Ehrnsten 2004). The wasp develops within the host larva and pupates inside the host integument the next spring. *Hyposoter horticola* is a large (12 mm body length) and extremely mobile wasp, and it parasitizes about a third of the host larvae in most populations, including very isolated and newly colonized host populations. In contrast to the spatially and temporally dynamic occurrence of *C. melitaearum*, the occurrence of *H. horticola* is stable, and it thereby uniformly suppresses host population sizes without directly causing local host population extinctions (van Nouhuys and Hanski 2002a, van Nouhuys and Ehrnsten 2004).

The two parasitoid species compete directly in those host populations in which they happen to occur together (section 8.3; Lei and Hanski 1998, van Nouhuys and Tay 2001). Additionally, each parasitoid has its own natural enemies. *Cotesia melitaearum* is host to the pseudo-hyperparasitoid *Gelis agilis* (Fabricius) (Ichneumonidae: Cryptinae) and less commonly to three other *Gelis* species (Lei et al. 1997). These species are all abundant wingless, generalist, ectoparasitic ichneumonids parasitizing insects developing in silken cocoons (Schwarz and Shaw 1999). *Gelis* aggregate in response to high host density and can greatly reduce the population size of *C. melitaearum*, probably even causing local extinctions (Lei and Hanski 1997, van Nouhuys and Tay 2001). An additional complication, supported by experimental results reported below, is the possibility that the dynamics of the primary parasitoid *C. melitaearum* is influenced, via apparent competition (Holt 1977), by the alternative prey of the generalist hyperparasitoid *G. agilis* (van Nouhuys and Hanski 2000). Finally, generalist predators kill some 40% of pupae in the early spring generation of *C. melitaearum*, leaving the populations small and prone to extinction (van Nouhuys and Tay 2001).

Hyposoter horticola is parasitized by the solitary ichneumonid hyperparasitoid *Mesochorus* sp. cf. *stigmaticus* (Brischke) (Ichneumonidae: Mesochorinae), which, like its host, is univoltine and very mobile and successfully parasitizes about a third of the host individuals throughout the Åland Islands

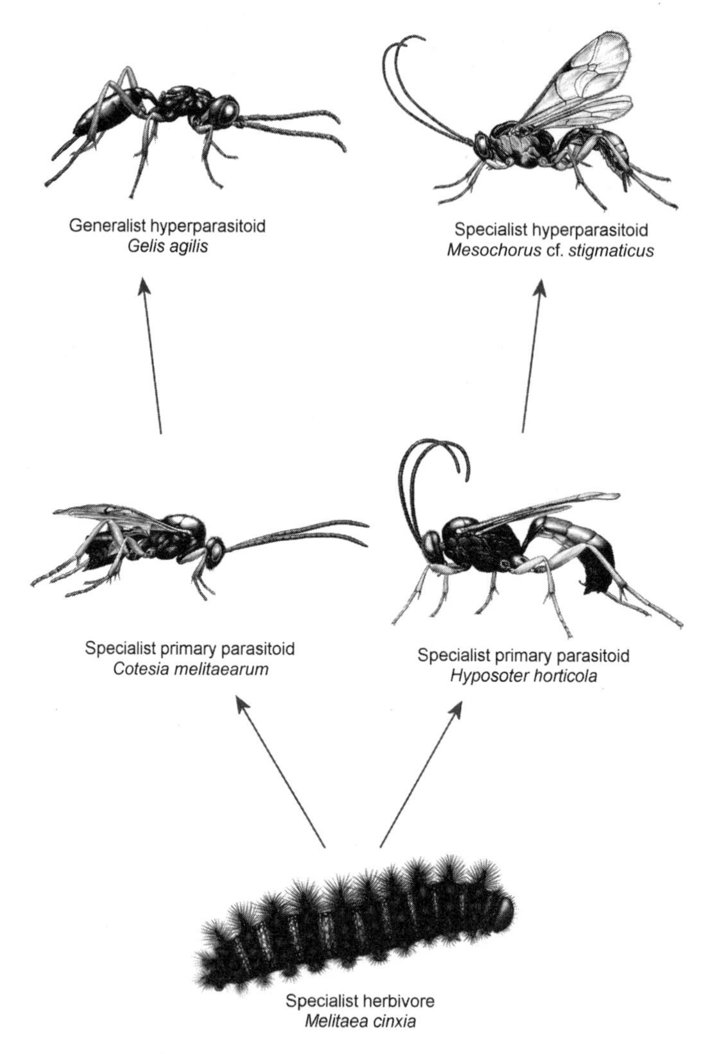

Figure 8.2. The parasitoid food web associated with *Melitaea cinxia* in the Åland Islands, involving two primary parasitoids (*Cotesia melitaearum* and *Hyposoter horticola*) and two hyperparasitoids (*Gelis agilis* and *Mesochorus* sp. cf. *stigmaticus*). Drawings by Zdravko Kolev.

(van Nouhuys and Hanski 2002b). *Mesochorus* sp. cf. *stigmaticus* also parasitizes *C. melitaearum*, though only occasionally (Lei et al. 1997). The systematics of the Palaearctic species of *Mesochorus* is in need of revision; hence we cannot be sure that the species is *stigmaticus* (Lei et al. 1997). However, *Mesochorus* are usually closely affiliated with the host of their primary parasitoid and are likely to have narrow host ranges. Thus, *M.* sp. cf. *stigmaticus* may be a specialist on *M. cinxia* in Åland (this is indirectly supported by distributional data; section 8.3).

There are at least four species of pupal parasitoids of *M. cinxia* in Åland: *Ichneumon gracilicornis* (Gravenhorst) (Ichneumonidae: Ichneumoninae), *Pteromalus apum* (Retzius) (Pteromalidae: Pteromalinae), *P. puparum* (L.) (Pteromalidae: Pteromalinae), and *Coelopisthia caledonica* (Askew) (Pteromalidae: Pteromalinae). All these species appear to have wide host ranges. Little is known of their impact on *M. cinxia*, apart from the observation that 14% of the pupae put out in a habitat patch used by *M. cinxia* during one season were parasitized (Lei et al. 1997).

Numerous samples of all larval stages suggest that there are no larval-pupal parasitoids in the community (S.van Nouhuys pers. obs.).

Biology of Checkerspot-Associated Parasitoids and Their Communities

Parasitoids in the genus *Cotesia* have been found to use most checkerspot species in Europe and North America. There are 4 recognized species in Europe using 13 different host species, and 2 species in North America using 6 host species (table 8.2). However, recent molecular studies suggest that there may be several more *Cotesia* species associated with checkerspots that have not been recognized by traditional taxonomy (M. Kankare pers. comm.). Additionally, only a few host species have been sampled extensively; thus the host records are likely incomplete. A few generalizations can nonetheless be drawn from the comparison of species in table 8.2. There is no clear pattern in the host ranges of the European *Cotesia* with respect to the phylogeny of the host butterflies (Wahlberg and Zimmermann 2000), nor is there an overall pattern in the use of host larvae on particular host plant species. The two generalist species of *Cotesia*, *C. melitaearum* and *C. acuminata*, share many host species (table 8.2), but they do not appear to coexist locally on the same host. One caveat here is that both *C. melitaearum* and especially *C. acuminata* may actually be aggregate names including two or more species each having a narrow host range (M. Kankare and M. Shaw pers. comm.).

Cotesia is a large genus parasitizing many families of Lepidoptera, but the *Cotesia* that parasitize checkerspots are not known to parasitize any other hosts. *Cotesia* species parasitizing *E. phaeton*, *E. aurinia*, and *M. cinxia* have specialized behaviors that are not characteristic of the genus *Cotesia* as a whole. These behaviors include moving freely inside the dense silken web of the host larvae, attending these webs for hours in the case of *E. phaeton* (Stamp 1981a, 1982b) or even days in the case of *M. cinxia* (Lei and Camara 1999), and often parasitizing only a fraction of the larvae available (Stamp 1981a, Porter 1983, Lei and Camara 1999). *Cotesia* that parasitize checkerspots also share many traits with other species in the genus, such as extreme variation in the number of eggs laid in a host depending on host size and flexible developmental time, both of which contribute to a potentially high rate of population increase. Additionally, at least some *Cotesia* that use checkerspots can control the rate of development of the host (e.g., Stamp 1984, Moore 1989b), and most pupate in clusters around the dead or dying host, as do other *Cotesia* (plate X).

Less is known about the behavior of the remaining parasitoids attacking checkerspots. Stamp (1982c) observed the ichneumonid *Benjaminia euphydryadis* making brief visits to *E. phaeton* larval groups in the field, where it successfully parasitized 4–6% of larvae and appeared to be deterred by larval defensive behavior. The ichneumonid *Hyposoter horticola* completely avoids any larval behavioral defenses of *M. cinxia* by parasitizing first instar larvae while they are still within the eggshell (plate X). This behavior leads to an extremely short window of opportunity for parasitism, only lasting for some hours per generation. It is at first surprising that the parasitoid nonetheless manages to parasitize about a third of the host larvae in most host populations, including the newly established ones. Part of the explanation is the great mobility of *H. horticola* (van Nouhuys and Hanski 2002a). But another reason for the high rate of parasitism, in spite of the short period of time during which the host is vulnerable, is the ability of the female wasps to learn the locations of the host groups before they are at the vulnerable stage for parasitism. By observing marked individuals of *H. horticola* in the field, we know that the wasps are long lived, and they actively forage for *M. cinxia* egg groups for several weeks before the eggs are due to hatch. During this time the wasps apparently memorize the locations of the egg clusters and return to parasitize them just before they hatch (figure 8.3; van Nouhuys and Ehrnsten 2004).

The better studied parasitoid complexes attacking checkerspots exhibit a clear pattern in species composition. Each host has a species of gregarious multivoltine *Cotesia* with complex behavior and extreme population fluctuations, as well as a single other important specialist parasitoid that is a large, solitary, univoltine ichneumonid (table 8.3; with the possible exception of *E. aurinia,* which appears to lack the ichneumonid) or a tachinid fly (table 8.2; Wahlberg et al. 2001). It may well be that coexistence of two *Cotesia* on the same host in the same locality is unlikely because of competition, while coexistence of a species of *Cotesia* and an unrelated parasitoid with contrasting behavior and phenology is possible (section 8.3). Unfortunately, the better studied systems all include just one common checkerspot host species, whereas there are locali-

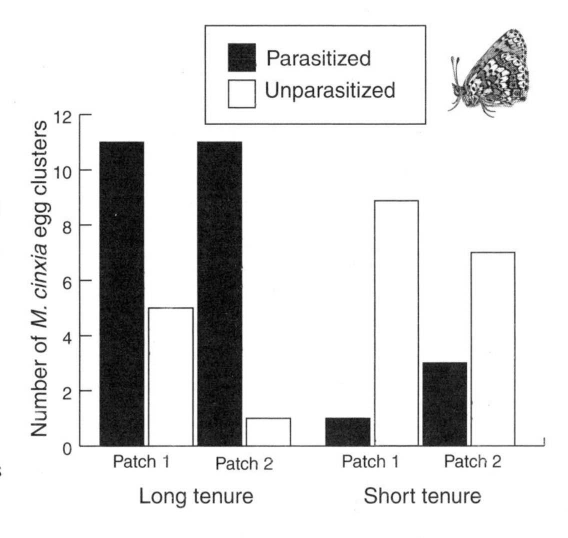

Figure 8.3. Parasitism of *Melitaea cinxia* egg clusters by *Hyposoter horticola* in a field experiment testing the hypothesis that *H. horticola* learn the locations of host egg clusters (van Nouhuys and Ehrnsten 2004). Two habitat patches occupied by both the butterfly and the parasitoid were divided into 144 and 132 2 × 2-m quadrants. Twenty quadrants in each patch were randomly assigned to receive transplanted food plants containing one-week old egg clusters (long-tenure eggs), and 10 quadrants in each were randomly assigned to receive transplanted plants with egg clusters on the day the eggs were to become susceptible (short-tenure eggs). Host egg clusters were laid by butterflies in the laboratory on potted food plants. Short tenure eggs were placed in the field by 9 AM when they began to slightly darken (see plate X), several hours before they became susceptible to parasitism. This assured that both egg cluster types were available for the entire time that they were vulnerable to parasitism. Egg clusters were checked every evening, and upon hatching returned to the laboratory, where the larvae were later dissected and scored as parasitized or unparasitized. The observed parasitism differs from the expected at $p < .01$ (Fisher's Exact test) in each habitat patch. Several of the long-tenure egg clusters in each patch were eaten by predators before the end of the experiment.

ties in Europe and Asia with up to 10 co-occurring checkerspot species. In these communities, several checkerspot species share the same habitat and even the same host plant species. For instance, in the steppe region of Buryatia, Russia, three checkerspot species, *Melitaea cinxia*, *M. latonigena,* and *M. didymoides,* all feed on *Veronica incana*, but they have not been observed to share the same species of *Cotesia* (*M. cinxia* has *C. melitaearum*, *M. latonigena* has *C. acuminata*, while no *Cotesia* was reared from *M. didymoides)*. In contrast, *M. phoebe* and *M. latonigena*, which feed on different host plant species in the same steppe habitat, share *C. accuminata* (Wahlberg et al. 2001). There is much scope for more work on the multihost communities, in which complex parasitoid communities with complex species interactions may occur.

Further complications occur due to varying host phenologies and development. Checkerspots in mild climates may have two or even three generations per year, in mountain regions host phenology changes with altitude, and in the north larvae may take more than one year to develop (Tolman and Lewington 1997). Examples include *M. cinxia,* which has two generations per year in much of southern France, but in the French Alps at higher altitudes there is only one generation per year (Tolman and Lewington 1997). *Euphydryas maturna* develops facultatively in two years in cool microclimates in Finland (Wahlberg 1998), whereas *E. gillettii* develops in two years at high altitudes in western North America (Stamp 1984). Such variation in host life cycles may have important consequences for host–parasitoid interactions.

8.3 Multitrophic-Level Interactions and Parasitoid Population Dynamics

The size and persistence of parasitoid populations at a local scale are affected by attributes of the host and the host food plants, which influence success

Table 8.3. Parasitoids in four checkerspot communities.

Butterfly[a]	Parasitoid	Host Stage Parasitized	Parasitoid Larval Behavior	Parasitoid Oviposition Behavior	Parasitoid Generations Per Year	% Host Population Parasitized
Melitaea cinxia	*Cotesia melitaearum*	Larvae	Gregarious	Specialist	2–3	Variable, 2–80%
	Hyposoter horticola	Larvae (in egg)	Solitary	Specialist	1	30% uniformly
	Coelopisthia caledonica	Pupae		Generalist		
	Coelopisthia extenta	Pupae		Generalist		
	Pteromalus apum	Pupae		Generalist		
Euphydryas editha	*Cotesia koebelei*	Larvae	Gregarious	Specialist	2	Variable, 0–66%
	Benjaminia fuscipennis	Larvae (?)	Solitary	Specialist	1	Very low, <1.0%
	Siphosturmia melitaeae	Larvae(?)	Solitary	Specialist		1–10%
E. phaeton	*Cotesia euphydridis*	Larvae	Gregarious	Specialist	2	6%
	Benjaminia euphydridis	Larvae			1	4–6% in
	Trichogrammatid	Eggs				<5%
E. aurinia	*Cotesia bignellii*	Larvae	Gregarious	Specialist	2–3	Variable, 0–80%

[a]*Melitaea cinxia* in the Åland Islands, 1993–2002 in a 50 × 70 km area (Lei et al. 1997, van Nouhuys and Tay 2001, van Nouhuys and Hanski 2002b); *Euphydryas editha* in California, USA, several years in coastal populations (Ehrlich 1965, White 1973, 1986), and several years in Sierra populations (Moore 1989a); *E. phaeton* in eastern USA, one population 1979 and 1980 (Stamp 1981a, 1982b); and *E. aurinia* in the United Kingdom, one site 1895–1926 (Ford and Ford 1930), and one site 1977–80 (Porter 1981, 1983).

of parasitism, and by parasitoid mortality caused by competition and natural enemies. At larger spatial scales parasitoid population dynamics may be constrained by the spatial dynamics of the species with which they interact, including factors that influence the extinction–colonization dynamics of the host and the relative range and the rate of dispersal of the interacting species (van Nouhuys and Hanski 2002b). Here we first discuss local tritrophic interactions among the butterflies, their food plants and the primary parasitoids, then local interactions among the parasitoids, and finally the spatial dynamics of the interacting species. All these interactions are primarily illustrated by results for *Melitaea cinxia* and its parasitoids in the Åland Islands, a system in which parasitoids play a significant role in the dynamics of the host butterfly.

Natural Enemies and Checkerspot Larval Host Plants

Natural enemies of herbivorous insects are influenced in various ways by the food plants of their host species (Godfray 1994, De Moraes et al. 1998). As reviewed in chapters 7 and 11, most checkerspots feed on plants containing iridoid glycoside or secoiridoid plant secondary compounds, which deter generalist herbivores. Checkerspot larvae sequester iridoids from their food plants (Bowers and Puttick 1986, Bowers 1988, Suomi et al. 2001), and predation by generalists would probably be higher if the hosts were not chemically defended. Specialist parasitoids appear to tolerate, overcome, or avoid iridoids sequestered by their hosts (Bowers 1981, Stamp 1982b, 1992). In a laboratory experiment using selected lines of *Plantago lanceolata* with high and low iridoid concentrations, Harvey and van Nouhuys (unpubl.) found that the rate of development and the size of *C. melitaearum* developing in *M. cinxia* larvae did not vary significantly with the iridoid content of the plants eaten by the host larvae. In contrast, in the field in the Åland Islands, where parasitoids may choose among hosts feeding on plants containing dissimilar iridoid concentrations, parasitism appears to be associated with low levels of the iridoid glycoside catalpol in the host plant (Nieminen et al. 2003). The apparent difference between the laboratory and field results might reflect a difference in parasitoid preference (which the field results largely or entirely reflect) and parasitoid performance (which the laboratory experiment measured). Alternatively, sequestration or metabolism of iridoid glycosides by host larvae in the laboratory may differ from that in natural conditions, or adult wasps may prefer larvae feeding on low-iridoid plants for reasons not related to the measures taken in the laboratory experiment. Finally, attributes of plants that are correlated with catalpol level may influence the adult oviposition behavior or success of *C. melitaearum* larvae. It would be worthwhile to compare the natural enemy communities associated with iridoid-defended versus undefended checkerspot butterfly species. For example, *Melitaea phoebe*, *M. scotosia*, *M. punica*, and *Chlosyne lacinia* all feed on plants in the family Asteraceae, which do not synthesize iridoids (Wahlberg 2001b). *Chlosyne lacinia* appears to have generalist predators, such as *Polistes* wasps, which have not been recorded for checkerspot larvae feeding on iridoid-producing hosts (table 8.1). The complication, of course, is that these noniridoid plants have a multitude of other secondary chemical compounds with unknown effects on checkerspot larvae and their parasitoids.

Production of toxic chemical compounds consumed and sequestered by herbivores is not the only factor related to plants that might influence natural enemies. For instance, volatile compounds released by herbivore-infested plants attract parasitoid wasps (Vet and Dicke 1992, Turlings et al. 1995). Furthermore, plants differ in their structure, phenology, and dispersion, which can all contribute to the success of a female wasp searching for hosts in which to lay eggs and to the subsequent success of her developing offspring (Godfray 1994). In the Åland Islands, larvae of *M. cinxia* feed on two host plant species, *Veronica spicata* and *Plantago lanceolata*. These plants differ in iridoid glycoside content and concentration (Suomi et al. 2002, Nieminen et al. 2003) and in their distribution within and among habitat patches (chapter 7). *Cotesia melitaearum* uses *M. cinxia* larvae on both host plant species, but observations on the rate of parasitism in natural populations over six years showed that larvae on *V. spicata* were more frequently parasitized than larvae on *P. lanceolata* (van Nouhuys and Hanski 1999). As a consequence, parasitoid populations are largest and persist longest in areas in which the host butterfly feeds predominantly on *V. spicata*. Colonization rate of host populations by the parasitoid also increased with the relative abundance of *V. spicata* in the habitat patch; hence the large-scale distribution of the two host plant species significantly influences the parasitoid metapopulation dynamics (van Nouhuys and Hanski 1999).

I. Palearctic Melitaeini (significant subgenera in parentheses)

Row 1. *Melitaea cinxia*, Hungary; *Melitaea cinxia*, Austria; *Melitaea cinxia*, Austria; *Melitaea diamina*, Austria.

Row 2. *Melitaea (Mellicta) deione*, France; *Melitaea (Mellicta) athalia*, Germany; *Melitaea (Mellicta) aurelia*, Germany; *Melitaea diamina*, Austria.

Row 3. *Melitaea (Didymaeformia) trivia*, Yugoslavia; *Melitaea (Mellicta) ambigua*, China; *Melitaea (Mellicta) asteria*, Austria; *Melitaea (Mellicta) britomartis*, Austria.

Row 4. *Melitaea (Cinclidia) scotosia*, Japan; *Melitaea (Mellicta) aurelia*, Italy; *Melitaea (Didymaeformia) didyma*, Germany; *Melitaea (Didymaeformia) didyma*, Germany.

Row 5. *Euphydryas desfontainii*, Spain; *Euphydryas aurinia*, Switzerland; *Euphydryas aurinia*, France; *Euphydryas aurinia*, Switzerland.

Row 6. *Euphydryas maturna*, Germany; *Euphydryas maturna*, Germany; *Euphydryas intermedia*, Austria; *Euphydryas intermedia*, Switzerland.

Row 7. *Euphydryas cynthia*, Switzerland; *Euphydryas iduna*, Sweden; *Euphydryas maturna*, Italy; *Euphydryas intermedia*, Italy.

II. Western Hemisphere Melitaeini

Row 1. *Euphydryas editha*, Santa Clara County, California; *Euphydryas editha*, Humboldt County, Nevada; *Euphydryas editha*, Humboldt County, California; *Euphydryas editha*, White Pine County, Nevada.

Row 2. *Euphydryas editha*, Douglas County, Nevada; *Euphydryas anicia*, Rovalli County, Montana; *Euphydryas anicia*, Apache County, Arizona; *Euphydryas anicia*, Teller County, Colorado.

Row 3. *Euphydryas anicia*, Yukon Territory, Canada; *Euphydryas anicia*, Nye County, Nevada; *Euphydryas chalcedona*, Eldorado County, California; *Euphydryas anicia*, Humboldt County, Nevada.

Row 4. *Euphydryas colon*, Washoe County, Nevada; *Euphydryas colon*, Siskiyou County, California; *Euphydryas chalcedona*, Los Angeles County, California; *Euphydryas chalcedona*, Clark County, Nevada.

Row 5. *Euphydryas phaeton*, Franklin County, Missouri; *Euphydryas gillettii*, Missoula County, Montana; *Chlosyne gabbi*, San Bernadino County, California; *Chlosyne neumoegeni*, Clark County, Nevada.

Row 6. *Chlosyne fulvia*, Gila County, Arizona; *Chlosyne californica*, Clark County, Nevada; *Polydryas arachne*, Apache County, Arizona; *Chlosyne guadealis*, Heredia Province, Costa Rica.

Row 7. *Phyciodes tharos*, Willacy County, Texas; *Phyciodes mylitta*, Cochise County, Arizona; *Anthanassa atronia*, Chiapas Province, Mexico; *Texola elada*, Sonora Province, Mexico; *Microtia elva*, Chiapas Province, Mexico.

Row 8. *Tegosa flavida*, Pichicha Province, Ecuador; *Eresia clara*, Limon Province, Costa Rica; *Eresia burchelli*, Rondonia Province, Brazil; *Dymasia dymas*, Pima County, Arizona; *Janatella leucodesma*, Punta Arenas Province, Costa Rica.

III. Selected butterflies from the Nymphalidae

Nymphalini (N), Kallimini (K), Coloburini (C), Melitaeini (M), Heliconiini (H)

Column 1. *Nymphalis antiopa* (N), Pershing County, Nevada; *Inachis io* (N), Germany; *Vanessa atalanta* (N), Elko County, Nevada; *Tigridia acesta* (C), Pichicha Province, Ecuador; *Polygonia interrogationis* (N), Rockingham County, Virginia; *Eresia eutropia* (M), Punta Arenas Province, Costa Rica.

Column 2. *Napeocles jucunda* (K), Rondonia, Brazil; *Smyrna blomfildia* (C), Chiapas Province, Mexico; *Siproeta steneles* (K), Rondonia Province, Brazil; *Heliconius erato* (H), Limon Province, Costa Rica; *Dryas julia* (H), Rondonia Province, Brazil.

Column 3. *Boloria graeca* (H), France; *Brenthis ino* (H), Germany; *Speyeria atlantis* (H), Newfoundland, Canada; *Mesoacidalia aglaja* (H), Austria; *Colobura anulata* (C), Rondonia Province, Brazil; *Junonia evarete* (K), Rondonia Province, Brazil.

IV. A checkerspot community in the Pyrenees (El Puig)
(studied for many years by C. Stefanescu)

Each row shows (from left to right) two males from the lower and upper side and a female from the upper side. The species are (from top to bottom): *Melitaea trivia, M. phoebe, M. cinxia, M. diamina, M. deione, M. athalia,* and *Euphydryas aurinia.* Plate by Jaakko Kullberg and Tapio Gustafsson.

V. *Euphydryas editha* sites

1. Aerial view of Jasper Ridge grassland island looking west. Bright floral displays are concentrated in serpentine soil areas. Arrows indicate area H in foreground, area C in upper right, and the small area G just above the center, with a small patch of light flowers in its lower corner. Photo by Paul R. Ehrlich.

2. Jasper Ridge H in April 1985 looking north. Note the substantial topographic complexity. Trail and isolated tree also clearly visible in aerial photo above. Photo by Paul R. Ehrlich.

3. Jasper Ridge area G in April 1961 looking north. Floral displays in this area were never as spectacular as those in C and H. Photo by Paul R. Ehrlich.

4. Jasper Ridge area C in April 1985 looking north. Note the lack of topographic complexity. Isolated clump of trees clearly visible in aerial photo above. Photo by Paul R. Ehrlich.

5. Morgan Hill, looking southwest, showing topographic complexity. Photo by Alan Launer.

VI. *Euphydryas editha* food plant and sites

1. *Plantago erecta*, the primary *Euphydryas editha* food plant in the Bay area coastal ecotype, in flower on Jasper Ridge, 1960.

2. Del Puerto Canyon. Photo by Paul R. Ehrlich.

3. Schneider's Meadow (1700 m). Photo by Michael Singer.

4. Rabbit Meadow (2500 m). Photo by Michael Singer.

5. Ebbetts Pass, high altitude site (2660 m). Cheri Holdren in foreground. Photo by Paul R. Ehrlich.

6. Gardisky Lake, high altitude site (3200 m). Photo by Paul R. Ehrlich.

VII. Nearctic *Euphydryas* adults

1. *Euphydryas editha luestherae* female from Del Puerto Canyon population. This ecotype was named after the late LuEsther T. Mertz, who has supported much of the Ehrlich group's research since 1970. Photo by Paul R. Ehrlich and Dennis D. Murphy.

2. *Euphydryas editha* male at Jasper Ridge. Photo by Paul R. Ehrlich.

3. *Euphydryas editha* female ovipositing at Edgewood Road, a population just north of Jasper Ridge. The egg mass is being deposited on stems of *Plantago erecta*. Photo by Paul R. Ehrlich.

4. *Euphydryas editha* male at Almont Summit (1984). Its genitalia are being dusted with fluorescent dye in an experiment to determine whether males on ridgelines are more likely to copulate than those on slopes below (Ehrlich and Wheye 1986). Photo by Paul R. Ehrlich.

5. *Euphydryas gillettii* female from the Gothic transplant population, ovipositing on the underside of a *Lonicera involucrata* shrub. Photo by Paul R. Ehrlich.

VIII. Nearctic *Euphydryas* early stages

1. *Euphydryas editha* prediapause larva on webbed plant. Morgan Hill. Photo by Jessica Hellmann.

2. *Euphydryas editha* postdiapause larva, near pupation at Edgewood Road, just north of Jasper Ridge. Photo by Paul R. Ehrlich.

3. *Euphydryas editha* larval phenotypes from California. Orange morph from Marron Valley, San Diego County (top); dark morph from Lake Skinner, Riverside County (bottom left). Black and white morph from Yucca Point, Kings Canyon National Park (bottom right); *E. editha* shows more color pattern differentiation from population to population and within populations as larvae than as adults. Photo by Rob Plowes and Michael Singer.

4. *Euphydryas chalcedona* larva on *Diplacus aurantiacus* at Jasper Ridge. Photo by Paul R. Ehrlich.

5. *Euphydryas editha* pupa at Edgewood Road, just north of Jasper Ridge. Photo by Paul R. Ehrlich.

IX. *Melitaea cinxia* adults and larvae in the Åland Islands in Finland

1. *Melitaea cinxia* female on the flower of *Allium schoenoprasum*, a favored nectar plant. Photo by Tapio Gustafsson.

2. A mating pair of *Melitaea cinxia*. Photo by Tapio Gustafsson.

3. An ovipositing female of *Melitaea cinxia*. Photo by Tapio Gustafsson.

4. An egg cluster of *Melitaea cinxia* underside a leaf of *Plantago lanceolata*. Photo by Tapio Gustafsson.

5. *Melitaea cinxia* fourth instar (prediapause) larvae. Photo by Saskya van Nouhuys.

6. *Melitaea cinxia* prediapause larval group within a web. Photo by Tapio Gustafsson.

7. *Melitaea cinxia* winter nest, inside which the larval group spends the winter in diapause. Photo by Tapio Gustafsson.

8. *Melitaea cinxia* postdiapause larvae basking. Notice the presence of some web. Photo by Marko Nieminen.

X. Parasitoids and host plants of *Melitaea cinxia*

1. A group of cocoons of *Cotesia melitaearum*, emerged from a large postdiapause larva of *Melitaea cinxia* in the spring. Photo by Saskya van Nouhuys.

2. *Cotesia melitaearum* adult. Photo by Niklas Wahlberg.

3. *Hyposoter horticola* female parasitizing larvae of *Melitaea cinxia* that are still inside the egg shell. Photo by Saskya van Nouhuys.

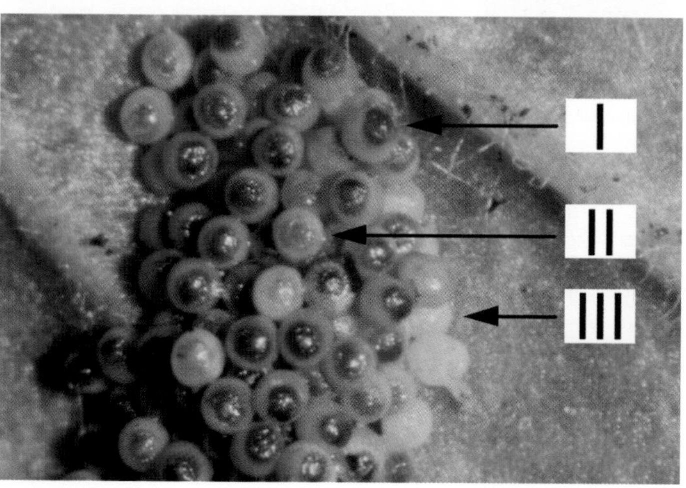

4. An egg cluster of *Melitaea cinxia* showing asynchrony of development. Three types of eggs are indicated: I, dark-topped eggs vulnerable to parasitism; II, light brown eggs soon to be vulnerable; and III, bright yellow eggs with no sign yet of the maturing larva inside (not vulnerable). Photo by Saskya van Nouhuys.

5. *Veronica spicata*, one of the two host plants of *Melitaea cinxia* in the Åland Islands (the photograph also includes some leaves of *Plantago lanceolata*). Photo by Ilkka Hanski.

6. *Plantago lanceolata*, one of the two host plants of *Melitaea cinxia* in the Åland Islands. Photo by Tapio Gustafsson.

XI. Postdiapause (sixth instar) larvae of 10 species of Eurasian checkerspots

Melitaea athalia from Finland.

Melitaea cinxia from Finland.

Melitaea deione from northern Spain.

Melitaea diamina from Finland.

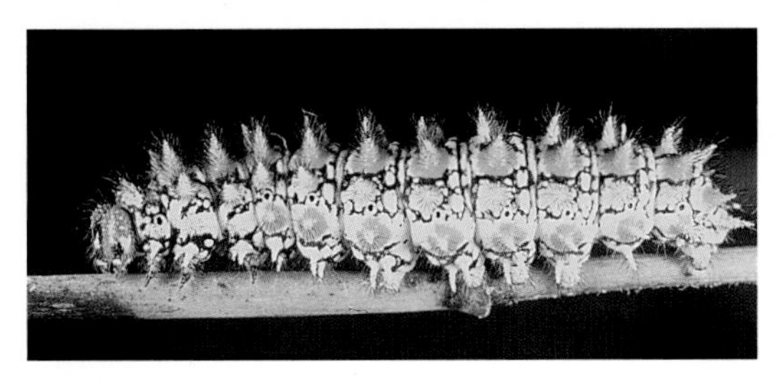

Melitaea didyma from southern France.

Melitaea didymoides from Buryatia (Siberia), Russia.

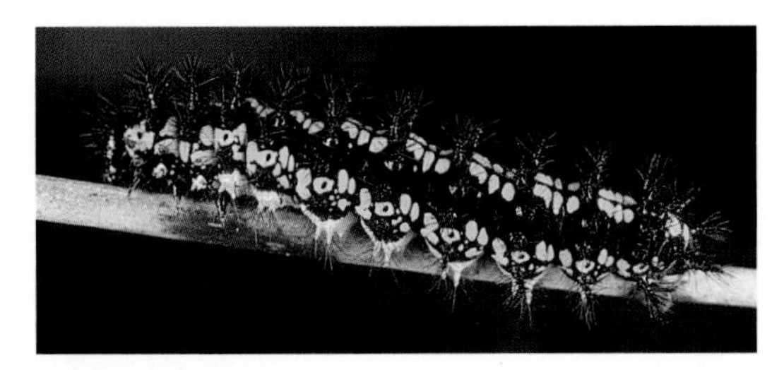

Euphydryas maturna from Finland.

Melitaea deione or *M. athalia celadussa* from southern France.

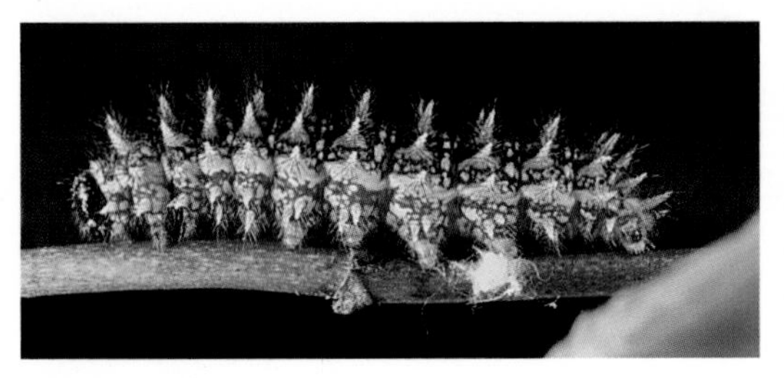

Melitaea parthenoides from southern France.

Melitaea trivia from Spain.

Photos by Niklas Wahlberg.

XII. Habitats of *Melitaea cinxia* and *M. diamina* in Finland

1. Habitat of *Melitaea cinxia* in the Åland Islands in southwestern Finland. Photos by Marko Nieminen and Tapio Gustafsson.

2. Habitat of *Melitaea diamina* near Tampere, southern Finland. Photos by Janne Heliölä.

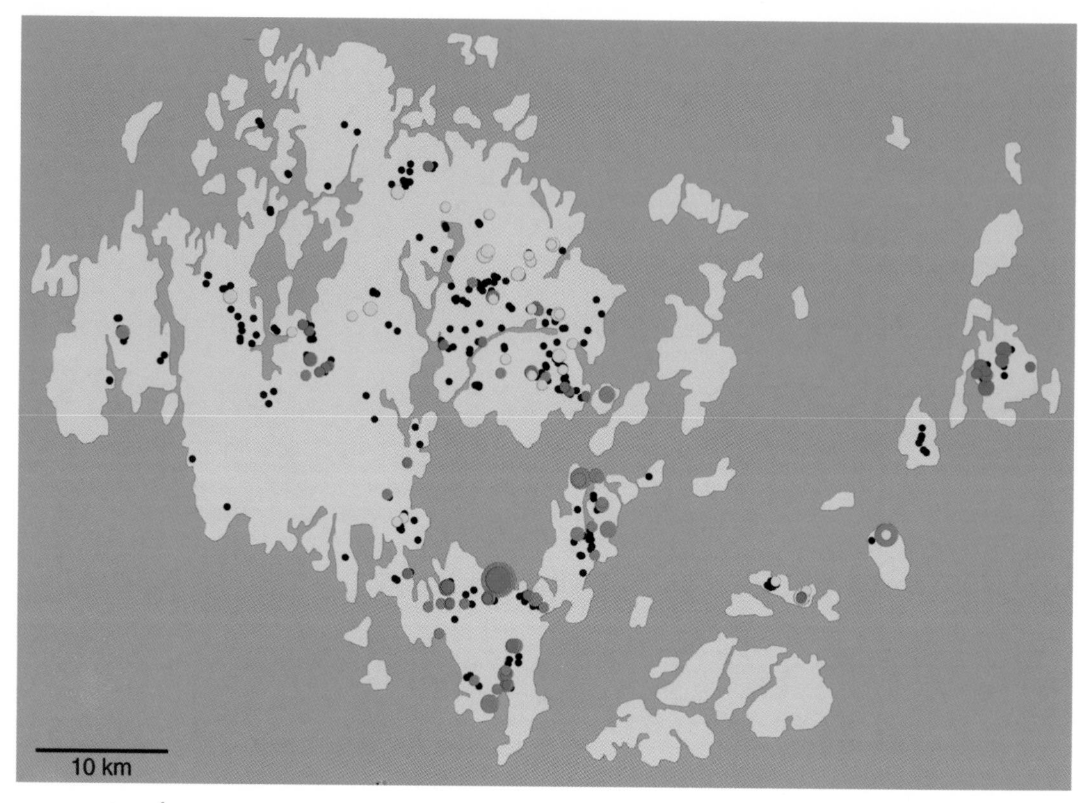

3. Map of the Åland Islands showing the distribution of four rare alleles of a microsatellite locus (CINX38) in *Melitaea cinxia*, superimposed on the distribution of local populations (black dots) in 1995. The size of the circle is proportional to the number of allele copies in the population. The identities of the four alleles are 2 (red), 7 (green), 8 (blue), and 9 (yellow) (see figure 10.2).

4. An aerial photograph of a small village in the Åland Islands showing two habitat patches for *Melitaea cinxia*.

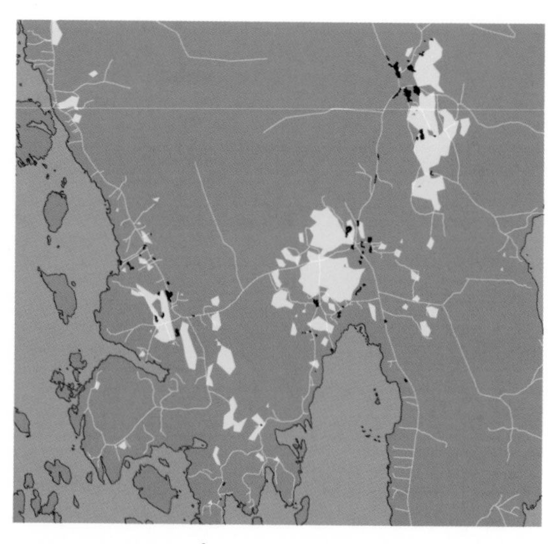

5. One part of the Åland Islands showing typical aggregated distribution of habitat patches for *Melitaea cinxia*. The habitat patches are black, cultivated fields are yellow, water (the Baltic) is blue, and the rest (mostly forests) are green.

The observed superiority of *V. spicata* as a host plant has many possible explanations. Host-occupied *V. spicata* may be easy for female parasitoids to find, or, once larvae on *V. spicata* are discovered, they may be easy to parasitize, or host larvae feeding on *V. spicata* may not be able to defend themselves against the parasitoid eggs or larvae as well as larvae feeding on *P. lanceolata*. Our results support the first explanation rather than the latter two explanations. Possibly, female wasps searching for host larvae find them more easily on *V. spicata* than on *P. lanceolata* because the larval nests on *V. spicata* are generally located closer together than are larval nests on *P. lanceolata* (van Nouhuys and Hanski 1999). Preliminary studies also suggest that *C. melitaearum* is more attracted to the odor of host-occupied *V. spicata* than to the odor of host-occupied *P. lanceolata* (figure 8.4; Anton and van Nouhuys, unpubl. data). Once the host larvae have been located by the wasp, the consequences of the host plant species for the parasitoid are ambiguous. Lei and Camara (1999) found in a laboratory study that host larvae reared on *P. lanceolata* were more readily attacked than larvae reared on *V. spicata* (the larvae were exposed to parasitoids without the host plants). In contrast, Anton and van Nouhuys (unpubl. data) found the opposite: larvae reared on *V. spicata* being more readily attacked. The clutch size and survival of wasp progeny was equal among larvae feeding on the two plant species.

Most other checkerspot butterflies also use several host plant species, which can influence the parasitoid community and population dynamics. For instance, *E. editha bayensis* populations use two host plant species that differ phenologically and that differ in their suitability for larvae depending on the season, weather, and topography. Larvae do not often stay in groups for long, but they move individually from senescent to healthy plant individuals and species (chapters 3 and 7). Such a complex and variable relationship with host plants may render plant-derived host cues unreliable, reducing the efficiency of foraging parasitoids and ultimately the sizes of their populations. This could contribute to the apparent small role of parasitoids in *E. editha* population dynamics (table 8.3). In *E. editha bayensis* in particular, there is typically high larval mortality induced by host plant senescence (chapter 3), which will inevitably lead to high parasitoid mortality as well.

Competition among Natural Enemies

Any predator or parasitoid that has an impact on the host population size is potentially involved in competitive interactions with other natural enemies. Thus the egg, larval, and pupal parasitoids and predators of checkerspots, which have at least occasionally a large impact on local host population size (Ford and Ford 1930, White 1973, 1986,

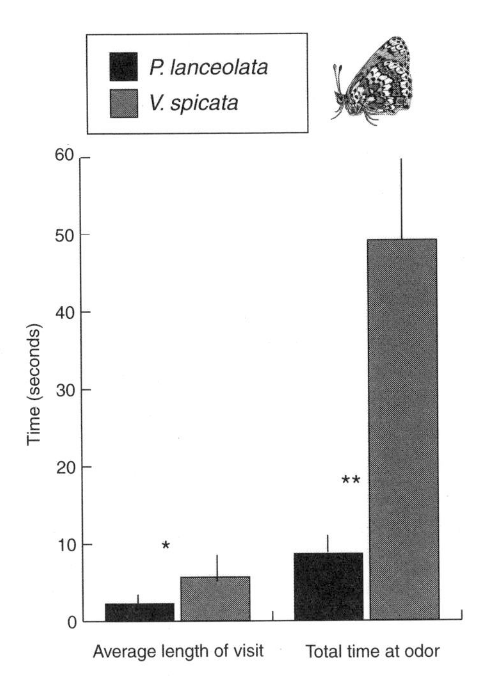

Figure 8.4. The relative attractiveness of the odor of herbivore-damaged *Plantago lanceolata* and *Veronica spicata* to the parasitoid *Cotesia melitaearum*. The gray bar is the host-damaged *V. spicata* + frass treatment. The black bar is the host-damaged *P. lanceolata* + frass treatment. The average visit length is the amount of time a wasp spent palpating and walking during one visit to an odor source. The total time at odor is the average sum of all visits to the odor source during the 5-minute observation period (n = 15 independent choice trials under laboratory conditions, $^{*}p < .05$; $^{**}p < .005$).

Moore 1989a, Lei and Hanski 1997), most likely compete with each other.

The two primary larval parasitoids of *M. cinxia* in the Åland Islands, *C. melitaearum* and *H. horticola*, coexist in about 10% of host populations, and the majority of the remaining host populations are occupied by *H. horticola* alone, as this species has a wider range of dispersal than the host and a much wider dispersal range than *C. melitaearum* (van Nouhuys and Hanski 2002a). Coexistence of the competing parasitoids would be enhanced if the superior disperser would be the inferior competitor locally (Nee et al. 1997). *Hyposoter horticola* is always the first one to parasitize a host individual because it oviposits in first instar host larvae while they are still within the eggshell (van Nouhuys and Ehrnsten 2004). *Cotesia melitaearum* does not appear to distinguish between unparasitized hosts and those parasitized by *H. horticola* (E. Punju unpubl. data). Therefore, upon hatching, host larvae already parasitized by *H. horticola* are still vulnerable to parasitism by two or three generations of *C. melitaearum*. In natural butterfly populations occupied by both wasp species, successful parasitism by *H. horticola* is lower than in populations in which it is alone, suggesting that *C. melitaearum* has a substantial competitive impact on *H. horticola* (Lei and Hanski 1998). Laboratory experiments have revealed that both parasitoids suffer from competition, but also that the outcome of competition depends on when, during the host life cycle, the larva is parasitized by *C. melitaearum*. Competition between the two parasitoids takes the form of both physical combat and physiological suppression (E. Punju, unpubl. data). Unexpectedly, it is only during the second (autumn) generation of the three *C. melitaearum* generations that *H. horticola* is the weaker competitor (E. Punju, unpubl. data; figure 8.5). One might have expected *H. horticola* larvae to perform poorly in competition in the smallest host larvae, in which the size difference between the parasitoids should favor *C. melitaearum*, but at this stage *H. horticola* is the superior competitor (figure 8.5).

Natural Enemies of the Natural Enemies

A major source of mortality of insect natural enemies is other natural enemies, in the forms of predation and hyperparasitism, intraguild predation and parasitism, and incidental consumption along with the host (Rosenheim et al. 1993, Godfray 1994, Stiling and Rossi 1994, Colfer and Rosenheim 2001, Snyder and Ives 2001). Because we know relatively little about generalist predators of checkerspots, we also know little about incidental and intraguild predation. What we do know about the hyperparasitoids and predators of *C. melitaearum* in Åland supports the notion that secondary natural enemies play an important role in insect population dynamics and community structure.

The two primary parasitoids of *M. cinxia* in the Åland Islands both suffer substantial parasitism, even though for one of them, *H. horticola*, such mortality is spatially and temporally relatively constant, while for the other one, *C. melitaearum*, mortality is variable and possibly density dependent. As we have already mentioned, *H. horticola* parasitizes about a third of the host larvae in Åland, of which about a third again are parasitized by the solitary hyperparasitoid *M.* sp. cf. *stigmaticus* (figure 8.2). This pattern has remained relatively stable over hundreds of local populations over several years (van Nouhuys and Hanski 2002b). In contrast, the generalist pseudohyperparasitoid *Gelis agilis* (and other less common *Gelis* species) attacking *C. melitaearum* appears to aggregate in response to high host density (Lei and Hanski 1997, van Nouhuys and Hanski 2000), and the fraction of *C. melitaearum* cocoons parasitized by *Gelis* varies from none to virtually 100% in different local populations (figure 8.6). Predation of *C. melitaearum* cocoons can also be high, and it varies greatly both spatially and temporally (figure 8.6). Together the parasitoids and predators of *C. melitaearum* probably cause local extinctions of high-density populations (Lei and Hanski 1997, van Nouhuys and Tay 2001). Because *Gelis* have many host species and aggregate, the local density of alternative hosts for *Gelis* can influence its impact on *C. melitaearum*. Such an indirect interaction, called apparent competition (Holt 1977), was demonstrated in a field experiment in which alternative hosts for *Gelis* (*Cotesia glomerata*) were placed in substantial numbers in natural *C. melitaearum* populations and the population sizes of *C. melitaearum* were compared with untreated control populations (van Nouhuys and Hanski 2000). All three treated populations declined, two of them to extinction, while the control populations remained relatively stable.

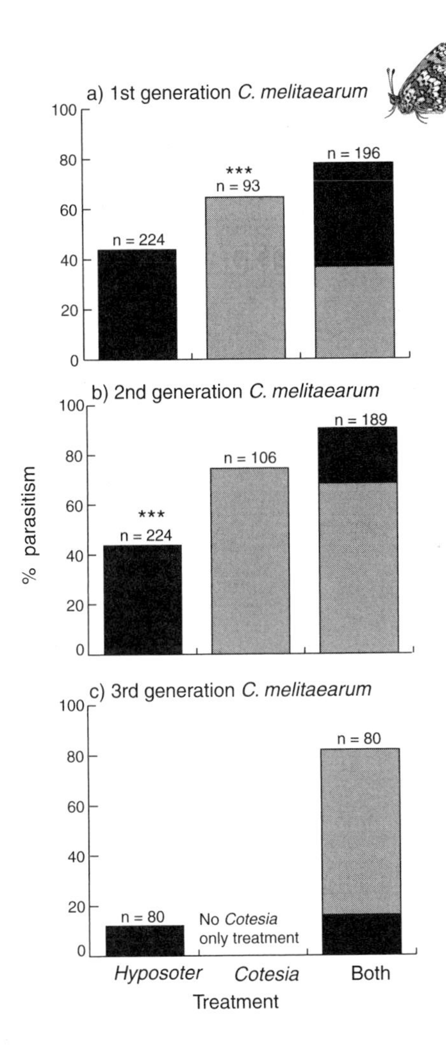

Figure 8.5. The outcome of direct competition between immature parasitoid wasps within *Melitaea cinxia* larvae in laboratory experiments. The gray bars represent successful parasitism by *Cotesia melitaearum*, and the black bars represent successful parasitism by *Hyposoter horticola*. (a) First-generation *C. melitaearum* attacks second instar *M. cinxia* larvae. If the host is already occupied by *H. horticola*, then *C. melitaearum* suffers ($\chi^2 = 12.66$, $p < .001$). (b) Second-generation *C. melitaearum* attacks fourth instar *M. cinxia* larvae. If the host is already occupied by *H. horticola* then *H. horticola* suffers ($\chi^2 = 20.25$, $p < 0.001$). (c) Third-generation *C. melitaearum* attacks sixth instar *M. cinxia* larvae. If the host is already occupied by *H. horticola*, then *C. melitaearum* dies (no significant difference in success of *H. horticola* with and without *C. melitaearum*). Data for panels (a) and (b) are from E. Punju (unpubl.), and for (c) from van Nouhuys and Tay (2001). Note that the initial rate of parasitism by *H. horticola* in (c) was lower than in (a) and (b).

If *Gelis* were absent, and not replaced by an equivalent other natural enemy, the population sizes of *C. melitaearum* would probably increase rapidly, and the butterfly populations would in turn decrease in size and be more prone to local extinction. The deletion of the hyperparasitoid *M.* sp. cf. *stigmaticus* from the species chain *M. cinxia–H. horticola–M. stigmaticus*, in contrast, would probably have no effect on host population dynamics because *H. horticola* is extremely efficient and parasitizes a fraction of almost every egg cluster in Åland. Eggs that escape parasitism have been found by *H. horticola*, but they remain unparasitized because the eggs in one batch mature over several hours (plate X) but are attended by an ovipositing wasp for less than one hour (van Nouhuys and Ehrnsten 2004; figure 8.3). Presumably, it would not pay for the wasp to spend several hours attending one egg cluster.

Spatial Population Dynamics of Interacting Species

At large spatial scales, species interactions may depend on both the structure of the landscape and the dispersal behavior of the interacting species. There is an extensive body of theory dealing with interspecific interactions in fragmented landscapes (Holt 1997, 2002, Nee et al. 1997, Hanski 1999b, Hassell 2000b). Prey and inferior competitors may escape enemies and superior competitors by dispersing to currently unoccupied habitat (temporary refuge). Over time, space may allow for sufficiently asynchronous local dynamics to be maintained to pre-

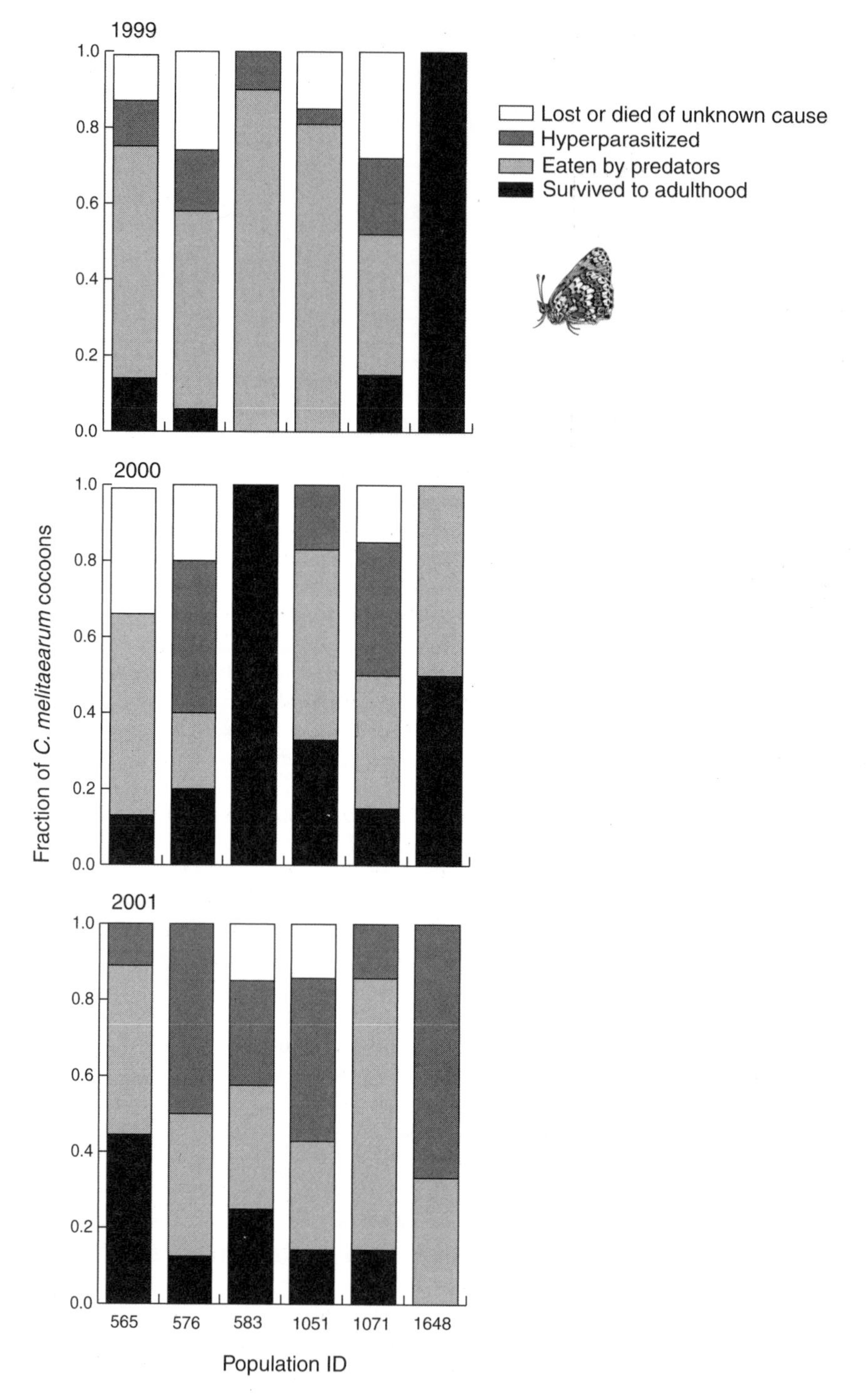

Figure 8.6. The fate of cocoons of the spring generation of *Cotesia melitaearum* in six populations in 1999, 2000, and 2001. Data for 1999 and 2000 are from van Nouhuys and Tay (2001).

vent metapopulation extinction. A critical issue is the rate and the scale of dispersal, which should be sufficient to allow for recolonization of currently empty habitat but not so extensive as to eliminate spatial asynchrony in local dynamics.

Butterfly movements within and among habitat patches have been much studied (Baguette and Nève 1994, Hill et al. 1996, Roland et al. 1999, Baguette et al. 2000, Ries and Debinski 2001), including the movement behavior of checkerspots (chapters 3 and 4). Checkerspots are convenient for the study of spatial population dynamics because the butterflies are relatively sedentary and because the habitat patches they use can often be delimited easily. In contrast, large-scale movements of insect predators and parasitoids are difficult to study and have rarely been measured in any system (Godfray, 1994; for exceptions, see Hopper 1984, Antolin and Strong 1987).

Studies conducted in the Åland Islands have produced unique data on the rates and ranges of colonization by *M. cinxia* and its two primary parasitoids, measured in field experiments and inferred from long-term observations on the presence and absence of the species in habitat patches (van Nouhuys and Hanski 2002a). This research shows that *M. cinxia* persists as a classic metapopulation in the Åland Islands (chapter 4). van Nouhuys and Hanski (2002a) found that *H. horticola*, the large ichneumonid, experiences the host as a single, patchily distributed population because of its high rate of dispersal and long colonization distances. In contrast, most of the local butterfly populations are temporally or more permanently inaccessible to *C. melitaearum*, the small braconid wasp, which has a limited dispersal range and therefore persists only in tightly clustered networks of host populations (figure 8.7). At the regional scale, the butterfly may escape *C. melitaearum* by colonizing empty habitat, but host dispersal does not limit parasitism by *H. horticola*, which consequently must be limited by local interaction. *Hyposoter horticola* suffers where it competes directly with *C. melitaearum*, but it mostly avoids direct competition because the majority of *H. horticola* populations occur outside the range of dispersal by current *C. melitaearum* populations.

The three-species chain *Melitaea cinxia–Hyposoter horticola–Mesochorus* sp. cf. *stigmaticus* forms a tightly linked group of interacting species. Hyperparasitoids are generally thought to be less specialized than primary parasitoids (Gordh 1981, Brodeur 2000), presumably because it is difficult to be a successful specialist at a high trophic level (Pimm 1991, Holt 1997). Both *H. horticola* and *Mesochorus* sp. cf. *stigmaticus* are very good dispersers, which facilitates metapopulation-level persistence. The island of Kumlinge, east of the main Åland Island (figure 2.3), is a notable exception. Here the hyperparasitoid is absent, but the butterfly and the primary parasitoid are present, most likely because the relatively small habitat patch network makes long-term persistence of the specialist hyperparasitoid unlikely.

Melitaea cinxia lives in a highly fragmented landscape in the Åland Islands, and hence its local populations are prone to extinction (chapter 4). In other localities, checkerspot butterflies and their natural enemies live in less fragmented landscapes and have

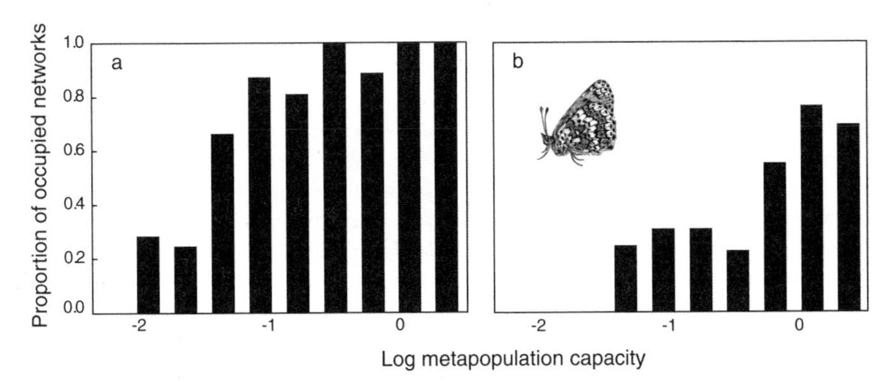

Figure 8.7. The fraction of habitat patch networks occupied by (a) *Melitaea cinxia* only and (b) those occupied by both *M. cinxia* and *Cotesia melitaearum* as a function of the metapopulation capacity of the network (van Nouhuys and Hanski 2002a). See box 12.1 for an explanation of metapopulation capacity.

larger populations. It is possible that *Cotesia* parasitoids in these less fragmented landscapes have generally larger populations and influence the host populations more strongly than in the Åland Islands. Checkerspot-associated *Cotesia* are relatively sedentary and typically have a high intrinsic rate of population increase because they are gregarious and have several generations per host generation. At high host density in an unfragmented habitat, this may lead to a parasitoid outbreak like the ones described by Ford and Ford (1930): a dramatic decline and increase, and then again a decline of *Euphydryas aurinia* populations near Carlisle, UK, in 1881–1929 (figure 1.5), accompanied by increase, decrease, and increase of parasitism by *C. bignellii*. Without having actual data on this, we may speculate that *Cotesia* populations are often kept in check by their own effective natural enemies, such as *Gelis agilis* in the Åland Islands, regardless of the habitat structure. Either way, comparison of the spatial population dynamics of the parasitoids in more and less fragmented landscapes would be welcome.

8.4 Conclusions and Outstanding Research Questions

Considering the numbers of species of predators and parasitoids in insect communities and their abundances, relatively little is known about them (Shaw 1994). This is in spite of the indisputable ecological importance of predators and parasitoids (Hochberg and Ives 2000, Shaw and Hochberg 2001), and their economic importance in pest control (Hochberg and Hawkins 1994, Barbosa 1998). Our lack of understanding of their taxonomy, natural history, and broader ecology is especially clear when we consider their absence from conservation plans. Invertebrate natural enemies, particularly parasitoid wasps, are prone to extinction because they are both high on food chains and specialized, yet they are part of very few conservation programs worldwide (Hochberg 2000, Shaw and Hochberg 2001).

The natural enemies associated with checkerspot butterflies represent a convenient system in which to learn more about the systematics, natural history, and population and community ecology of predators and parasitoids because we already have such a good understanding of the host ecology. In this final section we discuss four areas for further research, which are all relevant both for basic ecology and conservation as well as for the management of invasive species and biological control.

Role of Natural Enemies in the Dynamics of Checkerspots

We are aware of several examples of natural enemies having great influence on the dynamics of checkerspots, including the early observations by Ford and Ford (1930) of huge fluctuations in population size of *E. aurinia* over 30 years coinciding with great changes in the rate of parasitism by *C. bignellii*, and the metapopulation-level decline of *M. cinxia* in Åland after high levels of parasitism by *C. melitaearum* (Lei and Hanski 1997). Some species, like the ichneumonid *H. horticola* parasitizing *M. cinxia*, influence host population dynamics primarily by imposing a relatively constant rate of mortality in host populations (van Nouhuys and Ehrnsten 2004). The impact of yet other species remains poorly known. There is anecdotal evidence of generalist predation being occasionally very high (table 8.1), there are parasitoids that appear to be specialists but are rare, such as *Benjaminia fuscipennis* parasitizing *E. editha bayensis*, and there are pupal parasitoids, sometimes abundant, which we know little about.

Here we mention several directions for further research on the role of natural enemies in the dynamics of checkerspots. First, further study of the natural history and host ranges of the predators and parasitoids would yield useful information, fleshing out tables 8.1 and 8.3. Second, comparisons could be made of host–parasitoid interactions in communities structured around ephemeral versus more stable host plant species, among populations living in continuous versus fragmented landscapes, and among communities in which different numbers of checkerspot and host plant species co-occur. Third, population dynamics of host–parasitoid and competitive interactions could be analyzed using mathematical models parameterized for the species and ecological settings for which information is available. Two examples of this, though of limited ecological scope, are studies by Lei and Hanski (1997) and Hill and Caswell (2001). Finally, it would advance our understanding of the role of plant defensive chemistry on the natural enemies of herbivores if comparisons were available of the natural enemy communities associated with checkerspots feeding on host plants that do not confer chemical defenses (e.g., plants in the family Asteraceae) and those containing iridoid glycosides. The same question can be addressed experimentally, either by taking advantage of the natural variation in iridoid concentration (Stamp and Bowers 2000a, 2000b, Nieminen et al. 2001), by using artificial diet (Bowers and Puttick 1986, Lei and Camara 1999), or by using selected laboratory lines of the host plant species (Marak et al. 2000, Harvey and van Nouhuys unpubl. data).

Parasitoid Community Structure

Parasitoids are part of most terrestrial communities, in which they occupy several trophic levels and have complex direct and indirect interactions with many species (Hawkins and Sheehan 1994). Several theoretical expectations of community ecology can be illustrated using the parasitoid complex associated with *Melitaea cinxia* in Åland. First, as has been elucidated by theory (Nee and May 1992), the coexistence of the two competing primary parasitoids is at least to some extent facilitated by a negative association between competitive and dispersal abilities. *Hyposoter horticola* is clearly a better colonizer than *C. melitaearum*, but a worse competitor at least during one part of the host life cycle (figure 8.5; van Nouhuys and Hanski 2002a). Second, patterns in food chain length (section 8.3) support the notion that a minimum landscape area is needed to support high trophic levels (Holt 2002). The island of Kumlinge in Åland with a relatively small patch network (around 100 patches) that supports *M. cinxia* is also occupied by the primary parasitoid *H. horticola* but not by the hyperparasitoid *Mesochorus* sp. cf. *stigmaticus*. Third, a field experiment demonstrated apparent competition (Holt 1977, Bonsall and Hassell 1998) between two primary parasitoids sharing a common hyperparasitoid (van Nouhuys and Hanski 2000). A challenge for further empirical research is to analyze host–parasitoid relationships in communities of several coexisting checkerspot species and to relate such studies to predictions of the processes thought to maintain community structure.

Parasitoid Life-History Variation

Patterns of life-history variation within and among animal and plant populations reflect interactions and environmental conditions that are ecologically important. Parasitoids of checkerspots find their hosts in a range of habitats and use hosts with different phenologies. Some northern populations take several years to develop; most populations in central Europe and North America are univoltine, and southern populations are multivoltine (Tolman and Lewington 1997). There are checkerspot species that diapause in winter when it is cool and in summer during summer drought, and there are populations in which diapause is variable (chapters 7 and 11). The larvae feed on many plant species including herbs and shrubs, annuals and perennials, and the landscapes in which the host butterflies live range from vast open grasslands in Asia (I. Hanski pers. obs.) to highly fragmented meadows such as in the Åland Islands (chapter 4). *Cotesia* use conspecific host individuals ranging in size from tenths of milligrams to hundreds of milligrams, depending on the host instar. Therefore, although a parasitoid may have a narrow host range, it is able to use the host species under a variety of environmental and physiological conditions.

Characteristics of life history and behavior are likely to differ among populations of a single parasitoid species because natural selection on characters affecting longevity, fecundity, mating, and dispersal in each population may lead to local or regional specialization. Indeed, researchers have found genetic differences in life-history and behavioral traits among natural populations of parasitoids. For example, Ellers and van Alphen (1997) showed that the ratio of lipid storage to egg production differed among populations of the parasitic fly *Asobara tabida*, and Fleury et al. (1995) found differences in diurnal activity patterns among populations of the parasitoid *Leptopilina heteroma* collected from different latitudes. There are several examples of *Cotesia* populations differing in their searching behavior and attraction to the food plant species used by the host (Kester and Barbosa 1994, Benrey et al. 1997, van Nouhuys and Via 1999).

The extent to which checkerspot-specializing parasitoids have adapted to the particulars of their hosts deserves further study. For example, it would be interesting to know whether parasitoids of unrelated gregarious web-building caterpillars, such as the specialist parasitoid of the forest tent caterpillar *Malacosoma disstria* (Roland and Taylor 1997), behave like *Cotesia* attacking checkerspots. It is interesting that an unrelated specialist parasitoid of the forest moth *Hyphantria cunea* does attend the gregarious larval nests of their hosts for days (Morris 1976) like some *Cotesia* of checkerspots do. The costs and benefits of living in groups have been addressed to some extent using checkerspot larvae (chapter 7), but the costs and benefits for parasitoids of using a group-living host have not. We know that parasitism of *M. cinxia* and *E. phaeton* increases with the size of the host larval group (Stamp 1981a, Lei and Camara 1999), and *Cotesia* adults observed in the field appear to be unhindered by the silken web spun by the host larvae. However, inexplicably, they parasitize only a small fraction of the larvae available (Stamp 1982b, Lei et al. 1997). Finally, *Cotesia* parasitizing checkerspots often form very small local populations, which raises questions about mating behavior, possible inbreeding depression, and sex ratio (Crozier 1977, Kaitala and Getz 1992, Antolin 1999, Fauvergue et al. 1999).

The other side of the story is life histories of host butterflies evolving in response to their interaction with parasitoids (section 8.1). Although demonstrating host life-history evolution in experiments would be difficult, one can use the comparative approach to test predictions based on our knowledge of the host–natural enemy interaction and the biology of the species. For instance, the group-living behavior of checkerspot larvae, which shows great variation within and among the species (chapter 11) and which may have evolved in response to mortality caused by natural enemies, would be an interesting topic for comparative research. In another vein, parasitism of *E. aurinia* by *C. bignellii* late in the season has been shown to lead to a male-biased butterfly population (Porter 1984). How common such an effect might be is unknown, but butterfly dynamics are greatly influenced by adult sex ratio (chapter 5).

Parasitoid Phylogenies and Host Range

The host range of a parasitoid is influenced by the physiology, phenology, feeding niche, host food plant, habitat type, and evolutionary history of potential host species (Shaw 1994). Of the well-known Old World species, *Cotesia melitaearum* and *C. acuminata* use many checkerspot species feeding on different host plants over a large geographic area, whereas *C. bignellii* and *C. cynthia* appear to be entirely species specific (table 8.2). However, recent molecular results suggest that the picture is more complicated (M. Kankare and M. Shaw pers. comm.). For instance, a distinct haplotype of *C. melitaearum* has been reared from *Melitaea athalia* in the Åland Islands. These individuals have not been observed to parasitize *M. cinxia* (S. van Nouhuys pers. obs.); hence there may be two cryptic species using the two syntopic checkerspots in the Åland Islands. *Cotesia acuminata* is also likely to include two or several species using different host butterflies (M. Kankare and M. Shaw pers. comm.). Further studies will show the extent to which *Cotesia* associated with checkerspots have evolved or perhaps speciated in concert with their hosts. In other cases, parasitoids may have been specialized at least locally to host plant chemistry or habitat type rather than to host species. It will be particularly exciting to examine *Cotesia* in communities of many checkerspot species in which strong ecological dynamics may become intimately intertwined with local adaptation.

9

Dispersal Behavior and Evolutionary Metapopulation Dynamics

MICHAEL C. SINGER AND ILKKA HANSKI

9.1 Introduction

"Dispersal is the glue that keeps local populations together in a metapopulation" (Hansson 1991). Seen from the perspective of population biology, dispersal is also another kind of glue, a behavioral trait that couples ecological metapopulation dynamics with evolutionary dynamics. Indeed, we might describe our work on this topic as a study in either behavioral or evolutionary ecology, were it not that "behavioral ecology" is not regarded as proper ecology by ecologists, and "evolutionary ecology" is not really regarded as study of evolution by evolutionary biologists. Sadly, these researchers are usually correct; work that claims to be interdisciplinary at the interfaces between behavior, ecology, and evolution rarely achieves its aim. Yet metapopulation biology absolutely requires a truly behavioral understanding of the decision making that underlies dispersal, a truly ecological understanding of the interplay between dispersal and population dynamics, and a truly evolutionary understanding of the ways that dispersal responds to spatially and temporally varying environmental conditions. Studies of checkerspots have made substantial progress toward achieving this desirable combination of knowledge, and reciprocal interactions between the behavioral, ecological, and evolutionary facets of dispersal are becoming clear. For example, the influence of host plant preference (behavior) on dispersal drives population turnover (ecology) via biased immigration and colonization rates (Hanski and Singer 2001), while biased colonization rate, in turn, contributes to the evolution of host plant preference (Hanski and Heino 2003), bringing us back to the influence of preference on dispersal. Before reaching this level of complexity, however, we begin with a discussion of the classic approaches to the study of insect dispersal and how they have been applied in studies of butterflies in general and checkerspots in particular.

Studies of Insect Movement and Their Relevance to Butterflies

Both short-range dispersal, which characterizes all organisms, and long-distance migration, which characterizes only a few, have traditionally been studied as adaptations to spatial and temporal heterogeneity in habitat quality (Kennedy 1961, Southwood 1962, Hamilton and May 1977, Roff 1994, Mathias et al. 2001). The classic research on the responses of dispersers to habitat parameters was heavily, and understandably, biased toward species in which the likelihood of dispersal was easily assessed by morphological traits. These species include plants with bimodal distribution of seed size or shape (Venable 1979, Venable and Lawlor 1980, Olivieri et al. 1983, Imbert 2001) and insects with winged (or long-winged) and wingless (or short-winged) forms (Southwood 1962, Vepsäläinen 1973, Dingle et al. 1980, Roff 1986, 1994). Wing polymorphism in insects reflects genetic variation in some cases, while in other cases wing polymorphism is a direct response to environmental conditions (Roff 1986, Solbreck and Andersen 1989). In either case the influence of habitat parameters on dispersal can be surmised from comparisons of the frequencies of wing morphs under dissimilar environmental conditions.

In the majority of insect species, differences in dispersal tendency or capacity are not clearly reflected in morphology; rather, such differences are detectable only at the behavioral level. Lack of easily measured morphological traits presents an obstacle to the assessment of predilection to disperse, but occasionally the problem can be solved with simple behavioral assays. For example, a grasshopper held by a rod fixed to the thorax will vibrate its wings when its legs lose contact with the substrate. Variation among grasshoppers in the duration of such wing vibration reflects variation in the distance they would fly if at liberty (Rankin and Burchsted 1992). Such behavioral assays have permitted estimates of heritability of dispersal tendency in several species (Dingle 1968, 1991, Caldwell and Hegmann 1969, Roff and Faibairn 2001). These estimates have typically been high (Roff and Faibairn 2001), suggesting that dispersal is a heritable trait capable of rapid response to natural selection. Hence, dispersal behavior is an appropriate subject of evolutionary models (Ronce et al. 2000, Clobert et al. 2001, Heino and Hanski 2001).

How have butterflies figured in studies of insect movement behavior? References in Ford's (1945) classic book on butterfly evolutionary ecology deal exclusively with long-distance migration (e.g., C. B. Williams 1930), with no discussion of shorter movements. Yet butterflies have considerable potential for experimental and observational studies of local and regional dispersal. Butterflies can be obtained in large numbers, and they can be easily labeled as individuals (figures 5 and 3.5), which is one of the reasons Ehrlich initiated the long-term study of *Euphydryas editha* (chapters 3 and 12). For experimental studies of movement, butterflies offer advantages both over elephants, which are hard to obtain in sufficient numbers, and tend to resist experimental manipulation, and over *Drosophila*, which can be obtained in numbers but not tracked as individuals in the field.

As an example of the benefits and limitations of working with *Drosophila*, consider the experiments in which biologists have investigated learning of resource preference by newly emerged adults. Flies were exposed to different resources and those with different experience marked with different colors of fluorescent dye. They were then released and recaptured after they had been naturally attracted to different baits (Hoffman 1985). The question asked was whether flies that had different experiences of resource encounter were differently attracted to the baits under natural conditions. The experiments worked well, partly because of the large numbers of flies that could be used, but *Drosophila* cannot yet be individually marked, which is a disadvantage when we want to know not only where individuals end up but also how they get there. With butterflies, this can be achieved, and individuals can be tracked by being observed or captured repeatedly at multiple points during their lives (Ford 1945, Ehrlich 1965, Moore and Singer 1987). Tracking individuals and plotting their flight paths across a landscape (figure 9.1) allows deduction and modeling of rules of movement, which potentially leads to understanding of the mechanisms that generate distributions of individuals in space (Jones 1977, Mackay and Singer 1982, Kareiva 1983, Turchin 1991, 1998, Boughton 2000, Hanski et al. 2000, Ricketts 2001).

Our ability to observe butterfly movements directly allows us to record their behavioral responses to habitat boundaries, a key determinant of the distribution of individuals across landscapes. Different species show different responses to habitat boundaries, whether observed in the field (Ricketts 2001, Ries and Debinski 2001) or in outdoor cages large enough to allow seminatural behavior (Norberg et al. 2002). Ries and Debinski (2001) compared a habitat generalist and a specialist species. Both species were deterred to some extent from crossing habitat boundaries, but the specialist was deterred more than the generalist. Once the boundary had been crossed, the generalist showed no tendency to return, but the specialist did. Also, the specialist was deterred from leaving habitat patches of high conspecific density, whereas the generalist was not. Conradt et al. (2000) released butterflies at different distances from habitat patches and observed the distances from which patches could be detected. They also observed that butterflies released in nonhabitat performed systematic searches, describing oval flight paths of increasing size with repeated returns to the point of release.

Studies of butterfly movement have been beset by disagreements about the degree of control that fly-

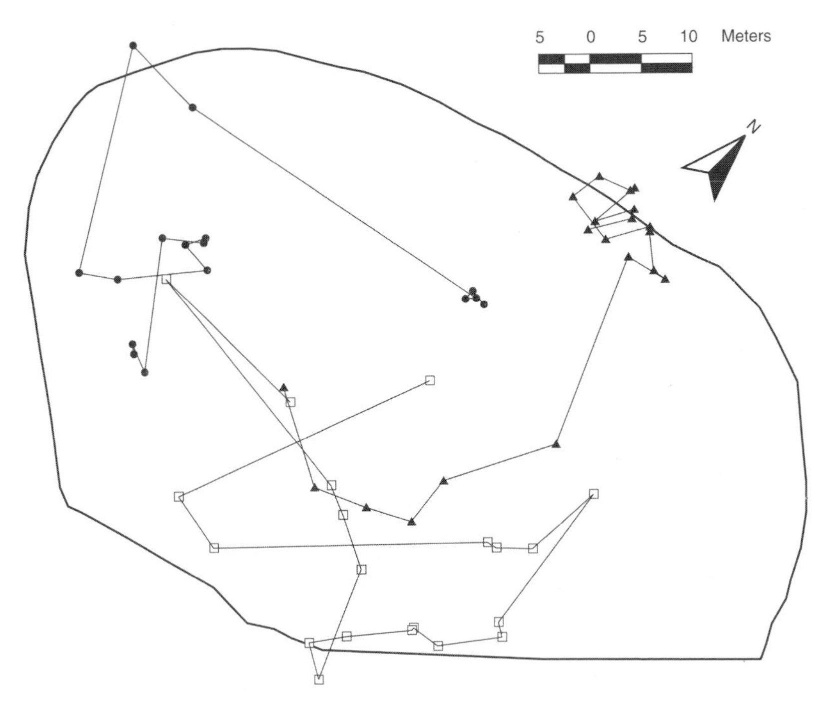

Figure 9.1. The flight paths taken by three individuals of *Proclossiana eunomia* within a habitat patch. Places of landing are indicated by an individual-specific symbol (M. Baguette and N. Schtickzelle, unpubl. data).

ing insects exert over their destinations. The possibility for such control depends on the mass and wing loading of the insect. Once a winged aphid has released its grasp on the substrate and left the boundary layer, the distance and direction that it travels are determined entirely by the wind (Johnson 1969). Only on reentering the boundary layer, which it apparently cannot do at will, is the aphid able to exert some control over its destination and express its preferences for alighting on particular substrates (Kennedy and Booth 1963). Aphids do become associated with particular resources in space, but they do so by repeated decisions to leave unacceptable habitats and by sampling new ones downwind, not by directed flight across the landscape. Even in larger insects, such as locusts and some butterflies and moths, the destinations of long-distance migrations are often predictable from wind speed and direction, regardless of the apparent heading of the migrants in a particular direction (Johnson 1969).

Early attempts to measure and model the evolution of butterfly movements assumed that butterflies were not aphids and that natural selection could act on the direction as well as on the distance moved. These attempts (e.g., Baker 1968) met vehement opposition from students of insect movement, who considered wind of paramount importance (Johnson 1969). If the insect cannot control its direction of flight, then natural selection cannot act upon it. Butterfly biologists have tended to think that butterflies are indeed sensitive to wind, but that, unlike locusts and aphids, they use this sensitivity to avoid flight under conditions where they would have no control over their destinations (Cullenward et al. 1979, Singer 1984). Recent advances in this field indicate that flying insects, including butterflies, can compensate for wind-imposed deviations in their flight paths, in some cases by using landmarks, but in other cases even while flying over water in the absence of landmarks (Perez et al. 1997, Oliviera et al. 1998, Srygley 2001, Srygley and Oliveira 2001). By some combination of avoiding flight in windy conditions and compensating for wind-induced displacement, migratory Monarch butterflies manage to arrive at exactly the same locations each winter. By the same means (and this brings us explicitly to our study organisms), discrete populations of checkerspot butterflies maintain their integrity despite winds that could scatter individuals across the landscape.

9.2 Checkerspots

Checkerspot butterflies lack traits for which insects were chosen in the early studies of movement by flight. Checkerspots have no distinct dispersing and nondispersing forms, they make no migrations to predictable destinations, and they rarely undertake long-distance movements (Ehrlich 1961b, 1965, Harrison 1989, Hanski et al. 1994, Kuussaari et al. 1996, Hanski 1999b, 2000). These features may seem to render checkerspots deficient as tools for studying insect dispersal. Indeed, checkerspots were passed over by researchers whose principal interests lay in either regional dispersal or long-distance migration. However, checkerspots are in all likelihood rather typical of winged insects. By analogy with Ehrlich's (1965) argument that he chose a checkerspot for an evolutionary study because it was such an ordinary insect and not known to be undergoing rapid evolutionary change, we could likewise argue that checkerspots are appropriately "normal" insects for understanding "normal" dispersal, lacking as they do the outstanding and relatively unusual traits that have drawn workers to wing-dimorphic species or to conspicuous migrants.

Checkerspot population biology was initially studied because their relatively sedentary nature made it feasible to repeatedly recapture and track individuals. Ford and Ford (1930) took advantage of the sedentary nature of checkerspots in their 55-year-long population dynamic study (chapter 1, figure 1.5), and the pioneering studies by Ehrlich in the 1960s used direct measures of butterfly movement as a means of defining population structure. These studies stimulated Turchin, Odendaal, Mackay, and others in the 1980s to measure the lengths of checkerspot flight paths, turning angles, and responses to conspecifics encountered in flight in order to model the processes that generate the spatial distributions of butterflies. More recently, the model system status that *Melitaea cinxia* has achieved in metapopulation biology has motivated studies by Kuussaari, Nieminen, Hanski, Setchfield, Ovaskainen, and others on the factors that influence decisions to emigrate from a habitat patch and on the criteria used to select a new patch. These studies rely on the ability to measure dispersal directly and to obtain large numbers of observations both of insects that move and of those that do not. Typically, a high proportion of marked individuals are resighted in the same habitat patch where they were marked, while many of those that change patches move sufficiently short distances to allow their new locations to be readily observed (more about this below).

The dispersal properties of checkerspots fit them perfectly for studies concerned with the dynamics of unstable local populations connected by dispersal (Ehrlich 1961b, Hanski 1999b). These studies provide practical examples of incorporating dispersal into testable models of real metapopulations (Hanski 1994b, Wahlberg et al. 1996), of estimating the rates of transfer among populations, and of estimating mortality during dispersal and of distinguishing disappearance caused by death from disappearance caused by emigration (Hanski et al. 2000, Petit et al. 2001, Ovaskainen 2003). Ecological knowledge of dispersal sets the stage for understanding the evolution of dispersal rate in response to ecological parameters such as the degree of habitat fragmentation (Heino and Hanski 2001, Hanski et al. 2002).

Manipulability of Checkerspots

A checkerspot is a manipulable insect, barely distracted by handling from its current task, whether it be courtship, oviposition, or dispersal. A male placed behind a female will court it; a female placed upon a host will assess it for oviposition; and a bevy of butterflies released in a habitat patch will form a population at least superficially resembling a natural one. Capturing and releasing other butterfly species may dramatically affect their movements, causing them to avoid the site of capture (Singer and Wedlake 1981, Morton 1984). Not so with the released checkerspot, which typically puts on a brief, rapid flight for two or three seconds and then departs the site of release at its normal pace and is most likely to be recaptured or resighted within 50–100 m (Ehrlich 1961b, Brussard et al. 1974). The influence of various variables on dispersal can thus be investigated experimentally in artificial populations. For example, releases of butterflies at experimentally determined densities have tested effects of density on movement (Kuussaari et al. 1996).

Checkerspot manipulability combines serendipitously with the large data set on metapopulation structure and dynamics that has been assembled for *Melitaea cinxia* in the Åland Islands in Finland. This data set makes it possible to obtain butterflies from populations of known current size, age, and dynamic history, and checkerspot manipulability allows their release in currently unoccupied habitat patches and allows observations of their residence times and movements among patches. This combination of opportunities has facilitated the investigation of the influence of population history on dispersal, as described in section 9.3.

There is an additional property of checkerspots that has allowed us to use them to tackle novel questions about dispersal. Host-plant oviposition preference, expressed as a response to tasting plants after alighting on them, can be measured either in the greenhouse or in the field and is often variable among individuals and populations (chapter 6). When particular host plants are concentrated in particular habitat types, insects that differ in their host plant preference may come to differ in habitat preference. In this case, different individuals will have dissimilar perceptions of the relative qualities of the same set of habitat patches. Thus it becomes possible to study not only the influence of habitat quality on dispersal but also the influence of the insects' dissimilar perceptions of that quality (C. D. Thomas and Singer 1987, Singer and Thomas 1996, Hanski and Singer 2001).

Measuring Checkerspot Movements

The movements of checkerspots, like those of many other butterflies, can be conveniently measured by mark–release–recapture (MRR; Ehrlich 1961b, 1965). The principal difficulty is to cover a sufficiently large area with sufficiently high effort to yield a capture probability around 0.3 or higher (if the capture probability is lower, the estimated parameter values have very wide confidence limits). Luckily, this can be done without an army of field assistants. Two students working full time for 28 days collected a data set on *Melitaea diamina* (Wahlberg 1997a) that has allowed effective modeling of dispersal in a metapopulation of 14 local populations within an area of 4 km^2. During this time, the students marked 842 butterflies and obtained 1908 recaptures. Capture probability for males was estimated to be 0.52 in an intensively studied large population and 0.26 in the other populations (Hanski et al. 2000).

Classic statistical models of MRR data estimate movement parameters for pairs of populations (Arnason 1973, Hestbeck et al. 1991, Hilborn 1991, Ims and Yoccoz 1997). This makes sense if there are only two or a few populations, but not if there are tens of populations like in typical checkerspot metapopulation studies. Hanski et al. (2000) solved this problem by first constructing a model of how the structure of the landscape is assumed to influence movements and then estimating the parameters of that model from the data. Specifically, Hanski et al. (2000) assumed that emigration and immigration scale as power functions of the habitat patch area; that dispersal is distance-limited; and that mortality during migration is a decreasing function of the connectivity of the source population. Therefore, rather than parameterizing the transfer rate of individuals among all pairs of patches, the idea was to parameterize the responses of individuals to the general features of the fragmented landscape, such as habitat patch areas and connectivities.

Fitting the model to males of *M. diamina* led to the following conclusions (Hanski et al. 2000). First, immigration and emigration scaled as patch area to powers 0.3 and –0.2, respectively, indicating that emigration rate is lower from and immigration rate is higher to large than small patches, though the scalings with patch area are far from linear. Second, the daily probability of leaving a patch of unit size (1 ha) was estimated to be 0.13, comparable in magnitude with the daily rate of mortality while in a habitat patch (0.11). Third, <1% of daily migration distances were >1 km, though a large fraction of individual lifetime dispersal distances would be a few kilometers. And fourth, 16% of all deaths were estimated to have occurred during migration, based on the model assumption that mortality during migration (but not mortality in habitat patches) depends on patch connectivity, which varies among the patches. Figure 11.2 shows the parameter estimates for five species of checkerspots in Finland.

Petit et al. (2001) have applied the same model to a long-term data set on *Proclossiana eunomia*, a non-checkerspot European fritillary. In this species the scaling of immigration with patch area was steeper (0.7) than in *M. diamina* (0.3), and the scaling of emigration was significantly different between the two sexes, –0.47 in males and –0.29 in females (Petit et al. 2001). Thus, emigration was much more sensitive to patch area in males than in females (figure 9.2), reflecting some fundamental differences in the movement behavior of the two sexes (for discussion, see Petit et al. 2001).

Ovaskainen (2003) developed a modeling approach based on a diffusion model. The advantage of this model is that it is based on a mechanistic description of individual movements, reflected in the model parameters that can be estimated from MRR

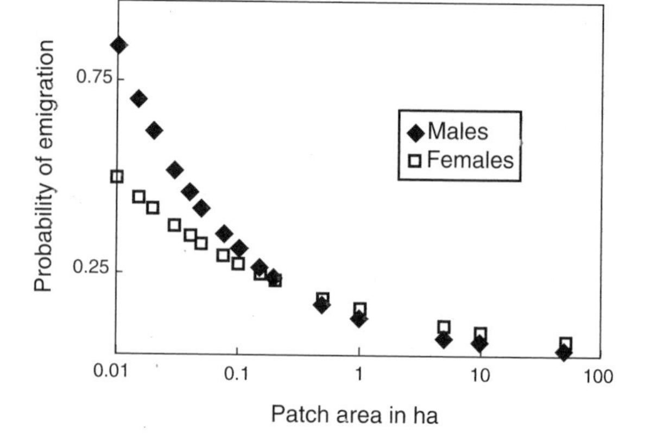

Figure 9.2. Daily probability of emigration of females and males of *Proclossiana eunomia* from a habitat patch of a given area based on the estimated parameter values of the model of Hanski et al. (2000). From Petit et al. (2001).

data. These parameters include habitat-specific mortality rate, which can be different in the breeding habitat and in possibly several matrix habitat types; habitat-specific diffusion coefficients; and a parameter that summarizes the behavior of butterflies at habitat patch boundaries as a consequent abrupt change in density (this parameter can be derived from the parameters of a mechanistic correlated random walk model describing individual movements; e.g. the probability that an individual will move toward the preferred habitat when located close to the habitat boundary). Ovaskainen (2003) fitted the model to the data set on *M. diamina*. He estimated the lifetime dispersal distance to be 1.5 km, which is consistent with the estimated daily distance of 0.2 km obtained by Hanski et al. (2000). Both models estimated the daily mortality rate at around 0.1. Ovaskainen's (2003) estimate of roughly 100-fold density difference between the habitat patches and the matrix suggests a strong tendency in the butterflies to stay in and to move to the habitat patches. The interesting novel prediction of Ovaskainen (2003) relates to the time-dependent location of individuals born in particular habitat patches in the network. Figure 9.3 gives two examples. To start with, the individual is located in the natal patch. As time passes, the probability of being in the natal patch declines, while the probabilities of being either in the matrix or in some other patch first increase, then decline, as the probability of the individual being dead gradually increases (figure 9.3). These results illustrate that a sophisticated tool box is available to extract much relevant information from the kind of observational

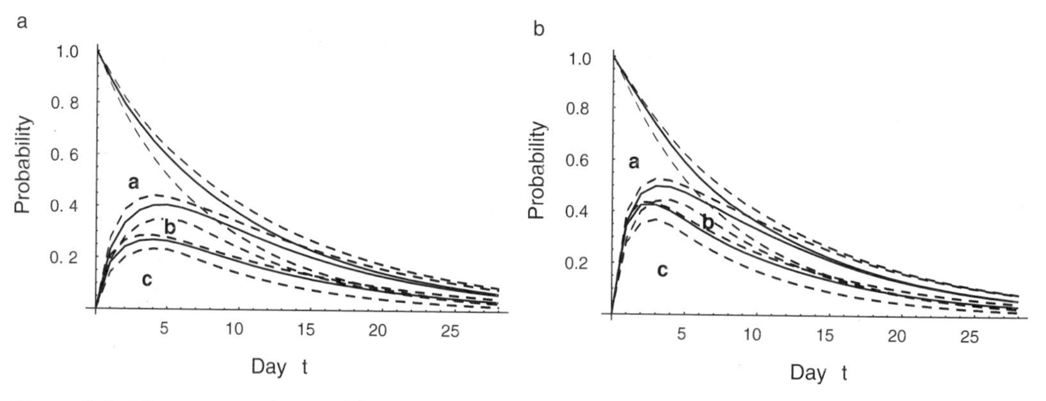

Figure 9.3. The two panels a and b in this figure show temporal changes in the probabilities that butterflies occur in (a) their natal habitat patch, (b) another habitat patch, and (c) the landscape matrix (that is, are dispersing). Panels a and b give results for two patches in the *Melitaea diamina* network discussed in the text. The continuous lines give the maximum likelihood estimates, dashed lines show the 95% confidence limits. From Ovaskainen (2003).

data that can be routinely collected from checkerspot metapopulations. One of the most significant achievements is the measurement of mortality during dispersal, which is a key factor influencing the evolution of dispersal rate (Hamilton and May 1977, Comins et al. 1980, Ronce et al. 2000, Heino and Hanski 2001) and a factor that has previously largely evaded empirical measurement.

The large-scale study of *Melitaea cinxia* in a network of some 4000 small meadows has allowed another approach to the study of movement distances complementing the MRR studies. Every autumn, after the dispersal of adult butterflies, the presence or absence of local populations in the habitat patches has been surveyed (chapter 4). These data allow one to estimate the distances across which female butterflies have successfully moved and established a new local population. Figure 9.4 shows that the probability of an empty patch turning to an occupied one in any one year dramatically declines with decreasing connectivity of the patch. Connectivity takes into account distances to all neighboring populations (potential sources of colonists) weighted by their sizes and distances to the focal patch (Hanski 1999b). The fact that well-connected, empty patches are much more likely to become colonized than poorly connected patches is reflected in the pattern of patch occupancy at any one point in time: the occupied patches are better connected than all patches on average (figure 9.4). In terms of distances to the nearest possible source population, which is not a good measure of connectivity but is easy to visualize, van Nouhuys and Hanski (2002a) found that, in most years, 95% of the colonizations occurred within 2.5 km of the nearest possible source population, and the longest recorded colonization distances were 4–5 km. These figures are entirely consistent with the modeling results of MRR data as we have described above.

9.3 Patterns in Checkerspot Dispersal

The discrete population structure and sedentary nature of checkerspots attracted Ford and Ford (1930) to use *Euphydryas aurinia* for one of the first studies employing MRR to investigate population dynamics. Later, Ehrlich (1961b, 1965), using *Euphydryas editha* at Jasper Ridge Preserve, was the first to use MRR to simultaneously study the dynamics of a set of populations and movement between them. Ehrlich's original intent had been to study the dynamics of a single population, but the numbering of butterflies as individuals allowed him to trace their movements and thereby to show that the presumed single population actually comprised three populations, among which movements were limited (figure 1.2), allowing their population dynamics to be partly independent. Although *E. editha* at Jasper Ridge do occasionally move several kilometers (Harrison 1989), they are remarkably sedentary both in relation to their physical capacity for flight and to the time spent flying, which would take them tens of kilometers per day in the absence of behavioral mechanisms to maintain discrete population structure. In a detailed study of within-population movements, Brussard et al. (1974) found that some individuals were so sedentary as to be repeatedly recaptured within the same 30-m quadrat, while others ranged quite freely around their habitat patch, which comprised about 20 such quadrats.

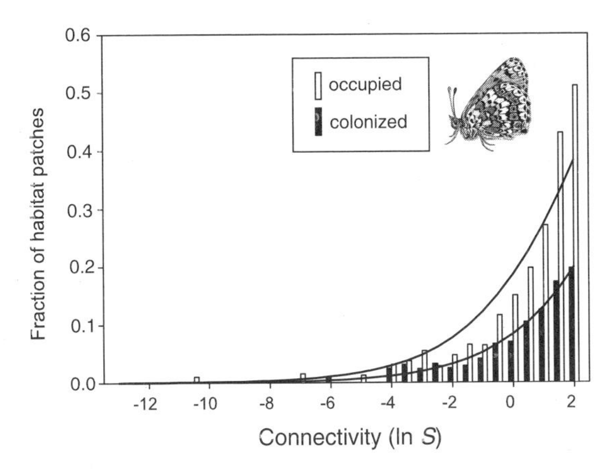

Figure 9.4. Habitat patch occupancy and colonization rate in *Melitaea cinxia* as functions of patch connectivity in a large data set collected in 1993–2000 (chapter 4). The horizontal axis gives the logarithm of *S*, a measure of connectivity (Hanski 1999b). The open bars give the proportion of patches occupied by the butterfly, the filled bars the proportion of empty patches that became colonized in one year. The two lines show the relationships based on fitting the incidence function model (Hanski 1994b) to these data. From van Nouhuys and Hanski (2002a).

Responses to Habitat Boundaries

The sedentary nature of *E. editha* at Jasper Ridge led Ehrlich (1961b) to propose the existence of intrinsic barriers to dispersal. These butterflies typically turned aside if chased to a chaparral boundary of their habitat, rather than rising the 1 or 2 m necessary to traverse the shrubs. When not chased, they were often unwilling to fly over vegetation more than about 10 cm tall. This behavior tended to keep them on the serpentine soils and rocky outcrops where the vegetation was short. The serpentine habitats occurred interspersed in a matrix of sandstone, where introduced grasses (principally *Avena fatua*) grew to a height of 20–30 cm. Butterflies arriving at the boundary between short and tall grass would normally turn around sharply. By this means they failed to leave occupied patches when it was physically easy for them to do so. As a consequence, they rarely colonized apparently suitable habitat just tens of meters from the occupied sites. In like manner, *Melitaea cinxia*, though not obviously sensitive to the height of grasses, are deterred both by trees and by the shade that trees create (Kuussaari et al. 1996, Norberg et al. 2002). Experimentally released *M. cinxia* resided longer in habitat patches surrounded by trees than in patches surrounded by open landscape (Kuussaari et al. 1996), a potentially important finding for conservation of this and similar species.

Montane *E. editha* at Rabbit Meadow, Sequoia National Forest, California (for the location, see figure 3.1), occupy a patchwork of two habitat types: rocky outcrops and woodland clearings (figure 9.5). Unlike *E. editha* at Jasper Ridge, these butterflies were not strongly inhibited from entering one habitat type from another. They entered the two habitat types at equal rates but achieved much higher equilibrium densities in the outcrop habitats because their emigration rate from that habitat type was lower (Boughton 2000). In this case the insects assessed habitat quality after arriving in a patch and

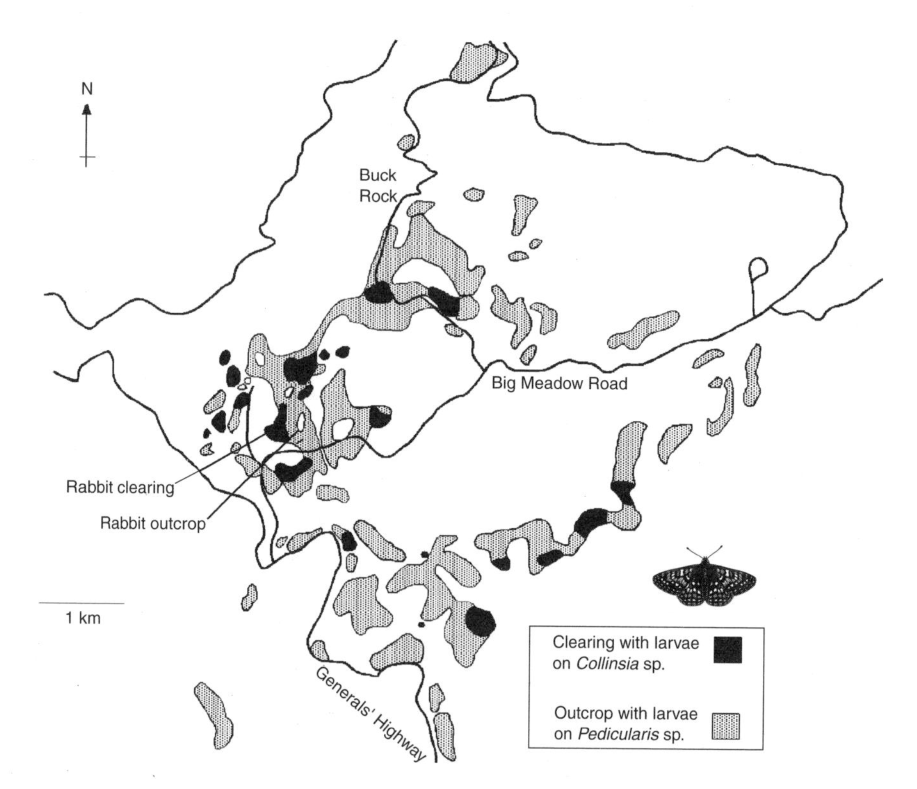

Figure 9.5. Map of the *Euphydryas editha* metapopulation with two patch types studied by Singer and Thomas (1996). Shaded patches had larvae in 1985. Patches shaded black were clearings with larvae feeding on *Collinsia*; those shaded gray were unlogged, open forest patches with larvae feeding on *Pedicularis*. The intensively studied pair of meadows is indicated.

used that assessment to decide whether to remain in the patch. Other ecotypes of *E. editha* also behave differently from Jasper Ridge *E. editha*. Butterflies in a population at Del Puerto Canyon (for the location, see figure 3.1) studied by Gilbert and Singer (1973) occupied a chaparral habitat, were undeterred by obstacles as tall as 3 m, and routinely moved much longer distances than *E. editha* at Jasper Ridge while commuting between oviposition sites and nectar sources, which were spatially separated (figure 2.6).

A larval transplant experiment investigated the difference in habitat choice between the Del Puerto Canyon and the Jasper Ridge butterflies, taking advantage of the fact that Jasper Ridge comprised a patchwork of grassland inhabited by local *E. editha* and chaparral resembling the habitat at Del Puerto Canyon, but into which the Jasper Ridge butterflies do not go. Postdiapause *E. editha* larvae from Del Puerto Canyon were placed at Jasper Ridge at a margin between the two habitat types. They generated a population of naturally emerging butterflies. These introduced butterflies were not only unintimidated by tall grass, they actually concentrated their activities in the chaparral, flying between 1 and 3 m above the ground and behaving entirely as though they were at Del Puerto. They made no genetic contribution to the local populations because they flew too late in the year to produce surviving offspring at Jasper Ridge (Gilbert and Singer 1973).

Response to Conspecific Density

Travis et al. (1999) stated that models of metapopulation dynamics and evolution of dispersal are almost exclusively density independent, despite empirical evidence to the contrary. Their own models suggest that density-dependent dispersal should almost always evolve and should be included in ecological metapopulation models. What information do we have for checkerspots? Gilbert and Singer (1973) observed that the proportion of marked *E. editha* at Jasper Ridge recaptured in a new patch tended to be higher in years when the density in the patch of origin was low. Reexamination of Gilbert and Singer's data shows that their conclusion was based on a small sample size and did not reach statistical significance. However, support for the same conclusion came from the observation by Brussard et al. (1974) that movements were longer at the beginning and end of the flight season, when densities were low, than at the peak of the season. Finally, convincing evidence of inversely density-dependent response to conspecific density by a checkerspot was provided by Kuussaari et al. (1996), who released *M. cinxia* at different densities and found that emigration by both males and females was reduced by high conspecific density. This influence of density on movement contributes to a demographic Allee effect in *M. cinxia* (Kuussaari et al. 1998), making low-density populations particularly vulnerable to extinction.

What mechanisms could underlie reduced emigration with increasing density? The mechanisms that have been suggested fall into two categories: mate seeking and copying habitat choice. Male *Euphydryas anicia* in a "featureless" landscape flew in straight lines until they encountered other butterflies, then they were likely to turn. The direction of flight by a male after an encounter was not correlated with the direction before it. By this mechanism, encounters caused males to increase their rates of turning and caused them to aggregate. Aggregation of males would therefore be expected to increase with increasing frequency of encounters, that is, with increasing density (Odendaal et al. 1988). This mechanism could explain reduced emigration of males at high density, though not that of females.

The second possible mechanism is copying habitat choice. Positive effects of density on aggregation could arise from a tendency to copy the habitat choices made by others or to use the presence of conspecifics to identify suitable habitat (Gilbert and Singer 1973). Animals often imitate each other, and copying of resource choice is well known, though the evolution of copying is not well understood (Dobson and Poole 1998, Mönkkönen et al. 1999). For example, an untrained octopus will, without reward, copy the food choices of trained individuals. A similar phenomenon could underly some of the observed tendencies of checkerspots to aggregate. Alternatively, if insects are not well able to assess important aspects of habitat quality, then the presence of conspecifics that have survived in the habitat may be a good guide (Gilbert and Singer 1973, Stamps 1987). Work on vertebrates suggests that reduced emigration from habitats with high density of conspecifics is frequent and has multiple causes (Stamps 1987, Reed and Dobson 1993, Dobson and Poole 1998).

Responses to density by *M. cinxia* were similar in males and females, but in *E. anicia* they appeared to be opposite. Odendaal et al. (1988) reported that female *E. anicia* increased their emigration at high densities as a consequence of avoiding males. In addition to avoiding males, females made sharper turns and flew for shorter distances in areas of high host density. However, male distribution was positively correlated with host distribution; therefore, the effects on female movement of male distribution and host distribution were opposed (Odendaal et al. 1989). Females of the checkerspot relative *Proclossiana eunomia* did emigrate more in a high-density than in a low-density year (Baguette et al. 1998), whereas males in the same study showed the same response to density as do checkerspots. The female response to density was at least partly mediated through encounters with males, since females released into sites with different male densities emigrated more from the high-density sites. In sum, reduction of male emigration with increasing population density is a well-established feature of the biology of checkerspots and related species. Females, in contrast, have shown a diversity of responses: responses of *M. cinxia* females resemble the males, and *E. anicia* (and *P. eunomia*) females increase their emigration rates at high densities in response to encounters with males.

The only study that did not show any effects of density on checkerspot movement comes from montane *E. editha* that maintained constant emigration rates of both sexes across a large range of densities (Boughton 1998). This study was facilitated by a natural extinction and subsequent recolonization of an extremely dense population in a forest clearing, while an adjacent population on a rocky outcrop remained stable. In 1984 butterfly density was an order of magnitude higher in the clearing than in the outcrop, but in 1996 the density in the outcrop was two orders of magnitude higher than in the clearing, which had been recently recolonized. Despite this dramatic change of relative densities, the probabilities that an individual in either the clearing or the outcrop would move to the other patch were not significantly different between the years for either males or females. In both years the two sexes expressed the same habitat preference, consistently moving from the clearing to the outcrop more than from the outcrop to the clearing.

None of the experiments or detailed observations that have been carried out so far account for the occasional reports of *Euphydryas* undertaking large-scale, unidirectional migrations away from sources of extremely high density. Such observations have been made close to the extreme northern and southern range limits of *E. editha*, in the Puget Sound area of Washington (D. Bauer, pers. comm.) and on the border between Baja California and the United States (D. D. Murphy, pers. comm.). The nature and cause of this phenomenon are mysterious. It is possible that instead of the classic migration syndrome in which all individuals take the same direction, when a population reaches extraordinarily high density, each individual chooses a direction to fly and holds that direction for some time. An observer outside the area of origin of the migrants would then have the false impression that they were all traveling in the same direction.

Morphological and Gender Correlates of Dispersal

The classic studies have documented how natural selection on insect dispersal influences the frequencies of short-winged (or wingless) and long-winged (winged) individuals (section 9.1). There is no wing polymorphism in butterflies. Researchers have nonetheless looked for more subtle morphological differences among groups of individuals that might bear the signature of selection on dispersal. Dispersal capacity of individuals might depend on the ratio of thorax weight (with wing muscles) to abdomen weight, on the ratio of thorax width to wing size (or some other measure of wing loading), or just on body size itself. The results have been conflicting. For instance, J. Hill et al. (1999a) found that individuals of *Pararge aegeria* were larger in newly colonized populations at an expanding range margin than in older populations, but no such difference was found in *Hesperia comma* (J. Hill et al. 1999c). Harrison (1989) found no differences in body size or in wing loading of *Euphydryas editha* between residents and migrants that had moved a short or a long distance. Lewis and Thomas (2001) made an interesting comparison between a wild population of *Pieris brassicae* and a conspecific population that had been kept in captivity for at least 100 generations. Females of the captive population had substantially higher fecundity, as could have been expected, but there was no correlated reduction in relative thorax mass (a surrogate of dispersal capacity), as might have been expected. Furthermore, butterflies in the captive population were significantly larger than butterflies in the wild population. It is reasonable to assume that individuals in the captive population had lost some of their dispersal capacity (as such behavior would only incur cost in the laboratory), yet there was no sign of this in the morphological measures. (However, flight capacity for purposes other than dispersal was still beneficial under laboratory conditions.) Like other butterflies, checkerspots have the physical capacity to fly very long distances, but most individuals of most species fail to do so. It is therefore likely that variation among individuals in the distances flown is due to some behavioral traits rather than to any obvious morphological traits.

Trends in body weight with age could potentially influence dispersal. Female *E. editha* lose weight rapidly as they age, but males do not (Moore 1987), and in some populations the high wing loading of newly emerged females obviously compromises their ability to fly. The most fecund females of *E. editha* at Jasper Ridge fly with difficulty and have trouble turning in flight until they have laid their first cluster of eggs. Some female *E. editha* in Oregon populations cannot fly at all until the first oviposition has occurred (C. Parmesan, pers. comm.). In any case, the maneuverability of females increases as they lose weight with oviposition. Under most circumstances the flight abilities of males change little over time, but at high density males interact frequently, and the contacts between wing-tips shorten the wings, compromising the ability to fly. In an unusually high-density population of *E. editha*, the result was a nice irony in the interactions between the sexes. In the beginning of the flight season the heavy, clumsy females were unable to fly very fast or change direction quickly and were continually pursued and harassed by the more agile males. Ten days later the females were the agile sex, and most of the males could hardly fly at all. Each female glided across the habitat followed by a sine wave of males jumping into the air, flapping their stubby wings, making no horizontal progress, and falling back to their approximate starting positions (M. Singer, pers. obs.).

Morphology and thermoregulation interact to influence butterfly flight (e.g., Heinrich 1993). Small butterflies tend to lose heat in flight and must pause to bask and reheat themselves to operating temperature. No studies have been done on checkerspots relating thermoregulation to body size, but such studies could prove useful in understanding patterns of dispersal. The extensive intraspecific variation in body size and wing loading might be helpful to such an enterprise: newly emerged female *E. editha* in the Sierra Nevada weigh about 320 mg at 600 m elevation, 130 mg at 2400 m and 100 mg at 3200 m (table 9.1).

Residence times of female *E. editha* at Jasper Ridge were much shorter than those of males in habitat patches (Ehrlich 1961b), indicating that females were more likely than males to either leave or die. Recent work shows that females have higher emigration rate than males, both in *M. cinxia* (Kuussaari et al. 1996) and in the non-checkerspot *P. eunomia* (Baguette and Nève 1994, Baguette et al. 1998). Why might this be so? Checkerspots do not normally take nectar as adults from plant species that are used as larval hosts. It is therefore not necessary for males to identify larval hosts in order to feed. If it were true that males do not respond to larval host plants, then females would be equipped to identify hosts and thereby assess habitats in terms of their suitability for larvae, whereas males would not be so equipped. This asymmetry would give migrant females an advantage over migrant males in distinguishing habitat from nonhabitat and could explain the greater willingness of females to emigrate. It could also explain why checkerspots differ from many other groups of animals, such as mammals (Dobson 1982), in which males are the more dispersive sex. But is it true that males do not recognize or respond to larval host plants? There are suggestions from *M. cinxia* that this is the case and suggestions from *E. editha* that it is not (see below). The situation in checkerspots needs clarification, especially in the light of recent evidence from another lepidopteran, the larch budmoth (*Zeiraphera diniana*), in which males from different host races have different host-alighting

Table 9.1. Variation in adult body size (teneral female mass) among populations in *E. editha* (from M. C. Singer unpubl. data).

Population	Latitude	Longitude	Altitude (m)	Mean Mass (mg)	Mass Range in Sample (mg)
Indian Flat	−119.82	37.65	610	325	291–358 (*n* = 10)
Rabbit Meadow	−118.87	36.22	2400	134	99–154 (*n* = 10)
Gardisky Lake	−119.25	37.97	3200	102	87–116 (*n* = 10)

preferences in the field (Emelianov et al. 2001). Where this occurs, the influences of host preference and host distribution on male movements could resemble those described for female checkerspots in section 9.4 below.

Dispersal of Individuals in New versus Old Populations

In the case of metapopulations consisting of unstable local populations experiencing frequent local extinctions, we would expect natural selection operating at the metapopulation level to generally favor increased dispersal, as high dispersal rate enhances recolonization. Assuming variation among individuals in the traits that influence dispersal rate, we would expect that individuals that have established new populations do not represent a random sample from the entire metapopulation but represent the more dispersive individuals. Further assuming that the relevant trait or traits are to some extent inherited by the offspring, we would predict that individuals originating from newly established local populations are more dispersive than individuals originating from old populations.

Hanski et al. (2002) conducted exactly such a comparison by sampling postdiapause larvae of *M. cinxia* originating from a large number of both newly established and old populations, rearing them to adult butterflies, marking the butterflies, releasing them into a natural environment, and collecting recapture observations both in the release patches and in their surroundings. The result of this experiment supported the prediction: females originating from the new populations were more dispersive than females originating from old populations (figure 9.6). There was no such difference in males (Hanski et al. 2002), indicating that selection on dispersal in the two sexes is different. The reasons for this difference were discussed in the previous section. Here we add another consideration: males are unlikely to gain any fitness by dispersing to currently empty habitat patches because any immigrant females that they might encounter there are most likely to be already inseminated, and the females usually mate only once in their life (chapter 5).

No morphological differences could be detected between females originating from new versus old populations, nor between individuals that dispersed away from the release patch versus those that did not (Hanski et al. 2002). These results support the conclusion drawn in the previous section that the relevant traits influencing dispersal are likely to be behavioral rather than morphological. There is, however, one indication that something else might be involved as well. Another comparison between a set of females from several new populations versus several old populations revealed that the total egg number at eclosion was roughly 10% lower in new populations (I. Hanski unpubl. data). This difference suggests an association between high dispersal capacity and reduced potential lifetime fecundity, which suggests that the physiological capacity to fly might be different among different individuals. There is much scope for further work in this area.

Timing of Dispersal and Its Relation to Gene Flow

Ehrlich (1965) pointed out that when individuals in a set of local populations have similar phenologies, immigrants must by necessity arrive, on aver-

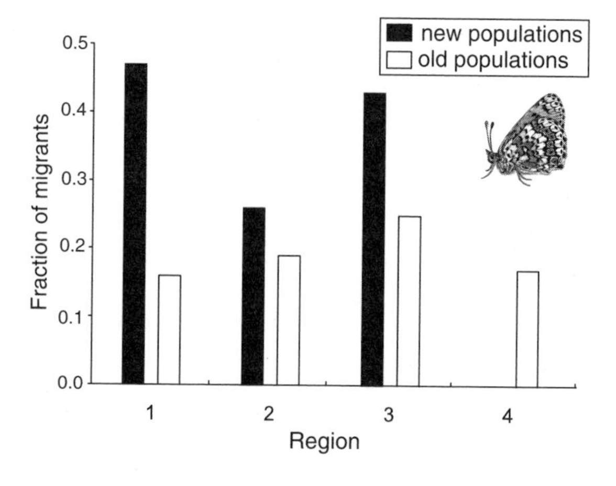

Figure 9.6. Fraction of female *Melitaea cinxia* observed to emigrate away from a release patch in an experiment. Filled bars refer to females originating from a large number of newly-established populations, open bars to females from old populations. Results are shown for four different regions (no new populations in region number 4). From Hanski et al. (2000), which gives further details and statistical tests.

age, later than local individuals eclose (as shown in figure 9.3), and so immigrants have to pay a cost relative to residents due to their late arrival (Morbey and Ydenberg 2001). Immigrant males are less likely than residents to encounter virgin females. Immigrant females will oviposit, on average, later than residents, which often leads to reduced offspring survival due to seasonal host senescence (chapter 3). (We describe in section 9.5 a surprising population dynamic consequence of this unavoidable delay in the reproduction of immigrants.) All these factors support the conclusion that dispersal of adult insects does not translate simply and directly into gene flow (Ehrlich 1965; see also chapter 12).

Ehrlich's ideas on this topic derived from his work on *E. editha* populations using an annual host plant (*Plantago erecta*) that was susceptible to rapid senescence. Rapid host senescence routinely killed a high proportion of larvae in each year, as the plant typically senesced before the larvae were large enough to respond to starvation by going into diapause (chapter 3). Boughton (1999, 2000) showed experimentally that the success of immigrant *E. editha* was sensitive to the timing of their arrival while entering a patch dominated by an annual host (*Collinsia*), but not when they entered habitat patches with perennial host (*Pedicularis*) not susceptible to rapid senescence. In this context it is interesting to compare *E. editha* with *M. cinxia*, as the latter species uses only perennial host plants. In the Åland Islands, the hosts of *M. cinxia* may occasionally be rendered inedible by summer drought (chapter 4), but they usually recover quickly, and they remain edible long after the larvae have entered diapause in August. In consequence, cinxiologists have been less impressed than edithologists by the importance of timing of dispersal. Another matter is that, as we described in the previous section, in the large metapopulation of *M. cinxia* consisting of unstable local populations, selection at the metapopulation level will necessarily favor substantial migration of females regardless of potential problems due to host plant withering during dry periods.

9.4 Relationship between Host Plant Preference and Female Movements

The study of dispersal in relation to oviposition host plant preference arose from an observation of the process by which a novel host plant, *Collinsia torreyi*, was incorporated into the diet of *E. editha* at Rabbit Meadow in Sequoia National Forest, California (chapter 6). After the disappearance of the traditional host, *Pedicularis semibarbata*, from a series of forest clearings created by logging, eggs were laid on the alternative host, *Collinsia*, in clearings. Butterflies that had developed on *Collinsia* in clearings showed a diversity of post-alighting oviposition preferences (for the definition and measurement of host preference, see chapter 6). Many butterflies still preferred the traditional host, *Pedicularis*, which was no longer present in the cleared habitat but which continued to be present in adjacent undisturbed habitat patches (Singer 1983). Some females preferred *Pedicularis* strongly, implying that they would search for a host plant to oviposit for more than a day before reaching the level of motivation at which the alternative host, *Collinsia*, would be accepted on encounter. A few butterflies that emerged from clearings with *Collinsia* would never accept this host at all, although *Pedicularis* was highly acceptable to them, and their larvae could survive well on both hosts (Singer 1983).

What's a mother to do when the combination of her host preference and the plant community in her habitat patch prevents her from ovipositing? The stark choice is either to emigrate or to die without reproducing. Of the two, emigration is surely the option that confers higher expectation of fitness, despite the cost of migration in terms of increased mortality (Hanski et al. 2000). The observation of locally maladaptive host plant preferences at Rabbit Meadow suggests the potential for a strong relationship between post-alighting preference and movement among habitat patches (and types), whenever preference is variable among individuals. To look for such a relationship, C. D. Thomas and Singer (1987) marked female butterflies in two adjacent habitat patches: a clearing where *Collinsia* was used and an outcrop where *Pedicularis* was used. They then tested the host preferences of recaptured butterflies that had stayed where they were marked and those that had moved between the patches. The result was clear: butterflies that preferred the host in the patch where they had been marked were less likely to move to the other patch than were butterflies with no preference, which were in turn less likely to move than butterflies whose preferred host was not present in the patch where they had been marked (C. D. Thomas and Singer 1987). There was biased reciprocal movement of females with dissimilar preferences among patches with different host plant composition (table 9.2).

Table 9.2. Oviposition preferences of Rabbit Meadow *E. editha* measured after mark–release–recapture in two habitat patches where different host plant species were used (from C. D. Thomas and Singer 1987).[a]

	Stay	Move
First recorded in patch with preferred host	35	2
First recorded in patch with less preferred host	9	11
Butterflies with no preference	52	24

[a]Statistics: preferred host versus less-preferred: $p = .00005$ (Fisher's Exact test); preferred host versus no preference: $p = .002$ (Fisher's Exact test).

The correlation between post-alighting host-plant preference and dispersal could be caused by an effect of preference on movement, as we had suspected, but it could also be caused by an effect of movement on preference. Movement could influence preference indirectly because movement influences host encounter, and encounter could affect preference. To test this possibility, we staged encounters between freshly captured butterflies and the two host plant species, then tested for an effect of encounter on preference. We found no sign at all for such an effect (table 6.4); hence it seemed safe to attribute the correlation between preference and movement to an effect of preference on movement. The following additional evidence from *E. editha* and *M. cinxia* further supports this conclusion. First, a difference in preference between the clearing and outcrop habitats at Rabbit Meadow was not reduced after the population in the clearing had been extirpated by a summer frost and then replaced by immigrants from the outcrop. After the extirpation, butterflies caught in the clearing (that were now all immigrants from the outcrop) still exhibited significantly weaker preference for the outcrop host, *Pedicularis*, than did butterflies that had remained in the outcrop habitat (Singer and Thomas 1996). Second, emigration of *E. editha* was higher from patches containing only *Collinsia* than from patches containing both *Collinsia* and the generally preferred host, *Pedicularis* (Boughton 1999, 2000). Third, in *M. cinxia* emigration of experimentally released females from meadows containing only *Plantago* was higher in the case of females originating from a *Veronica*-preferring population than in the case of females originating from *Plantago*-preferring populations (Hanski et al. 2002). Fourth, *Plantago*-preferring and *Veronica*-preferring genotypes of *M. cinxia* were released in a habitat patch containing *Plantago* and in another nearby patch containing *Veronica*. Differential rates of loss of the insects from the two habitat patches were as predicted if there were preference-biased emigration (Hanski and Singer 2001).

This series of studies on *E. editha* and *M. cinxia*, though none is entirely conclusive alone, collectively supports the hypothesis that host plant preference strongly influences female dispersal. In the next section we turn to a metapopulation-level phenomenon that is predicted to occur because of preference-biased movements and which has been observed empirically. We also discuss issues in which individual behavior, population dynamics, and evolutionary change are all closely linked.

9.5 Metapopulation-Level Consequences of Preference-Biased Movements

Source–Sink Dynamics in Euphydryas editha

As we have already explained, ovipositing females of *Euphydryas editha* at Rabbit Meadow originally preferred *Pedicularis semibarbata*, which grows in the traditional outcrop habitat of the butterfly. Another habitat type was created by forest logging in the 1980s, and this habitat type has another host plant, *Collinsia torreyi*. At that time, the host plant that conferred the highest fitness by far was the novel host *Collinsia* in the clearings, though most butterflies still preferred *Pedicularis* (Singer and Thomas 1996, C. D. Thomas et al. 1996). Large numbers of butterflies emerged in clearings, which lacked the preferred host. These butterflies tended to emigrate (Boughton 2000), as described in the previous section. In the case of small patches, the loss of butterflies due to emigration could be expected to be so rapid that no local populations might be established. In contrast, the larger the habitat patch, the greater the proportion of females that would reach the level of motivation at which *Collinsia* would be accepted before they had emigrated. We therefore expected a positive relationship between clearing size and larval density. Such a relationship was, indeed, detected (C. D. Thomas and Singer 1998; figure 9.7). Furthermore, examining the same relationship at a different site, Tamarack,

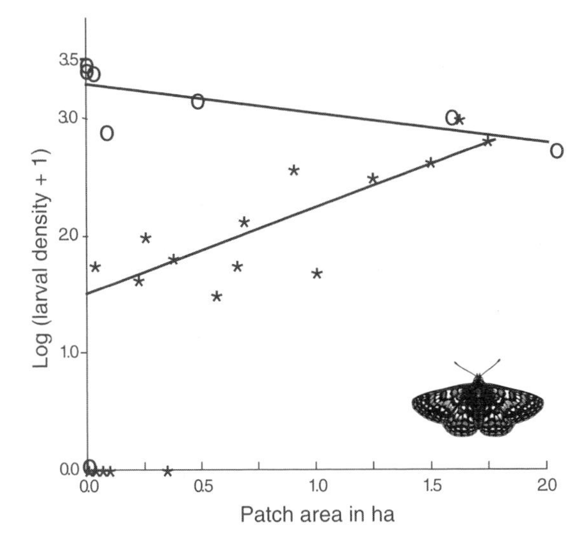

Figure 9.7. Relationship between habitat patch size and *Euphydryas editha* larval density at two sites, Rabbit Meadow (stars) and Tamarack (open circles). The Rabbit Meadow data are expressed as numbers of prediapause larval webs per hectare; the Tamarack web density has been divided by 10 because the numbers of larvae per web are almost exactly 10 times higher at Rabbit Meadow than at Tamarack (mean egg cluster size at Rabbit Meadow 52, at Tamarack 5.5; M.C. Singer unpubl. data).

where *Collinsia* was preferred for oviposition, failed to show a positive relationship between patch size and larval density (figure 9.7; M. Singer and D. Boughton, unpubl. data). These results demonstrate clearly how the behavior of ovipositing females greatly influences their dispersal, with important consequences for population dynamics.

Emigration of butterflies from the *Collinsia*-dominated habitat patches in the clearings was so intensive that densities in the nearby undisturbed outcrop patches became elevated (C. D. Thomas et al. 1996). The clearing populations acted as sources, while the outcrop populations were sinks. In the early summer 1992, however, the source populations feeding on *Collinsia* were extirpated by a summer frost. The densities in the sink populations using *Pedicularis* did not decline to extinction, however, as true sinks would have done, but settled to a lower level. This was not unexpected, as the butterfly had happily lived in the outcrop habitats before the formation of the clearcuts. But now the relationship between larval density on outcrops and isolation of the outcrop population from clearings, which had been a marked relationship before 1992, disappeared completely. The change in this relationship was statistically significant. C. D. Thomas et al. (1996) described these observations as the first empirical demonstration of a source–pseudosink system. A pseudosink is a population in which density is elevated due to immigration but which would settle to a lower stable state, rather than go extinct, in the absence of immigration (Watkinson and Sutherland 1995). It follows that, in the presence of immigration, breeding individuals in a pseudosink are not able to replace themselves, though they would be able to do so without the competition of the immigrants.

Boughton planned to study the recolonization of the set of clearing habitats following the catastrophic extinction of their populations due to the summer frost in 1992. However, he was initially unable to do so because, although migrants from the occupied outcrop habitats did move into the empty clearing habitats, they consistently failed to establish populations there. Experimental placement of larvae in the field showed that individuals in the outcrops grew slowly and emerged as butterflies too late to fly to the clearings and produce surviving offspring. Insects developing in the clearings emerged on average 10 days earlier and had high offspring survival whether they stayed in the clearing or emigrated to the outcrop habitat. There was thus a unidirectional phenological barrier that made it easy for butterflies from the clearings to colonize outcrop habitats, but not vice versa (Boughton 1999, 2000). These observations suggest that the system had alternative stable states, with source–sink relationships in the opposite directions. Before 1992 the system existed as a source–pseudosink system, with sources in the clearings, whereas after 1992 it turned to a true source–sink system with sources in the outcrops. The switch between these two states occurred when the system was disturbed by the summer frost that killed, for one year, the host plants in the clearing (Singer and Thomas 1996, C. D. Thomas et al. 1996, Boughton 1999, 2000). This switch involved a dramatic shift in resource use by the butterfly and has stimulated modeling of the evolution of resource use in source–sink systems (Ronce and Kirkpatrick 2001).

The "Colonization Effect" in Melitaea cinxia

Preference-biased female movements also occur in *Melitaea cinxia*, and in this case the biased movements have population dynamic consequences at the level of metapopulations. Our research has been greatly facilitated by high statistical power resulting from the small number of host species used by the butterfly (two hosts) and the large number of habitat patches in the Åland Islands (around 4 kilopatches). To characterize the regional host plant use in the surroundings of each habitat patch, we divided the familiar measure of patch connectivity (Hanski 1999b; chapter 4) into two components: connectivity to larval groups using *Plantago lanceolata* and connectivity to larval groups using *Veronica spicata* (Hanski and Singer 2001). Thus, a habitat patch that is well connected to larvae found on *Veronica* and poorly connected to larvae found on *Plantago* is a patch situated in a region with high use of *Veronica* relative to *Plantago*. Using this measure, we found that colonization of empty patches was strongly related to regional host plant use (figure 9.8). The rate of colonization of empty patches containing principally *Veronica* increased with increasing regional use of *Veronica*. Independently, the rate of colonization of empty patches containing principally *Plantago* decreased with increasing regional use of *Veronica*. We termed this effect, which was documented based on 317 observed colonization events in the years 1993–97, the "colonization effect" (Hanski and Singer 2001).

The colonization effect could be due to spatially variable traits of host plants, such as acceptability to ovipositing butterflies or suitability for growth and survival of larvae. In this case, a currently empty habitat patch in a region in which *Plantago* was mostly used as a host plant would have low-quality *Veronica* and hence the rate of successful colonization would be low. By "low-quality" *Veronica*, we mean plants that are unacceptable to ovipositing butterflies or resistant to attack by their larvae. If the quality of *Veronica* were spatially autocorrelated, a target patch containing low-quality *Veronica* would be surrounded by patches with low-quality *Veronica*, causing the patch to remain uncolonized and the larvae in nearby patches to have been found on *Plantago*. Alternatively or additionally, the colonization effect could be caused by spatially variable insect oviposition preference or larval performance. For example, an empty target patch containing principally *Veronica* could remain uncolonized by the butterflies because the

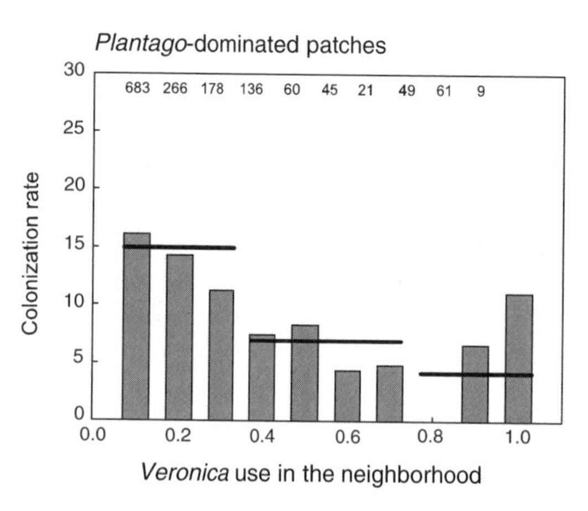

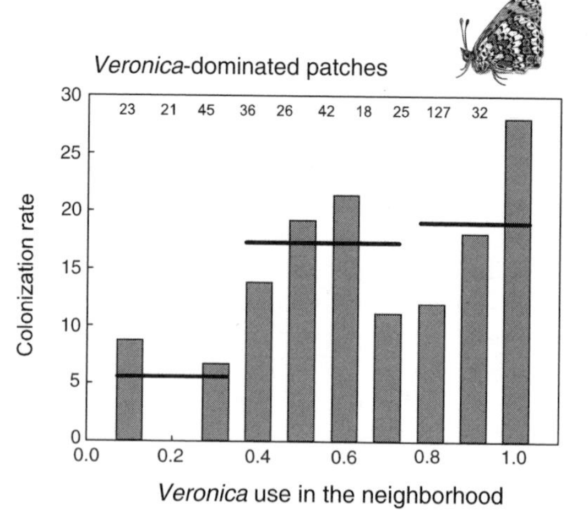

Figure 9.8. Colonization rates of empty habitat patches by *Melitaea cinxia* as a function of host composition in the target patch and relative isolation of the patch from larvae previously found on the two host species, *Plantago lanceolata* and *Veronica spicata*. The numbers at the top of the bars give sample size. The horizontal lines give the means for high, intermediate and low use of *Veronica*. From Hanski and Singer (2001).

migrating butterflies in the region mostly preferred *Plantago*.

The above arguments make it clear that we need to describe the patterns of spatial variation of butterfly and plant traits to identify the mechanistic causes of the colonization effect and thereby begin to understand its significance in the system. Hanski and Singer (2001) undertook this task in the Åland Islands in summer 1998 by gathering and testing plants and butterflies in selected habitat patches. We preference tested butterflies on freshly gathered plants from different areas to find out whether insect preference and/or plant acceptability varied spatially in a manner that could generate the colonization effect. Preference did, but acceptability did not. We manipulated butterflies to lay eggs in their habitat patches of origin on experimentally chosen plants and measured survival of eggs and larvae. These experiments showed that the combined variation of insect performance and plant suitability did not take a form that could produce the colonization effect. For that reason we were left with the conclusion that spatial variation in adult host plant preference occurred at the relevant spatial scale and that the other possible causes of the colonization effect could be discounted. We concluded that the colonization effect is generated by host preference influencing the movement and/or residence and oviposition patterns of female *M. cinxia*. Migrant females with dissimilar host preferences had different perceptions of relative patch quality, which influenced their likelihood of colonizing patches with particular host composition (Hanski and Singer 2001). Note that this explanation is based on the very phenomenon—preference-biased movements—that had previously been documented for *E. editha* in California and which we described in the previous section.

The bias in population establishment that comprises the colonization effect suggests that selection on host-related adaptations should occur among populations in a manner that can be predicted from the spatial arrangement and host composition of habitat patches. For example, in a patch network in which every patch contains 80% *Veronica* and 20% *Plantago*, we might expect that selection among patches would not be different from selection within them. Now consider a patch network where 80% of the patches contain only *Veronica* and 20% contain only *Plantago*. Selection on butterflies occupying patches of *Plantago* may favor adaptations to this host, including oviposition preference for it. However, the colonization effect implies that *Plantago*-preferring butterflies would be poor colonizers of empty patches, and selection at the metapopulation level would disfavor *Plantago* preference, especially if the 20% of patches containing *Plantago* occurred in spatial locations that make them unimportant for the dynamics of the metapopulation as a whole (Hanski and Ovaskainen 2000). *Plantago*-preferring females would also be disadvantaged relative to residents if they immigrated to patches containing extant populations, most of which occupy patches of *Veronica* (see Kuussaari et al. 2000).

Evolution of resource use should differ between the two scenarios outlined above. Modeling this situation leads to metapopulation-level predictions of the evolution of oviposition preference and butterfly diet (Hanski and Heino 2003). How can such predictions be tested? Preference cannot be measured in large numbers of populations, for practical reasons, but a reasonable surrogate for preference is electivity, the proportional use of the two hosts as a function of their availabilities in the habitat (Singer 2000). Hanski and Heino (2003) regressed metapopulation-level electivity on the proportion of the two hosts as estimated from plant census data. The regression was, as expected, positive. What was less expected was that the residuals from this regression were significantly explained by spatially explicit models incorporating the colonization effect. In these models, preference evolves in response to regional plant availabilities and population establishment is influenced by the match between the host composition of habitat patches and the oviposition preferences of migrating butterflies. The finding that metapopulation-level electivity is significantly explained by these models implies that the metapopulation-level evolution of resource use is influenced by the colonization effect and hence by extinction–colonization dynamics. The behavior of butterflies, their ecology, and the evolutionary change in a key trait are all dynamically coupled.

9.6 Conclusions

The study of dispersal in checkerspots started with Ehrlich's demonstration that clear-cut responses to habitat patch boundaries and limited dispersal were responsible for the typically discrete structure of local populations of *Euphydryas editha*. This expanded to studies of the roles of dispersal in generating spatial distributions of *Euphydryas anicia* and then to metapopulation dynamics of *Melitaea cinxia*, in a system in which movement is critical because all populations exhibit a considerable risk of extinction. Some degree of movement among the ephemeral local populations is a necessary condition for long-term persistence at the metapopulation level (chapter 12). Manipulability of checkerspots has been an invaluable tool, facilitating the study of emigration from experimental populations established by releasing butterflies of known origin at experimentally determined densities. This type of experiment has shown that butterflies, especially males, are more likely to emigrate when their density is low. The experiments have shown that females (but not males) originating from newly established populations have a higher dispersal tendency than females originating from older populations, indicating that extinction–colonization dynamics select for increased dispersal rate. The experiments have also documented the influence that the surrounding landscape structure has on immigration, which knowledge can be profitably used in spatially realistic metapopulation models (Hanski 1994b, 2001; box 12.1).

Much of the metapopulation research on checkerspots and other taxa has been concerned with the influence of landscape configuration on population processes (chapter 12). Habitat quality is harder to measure than patch area and has taken a back seat. Yet research on checkerspots has also produced novel insight to the influence of habitat quality, demonstrating not only that quality (e.g., host plant species composition) varies spatially but that the perception of that quality also varies among individuals (due to variation in individual preference). We have examined the influence of this individual perception of habitat quality on dispersal. Checkerspots frequently occupy habitat patch networks in which patches vary drastically in host plant species composition. Regional oviposition preferences evolve in response to regional host availabilities, but these preferences also interact with local host composition to affect the insects' assessment of patch quality and hence their decisions that influence patterns of movement and colonization. The result of all this is that biased colonization rates can be predicted from the match between the host plant composition of the patch and the regional oviposition preferences of migrating butterflies. In turn, the biased colonization rates influence the evolution of resource use at the metapopulation level and the regional oviposition preferences from which our argument started.

It is clear that emigration rates of checkerspots increase with low population density and when the butterflies perceive that habitat quality is poor. Where do we go from here? Further advances in our ability to understand the role of dispersal should come from more quantitative understanding in two areas. First, how does dispersal rate evolve in response to population history? Second, how do insects make decisions to disperse? The first question is amenable to modeling. It can in parallel be tackled empirically by a combination of longitudinal studies, geographical comparisons, and manipulative experiments. These could take advantage of the regional variation in population size and persistence that exists within most of our study species. Answers to the second question could start with the established conclusion that emigration and colonization decisions depend in part on the insects' perception of patch quality. We need to know more about what factors influence those perceptions. For example, we know nothing about whether or how the butterflies can detect important natural enemies such as parasitoids. We also need to know more about the ways in which the insects' perception of habitat quality is influenced by their own activities. For example, plants on which eggs are laid or that are attacked by *M. cinxia* larvae become more attractive to adults (M. Singer, L. Ramakrishnan, and M. Kuussaari, unpubl. data). Both these effects should strengthen the known tendency for emigration to decrease at high density.

To incorporate these and other factors into models, we need to be able to translate quantitative measures of preference into predictions of emigration rates from patches where there are known encounter rates with specific resources. This we cannot yet do. But pending additional information, we do have sufficient documentation to assert that the interplay among resource availability, resource-biased dispersal, population establishment, and evolution of resource choice shows how evolution of individual behavior can have major consequences for population dynamics, and indeed how population dynamics can influence individual behavior and its evolution.

10

Genetics of Checkerspot Populations

ILIK J. SACCHERI, CAROL L. BOGGS, ILKKA HANSKI,
AND PAUL R. EHRLICH

10.1 Introduction

The integration of ecological and evolutionary processes into a common framework has been a long-standing goal in population biology, to which E. B. Ford's pioneering studies on the European checkerspot *Euphydryas aurinia* made distinct contributions (Ford and Ford 1930, Ford 1975). Subsequently, Sheppard and Ford's work (Sheppard 1953, 1956, Ford and Sheppard 1969) on the factors controlling the *medionigra* wing pattern polymorphism in the scarlet tiger moth, *Panaxia dominula*, was central to the development of ecological genetics, in which theory, field observations, and experimentation were combined to analyze the genetics of natural populations. This integration has also been one of the key themes in our own research on *Euphydryas editha* and *Melitaea cinxia*. Merging ecological and evolutionary studies with their dissimilar perspectives and techniques is necessary for several reasons. In the first place, the key demographic parameters, including variation in reproductive success and migration, are studied much more effectively through a combination of direct observation and genetic marker analysis than with just one approach alone (Avise 1994). Additionally, a mechanistic understanding of variation in individual behavior and fitness that underlies population-level phenomena must include the interaction of ecological and genetic factors. Not only do genetic constraints have the potential to influence the ecological dynamics of populations and metapopulations, but population and metapopulation dynamics are expected to influence patterns of genetic variation and the evolution of life-history traits. Two examples are the evolution of migration rate and oviposition host plant preference in checkerspots, discussed in chapters 6 and 9.

Much of the success of the combination of ecological and genetic approaches to checkerspot biology has derived from the large body of information on their natural history and population dynamics. Without such knowledge, the significance of quantitative genetic variation in a given trait (e.g., oviposition host plant preference and egg hatching rate) to individual fitness could not be appreciated. Neither would it be possible to interpret patterns of genetic variation within and between populations in a way consistent with the migration behavior and population dynamics of the species.

Two broad questions, both aimed at better understanding of population dynamics and the evolutionary process, have dominated the research on checkerspot population genetics. What is the population structure? And, to what extent do interactions between genetic and ecological factors influence population dynamics? In its broadest sense, 'population structure' can be taken to mean all the factors governing the pattern in which gametes from various individuals unite with each other (Wright 1951, Ehrlich and Holm 1963). Typically these factors include the geographic distance separating two individuals relative to the species' capacity for movement, physical barriers to dispersal, habitat preferences and phenologies, and mate choice. In practice, the first step in studies of population structure is to describe the pattern of interactions among individuals and local populations, using both direct (ecological) methods, such as mapping the location of individuals and mark–release–recapture (MRR) experiments, and indirect (genetic) methods, which infer demographic parameters from patterns in the distribution of genetic markers. This approach has been applied to several species of checkerspots, revealing a diversity of population structures in a group of insects known for their habitat specificity and restricted dispersal (section 10.3). The approach has also been used in checkerspots to address questions within the selectionist/neutralist controversy (section 10.2), which can only be done in the context of a mechanistic understanding of population structure.

By extending the concept of population structure to include the entire range of a species, we enter the realm of phylogeography (Avise et al. 1987), describing the history of relationships among regional populations and metapopulations (section 10.4). These studies provide insight into the impact of geography, primarily climate and topography, on migration routes during range expansions and the formation of regional clades (Hewitt 2001). Molecular genetic markers have been used to study the genesis of intraspecific biodiversity in *Euphydryas* species in western North America (Brussard et al. 1989, Baughman et al. 1990a) and in *Melitaea cinxia* across Eurasia (I. Saccheri and N. Wahlberg, in prep.). The different selection regimes associated with a wide range of habitats, coupled with geographic isolation, present great scope for evolutionary divergence, as reflected by the evolution of oviposition host plant specialization in checkerspots (chapter 6; Radtkey and Singer 1995, Wahlberg and Zimmermann 2000, Wahlberg 2001b). Checkerspot research has also made major contributions to the study of the evolution of migration rate in fragmented landscapes (chapter 9; Heino and Hanski 2001).

The other major area of inquiry has been motivated by the fundamental assumption that differences in survival and reproduction among individuals are partly determined by genetic differences, and that, contingent on density-dependent responses, these may influence the growth and persistence of populations. There is, however, limited empirical evidence demonstrating a causal link between any genetic characteristics of a population and its persistence, at least in natural populations. The need to consider the effect of genetic factors on the risk of population extinction was highlighted during the early development of conservation biology (Frankel and Soulé 1981). Two related genetic processes, both a consequence of small population sizes, are potentially important. The first one is inbreeding, which is usually associated with decline in the fitness of individuals, known as inbreeding depression. The second process is loss of genetic diversity through random genetic drift, which may compromise a population's ability to adapt to environmental change. These genetic factors are sometimes referred to collectively as "genetic erosion" (Ouborg and Vantreuren 1994) or "genetic stochasticity." A major debate in conservation biology has revolved around the importance of these genetic factors to extinction risk relative to environmental and demographic factors (Soulé 1986).

The essential argument against an important role for the genetic factors is that substantial losses in fitness due to inbreeding depression and/or reduced genetic diversity are believed to occur only at effective population sizes of 10 or fewer individuals. Such small populations are vulnerable to unfavorable environmental conditions ("environmental stochasticity"), and their fate strongly depends on random variation in individual survival and reproduction ("demographic stochasticity"), as illustrated by relevant demographic models (Lande 1993, Foley 1994). Very small populations are therefore most likely to be pushed relatively quickly to extinction by one of the nongenetic factors, which are perceived to operate largely independently of variation in fitness or genotype. Furthermore, once exposed by inbreeding, the frequency of deleterious recessive alleles may be significantly reduced by directional selection (genome purging). There is evidence for genome purging (section 10.5), but this is likely to be only partial due to the ineffectiveness of selection in relatively small populations, particularly with respect to minor effect mutations that nevertheless collectively account for a large part of inbreeding depression.

The counter-arguments are that there are strong genotype-by-environment interactions determining viability and reproductive success, particularly under environmentally stressful conditions; hence genetics plays a role at all population sizes. It should also be noted that high variance in reproductive success, population substructure, and historical population bottlenecks may often lead to effective population sizes being an order of magnitude smaller than the census population size (Frankham 1995), amplifying the role of genetic relative to purely environmental and demographic factors. In metapopulations composed of local populations that are small and prone to extinction, there is especially great scope for drift and inbreeding to have consequences for population dynamics. Lande (1995) argued that mutations involving only mildly deleterious alleles also threaten population persistence due to drift and inbreeding, even at effective population sizes as high as 5000. This further opens the door for the potential importance of drift and inbreeding in metapopulation dynamics.

The consequences of reduced genetic diversity for population viability may be subtle, at least within ecological time frames, in all but extreme or otherwise rare environmental conditions. In contrast, the effects of inbreeding per se (i.e., increased homozygosity) are typically less environmentally conditioned and have a large impact on major components of fitness that are under consistently strong directional selection (e.g., fertility and early zygote viability). Our research on this topic using *Melitaea cinxia* has therefore been focused on inbreeding depression rather than loss of genetic variation (though in observational studies it is difficult to distinguish between the two). A major finding has been that levels of inbreeding common in natural populations of *M. cinxia* do indeed increase the risk of local population extinction (section 10.4). Genetic drift and inbreeding are thereby features of small checkerspot populations that may influence their population dynamics and evolutionary trajectories.

In this chapter we review research in which population genetic principles and techniques have been applied to the study of checkerspot population structure and dynamics. The early work on wing pattern variation (phenetics) in *E. editha* was followed by studies of allozyme variation in a wider set of *Euphydryas* populations, raising fundamental questions about the forces acting on enzyme loci (section 10.2). The analysis of genetic markers has provided valuable insights into the scale of genetic differentiation, spatial genetic structure, and genetic consequences of population turnover (section 10.3). The phylogeography of *Melitaea cinxia* and *Euphydryas* species inferred from mitochondrial DNA sequence data and allozymes places our study populations in the Åland Islands and on Jasper Ridge within a broader historical context (section 10.4). Section 10.5 describes laboratory and field studies designed to quantify the level of inbreeding depression in *M. cinxia* and its effect on the extinction risk of local populations. We close the chapter with a discussion of the ongoing and future work in this expanding area of checkerspot research.

10.2 Phenetics and the Neutrality Controversy

In the early 1960s, when the techniques to study molecular genetic variation had not yet been developed, Ehrlich had some important questions that he thought might be answered with the *Euphydryas* system. Among these questions were the amount and spatial scale of genetic differentiation and the relationship between population size and the degree of heterozygosity. For example, would populations become more homozygous as population size dropped, or would selection favoring heterozygotes possibly counter the influence of inbreeding and drift? Initially, Ehrlich and co-workers (Ehrlich and Mason 1966, Mason et al. 1967) took a "population phenetic" approach to such issues, studying the variation in wing pattern elements in *E. editha*. They attempted to infer the extent to which the butterflies found in the three Jasper Ridge areas (figure 1.2; chapter 3) were genetically differentiated, and by extension isolated, from one another. Ehrlich expected that phenotypic differences would be found between the populations C and H at Jasper Ridge because of marked differences in their slope exposures and population dynamics. However, no differences were detected in these studies, though the results suggested that some wing characters were changing in response to selection (Ehrlich and Mason 1966). Overall the analyses proved very difficult, and the population phenetic studies did not permit answers to the main questions of interest. The studies were abandoned in part because the underlying assumptions for calculating heritabilities could not be met in natural populations.

Following the seminal work of Lewontin and Hubby (1966) on patterns of enzyme variation in *Drosophila*, Ehrlich's group began to investigate such variation in *Euphydryas editha* and *E. chalcedona*. They originally thought that, at last, it would be possible to get a more-or-less random sample of the genome and to answer a wide array of evolutionary questions of the sort posed above. But nature was not cooperative, and the "neutrality" controversy reared its ugly head (e.g., Hartl and Clark 1997), complicating the analysis of *Euphydryas* allozyme patterns in nature (McKechnie et al. 1975).

After considerable difficulty, Ehrlich's group developed methods to evaluate enzyme variation at eight polymorphic loci, and that variation was surveyed in 21 populations of *E. editha* and 10 populations of *E. chalcedona* throughout California. The level of genetic variability in these two *Euphydryas* species turned out to be similar to that found in other animal populations (McKechnie et al. 1975). One result from the allozyme studies confirmed Ehrlich's preconceptions: at one of the eight loci, *Bdh* (β-hydroxybutyric acid dehydrogenase), the C and H populations at Jasper Ridge were significantly differentiated in two consecutive years. Overall, though, the results for the 21 populations distributed across an area of approximately 300,000 km² suggested sufficient gene flow to maintain relative uniformity at seven of the eight loci. In striking contrast, at the eighth locus, *Hk* (hexokinase), there were great differences among the 21 populations, interpreted as evidence of strong selection, either on that locus or on one closely linked to it (Kaplan et al. 1989). The observed pattern is shown in figure 10.1.

Analysis of allozyme frequencies in a population of *Euphydryas editha* in Gunnison County, Colorado (Pioneer Resort), which had been isolated from the West Coast populations for at least some 7000 generations, showed interesting relationships to California *E. editha* (Ehrlich and White 1980). At seven of the eight loci the similarities of the Colorado population to the others was striking. At the eighth locus, this time *Pgm* (phosphoglucoisomerase), the frequencies were strikingly different (table 10.1). Here, as

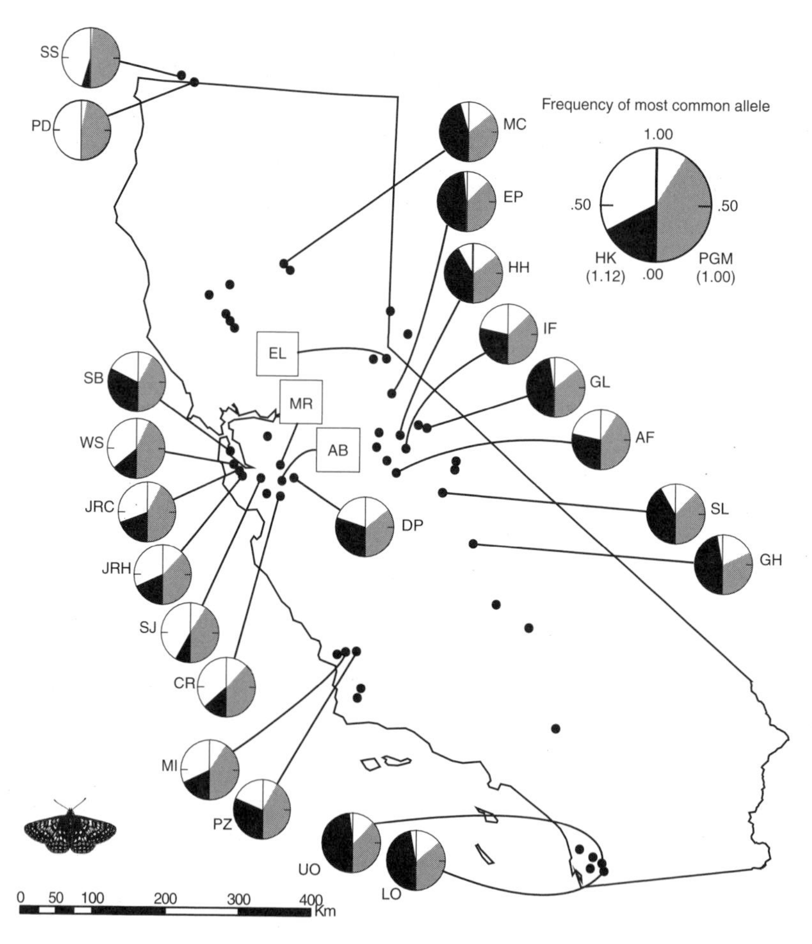

Figure 10.1. Allele frequencies at phosphoglucomutase (*Pgm*) and hexokinase (*Hk*) loci in *Euphydryas editha* in California. For names of the sites, see figure 3.1.

Table 10.1. Allele frequencies for eight polymorphic allozyme loci in *Euphydryas editha* populations at Sulphur Springs, Oregon (SS); General's Highway, California (GH); and Pioneer Resort, Colorado (PR) (from Ehrlich and White 1980).

	Allele	SS	GH	PR
Phosphoglucomutase (*Pgm*)	0.87		0.25	
	0.94	0.07		0.78
	1.0	0.93	0.69	0.22
Phosphoglucose isomerase (*Pgi*)	0.80	0.22	0.07	0.11
	1.0	0.57	0.84	0.87
Hexokinase (*Hk*)	1.0	0.97	0.03	
	1.12	0.03	0.96	0.97
	1.24			0.03
Glutamic acid oxaloacetic acid	0.6	0.05	0.10	
transaminase (*Got*)	1.0	0.90	0.84	0.99
	1.4	0.05	0.06	0.01
β-hydroxy butyric acid dehydrogenase (*Bdh*)	0.58	0.03	0.04	
	1.0	0.97	0.82	1.0
	1.4		0.14	
Tetrazolium oxidase (*To*)	0.26		.04	
	1.0	1.0	0.96	1.0
Adenylic acid kinase (*Ak*)	1.0	0.96	1.0	1.0
	1.2	0.02		
Malic acid dehydrogenase (*Mdh*-1)	0.64	0.06	0.06	
	1.0	0.94	0.88	0.97
	1.3		0.06	0.03

Alleles are numbered by mobility, with 1.0 being the most common. Only data for the two most common alleles in each population are presented here. Sample sizes are 57 individuals for *Got* and 58 for all other allozymes for Sulphur Springs; 48 for *Mdh*-1, 50 for *Pgi*, *Bdh*, and *To*, and 51 for all other allozymes for General's Highway; and 119 for *Hk*, 120 for *Pgm* and *Pgi*, and 121 for all other allozymes for Pioneer Resort.

in the U.S. West Coast data, frequencies at different loci present strikingly different patterns, leading to the conclusion that selection is operating either directly on some loci or on closely linked loci. A subsequent study of allozyme frequencies at six polymorphic loci over eight generations in the C and H populations of *E. editha* at Jasper Ridge showed patterns suggestive of fluctuating selection (Mueller et al. 1985). There is similar indication of such selection in multiyear analyses of nine allozyme loci in an alpine Colorado population of a close relative of *E. editha*, *E. anicia* (Cullenward et al. 1979).

Slatkin (1987) reanalyzed the West Coast data on *Euphydryas editha* allozyme frequencies and concluded that the results of his indirect method of estimating gene flow were inconsistent with the direct evidence of movement based on MRR studies and the asynchronous flight periods of the different ecotypes. His analysis suggested that gene flow maintained genetic similarity in populations at seven of the eight loci, in which variation was more or less neutral, but he also concluded that selection was involved in creating the *Hk* pattern (figure 10.1). Direct studies of movement, in contrast, have shown that the gene flow required to maintain similarity at the seven loci is currently nonexistent. The resolution of the apparent paradox probably lies in dramatic changes in the distribution of *E. editha* as a result of the invasion of California by European weeds (chapters 2 and 13) or in consistent selection pressures across the western United States for those seven loci.

The neutrality controversy is much closer to resolution today, nearly four decades after it first appeared in its molecular form. Its roots go much farther back to disagreements between the schools of Herman Muller (1950) and Theodosius Dobzhansky (1955). Muller and his colleagues believed that virtually all loci in an individual were homozygous for "wild-type" genes and heterozygous at a few loci for deleterious recessive mutants. Those mutants, in turn, were said to produce a "genetic load" in a population by reducing its average fitness (e.g., Morton et al. 1956, Spiess 1977), and they were constantly being removed by "purifying selection" (directional selection). Dobzhansky and his colleagues thought that most individuals in populations were heterozygous at most loci, maintained that way by "balancing selection" (overdominance for fitness). At first, the classic paper by Lewontin and Hubby (1966) appeared to settle this argument in favor of the school of Dobzhansky by revealing high levels of genetic variation. The entire discussion was then made more complex by the appearance of the "neutralist" school (e.g., Kimura 1968, King and Jukes 1969, Kimura and Ohta 1971a, 1971b), which claimed that most of the newly observed variation was not under the control of selection at all—either directional or balancing.

The issue of the proportion of protein coding sequences whose allele frequencies are significantly influenced by selection remains unresolved, although the consensus now is that a great deal of the variation observed is either neutral or under very weak selection (Ohta 1992). Genetic load due to strongly deleterious alleles is therefore much less than previously assumed under the purely selectionist models. But at least some loci are under strong selection (see Gillespie 1991, Watt 1994, for summaries). To actually document selection operating on allozymes requires careful, detailed work focused on one or a few loci, an approach employed with great success by Ward Watt in the process of building another butterfly model system, the pierid genus *Colias* (e.g., Watt 1977, 1992, 2003, Watt et al. 1983, 1986, 1996).

In *Euphydryas* and many other species (e.g., Christiansen and Frydenberg 1974, Kreitman and Akashi 1995, Watt and Dean 2000), it has proved possible to detect the effects of selection, but it has not yet proved possible to answer all the sorts of questions Ehrlich originally wished to answer. Since much of the observed allozyme variation may be quasi-neutral, its elucidation has proved useful in answering questions about the genetic structure of populations, as we see below.

10.3 Population Structure

How much movement occurs between local populations in a metapopulation? This is the first question we ask when faced with a spatially structured population. As discussed in chapters 3, 4, and 9, rate of movement (migration) is a crucial determinant of the ecological and evolutionary dynamics of a metapopulation. As a result, much effort has gone into studying the movement behavior of individuals within and between suitable habitat patches using MRR techniques in *E. editha* (Gilbert and Singer 1973, Ehrlich and White 1980), *M. cinxia* (Hanski et al. 1994, Kuussaari et al. 1996), related checkerspots (Schrier et al. 1976, Cullenward et al. 1979, Brown and Ehrlich 1980, White 1980, Munguira et al. 1997, Hanski et al. 2000, Wahlberg 2000b, Wahlberg et al. 2002b), and other butterflies (e.g., Petit et al. 2001). This approach, when undertaken on a large enough spatial scale and with sufficient effort, is effective for describing the key parameters of movement behavior: emigration from and immigration to specific habitat patches, scaling these rates with patch area, and even the rate of mortality during migration (Hanski et al. 2000, Ovaskainen 2003; see chapter 9).

A major contribution of studies on *E. editha* has been to show that the movement behavior of individuals can differ radically in different habitats, depending on the spatial distribution of larval and adult resources (e.g., figure 2.6) and the associated selection history (Gilbert and Singer 1973). Similar results have been obtained for *M. cinxia* (Kuussaari et al. 1996). However, MRR alone gives no information on the reproductive success of immigrants in the new patches, only that individuals have arrived at these patches. The most clear-cut demographic signals indicating that many immigrants do have successful offspring in the new patches are recolonization of previously empty patches and the rescue effect, which are both abundantly documented for *M. cinxia* (chapter 4) and to a lesser extent for *E. editha* (chapter 3). In the case of *M. cinxia*, MRR studies indicate the same spatial scale of migration as inferred from studies of recolonization (van Nouhuys and Hanski 2002a).

There are several benefits of applying neutral molecular genetic markers to study structured populations. First, the pattern of variation in such allele frequencies over space and time is of interest in itself because it describes the consequences of demographic history, population dynamics, and individual behavior for the raw material of evolution: genetic variation. The same information on allelic patterns, used in conjunction with ecological data, may be probed to make inferences about demographic history and population dynamic processes, typically on a longer time scale than would be possible with ecological data alone. In particular, we are interested in knowing how gene flow, genetic drift, and population turnover interact to generate the observed patterns of genetic variation.

The *E. editha* and *M. cinxia* studies exemplify how the dynamics of populations can alter extrapolation from contemporary patterns of movement to patterns of genetic differentiation. Boughton's (2000) research on *E. editha* populations in the Sierras in California demonstrates convincingly that phenological differences among butterfly populations and in their host plants can mean that butterfly dispersal into new populations may practically never translate into gene flow. Among the three Jasper Ridge populations, very little genetic differentiation at allozyme loci was observed, in contrast to expectations based on MRR data showing little migration. The case of *Melitaea cinxia* in the Åland Islands is very different. Here we have high rates of movement between neighboring habitat patches, but also complex spatiotemporal dynamics at the scale of individual patches (including frequent extinctions and recolonizations), patch networks, and even larger regions (chapter 4; Hanski 1999b). The expectation might be for little differentiation, both among local populations (occupying individual habitat patches) within networks and among metapopulations (occupying networks of habitat patches) within regions. But, in fact, we have found substantial genetic differentiation at a relatively small spatial scale and evidence that population turnover (extinction–colonization dynamics) promotes it. In this section we review the findings of several studies of population structure in the American checkerspots and then showcase the analysis of spatial patterns of molecular genetic variation in the Åland Islands 'megapopulation' of *Melitaea cinxia*.

Spatial Genetic Structure of American Checkerspot Populations

A series of studies have been carried out to examine the genetic structure of Nearctic *Euphydryas* and *Chlosyne* species. These investigations have shown that, not unexpectedly, patterns are strongly influenced by the structure of habitats. For example, for *E. editha* in the subalpine zone of the Rocky Mountains, where suitable habitat is extensive and often essentially continuous (although not continuously occupied; see chapter 3), genetic differentiation at 13 allozyme loci among geographic samples was readily explained by isolation by distance (F_{ST} population-local area and local area-regional < 0.03). Clustering of population samples was similar whether based on their genetic similarities or on geographic locations (Britten et al. 1995). In contrast, in Great Basin samples from Nevada the population-local area F_{ST} was 0.056, indicating genetic structuring at a relatively small geographic scale. In the Great Basin area, populations are located in relatively discrete habitat patches in mesic mountain ranges, isolated not by distance but by dry intermountain-range valley devoid of *Euphydryas* habitat. In short, hostile barriers to migration appear to be the dominant structuring mechanism in the Great Basin drift in Colorado.

In *Euphydryas anicia*, Cullenward and colleagues (1979) found only limited differentiation at nine allozyme loci when comparing samples from populations of alpine and high plateau ecotypes. This could indicate more movements than in either *E. editha* or *E. chalcedona*, or similar selection pressures, as suggested by Ehrlich and Raven (1969) as a general explanation for similar gene frequencies in isolated populations and documented for a pierid in similar habitat (Watt et al. 2003). Similarly, little differentiation was found at three polymorphic allozyme loci between three populations of *Euphydryas phaeton* in eastern North America (Opler and Krizek 1984). These populations were 200 m to 18 km apart in upstate New York (Brussard and Vawter 1975). This species has the reputation of being highly colonial (e.g., Scudder 1889, Macy and Shepard 1941), and the populations studied had sizes of 50–200 individuals. The hypothesized reason for the genetic similarity of the populations was occasional transfer of mated females, which anecdotally are often seen far from known populations (Macy and Shepard 1941, Brussard and Vawter 1975, Ehrlich unpubl. data).

Allozyme variation in the most distinct Nearctic *Euphydryas*, *E. gillettii* (the North American species with Palearctic affinities; chapter 2), showed the expected pattern: samples from different units of the Glacier National Park metapopulation were more similar to each other than to populations from central Idaho and central Wyoming (F_{ST} 0.041 vs. 0.325; Debinski 1994). One of the interesting results of this study was marked genetic differences in the same population in consecutive years. This is consistent with other evidence that *E. gillettii* can have a two-year development or have a less well-defined multiple-year diapause (E. H. Williams et al. 1984); thus there may be partially isolated *E. gillettii* populations using the same habitat in alternate years.

Schrier and co-workers (1976) compared the genetic population structure of a checkerspot in the genus *Chlosyne* with that of *Euphydryas editha* under similar circumstances. Three montane *Chlosyne palla* populations in Gunnison County, Colorado, showed no detectable differentiation at 11 allozyme loci among sites 5–12 km apart. This result, combined with information on movement patterns, implies that all three sites belong to what is essentially a single panmictic population (Schrier et al. 1976). The Pioneer Resort *E. editha* population, which is sympatric with that of *C. palla*, appeared to occupy an area close to an order of magnitude smaller than that of *C. palla*. The reason for the different distributions lies in the respective host plants, that of *C. palla* (*Aster occidentalis*) being more widespread than the local host plant of *E. editha* (*Castilleja linariifolia*).

Spatial Genetic Structure of Melitaea cinxia *in Åland*

The effect of local population turnover, that is, extinctions and recolonizations, on the genetic variance among local populations depends on several factors (Whitlock and McCauley 1990, Whitlock 1992). In structured populations with minimal turnover, a high rate of gene flow among local populations results in genetic homogenization to a scale dependent on the average migration distance. The processes of local extinction and colonization are expected to increase genetic differentiation when the average number of colonists to an empty patch is small and when the relatedness of colonists is high within patches but low between patches. Thus, the genetic variance among local populations is expected to increase when any given patch is colonized by immigrants from a single local population, but different patches are each colonized by migrants from different local populations. The effect of the rate of population turnover on genetic variance among local populations depends on how extinction and colonization influence the pattern of migration (and hence gene flow) and local population sizes (genetic drift), gene flow tending to decrease the variance and drift tending to increase the variance.

In the Åland metapopulation of *M. cinxia*, most local populations are small (chapter 4), and the average number of colonists to an empty patch is also very small, often consisting of a single mated female. Both of these factors tend to increase genetic differentiation. The maximum number of colonists is given by the number of larval nests found in previously empty patches, which was 1 and ≤3 in 55% and 86%, respectively, of the newly colonized populations in 2001. However, we have currently little information about how many source populations the colonists typically come from, and we know that migration between neighboring habitat patches and populations is high (chapter 4; Hanski 1999b), making it difficult to predict what level of genetic differentiation to expect. At a larger scale, there is the added complication of spatially correlated changes in population sizes (figure 4.12), primarily driven by spatially correlated weather phenomena (chapter 4). Such spatially correlated dynamics are likely to play an important role in shaping genetic differentiation among regions. For instance, regional crashes and recoveries could affect genetic differentiation among regions by increasing genetic drift if recoveries occur within regions, but the regional crash–recovery episodes could also promote greater genetic mixing if pronounced density differences among regions would enhance the reproductive success of migrants from one region to another. With these qualitative predictions in mind, our studies have been aimed at answering a number of basic questions. How much spatial genetic differentiation is there, and at what scale? Is there a relationship between genetic and geographic distances? Does population turnover increase genetic differentiation? To what extent can variation in gene diversities be explained by recent population sizes and connectivities?

An extensive sample was collected in 1995 at the level of one larva per every larval group (most likely to be full sib families) found during the annual survey (chapter 4). The sample size was restricted to one larva per family to minimize damage to the larval nests and impact on local populations. The genotype of 1216 larvae representing 369 local populations was scored at six polymorphic allozyme loci (peptidase A, peptidase B, glucose phosphate isomerase, glutamate oxaloacetate transferase, adenylate kinase, 6-phosphogluconate dehydrogenase) and two polymorphic microsatellite loci (referred to as CINX22 and CINX38).

Scale of Genetic Differentiation Strong spatial genetic structuring is evident not only among local populations and networks across the entire Åland 'megapopulation', but also among local populations within networks (Saccheri et al. in prep.). The clustering of rare alleles is evident from maps showing their locations, particularly at CINX38, the most polymorphic of the loci assayed (plate XII). At the largest scale, it is clear that the pattern of clustering has been strongly influenced by the effect of large water bodies acting as barriers to migration. The presence of a few scattered copies of a rare allele far away from any core region of that allele may represent the chance survival of a migrant lineage carrying the rare allele, with other members of the lineage having gone extinct along the way. Rigorous statistical analysis (Saccheri et al. in prep.) shows that rare alleles are typically strongly aggregated at spatial scales varying from a few kilometers to 10–15 km (figure 10.2). Compare these statistical results for the locus CINX38 with the actual spatial distribution of the respective alleles in the map on plate XII.

Further analysis, incorporating all alleles at allozyme and microsatellite loci (Goudet et al. 1996), reveals highly significant differentiation of genotypes among local populations within all the 17 regional networks of habitat patches with at least 20 individuals in the 1995 sample, consistent with the view that these networks are metapopulations. Hierarchical analysis of variance (Excoffier 2001) indicates that, on average, the total genetic variance is partitioned within local populations, between local populations within networks, and between networks in the ratio 89:8:3. The relatively low proportion of the total variance accounted for by variance among regional networks is likely to reflect a high level of gene flow among neighboring networks, masking greater differentiation seen at larger spatial scales (plate XII, figure 10.3). Global F_{ST} among all local populations is close to 0.1, which is somewhat higher than found for two other fritillaries, *Proclossiana eunomia*, $F_{ST} = 0.08$ (Nève et al. 2000) and *Melitaea didyma*, $F_{ST} = 0.06$ (Johannesen et al. 1996), particularly when one takes into account that the spatial scale covered was larger in the latter two studies. The greater degree of genetic structuring in Åland must in part reflect the highly fragmented distribution of the breeding habitat, but it may also be that the frequent extinctions and recolonizations promote differentiation among populations in Åland.

Isolation by Distance A large body of theory has been developed to use F_{ST} (Wright 1951), a standardized measure of genetic variance among local populations, to draw inferences about gene flow and genetic drift (Kimura and Maruyama 1971, Slatkin 1985). It is now widely accepted that it is inappropriate to use F_{ST} to estimate gene flow per se ('the effective number of migrants') because of discrepancies between the underlying model and real meta-

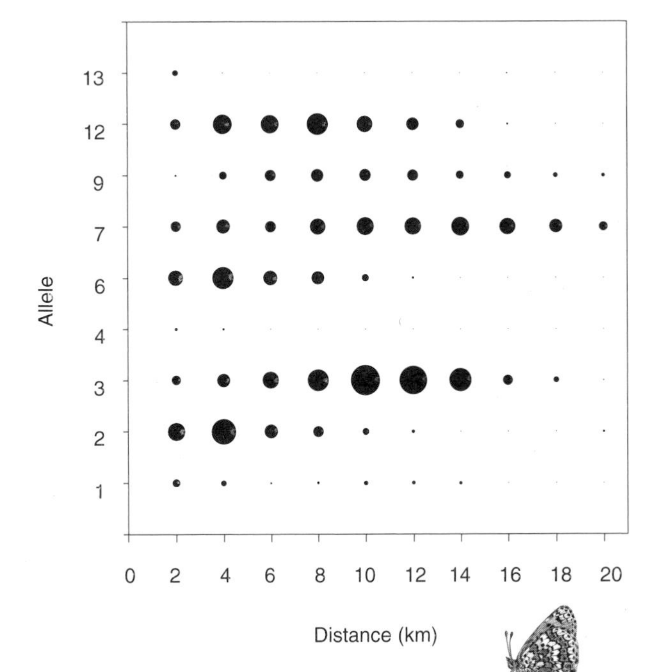

Figure 10.2. The scale of spatial aggregation of uncommon alleles at microsatellite locus CINX38 (on the rows). The significance of spatial aggregation at different spatial scales is shown by the size of the dot (larger the dot, more significant the aggregation). Significance of spatial aggregation was tested with a modified K-function approach (Diggle and Chetwynd 1991). The null hypothesis assumes that copies of the allele are randomly distributed among the individuals in the sample, regardless of their spatial location. Allele 8 (blue in the map on plate XII) is missing from the graph, as the outer islands were excluded from this analysis.

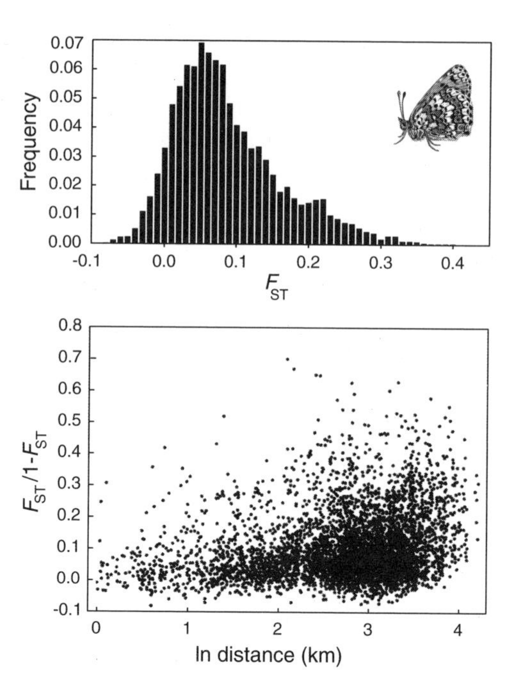

Figure 10.3. Distribution of pairwise F_{ST} for local populations (> 3 larval groups) of *M. cinxia* in Åland (upper panel). Pattern of pairwise genetic distance versus geographic distance (lower panel).

populations (Whitlock and McCauley 1999). One can nonetheless gain valuable insight into population structure by analyzing the relationship between F_{ST} and geographic distance for pairs of populations (Rousset 1997, Hutchinson and Templeton 1999).

The distribution of pairwise F_{ST} (figure 10.3, upper panel) illustrates the wide range of interpopulation co-ancestries existing in the metapopulation and the limitations of point estimates as descriptors of genetic structure of metapopulations. The pattern of isolation by distance (figure 10.3, lower panel) indicates a strong association between genetic and geographic distances between local populations. At small interpopulation distances gene flow tends to homogenize allele frequencies. With increasing distance separating pairs of populations, both the mean and the variance of F_{ST} increase, reflecting the greater role of genetic drift relative to gene flow.

Turnover, Genetic Variance, and Gene Diversity The effective population size is essentially determined by the number of reproducing residents and the number of reproducing immigrants, as well as the co-ancestries of these individuals. By increasing the genetic variance among local populations, population subdivision and turnover can increase the effective size of a metapopulation (Nichols et al. 2001). This appears to be the case with *M. cinxia* in Åland, where the slope of the regression of pairwise genetic distance against geographic distance is twice as great for young populations (less than two years old) as for old populations (greater than two years old; Saccheri et al. in prep.).

There are no distinct geographic trends in gene diversity (average heterozygosity) within patch networks. There is, as expected, both increased variance in gene diversity with decreasing network size and a weak negative relationship between average gene diversity and network size. Ongoing work is aimed at further quantifying the relative effective sizes of young and old populations to understand the influence of extinction–colonization dynamics in maintaining genetic variation in this metapopulation. There is also a need to explain variation in effective population size in terms of demographic population size and isolation.

10.4 Phylogeography

Research on *E. editha* and *M. cinxia* was initially focused on single populations and metapopulations within restricted geographical areas: the three nearby populations at Jasper Ridge and the large metapopulation in the Åland Islands. Later, the need arose to compare certain features of the focal populations with other geographically distinct populations, to evaluate hypotheses about ecological and evolutionary mechanisms, and to search for generalities. In the case of *E. editha*, studies of other populations revealed complex differences in host plant use and associated behavior of larvae and adults (chapters 3, 5, and 6). For *M. cinxia*, there was an interest to examine the strategies of parasitoids that are specialists in Åland in other locations, (chapter 8); to find out whether metapopulation dynamics affected the genetic load; and to study variation in oviposition host plant preference (chapter 6). Such comparisons require a frame of reference to separate recent from ancient effects of environment and demography. Intraspecific phylogenies (e.g., Hewitt 2001) provide an elegant way to establish this historical perspective. By informing us about evolutionary history, the geographical distribution of genealogical lineages, or phylogeography (Avise et al. 1987, Avise 1998), has also become an invaluable tool for setting management priorities aimed at maximizing the conservation of the evolutionary past (and possibly future) for single species and regional communities (Avise and Hamrick 1996).

Melitaea cinxia is widely distributed across Eurasia, from the steppes of Buryatia and the Tian Shan mountains in China in the east to meadows in France and the Iberian Peninsula in the west, extending northward to Åland in Finland, and southward to the Atlas Mountains in North Africa and to Israel. Across this vast area it occupies a wide variety of habitats (chapter 4), feeding on different species of *Plantago* and *Veronica* (table 7.2). It is perhaps this host plant specialization, coupled with its relatively sedentary behavior, that has led to the genesis of a highly structured phylogeography for this species.

The phylogeny of *M. cinxia* inferred from partial sequences (600 bp) of the mitochondrial gene cytochrome oxidase I is shown in figure 10.4. Parsimony analysis of the available samples supports the existence of three main clades: (1) eastern and central Eurasia; (2) Iberia and France; and (3) a southern/central European group including southern Scandinavia, Germany east to Ukraine, northern Italy, the Balkans, Greece, Turkey, and the Levant. The relationships between these clades, which are currently unclear, will be resolved with more sequence data (I. Saccheri and N. Wahlberg in prep.). Within clades there are several interesting patterns. It was at first surprising to discover that the Åland metapopulation shares such high mitotype similarity with populations from the Lake Baikal region, some 5000 km away. Glacial refugia and postglacial colonization routes are less well established for central and eastern Eurasia than for western Europe (Hewitt 1999), and further study is required to reconstruct the history of this extensive clade. The northern half of Italy groups with the central European clade, whereas the southern half, including Sicily, is more closely related to the Iberian/French clade. It seems likely that a population expanding northward from a southern Italian refugium has come into contact with a southward expansion, possibly originating in the Balkans. There also appears to be a contact zone in northeastern France, where a mixed population occurs (Mourmelon), containing individuals with Iberian/French mitotypes and other with central European mitotypes. The density of samples in eastern Europe is as yet too low to localize zones of contact between central european and Eurasian clades.

Euphydryas editha's population phylogeny in the western United States has been examined using morphology, allozyme loci, and mtDNA. Baughman et al. (1990a) studied 41 populations, using 19 allozyme loci, almost all of which were in Hardy-Weinberg equilibrium. Using a variety of analyses, they identified six phylogeographic groupings: (1) the southern Rocky Mountains; (2) the northern Great Basin; (3) the southern Great Basin and mostly eastern Sierra Nevada; (4) California west of the Sierra Nevada; (5) a lone north-central California population, and (6) a lone south-central California population (figure 10.5). The phylogeographic pattern of *E. editha* resulting from allozyme studies is correlated with that for *E. chalcedona*, for areas where the two species occur sympatrically (Brussard et al. 1989).

The groupings based on allozyme loci do not correlate well with the morphological variation historically used to define subspecies (Baughman et al. 1990a), and they also differ from patterns revealed by mtDNA data. In particular, Radtkey and Singer (1995) examined mtDNA haplotypes using restriction enzymes in 24 populations in California and Nevada in a study aimed at exploring the evolution of host plant oviposition preference. Their data overlap a number of the populations surveyed earlier using allozymes. The pattern of mtDNA haplotypes among populations suggests first that populations vary in age and gene flow patterns (M. Singer pers. comm., Radtkey and Singer 1995). The data also lend some support to grouping of the northern Great Basin as a clade distinct from other groups in California and Nevada and to east-

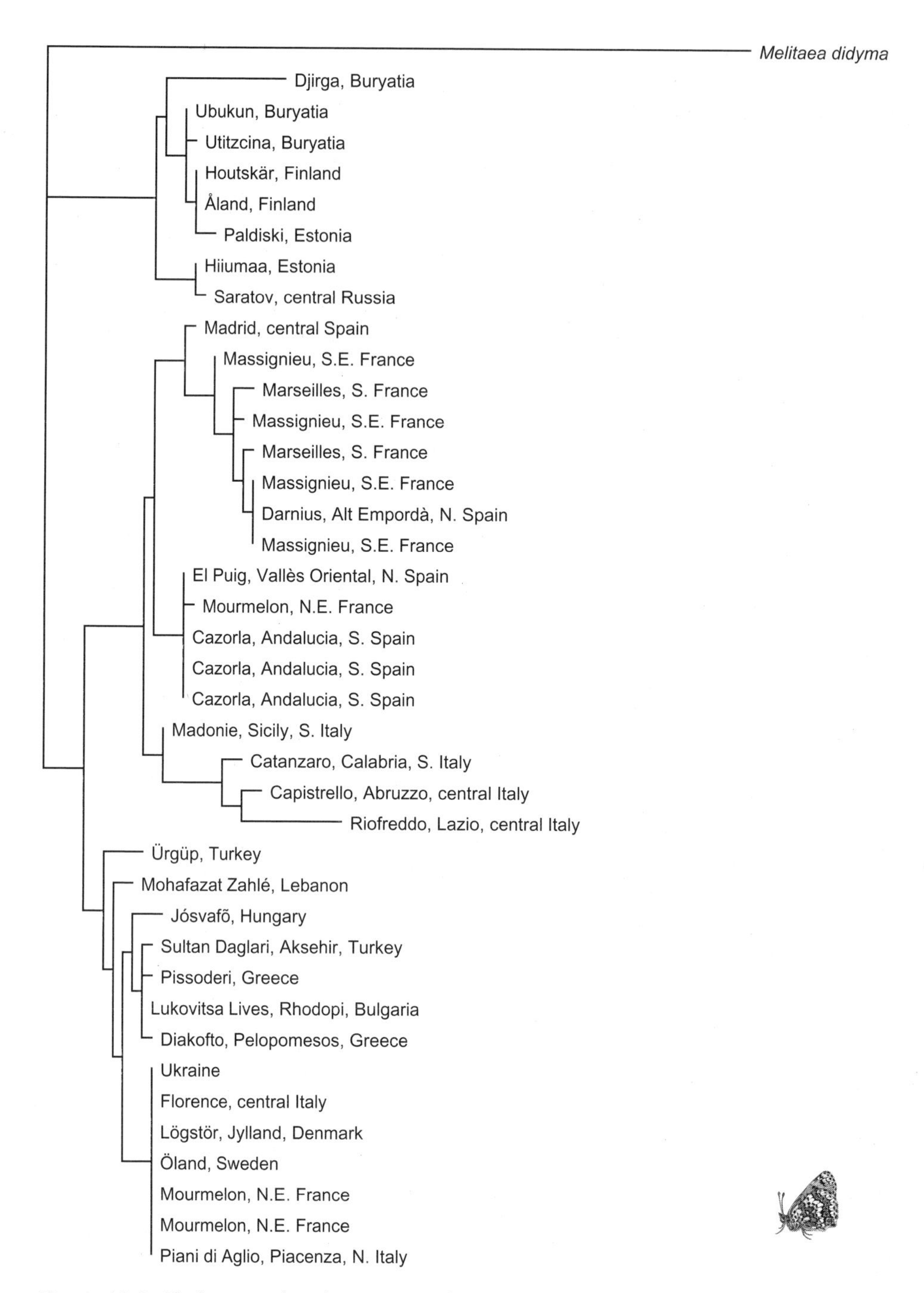

Figure 10.4. Phylogeny of *Melitaea cinxia* based on partial sequence of COI mitochondrial DNA.

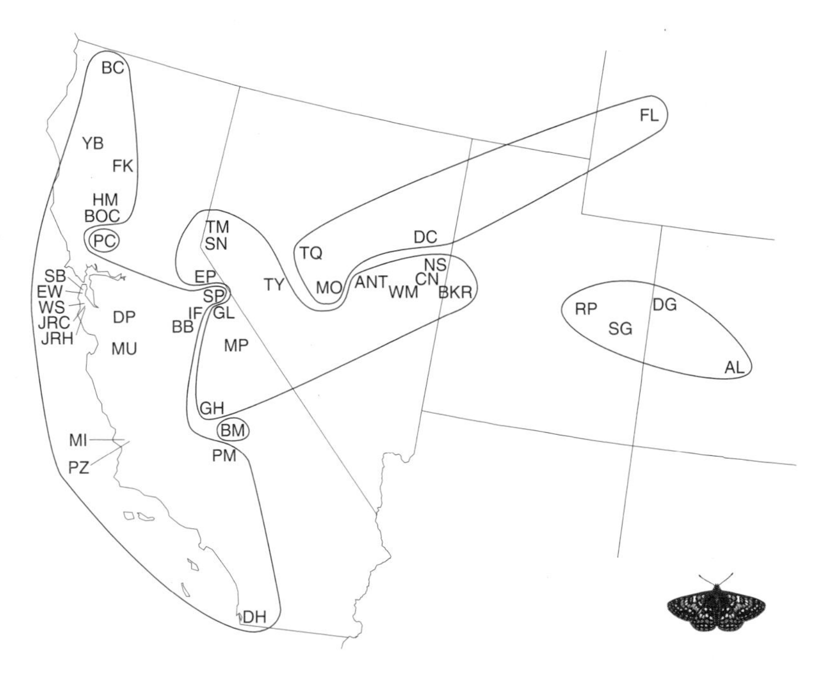

Figure 10.5. Phylogenetically distinct groups of *Euphydryas editha* in the western United States based on allozyme markers, modified from Baughman et al. (1990a).

ern versus western Sierra Nevada clades, although some Sierran population affinities differ from that seen using allozymes. These differences are not surprising, as mtDNA and allozyme loci experience different dynamics, which have led researchers in more recent years to utilize DNA for such phylogenies.

Given these current data, there remain multiple possible causal explanations for the groupings, including historic and recent patterns of gene flow influenced by geography and climate (including glaciation patterns), and congruent selection pressures operating at the allozyme loci. It is possible that further DNA data will shed more light on the relationships among these groups, allowing tests of causal hypotheses for their distributions, but those data have not yet been collected.

10.5 Genetic Load and Inbreeding Depression in Metapopulations

The phenomenon of inbreeding depression—the reduced viability and fertility of the offspring produced by the mating of related individuals—has been documented for many animals and plants (Charlesworth and Charlesworth 1987). In the past, there has been considerable debate about the underlying causes, with the two competing hypotheses being loss of heterozygosity per se (the overdominance hypothesis) versus increase in deleterious recessive homozygotes (the dominance hypothesis; Crow 1952, 1993). Nowadays, it is widely accepted that by far the most important factor is the expression of deleterious recessive alleles in homozygotes (Charlesworth and Charlesworth 1999), the frequency of homozygous loci within individuals increasing with inbreeding.

The concept of "genetic load" (Haldane 1937, Crow 1948, Muller 1950) was introduced to describe the reduction in fitness due to deleterious mutations, the majority of which are recessive. This concept quantifies the potential magnitude of inbreeding depression in a population. As noted in section 10.2, it also encourages analysis of the mechanisms and processes causing genetic loads to differ among populations: mutation, selection, genetic drift, and gene flow (Kimura et al. 1963, Whitlock 2002).

Although most of the data on inbreeding depression come from laboratory and zoo populations, an increasing number of studies have described inbreeding depression in natural populations (Hedrick and

Kalinowski 2000). Furthermore, laboratory and field experiments that have looked at the interaction between environmental stress and inbreeding depression have often found that inbreeding depression is greatly increased under less favorable conditions (Bijlsma et al. 1999, Meagher et al. 2000, Keller et al. 2002), though the mechanisms for the interaction are poorly known. Comparison of estimates of inbreeding depression measured under natural versus benign laboratory environments (Crnokrak and Roff 1999) confirm that laboratory estimates of inbreeding depression are likely to underestimate the impact of genetic loads in natural populations, particularly when one considers that most studies do not measure inbreeding depression for total fitness but for a component of it (Keller 1998).

At present, there is convincing evidence that, for many species, inbreeding can cause large reductions in individual fitness in nature (Keller and Waller 2002). What has been much more difficult to establish is whether such reductions in individual fitness have any significant impact on the dynamics, and in particular on the risk of extinction, of natural populations. The contribution of the *M. cinxia* project has been to demonstrate that inbreeding can indeed increase the extinction rate of small populations in nature. This conclusion was reached through a combination of an observational study of natural populations (Saccheri et al. 1998), laboratory experiments measuring inbreeding depression under controlled conditions (Haikola et al. 2001), and field experiments with artificially inbred populations to test the results of the original observational study (Nieminen et al. 2001).

Inbreeding Depression in Melitaea cinxia

It is likely that many species of butterfly are sensitive to inbreeding. *Bicyclus anynana*, an African satyrine butterfly, has been the most intensively studied species in this respect, and it suffers a 25% decline in fertility (measured as the proportion of eggs that hatch) per 10% increase in inbreeding, equivalent to a 60% decline in the offspring of full sibs (Saccheri et al. 1996). The handful of other Lepidoptera for which there are published data also exhibit large declines in egg hatching rate with inbreeding (e.g., Higashiura et al. 1999, Cassel et al. 2001). In contrast, inbreeding depression in *Drosophila melanogaster* (3% decline in egg-pupa viability per 10% increase in inbreeding) is unusually low for an insect (Saccheri et al. 1996). The causes of such differences in genetic load among species are little understood, but undoubtedly include history of inbreeding and purging, developmental architectures for the relevant traits, and variation in the mutation process.

The focus on egg hatching rate is partly because it is a convenient trait to measure, but also because it is a major component of fitness. In *B. anynana*, number of eggs hatching is highly correlated with reproductive success and total fitness, at least in the laboratory environment where larval mortality is very low (Brakefield et al. 2001). Results for *M. cinxia* demonstrate that under conditions where the environment and life history are more complex, there is greater scope for inbreeding depression to manifest itself in other components of fitness, such as larval viability, mating success, and adult longevity.

Conducting breeding experiments with *M. cinxia* has been a challenging and frustrating enterprise, reflecting the unfortunate fact that checkerspots do not make good laboratory organisms. The species succumbs to viral infections under conditions of laboratory mass culture, artificial breaking of the obligatory diapause often leads to increased larval mortality, and mating under artificial light is problematic. In spite of these constraints, much useful information on the effects of inbreeding on components of fitness has been obtained through the determined efforts of several researchers.

Several of these experiments have looked at the effects of inbreeding on egg hatching rate. A complication with measuring inbreeding depression for egg hatching rate is that it is affected by the fertility of both parents as well as the viability of the zygote (in the eggs that are fertilized). Not only is inbreeding likely to affect each of these components to a different degree (in *B. anynana* male fertility is the major component of inbreeding depression for egg hatching rate; Saccheri et al. unpubl. ms), but the degree of inbreeding may differ substantially between the parent and the offspring. For example, siblings whose parents are not closely related will have an inbreeding coefficient, F, of zero regardless of how inbred the parents were; their fertility cannot therefore be affected by inbreeding. Offspring produced by the mating of these sibs are inbred ($F = 0.25$), which may manifest through zygote, larval, and adult viabilities, mating success, fecundity, and fertility. If these offspring interbreed, parents (second filial generation F_2) and offspring (F_3) are inbred, in which case both the fertility and viability components are likely to affect inbreeding depression of egg hatching rate.

Three independent laboratory experiments (Saccheri et al. 1998, Haikola et al. 2001, Nieminen et al. 2001) with different population samples from the field have found that brother–sister matings ($F = 0.25$) lead to an 8–46% decline (average 26%) in egg hatching rate. For reasons just given, in the absence of information about previous inbreeding, these results are interpreted as measuring inbreeding depression on the zygote viability component of egg hatching rate, but not on the fertility component. In the single experiment that has been successfully continued to the next adult generation (F_2), inbreeding depression for egg hatching rate was 31% greater than in the equivalent lines in F_1 (S. Haikola pers. comm.).

Several other fitness components apart from egg hatching rate are adversely affected by inbreeding. Nieminen and co-workers (2001) found that inbred larvae often construct poor winter nests (chapter 4), possibly reflecting reduced larval activity, and that inbred larvae suffer high overwinter mortality during diapause. In one laboratory experiment, the proportion of mated females in laboratory populations derived from small (1–2 larval families) and isolated populations was 85% lower than in equivalent laboratory populations derived from large (9–22 larval families) and well-connected populations (Haikola et al. 2001). Such an effect of inbreeding on mating behavior, which has also been documented in *B. anynana* (M. Joron pers. comm.), is likely to have contributed to the lower frequency of mated females found in small and isolated populations than in large populations in the field (Kuussaari et al. 1998). There is also evidence suggesting that adult longevity is compromised by inbreeding (Saccheri et al. 1998, van Oosterhout et al. 2000). The compound effect of inbreeding depression for these components of fitness has a major impact on survival and reproductive success. It is not that surprising, therefore, that inbreeding should have significant consequences for population persistence.

Inbreeding and Extinction in the Melitaea cinxia *Metapopulation*

An inherent problem of critically studying the causes of extinction is that one needs to observe several populations going extinct. If one is additionally interested in examining the effect of inbreeding on extinction risk, there needs to be substantial variation in the degree of inbreeding among local populations. With its high rate of local population extinction and local inbreeding (chapter 4), the Åland metapopulation of *M. cinxia* presents excellent opportunities to study the impact of inbreeding on local dynamics and extinction.

The frequency of matings among close relatives is predicted to be very high in the Åland metapopulation of *M. cinxia* because 10–40% of all individuals in the entire metapopulation occur in local populations made up of just one to three larval families. In the majority of cases, larval groups consist entirely of full sibs. A small fraction of females (<10%) are doubly mated (chapter 5), in which case some of the larvae in a group might be half sibs. In such small local populations, there is a high probability of sib matings, giving rise to inbred offspring with an inbreeding coefficient of 0.25. Individuals with this level of inbreeding are expected to suffer measurably reduced viability and fertility. Allozyme and microsatellite markers showing a high degree of homozygosity within local populations provide further evidence for high rate of inbreeding. The process of inbreeding is therefore of direct relevance to the population biology of *M. cinxia* in Åland.

To examine the likely impact of inbreeding on the dynamics of small populations in the field, a sample of adult females was taken from 42 local populations, scattered across Åland and varying in population size and degree of isolation (Saccheri et al. 1998). The autumn survey of the same year established that seven of these populations had gone extinct (figure 10.6). Having obtained local population estimates of heterozygosity at six allozyme loci and one microsatellite locus, we found that average heterozygosity was significantly negatively correlated with population extinction (Frankham and Ralls 1998, Saccheri et al. 1998). By including in the statistical model of extinction risk the ecological factors known to influence extinction (most importantly, population size and connectivity), we were able to rule out the possibility that the heterozygosity effect was due to a correlation between heterozygosity and the size and isolation of local populations. These results imply that the mechanism underlying the observed heterozygosity–extinction correlation is inbreeding depression, variation in average heterozygosity reflecting relative differences in the degree of inbreeding among local populations. The result suggests that even in situations where the ecological factors predominate, genetic factors can substantially magnify the extinction risk (figure 10.7).

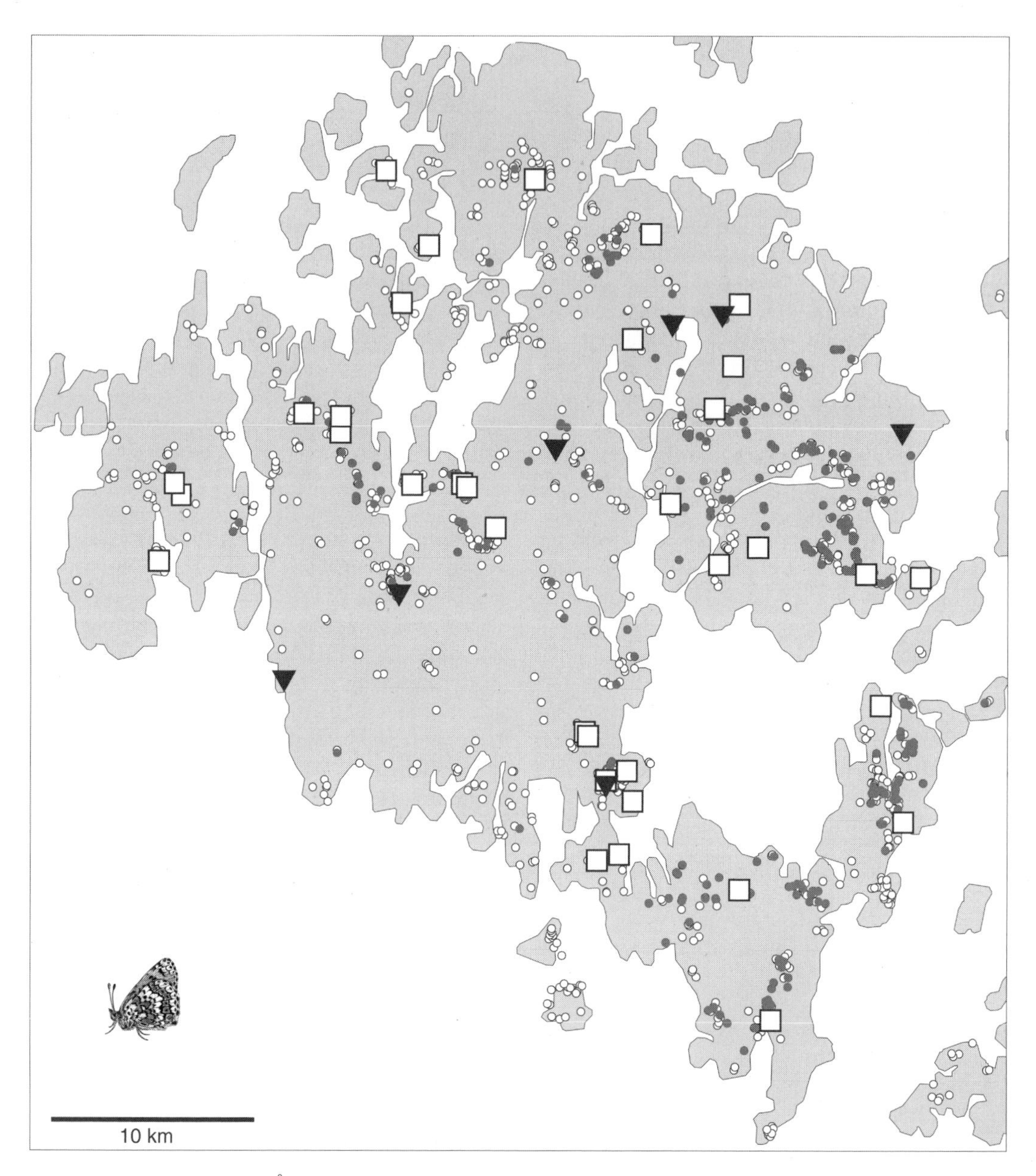

Figure 10.6. Map of the Åland Islands showing the locations of the 42 local populations from which adult female butterflies were sampled in summer 1996 (large symbols). All known suitable meadows are shown as small circles, with meadows in which Glanville fritillary larvae were present in autumn 1995 shown by gray circles (and large symbols) and unoccupied meadows shown by white circles. Of the 42 local populations sampled, the 35 that survived to autumn 1996 (white squares) are distinguished from the seven that went extinct (black triangles). From Saccheri et al. (1998).

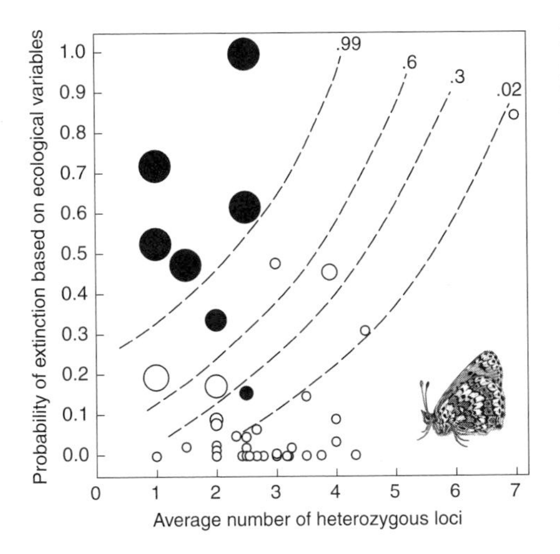

Figure 10.7. The probability of extinction in *M. cinxia* is influenced both by ecological factors and by heterozygosity, which is used here as a measure of the level of inbreeding. The vertical axis gives the probability of extinction for the 42 populations in figure 10.6 as predicted by a model including several ecological factors. The horizontal axis gives the average number of heterozygous loci per individual in a sample of eight polymorphic enzyme and microsatellite loci. The size of the symbol is proportional to the probability of extinction predicted by a model including both the ecological factors and heterozygosity (the isoclines of equal extinction risk were drawn by eye). Of the 42 populations studied, seven populations (black circles) went extinct in one year, as indicated in figure 10.6. From Saccheri et al. (1998).

To further test the prediction that inbreeding can significantly increase the risk of population extinction, a field experiment was undertaken comparing the survival of controlled inbred versus outbred populations (Nieminen et al. 2001). The parental generation for this experiment was drawn from two widely separated populations in Åland. Butterflies were mated at random in the laboratory for one generation, restricting matings to within source locality. In the F_1 (the first filial generation), butterflies were mated either to a full sib or to a butterfly from the alternate source population, producing groups of F_2 larvae with inbreeding coefficients of 0.25 or 0. After the winter diapause, these larvae were introduced into 12 unoccupied patches of suitable habitat, creating 6 inbred and 6 outbred populations, each consisting of 3 equally large larval groups. Surveys in early and late summer suggested little difference in survival to adulthood between the two treatments but clearly showed lower reproductive success of the inbred populations, only two of which left surviving offspring, compared to all six outbred populations. A survey the next spring revealed that the two remaining inbred populations had perished during the winter diapause but that larvae had survived in four of the six outbred populations. This experiment supports the interpretation that the correlation between average heterozygosity and local population extinction in the study of Saccheri et al. (1998) was indeed due to inbreeding depression.

Given the substantial impact of inbreeding on individual fitness and on local population dynamics in *M. cinxia*, one might expect that females have evolved to discriminate against close kin as mates. Two experiments showed that this is not the case, however: mating was equally as likely with brothers as with unrelated males in one laboratory experiment and in one field experiment (Haikola et al. 2004). The same result was obtained with females from the Åland metapopulation, where inbreeding is frequent, and with females from southern France, where inbreeding is thought to be infrequent. In the latter case, there has perhaps been little selection for inbreeding avoidance to evolve. In contrast, in the small local populations in Åland the cost of inbreeding avoidance might be prohibitive because often the females would have no choice, and it is clearly better to mate with a brother regardless of inbreeding depression than to remain unmated. A more detailed analysis is needed to assess the benefits and costs of inbreeding avoidance in different kinds of populations.

Dynamics of Genetic Load

The simplest model for inbreeding depression assumes a large, randomly mating population, in which deleterious alleles arise through spontaneous mutation and are eliminated at a rate determined by their selective disadvantage and dominance (Charlesworth and Charlesworth 1999). In small populations, or where inbreeding occurs for other reasons (such as spatial or social structuring), deleterious recessive alleles should be eliminated more effectively due to their greater exposure in the homozygous state. This process, known as purging (Barrett and Charlesworth 1991), has greatest impact on alleles with strongly detrimental effects (Hedrick 1994). However, small population size also hinders purging because the probability that deleterious alleles may drift to high frequency by chance is increased (Kimura et al. 1963, Wang et al. 1999).

In a closed population, a deleterious allele that has become fixed can only be replaced once an alternative allele has entered the population through mutation, which is likely to take hundreds to thousands of generations. A solution to this problem, used extensively in agricultural breeding programs, is to introduce alternative alleles from another population or genetic lineage. Repeated cycles of inbreeding and genetic mixing between populations have much in common with the genetic processes taking place in metapopulations and are predicted to lead to more effective purging of deleterious recessive mutations relative to equivalent panmictic populations (Wang 2000). We tested this prediction with *M. cinxia* by comparing the inbreeding depression on egg hatching rate in Åland with that in a more continuously distributed population in southern France (figure 10.8; Haikola et al. 2001). After one generation of full-sib mating, inbreeding depression in the French population was 42%, compared to 8% in the Åland metapopulation (taking into account two other experiments, the average for Åland is 26%, but because the experimental results are influenced by environmental conditions, it is best to compare experiments run concurrently). Partial purging of deleterious recessive alleles affecting egg hatching rate has also been observed in inbred laboratory populations of *B. anynana* (Brakefield and Saccheri 1994, Saccheri et al. 1996).

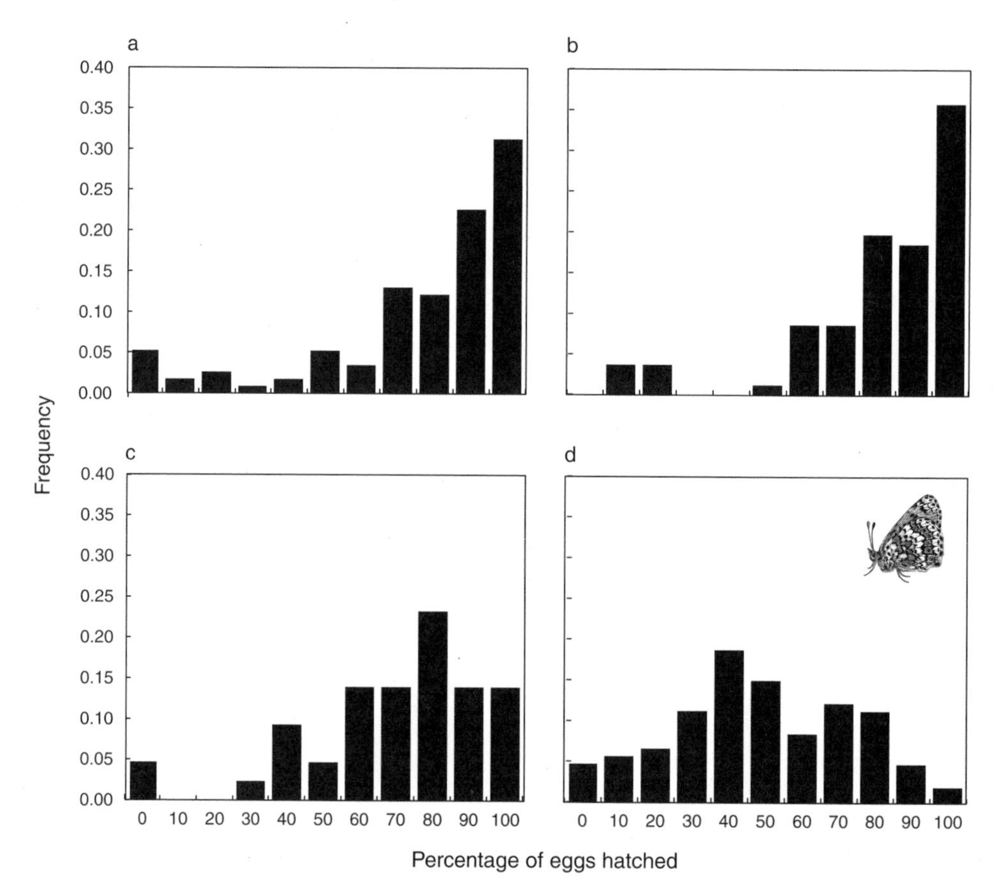

Figure 10.8. Comparison of inbreeding depression for egg hatching rate in *M. cinxia* from Åland and France. Each panel shows the frequency distribution of the proportion of eggs hatching in batches laid by different *M. cinxia* females, mated to nonsibs or full sibs. The four panels give the results for between-family crosses in (a) Åland and (b) France and for sib-crosses in (c) Åland and (d) France. The reduction in egg hatching rate due to inbreeding is substantially greater in the French than in the Åland population samples (Haikola et al. 2001).

Whitlock (2002) made quantitative predictions about the relationship between metapopulation F_{ST} and equilibrium inbreeding depression relative to a panmictic population. The observed 38–81% lower inbreeding depression in Åland, with an average F_{ST} of 0.1, is qualitatively consistent with the prediction, assuming small dominance coefficients, but suggests more effective purging than expected. It is possible that the efficiency of selection is further enhanced by the high level of metapopulation structuring in Åland (a population of metapopulations; chapter 4) and that this multilevel interaction between inbreeding and selection is not adequately captured by global F_{ST}, which simplifies the structure to two levels. The result also implies that progressive accumulation of deleterious mutations in small populations is unlikely to lead to "mutational meltdown" in metapopulations (Lynch et al. 1995, Gilligan et al. 1997, K. Higgins and Lynch 2000). The explanation that the lower genetic load in Åland than in France is entirely due to differences in contemporary population structures should be treated with caution, however, because populations in Åland and in southern France belong to distantly related lineages (section 10.5; figure 10.4). Differences in effective population sizes associated with range expansions are also expected to influence genetic loads, though there is little empirical evidence for genetic loads being lower at range margins (Sperlich et al. 1977, Brewer et al. 1990). Regardless of these issues, the fact remains that reduced genetic load in Åland has not removed the negative impact of inbreeding on population persistence.

Another dynamic consequence of inbreeding depression in metapopulations is that offspring of immigrants mated to local residents are expected to show hybrid vigor (heterosis), assuming that the set of deleterious recessive alleles carried by immigrants differs from the alleles occuring at high frequency in the inbred local populations (Whitlock et al. 2000) and that alleles at interacting loci are not locally co-adapted (Templeton 1986). Partly inspired by the Åland system, a laboratory experiment was conducted with *Bicyclus anynana* (Saccheri and Brakefield 2002) that demonstrated that heterosis can greatly promote the spread of immigrant genes into inbred populations. Thus, from a genetic perspective, the effective migration rate (gene flow) may be much greater than that implied by the number of migrants (Ingvarsson and Whitlock 2000). Heterosis may also reduce the risk of population extinction, giving rise to a genetic rescue effect (C. M. Richards 2000), which complements the well-established demographic rescue effect (Brown and Kodric-Brown 1977).

Preliminary experiments comparing between-family crosses within and between populations have not detected significant heterosis for egg hatching rate in *M. cinxia* (Haikola et al. 2001). This is surprising given the strong evidence for inbreeding depression in local populations coupled with genetic differentiation, but the result may reflect that the families used in this experiment came from relatively large and well-connected populations that are unlikely to have undergone much inbreeding. In contrast, strong heterosis has been found after immigration to seminatural populations in a *Daphnia magna* metapopulation (Ebert et al. 2002), with similar structure and dynamics to the Åland system.

10.6 Challenges for Future Work

It is most unfortunate that many families of Lepidoptera seem to present formidable challenges to the development of informative microsatellite markers (Nève and Meglécz 2000), apparently due to a combination of their low frequency in the genome and complex mutation processes. *Melitaea cinxia* has not been an exception to this rule, but after several years of work, Hanski's group is accumulating a sufficient number of polymorphic microsatellites that will allow more detailed analyses of metapopulation genetics. The greater density of genetic information provided by these markers will yield a higher resolution of the spatial genetic structure than has been possible to achieve to date. The markers may also be used to study the variance in reproductive success among individuals, which is a critical (short-term) determinant of the ratio of census to effective population size. Given sufficient variation, reliable assignment of parentage to larval groups within networks of local populations would be a powerful means for detecting gene flow directly.

A fundamental remaining issue is to understand fully the capacity of complex metapopulations to maintain genetic variation, which sets the effective metapopulation size. A challenge for the future is to combine the existing extensive demographic information (chapter 4) with high-resolution genetic information, most effectively, perhaps, in the context of individual-based simulation models, such as developed by Heino and Hanski (2001). This framework would provide a natural way of predicting the role of local and regional turnover of populations on spatial and temporal patterns in genetic variation, and predictions could subsequently be validated with microsatellite data. We expect that monitoring changes in microsatellite allele frequencies through time will give particularly valuable insight to the way in which local and regional turnover shape the genetic structure. Longer term ventures may include the study of the genetic basis of observed oviposition host plant preference (chapter 6) and the behavioral, physiological and ultimately genetic mechanisms that underpin the observed variation in migration rate (chapter 9; Hanski et al. 2002).

Another dimension to the genetic study of the *Melitaea cinxia* model system relates to genetic population structure of the two primary parasitoids and their respective hyperparasitoids associated with the butterfly in Åland (chapter 8). It is now known that the two primary parasitoids exhibit very different powers of migration, which are reflected in their typical migration distances (van Nouhuys and Hanski 2002a) as well as in the spatial patterns of parasitism (Lei and Hanski 1998). One would expect that these differences are reflected in the spatial genetic structure of the parasitoid populations. Equally interesting are the preliminary results on the braconid wasps in the genus *Cotesia* that are associated with checkerspot butterflies (M. Kankare pers. comm.). These results clearly indicate that the species recognized in traditional taxonomy are in fact clusters of closely related taxa, each specializing on a different host species or a group of closely related host species. There is as yet no clear evidence for coevolutionary changes in the butterflies and their parasitoids, but the parasitoids appear to have closely evolved with their host species.

With regard to *Euphydryas*, a number of significant opportunities now exist to explore the interaction of habitat and population demographic structure with population genetic structure. First, as noted in chapter 3, the large south San Francisco Bay population at Kirby Canyon is spread across a diverse topography, and the larval populations occupy different slopes to differing degrees among years, depending on weather conditions. The structure is thus intermediate between a classical open population and a metapopulation and presents an excellent opportunity for a combination of modeling and empirical work using markers such as AFLPs (amplified fragment length polymorphism) or possibly microsatellites for examining the effects of habitat structure on the population genetic structure of such a system. Second, intraspecific genetic analysis based on coalescence theory is gaining popularity because of its potential to shed light on demographic and historical processes. Insight into the characteristics of the habitats over which these demographic and historical processes operate may be gained from geographical information systems, currently used extensively in landscape ecology studies. Such tools, when used in conjunction with behavioral and physiological experiments, offer a means to further investigate the relative roles that behavior, physiology, and habitat characteristics play in the genetic structuring of natural populations. The extensive ecotypic variation shown by checkerspot butterflies, especially *E. editha* and *E. aurinia*, has paved the way for ongoing investigations of the relative roles of local adaptation and landscape features as influences on patterns of gene flow (B. Wee and M. Singer in prep.).

Exciting opportunities exist in both the *E. editha* and *E. gillettii* systems to explore the population genetic dynamics of colonization and metapopulation establishment in areas that are more isolated relative to species' movement patterns than occurs in *M. cinxia* in the Åland Islands. The extinction of *E. editha* at Jasper Ridge (figure 3.7) opens the possibility of reintroduction experiments designed to explore the dynamics of allozyme loci previously suggested to be selected or neutral under monitored environmental and population demographic conditions. The recent explosion of *E. gillettii* in Colorado, both in numbers and area, offers a chance to look at genetic dynamics associated with the establishment of a metapopulation system, assuming that the species will eventually establish similar demographic dynamics to those in the northern Rocky Mountains.

This short summary of challenges for future work is not inclusive, but rather gives a flavor of the opportunities that exist within the checkerspot system to further our understanding of the interplay between ecology and evolution through understanding the interactions among population ecological, demographic, behavioral, and genetic structures. As new technologies develop, our ability to probe the system will mature as well. But we have indeed come a considerable distance since Ehrlich first picked up a marking pen and marked butterfly #1 on Jasper Ridge in 1960.

11

Bay Checkerspot and Glanville Fritillary Compared with Other Species

NIKLAS WAHLBERG, PAUL R. EHRLICH, CAROL L. BOGGS,
AND ILKKA HANSKI

11.1 Introduction

The long-term intensive studies on *Euphydryas editha* and *Melitaea cinxia* have produced substantial knowledge about their population biology. It is important to ask how much of this knowledge can be applied to other checkerspots, to other butterflies, and, indeed, to other taxa apart from butterflies. To answer these questions requires judicious comparisons among species, following in the steps of hundreds of population biologists who have used the "comparative approach" since Darwin (1859) formulated the theory of evolution by natural selection.

The Comparative Approach

The comparative approach as currently practiced in population biology is an important tool to study adaptation. Working on butterflies, analyses examine characters such as wing length, host plant use, number of eggs per batch, or adult movement ability in fragmented landscapes and how these characters vary among species. Through comparative study, we can infer whether the character of interest may be adaptive and whether natural selection has produced similar solutions to ecological challenges in species that live in similar environmental circumstances.

Before making comparisons, however, we must consider the pitfalls associated with comparing groups of species. The main difficulty is statistical non-independence of data points (character values for different species) due to common ancestry of the species (Wanntorp et al. 1990, Harvey and Pagel 1991, Harvey 1996). This relates to the fundamental observation that different species may have the same adaptations for two entirely different reasons: they may share a common ancestor (identity by descent), or natural selection may have resulted in similar characters in evolutionarily unrelated species (parallel or convergent evolution; Harvey and Pagel 1991). For example, all checkerspot larvae are gregarious at least in the early larval instars (chapter 7). Is this a result of similar selection pressures favoring larval gregariousness or has gregarious behavior been inherited from a common ancestor and remained a common character though it may carry no selective advantage in some of the species? It appears that the latter explanation is correct (chapter 7), though it may be that selection also favors gregariousness (see below). As a second example, local populations of checkerspots are well delimited from each other (chapters 3 and 4). In this case it is much less clear to what extent the similarity is due to selection pressures in similar environments and to what extent it is due to common ancestry. There is no reason, of course, for only shared selection pressures or common ancestry to explain observed similarity: both selection and ancestry may act congruently.

Why Are Related Species Similar?

Similarity due to common ancestry requires some further thought. Harvey and Pagel (1991) list three important processes affecting the similarity of related species: phylogenetic niche conservatism, phylogenetic time lags, and adaptive responses. The first process, phylogenetic niche conservatism, is purely adaptationist: a vacant niche is occupied by a species that is already adapted to a similar niche. For instance, a butterfly adapted to open, moist habitats is more likely to invade a new open but dry habitat than a butterfly adapted to moist forest habitat. Thus checkerspots may be mostly found in patchily distributed meadows because they are all descended from an ancestor that was adapted to that niche.

Second, phylogenetic time lags also maintain similarity among related species. A character that has become redundant in a species will not necessarily disappear quickly. For instance, it would be difficult to understand why whales have hip bones (which are not used in any way as far as is known) if one did not realize that whales are descendants of terrestrial mammals. Evolution has not had time to completely remove the hip bones from the now purely aquatic whales.

The third process, adaptive responses, results from the fact that close relatives are more likely to share similar genetic variation. When such variation is acted on by natural selection, more closely related species are likely to respond similarly to similar selective pressures than are less closely related species. This process may be important in comparisons among checkerspots and other groups in the superfamily Papilionoidea (chapter 2). However, even closely related species can differ in their response to the same selective pressure (Harvey and Pagel 1991). The nature and amount of genetic variation within species depends on its population dynamic history (Futuyma 1986).

Phylogenies and Their Uses in Comparative Study

To address the issue of similarity by descent, one needs to know the evolutionary history, or phylogeny, of the group of species. Unfortunately, the evolutionary history of most species in terms of sequences of divergence (speciation events) is at best sketchy because the fossil record is inadequate (and grossly so for butterflies). Thus, what one observes is a snapshot of the current state of an unknown evolutionary sequence, whereas what one would like to observe is a motion picture of the entire process of evolutionary development. Nonetheless, in the past decade new molecular systematic methods have facilitated the construction of phylogenetic trees, which show the best approximation of the times of divergence for a group of species given their current character states.

Phylogenies are used by comparative biologists to study common evolutionary patterns across species and to infer which characters may have evolved in particular species as adaptations to their environment (Harvey and Pagel 1991). The definition of an adaptation has been the subject of debate (e.g., Ehrlich and Holm 1962, Mayr 1982, Rose and Lauder 1996). Most simply, an adaptation can be defined as "a derived character that evolved in response to a specific selective agent" (Harvey and Pagel 1991, p. 13). Methods have been developed based on explicit evolutionary and statistical models that allow one to study the processes of evolution leading to adaptations (for a review, see Harvey and Pagel 1991). These methods are powerful in that they take into account the phylogeny of the species in question, allowing one to weed out cases where different species have similar adaptations simply because they are derived from a common ancestor.

Evolutionary trends are usually identified by comparing values of characters or characters and environments across taxa (Harvey and Pagel 1991). Correlated evolution among characters or between characters and environments is evidence of adaptive evolution. A good example of two phylogenetically correlated traits is the frequent gregarious behavior of aposematically colored butterfly larvae (Sillén-Tullberg 1988), a phenomenon exhibited by checkerspots (chapter 7). In this case it has been hypothesized that gregariousness is adaptive because it amplifies the aposematic signal to potential predators.

The comparative approach can also be used to study the process of adaptive radiation (Harvey and Nee 1997). Adaptive radiation may be preceded by a "key innovation" in the ancestral species (Futuyma 1986). Key innovations can release an ancestral species from some developmental or other constraint and thereby allow adaptive radiation to occur (Futuyma 1986). Berenbaum et al. (1996) proposed one such key innovation in swallowtail butterflies of the genus *Papilio*. A single gene, *P450*, whose product metabolizes a class of secondary compounds (furanocoumarins) in the larval host plants, allows the larvae to develop on a range of plant families containing these compounds (Berenbaum 1983, Berenbaum et al. 1996). Although this hypothesis is so far based on studies of only three species of *Papilio*, a more comprehensive study may well support the claim. Similarly, in checkerspots, a possible key innovation has allowed the larvae to feed on plants containing iridoid glycosides (see section 11.7 for further discussion). This may have allowed a radiation onto plants containing these compounds in the plant subclass Asteridae, a group of plants that is otherwise not much utilized by butterflies.

Comparative studies on butterflies are still uncommon because well-resolved phylogenies have not been widely available, even though butterflies are otherwise so well studied. The main reason for the dearth of phylogenetic studies on butterflies has been the difficulty of finding and coding homologous morphological characters (de Jong et al. 1996). This is now changing rapidly with the advent of molecular systematics. In the past few years phylogenetic hypotheses have been advanced for the Papilionoidea (Weller et al. 1996, Ackery et al. 1999), for some of the butterfly families (Brower 2000, Campbell et al. 2000, Caterino et al. 2001), and for a range of smaller groups (e.g., Brower and Egan 1997, Brunton and Hurst 1998, Pollock et al. 1998, Aubert et al. 1999, Penz 1999, Martin et al. 2000, Morinaka et al. 2000, Rand et al. 2000, Monteiro and Pierce 2001, Nylin et al. 2001) including the checkerspots (Wahlberg and Zimmermann 2000, Zimmermann et al. 2000; chapter 2).

Organization of Chapter

Previous chapters in this volume have compared particular aspects of the biology of *E. editha* and *M. cinxia*, including their reproductive biology (chapter 5), larval biology (chapter 7), movement behavior and migration (chapter 9), and population genetics (chapter 10). In sections 11.2–11.5, we focus on population and metapopulation structures and dynamics, and we extend the comparisons beyond checkerspots. In these sections, the evolutionary histories of the species are taken into account for the most part only to the extent that comparisons are made between *Euphydryas* and *Melitaea* (section 11.2), with other checkerspot species (section 11.3), with other butterflies (section 11.4), and with other taxa apart from butterflies (section 11.5). The purpose of these comparisons is to find common elements in the population structure and dynamics of species living in patchily distributed or fragmented habitats. In sections 11.6 and 11.7 we consider the evolution of host plant use in checkerspots, with an explicit phylogenetic-based approach used in section 11.7. This analysis represents the first attempt to elucidate broad evolutionary patterns in checkerspots using phylogenetic information.

11.2 Comparisons between *Euphydryas editha* and *Melitaea cinxia*

Two tentative conclusions can be drawn concerning the population structure and distributions of *Euphydryas editha* and *Melitaea cinxia*, even though the research groups of Ehrlich and Hanski have targeted somewhat different issues in population biology, used different techniques, and faced different logistical constraints. First, *E. editha* generally appears to have a more patchy population structure throughout its range than does *M. cinxia*. Even in Colorado and the Great Basin, where there are large expanses of apparently suitable habitat, the distribution of *E. editha* is extremely patchy, and, as explained in chapter 2, we believe it was colonial in California even before the disruption of the native flora by European invasives. There has not been sufficient study of *M. cinxia* in continuous habitat to be sure of its population structure in the absence of habitat fragmentation. Russian *M. cinxia* is described as "rather common everywhere, flying over dry meadows" (Tuzov and Churkin 2000). In Lebanon, it is reported as "somewhat localized; no particular type of terrain seems favored" (Larsen 1974). In Morocco and Algeria *M. cinxia* flies in the mountains on flowery, grassy meadows and slopes between 1500 and 2600 m (Tennent 1996). More continuous populations may occur in the extensive steppe regions in Asia (I. Hanski and N. Wahlberg pers. obs.), but no detailed studies have been conducted there.

Second, *E. editha* appears to be tolerant of more diverse environmental conditions than does *M. cinxia*, as suggested by its much wider elevational range. In southern England *M. cinxia* is now primarily restricted to a few coastal landslip sites on the southern shores of the Isle of Wight, the adjacent Hampshire coast, and the Channel Islands, where the warm local microclimates that the butterfly requires in the early spring are present (Asher et al. 2001). Previously *M. cinxia* was more widely distributed in the warmer areas of southeastern England, and the reason for its present restriction is not clear (Watson 1969, Dennis 1977).

With respect to local population dynamics, weather-dependent phenological synchrony between the host plants and larval development is an important factor for both *E. editha* (chapters 2 and 3) and *M. cinxia* (chapters 4 and 7). There are some differences between the species, however, as *Melitaea cinxia* larvae in Åland may suffer more from intraspecific competition (chapters 4 and 7) than do larvae of Bay area *Euphydryas editha* (chapter 3). Parasitoids also have a much greater impact on *M. cinxia* than *E. editha* (chapter 8).

11.3 Comparisons with Other Checkerspots

Population Dynamics

Most checkerspots live in small local populations. Small populations are influenced by several types of stochastic processes that affect their risk of extinction and which can be grouped into demographic, environmental, and genetic stochasticities (Harrison 1991). The significance of demographic stochasticity is elevated in those species of checkerspot, such as *M. cinxia*, in which larvae spend a long time in a sib group (Hanski 1999b) (table 11.1). Additionally, in very small populations, not all females find a mate quickly (Kuussaari et al. 1998), which together with inversely density-dependent emigration (chapter 9) decreases the growth rate of small populations and increases their risk of extinction. Group living also increases the incidence of intraspecific competition (White 1974, Kuussaari 1998) even in low-density populations because the prediapause larvae, especially, have limited dispersal capacity (chapter 7). One could therefore expect that checkerspot species would exhibit especially large-amplitude population oscillations. Unfortunately, adequate data do not yet exist to test this prediction.

The impact of environmental stochasticity has been discussed at length in the case of *Euphydryas editha* (chapter 3) and *Melitaea cinxia* (chapter 4). There are no extensive studies on the other checkerspot species that would allow informative comparisons (although some short-term studies do exist on, e.g., *E. chalcedona*), but there are no reasons to assume that the results for our two focal species would not broadly apply to checkerspots in general. Thus we expect that the interaction between variable weather conditions and spatial variation in habitat quality is a key buffering mechanism that allows metapopulations to occur in the face of strong environmental stochasticity. Large-scale exceptional events, such as the two-year drought in California that resulted in the extinction of many *E. editha* populations (figure 3.11; Ehrlich et al. 1980), may nonetheless have long-lasting consequences. Parmesan (1996) attributed the observed northward range shift of *E. editha* in California to increasing frequency of drought years with climate change. The responses of some *E. chalcedona* populations to the California drought were similar to those of nearby *E. editha* populations, but at Jasper Ridge and San Bruno in the Outer Coast Range, populations showed slight increases rather than the decreases suffered by local *E. editha*. The large, bushy host plants of the coastal *E. chalcedona* populations were not noticeably affected by the drought, suggesting that host plant characteristics may play an important role in buffering environmental stochasticity in checkerspot populations.

Genetic stochasticity is primarily associated with populations with small effective population size. Therefore one could expect it to play an especially significant role in checkerspots with gregarious larvae, which leads to substantial correlation among families in survival (Hanski 1999b) and thus to small effective population size. *Melitaea cinxia* exhibits substantial inbreeding depression after just one generation of sib-mating (Haikola et al. 2001). Given the metapopulation structure of the species, with a high incidence of very small local populations (Hanski 1999b), it is not surprising that recent studies have documented a significant impact of inbreeding on population dynamics and extinction (Saccheri et al. 1998, Nieminen et al. 2001; chapter 10). Comparative studies on other species are badly needed. It would be especially interesting to examine the role of inbreeding in species with larvae that tend to spend most of their lives in sib groups versus those that disperse during the early larval instars.

Movement Behavior

Movement behavior plays an important role in both population and metapopulation dynamics, as well as in connecting the two. Mark–release–recapture (MRR) studies have been used extensively to study the movements of checkerspots, beginning with Ehrlich's (1961b, 1965) early work. In general,

Table 11.1. The average size of egg clusters and degree of larval gregariousness in checkerspot species as reported in the literature.

Species	Mean No. of Eggs Per Cluster	Degree of Gregariousness[a]	Reference
Melitaea cinxia	169	High	Wahlberg (1995)
Melitaea diamina	97	Intermediate	Wahlberg (1997a)
Melitaea didyma	20	Low	Vogel and Johanessen (1996)
Melitaea didymoides	~15	Low	Wahlberg et al. (2001)
Melitaea athalia	53	Low	Wahlberg (1997b)
Melitaea parthenoides	201	Intermediate	N. Wahlberg (pers. obs.)
Euphydryas aurinia	273	High	Klemetti and Wahlberg (1997)
Euphydryas maturna	205	Intermediate	Wahlberg (1998)
Euphydryas gillettii	146	High	E. H. Williams et al. (1984)
Euphydryas phaeton	274	High	Stamp (1982d)
Euphydryas chalcedona	160	Intermediate	Rutowski and Gilchrist (1987)
Euphydryas editha	45	Intermediate	Labine (1968)
Euphydryas anicia	~70	Intermediate	White (1979)
Poladryas minuta	38	Low	Scott (1974)
Chlosyne lacinia	139	Low	Drummond et al. (1970)
Chlosyne harrisii	200	Intermediate	Scott (1986)
Chlosyne nycteis	121	Low	Scott (1986)
Chlosyne janais	183	Intermediate	Denno and Benrey (1997)
Phyciodes cocyta	40	Low	Scott (1986)
Phyciodes tharos	63	Intermediate	Scott (1986)
Phyciodes picta	~70	Low	Scott (1986)
Eresia alsina	61	High	Young (1973)
Anthanassa ardys	42	Low	Feldman and Haber (1998)
Anthanassa frisia	~50	Low	Scott (1986)
Anthanassa tulcis	72	Low	Feldman and Haber (1998)

[a]Degree of larval gregariousness: high = larvae gregarious until final instar; intermediate = larvae gregarious until diapause (or mid-instar if there is no diapause); low = larvae gregarious only in first to second instar.

checkerspots are fairly sedentary butterflies, with migration during the lifetime of individual butterflies rarely exceeding 2–3 km, as documented for *Euphydryas editha* (Ehrlich 1965), *Chlosyne palla* (Schrier et al. 1976), *Euphydryas anicia* (Cullenward et al. 1979, White 1980), *Euphydryas chalcedona* (Brown and Ehrlich 1980), *Melitaea athalia* (Warren 1987b, Wahlberg et al. 2002b), *Melitaea cinxia* (Hanski et al. 1994), *Melitaea didyma* (Vogel and Johannesen 1996), *Euphydryas aurinia* (Munguira et al. 1997, Wahlberg et al. 2002b), *Melitaea diamina* (Wahlberg et al. 2002b), and *Euphydryas maturna* (Wahlberg et al. 2002b). The most extensive data are available for *Melitaea cinxia* in the Åland Islands, where 95% of observed recolonizations of empty but suitable habitat patches have occurred within 2.3 km from the nearest source population, and the longest recorded colonization distance from the nearest population was 6.8 km (van Nouhuys and Hanski 2002a). Comparable results have been obtained for *Euphydryas editha bayensis* (Harrison 1989). Although infrequent, successful long-distance migration events can critically affect the genetic structure of populations and metapopulations (chapter 10) and metapopulation dynamics in sparse patch networks.

What the observed migration distances mean for the occurrence of species in fragmented landscapes can be illustrated with the results for *Melitaea diamina* (figure 11.1; Wahlberg et al. 1996, Hanski et al. 2000), a species that is endangered in Finland. A survey of suitable habitat patches for this species showed that there were three clusters of patches in the study area and a scattering of more isolated individual patches (Wahlberg et al. 1996). The butterfly was found primarily in the areas with clustered habitat patches (figure 11.1), illustrating the consequences of limited migration distances for spatial occurrence. Limited migration distances also mean that the chance of encountering a previously unoccupied environment is limited. Thus Nève et al. (1996) reported that the European fritillary *Pro-*

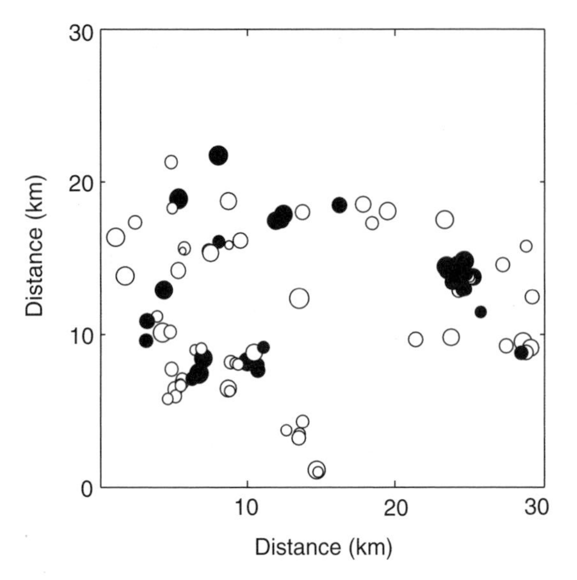

Figure 11.1. Left: A map of the observed patch occupancy of *Melitaea diamina* in the Tampere region in Finland in 1995. Filled and open circles indicate occupied and empty habitat patches, respectively. The size of the circle is proportional to the area of the patch. Right: *Melitaea diamina*. Photo by Janne Heliölä.

clossiana eunomia, which is morphologically similar to temperate checkerspots, expanded its distribution at the average rate of 0.4 km per year during 25 years after its introduction to an area in central France. The species successfully crossed gaps of 3–4 km across unsuitable habitat, but suitable habitat isolated by >10 km remained uncolonized, suggesting that a jump >5 km is unlikely, in agreement with the results for *M. cinxia* (van Nouhuys and Hanski 2002a).

Factors influencing movement behavior in checkerspots are discussed in detail in chapter 9. In brief, the quality of the habitat patch in general has a big influence (White and Levin 1981, Murphy and White 1984, C. D. Thomas and Singer 1987, Kuussaari et al. 1996). Butterflies tend to leave patches that have few nectar sources (Kuussaari et al. 1996) or that do not have the preferred host plant (C. D. Thomas and Singer 1987, Hanski and Singer 2001, Hanski et al. 2002). Results on *E. anicia* (Cullenward et al. 1979) and *E. chalcedona* (K. S. Williams 1983) suggest that butterflies tend to leave areas that are not thermally satisfactory, and butterflies are likely to emigrate from patches that are susceptible to drought in dry years (White and Levin 1981, Murphy and White 1984). A negatively density-dependent emigration rate has been found in *E. editha* (Gilbert and Singer 1973), *E. chalcedona* (Brown and Ehrlich 1980), and *M. cinxia* (Kuussaari et al. 1996). The study by Kuussaari et al. (1996) found weak evidence that larger females are more dispersive than small ones in *M. cinxia*, but a more comprehensive study by Hanski et al. (2002) found no evidence for butterfly morphology influencing emigration in the same species.

The degree of adaptation of checkerspot movement behavior to the structure of their often fragmented landscapes is an important question because it reflects the role of evolution in shaping population and metapopulation structure. For instance, if *E. editha* were transplanted to an environment in which its suitable habitat was configured as is the habitat for *M. cinxia* in the Åland Islands, would the butterfly disperse in the same way as it does in its native landscapes in California? Would it do so in 10 or 50 generations? These questions relate to the factors that influence the fitness of dispersing versus nondispersing individuals and how fast selection might change the behaviors of individuals if a substantial change in landscape structure were to occur. We know that in another key trait, host plant oviposition preference, substantial evolutionary changes may occur within only 10 generations (Singer et al. 1993; chapter 6). In the case of migration, several factors are likely to influence the evolution of relevant behavioral, physiological and/or morphological traits. The selective factors include spatial and temporal variation in fitness (well documented for checkerspots; chapters 3, 4, and 13), mortality during migration, inbreeding (chapter 10), and local Allee effects (chapter 4). A modeling study by Heino and Hanski (2001), based on empirically estimated parameter values for *M. cinxia* and *M. diamina,* found that the observed emigration rate (0.130 per day from a patch of 1 ha) is close to the model-predicted value (0.104). This finding suggests that the migration behavior is adapted to the current structure of the landscape (for further discussion, see chapter 9). Heino and Hanski (2001) also demonstrated with their model an evolutionary response in emigration rate to habitat loss and fragmentation, but the predicted changes were so small that it would require very large sample sizes to demonstrate such changes in practice.

A study by Wahlberg et al. (2002b) is the most ambitious attempt so far to ascertain whether the seemingly similar movement behavior in checkerspots is due to the species having adapted to similar landscapes or whether there is a phylogenetic component to the propensity to emigrate. Wahlberg et al. (2002b) used the Virtual Migration (VM) model of Hanski et al. (2000), which can be parameterized with MRR data collected from multiple populations. The parameters of this model describe the daily survival probabilities within habitat patches and during migration, emigration propensity, the scaling of emigration and immigration rates by patch area, and the effect of distance on migration. MRR data for five species of Finnish checkerspots were used to parameterize the VM model. The results indicate that the five species are more similar to each other than any is to several unrelated species (Wahlberg et al. 2002b). However, a closer look revealed some variation among the species. For instance, *E. aurinia* tends to move farther than *E. maturna,* and *M. cinxia* females have a greater propensity to emigrate than all the other species and sexes (figure 11.2). The estimated parameter values for each species and sex implied about 10–20% mortality during migration. These results imply that although there is no phylogenetic component to migration at the level of the five species, there may be phylogenetic constraints in the group as a whole. Checkerspots are rather sedentary, and, indeed, there are no records of checkerspots migrating very long distances, unlike for some species in the related Nymphalini, such as *Vanessa atalanta, V. cardui,* and *Aglais io.*

Something further can be learned about long-distance movements and checkerspot distributions from experiments carried out on *Euphydryas gillettii.* The genus *Euphydryas* appears to have evolved in the Nearctic region, then invaded the Palearctic where it speciated. *Euphydryas gillettii* is the only species from the Palearctic radiation that has reinvaded the Nearctic (Wahlberg and Zimmermann 2000). This species occurs as far south as the inhospitable Wyoming Basin desert, where its food plants, including *Lonicera involucrata,* are absent. In 1977, individuals were transplanted from Wyoming to central Colorado south of the barrier (Holdren and Ehrlich 1981; figure 11.3). One of the transplanted populations persisted through the year 2000 at a very small size, fewer than 100 individuals in most years. However, in 2002, this population exploded in size by an order of magnitude, reaching an estimated 1000 adults, and in area occupied by larvae to approximately 30 times the size of the original transplant population site. An individual female was caught as far as 6.4 km from the original introduction site. This population explosion may have been due to a series of three dry winters in a row, which extended the phenological window for larval growth, allowing larvae to diapause as fourth instars rather than as second instars and potentially to complete development in one year rather than in two (Boggs et al. unpubl. ms). Previously, the butterfly had not colonized areas with the requisite resources and located only within a few meters, let alone kilometers, of the transplanted colony. These results suggest that the species' range was restricted simply due to lack of sufficient time to spread farther into satisfactory habitats.

Metapopulation Structures

The generally patchy spatial population structure of checkerspot populations was noted early on (Ford and Ford 1930, Ehrlich 1961b). This patchiness is highlighted by the often sedentary behavior of individuals within a habitat patch (Ehrlich 1961b, 1965, Warren 1987b, Harrison 1989, Hanski et al. 1994), which has the consequence that populations occupying different habitat patches often fluctuate in size relatively independently of each other (e.g., Ehrlich et al. 1975, Ehrlich and Murphy 1981b, Hanski et al. 1995a). Local populations may go extinct, but the extinctions may be balanced by colonizations from currently occupied patches (Harrison et al. 1988, Hanski et al. 1995a). Such a metapopulation structure is commonplace in checkerspots, as documented

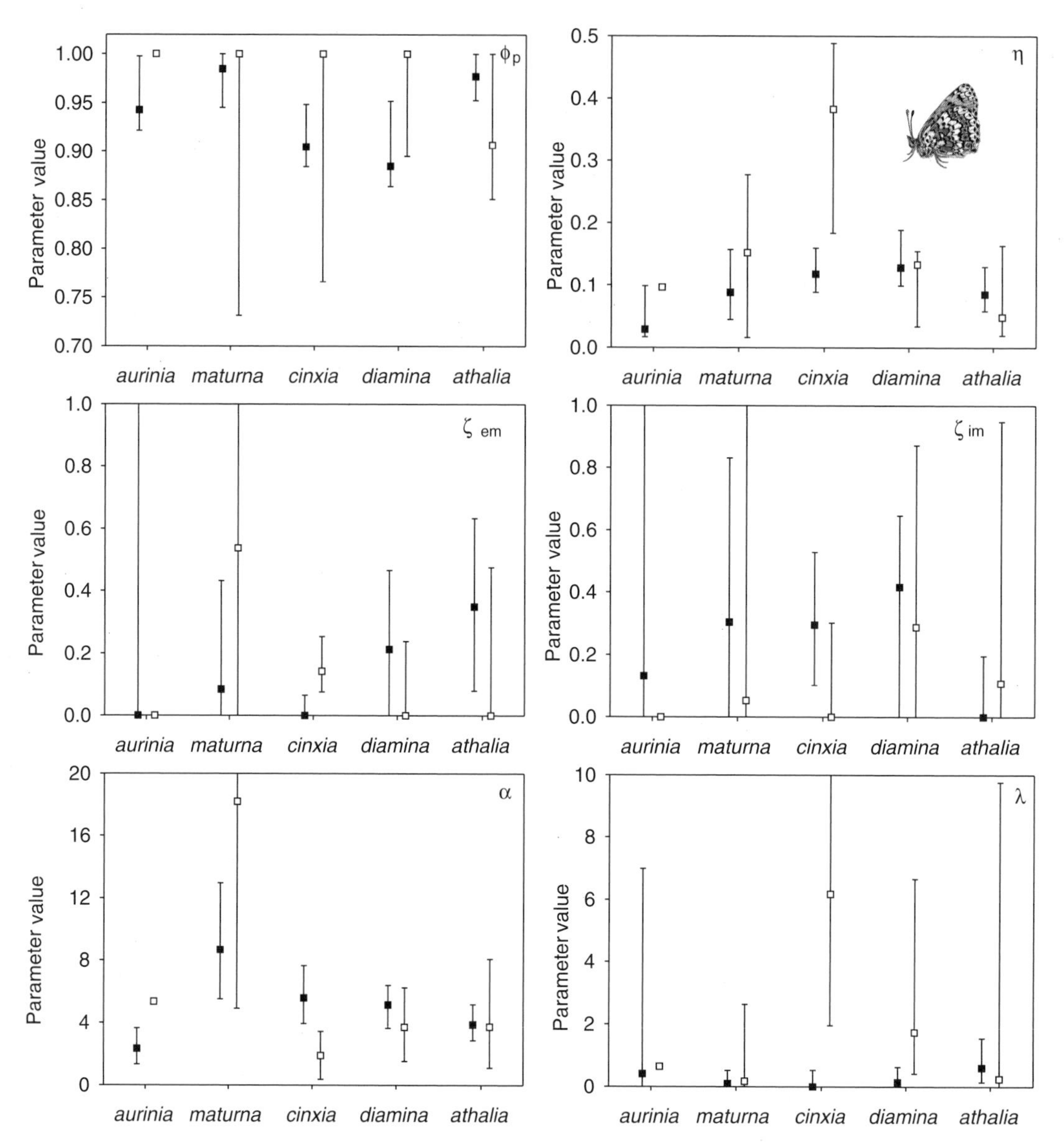

Figure 11.2. The parameter values of the VM model (squares) and their associated 95% confidence limits for the two sexes in five species of checkerspot butterflies. Filled and open symbols give the results for males and females, respectively. Parameter ϕ_p gives the daily survival probability of individuals, $\eta > 0$ is a parameter describing propensity to emigrate (daily emigration rate from a patch of 1 ha), $-\zeta_{em} > 0$ is a parameter scaling emigration to patch area, $\zeta_{im} > 0$ scales immigration to patch area, λ determines mortality during migration, and parameter $\alpha > 0$ determines the effect of distance on migration. See Wahlberg et al. (2002b) for more detailed interpretation of the parameter values. Note the very large confidence limits, reflecting the fact that large sample sizes are needed for precise parameter estimates.

Figure 11.3. Top: Habitat opposite to the Rocky Mountain Biological Laboratory, Gothic, Colorado (ca. 3000 m), to which Wyoming *Euphydryas gillettii* were transplanted. Bottom: Marked male *E. gillettii* of the fourth Gothic generation (1981). Photos by Paul R. Ehrlich.

for *Euphydryas editha* (Ehrlich et al. 1975, C. D. Thomas et al. 1996), *E. chalcedona* (Brown and Ehrlich 1980), *E. gillettii* (Debinski 1994), *E. anicia* (White 1980), *E. phaeton* (Brussard and Vawter 1975), *E. aurinia* (Warren 1994, Lewis and Hurford 1997, Wahlberg et al. 2002a), *E. maturna* (Wahlberg et al. 2002b), *Melitaea cinxia* (Hanski 1999b), *M. diamina* (Wahlberg et al. 1996), *M. didyma* (Vogel and Johannesen 1996), and *M. athalia* (Warren 1987b, Wahlberg et al. 2002b).

In spite of the overall similarities, there are also differences among the checkerspot species and their metapopulations in the rate of population turnover, for example. Frequent local extinctions and recolonizations of previously empty habitat patches have been reported for several species, including *M. cinxia* (Hanski et al. 1995a, Hanski 1999b) and *M. diamina* (Wahlberg et al. 1996, Moilanen and Cabeza 2002) in Finland, *M. athalia* (Warren 1987b) in Britain, *E. aurinia* (Warren 1994, Lewis and Hurford 1997, Wahlberg et al. 2002a) in Finland and Britain, and *E. editha* (C. D. Thomas and Singer 1998) in California. However, in some cases the patch network is so dense that any extinctions are immediately negated by colonizations. *Melitaea didyma* occurs as a metapopulation at the northern edge of its range in Germany (Vogel and Johannesen 1996), where the species apparently occupies practically all suitable habitat, and no extinctions or colonizations have been reported. The same is largely true for many Finnish populations of *M. athalia* and *E. maturna*. For these species suitable habitat occurs densely across many landscapes, and hence even though the butterflies are not more vagile than *M. cinxia* (Wahlberg et al. 2002b), for example, they appear to occupy practically all available habitat. In other cases, the species differ in their mobility with significant consequences for spatial population structure. The best example comes from China, where Rongjiang Wang and collaborators (pers. comm.) have studied the metapopulations of *Euphydryas aurinia* and *Melitaea scotosia* occupying the same set of habitat patches in a valley north of Beijing. MRR studies showed that *M. scotosia* is much more vagile than *E. aurinia*, which agrees with the significantly greater population genetic structuring within the metapopulation of *E. aurinia* than within that of *M. scotosia*.

Dissimilar spatial population structure in two species living in the same landscape but using different resources can be seen in *Euphydryas chalcedona* (figure 13.2) and *E. editha*, which co-occur at Jasper Ridge (figure 11.4) and elsewhere. At Jasper Ridge, there have been two populations of *E. chalcedona* in the chaparral and oak-woodland edges surrounding the serpentine patches supporting *E. editha*. Although adult *E. chalcedona* tend to remain clustered in the more favorable areas of the Jasper Ridge chaparral, they do move more than *E. editha* (figure 11.4), frequently visiting nectar sources that are not near larval host plants (the nectar plants include shrubs, such as *Eriodictyon californicum*, *Prunus ilicifolia*, and *Aesculus califor-*

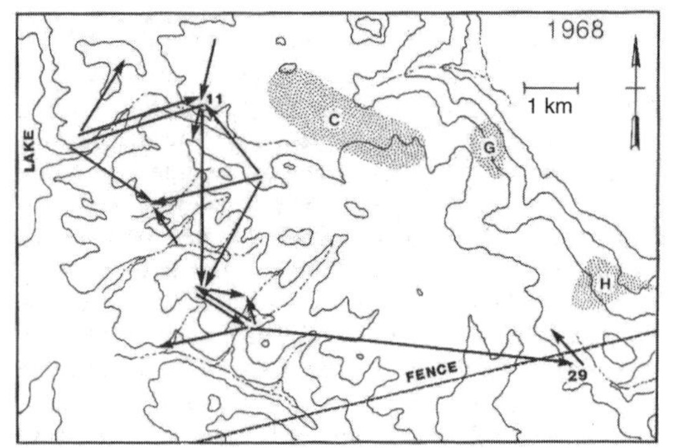

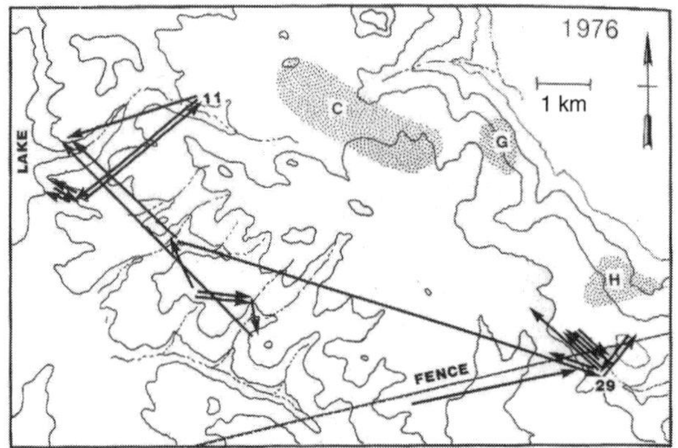

Figure 11.4. Movements of *Euphydryas chalcedona* at Jasper Ridge in 1968 and in 1976. Each arrow indicates a capture and release of a single butterfly (at its origin), and another capture and release of the same butterfly (at its point). The stippled areas C, G, and H indicate the positions of the three *Euphydryas editha* populations discussed by Ehrlich (1965) and McKechnie et al. (1975) and shown in Figure 2.6. The major contour lines represent 90, 120, 150, and 180 m.

nica, as well as various herbaceous composites). As in *E. editha*, the sizes of the two *E. chalcedona* populations have changed relatively independently, in this case in response to changes in the abundance of larval host plants. It would appear that *E. chalcedona* at Jasper Ridge has a metapopulation structure more similar to that of *Melitaea cinxia* than to that of its close relative *E. editha*, with the habitat patches defined primarily by the co-occurrence of larval food plants and adult nectar sources.

Another North American species, *Euphydryas anicia*, is best considered as a part of a superspecies with *E. chalcedona* and *E. colon* (Scott 1986, Wahlberg and Zimmermann 2000), and its widespread populations are, like those of *E. chalcedona*, found in a wide variety of habitats (Scott 1986). The population structure of *E. anicia* has been most thoroughly studied in areas of high altitude tundra (~3700 m) of the Rocky Mountains of Gunnison County, Colorado (see figure 11.5). In one extensive tundra area at Cumberland Pass, alpine *E. anicia* was relatively sedentary (Cullenward et al. 1979), although not as sedentary as some high altitude *E. editha* (White 1973). Larval and adult resources were both present in only a few areas, and adults were found where both resources were scarce, including an area at an altitude 150 m below the main colony area. Males showed a tendency to aggregate on hilltops (chapter 5). One interesting suggestion that might explain the presence of adults in the lower area was put forward by Watt and his colleagues (1977) in a parallel study of *Colias meadii*. They suggested that butterflies tend to spend some time in warmer areas where they can reach temperatures suitable for activity. White (1980) was able to mark 960 of an estimated 5,000–10,000 *E. anicia* on four peaks higher than 3700 m on both sides of the upper East River Valley near the Rocky Mountain Biological Laboratory at Gothic, Colorado. He obtained 227 recaptures and recorded two interpeak transfers, across a distance of 2 km. Three of the 85 individuals displaced to lower altitudes were observed to make their way back to the peaks (distances of ~3 km). White's results suggest that the montane populations of *E. anicia* have a classic metapopulation structure.

So far the population structures of *Chlosyne* or *Phyciodes* species have not been studied in any detail, and nothing is known about widespread multivoltine species like *Melitaea trivia*, which seem to inhabit weedy areas (Larsen 1974, P. Ehrlich pers. obs.). Dethier and MacArthur (1964) report that *C. harrisii* inhabits a network of old fields in a forested region, suggesting that this species might occur as a metapopulation. *Chlosyne lacinia* in Texas also appears to have a metapopulation structure, with local populations occurring around stands of the composites on which the larvae feed, but with individuals moving frequently among small local populations (Neck 1973). In another *Chlosyne* species, *C. palla*, the pattern of movements was similar to that of *E. editha* in one locality where they co-occur in Colorado (Schrier et al. 1976, Ehrlich and White 1980). Both species move downslope to a moist seep where males "mud puddle" and then return to higher areas containing larval host plants and adult nectar sources. But in contrast, an MRR study on *C. palla* (Schrier et al. 1976) suggested that this species occupies much larger areas of habitat and moves more than the well-studied *Euphydryas* and *Melitaea* species.

In conclusion, it appears that at least for *Melitaea* and *Euphydryas* species, a patchy population structure is nearly ubiquitous. This may not be an intrinsic feature of the butterflies, but is largely dependent on the distribution of larval host plants and adult nectar sources. Whether all suitable habitat patches are constantly occupied or whether the species exists in a stochastic balance between extinctions and colonizations depends on the spatial configuration of the habitat patches and on their number. Most of the European studies have been conducted close to the northern range limit of the species, where they may have a more fragmented population structure than in areas closer to the center of the species' range.

11.4 Comparisons with Other Butterflies

The large body of data that has been gathered in studies of *E. editha* and *M. cinxia* has contributed to some fundamental changes in our perception of animal population structure. For instance, the early comparisons of *E. editha* with the satyrine butterfly *Erebia epipsodea* (Brussard and Ehrlich 1970b) showed that similar insect species can have very different population structures (figure 11.6). These findings led researchers to classify populations of butterflies as "open" or "closed" (e.g., J. A. Thomas 1984), referring to the degree of exchange between local populations (not to be confused with another concept of open versus closed populations in terms

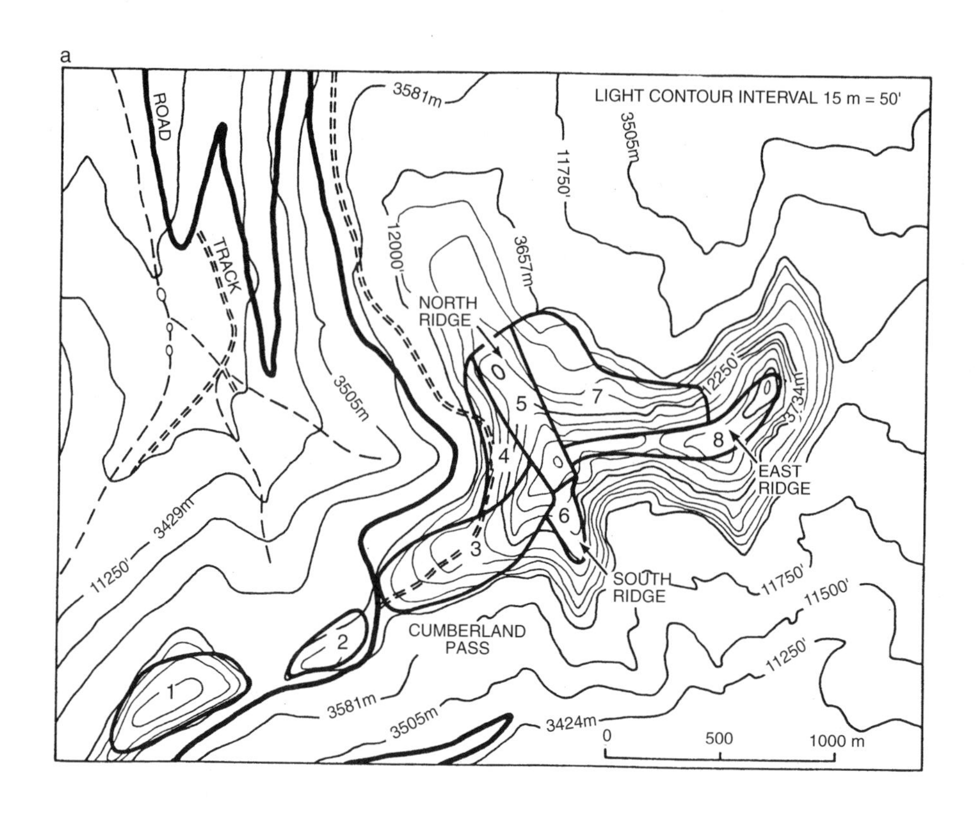

a
ROAD
TRACK
3581m
LIGHT CONTOUR INTERVAL 15 m = 50'
3505m
11750'
12000'
3657m
NORTH RIDGE
3505m
12250'
3734m
1
2
3
4
5
6
7
8
EAST RIDGE
3429m
11250'
SOUTH RIDGE
11750'
11500'
CUMBERLAND PASS
11250'
3581m
3505m
3424m
0
500
1000 m

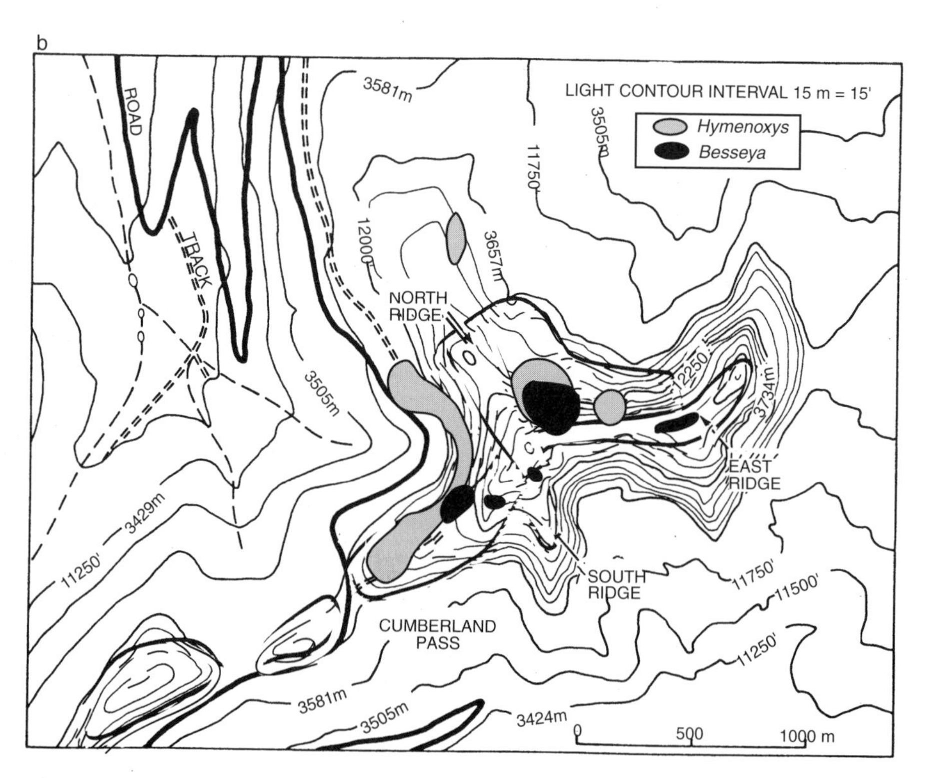

b
ROAD
TRACK
3581m
LIGHT CONTOUR INTERVAL 15 m = 15'
3505m
Hymenoxys
Besseya
11750
12000'
3657m
NORTH RIDGE
3505m
12250'
3734m
EAST RIDGE
3429m
11250'
SOUTH RIDGE
11750'
11500'
CUMBERLAND PASS
11250'
3581m
3505m
3424m
0
500
1000 m

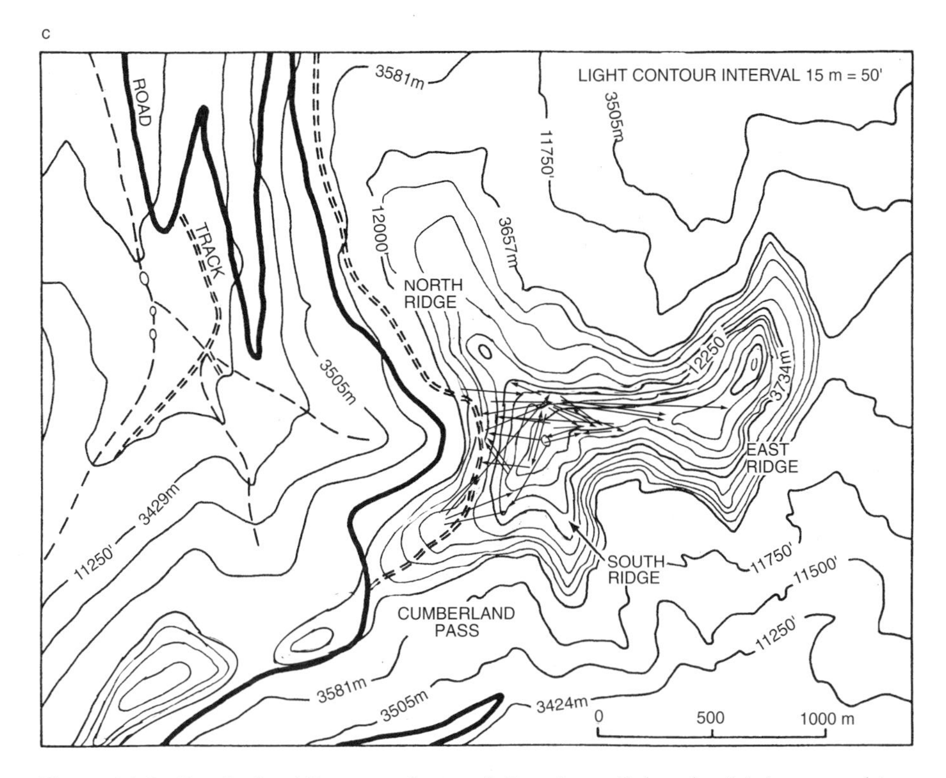

Figure 11.5. Cumberland Pass, northeast of Gunnison, Colorado. (a) Areas used in mark–release–recapture studies of *Euphydryas anicia*. Over two seasons (1976 and 1977), 3483 males and 1122 females were marked and released. (b) Distribution of *E. anicia* larval host plant, *Besseya alpina*, and adult nectar source, *Hymenoxys grandiflora*. (c) Patterns of movement illustrated by male transfers (showing only movements of > 100 m). No individuals were captured in areas 1 or 2. Males were commoner on ridgetops, females were commoner in gullies, especially in the area directly around the numeral 7 in (a), which coincides with the location of the highest density of the oviposition plant *Besseya alpina*. After Cullenward et al. (1979).

of whether, during a period of study, there is significant recruitment or mortality; Seber 1982). Further work with *M. cinxia* consolidated the notion that there is, in fact, a continuum from "closed" to "open" population structures (Hanski et al. 1994), and this research provided a paradigmatic empirical example of the classic metapopulation structure (Hanski et al. 1995a, Hanski 1999b).

Are checkerspots more prone to a metapopulation structure than other butterflies? Apparently not: metapopulation structure is common in butterflies in general (Hanski and Kuussaari 1995, C. D. Thomas and Hanski 1997). Hanski and Kuussaari (1995) estimated that 65% of the 94 species of butterflies living in Finland have a metapopulation structure. So what characteristics of butterflies might make them especially likely to be structured as metapopulations? Two major factors are the occurrence and permanence of discrete local populations and the magnitude of adult movements among such populations. Suitable habitat for many butterflies, which is often determined by the presence of just one or a small number of uncommon larval host plants, is generally rather well delimited from the rest of the landscape, and it often occurs in relatively small patches. In the case of our model species, *E. editha* and *M. cinxia*, suitable habitat is not susceptible to rapid changes, though more gradual changes do occur in response to global environmental changes (chapter 13). Many other butterflies live in more

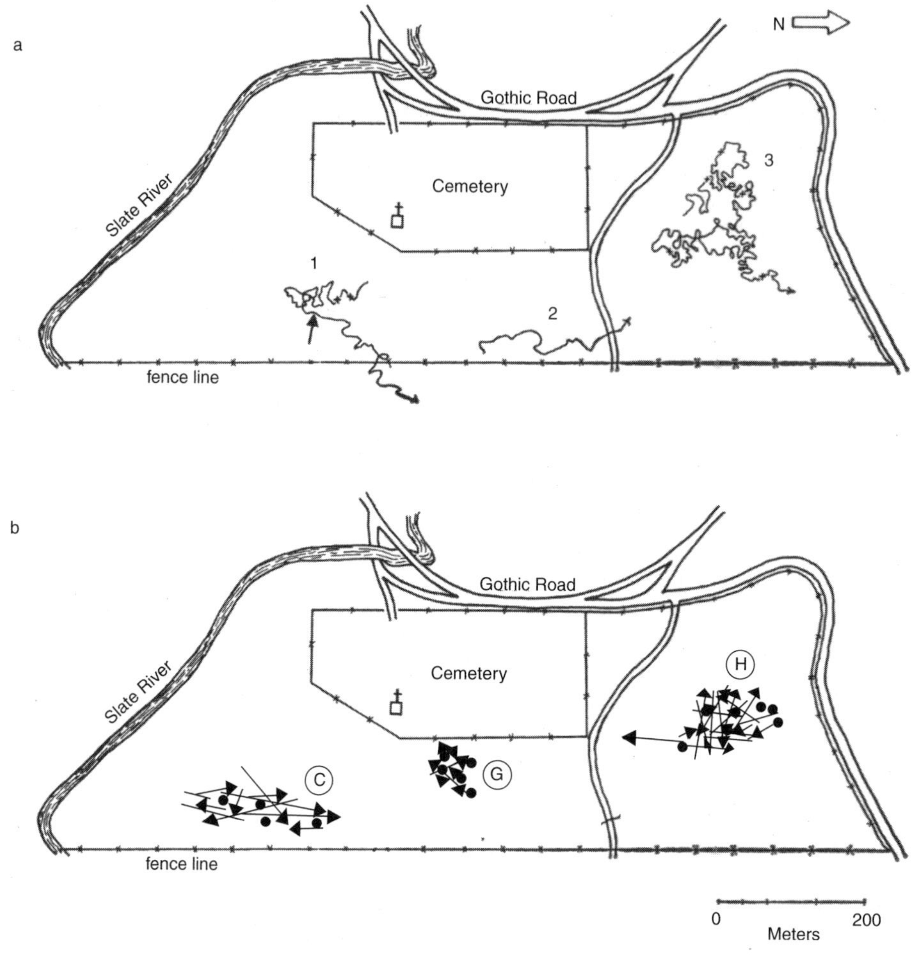

Figure 11.6. (a) Individual movements in an open population of *Erebia epipsodea*: three examples of flight paths in the cemetery study area, Crested Butte, Colorado, 1968 (after Brussard and Ehrlich 1970a). These sample tracks are representative of different patterns recorded in most cases for much less than an hour of tracking. Movements in this population recorded by mark–release–recapture were frequently > 600 m, and some butterflies were found as far as 10–13 km from the release site (Brussard and Ehrlich 1970c). (b) *Euphydryas editha* movements in the closed Jasper Ridge population (after Gilbert and Singer 1973) superimposed on cemetery site for comparison. This figure shows maximum movements recorded by mark–release–recapture over an entire flight season.

successional habitats, in which, to persist regionally, the populations must colonize newly appearing habitat patches. Two examples are the checkerspots *Melitaea athalia*, occurring in ephemeral clearings in woodlands in Britain (Warren 1987b), and *Euphydryas aurinia*, inhabiting a network of forest clearcuts and grazed meadows in Finland (Wahlberg et al. 2002a). Nonetheless, though in the latter cases the successional nature of the habitat adds to the local extinction rate, the type of population structure and dynamics is broadly the same as in the case of species living in networks of more permanent habitat patches. The overall similarity in the dynamics of butterfly species occurring in fragmented landscapes is underlined by Hanski's (1994b) modeling of extinction–colonization dynamics in three different European species, the checkerspot *Melitaea cinxia*, the skipper *Hesperia comma*, and the blue butterfly *Scolitantides orion*. The estimated parameter values describing the extinction proneness and colonization ability of the three species were broadly similar, suggesting that one set of parameters would give a reasonable qualitative picture of the dynamics of many butterflies living in fragmented landscapes (Hanski 1994b).

There are, however, other butterflies with population structures that do not fit into the metapopulation paradigm. One well-studied North American example is the satyrine *Erebia epipsodea* (Brussard and Ehrlich 1970b) mentioned above. In terms of nonpest butterfly species that have been thoroughly investigated, this species may represent a rather extreme case with an "open" population structure (the truly extreme cases include the migratory monarch, *Danaus plexippus*, and some large migratory tropical pierids, among others). *Erebia epipsodea* is one of the commonest members of its genus in North America, flying from Alaska east to Saskatchewan and south to northern New Mexico, in the subalpine and alpine zones of mountains, and on prairies in the north. It has been studied intensively in subalpine central Colorado, where it forms large, essentially panmictic populations over areas as large as hundreds of square kilometers. Individuals move at random within areas of suitable habitat but depart from those that are unsuitable (Brussard and Ehrlich 1970a, 1970c). Nonetheless, the latter areas are not barriers to movement and do not subdivide *E. epipsodea* populations. A similar population structure has been documented for *Speyeria mormonia* in Colorado in the same habitat (Boggs 1987b); this species is more closely related to the checkerspots than is *E. epipsodea*.

Neither *Erebia epipsodea* nor *Speyeria mormonia* have been investigated in other localities, where they might show a different population structure. Such differences have been recorded in another satyrine, *Maniola jurtina*. On the deserted Isle of Tean, in the Scilly Isles off the tip of Cornwall, southwestern England, *M. jurtina* shows an extremely tight population structure, reminiscent of *Euphydryas editha* at Jasper Ridge, with only 3 of 183 recaptured individuals moving between areas of suitable habitat only a few hundred meters apart (Dowdeswell et al. 1949, Dowdeswell 1981). In contrast, in a heavily farmed area of northeast Hampshire, *M. jurtina* showed movements of >1 km and may have had an overall population structure more similar to that of *E. epipsodea* (Dover et al. 1992).

Beyond similarities in gross population structures, we can ask whether factors controlling the presence or absence of our focal species within habitat patches are representative of such factors in other butterfly species. In our research, we have often emphasized the significance of habitat patch areas and connectivities to existing populations as key determinants of habitat occupancy, both for *M. cinxia* (Hanski 1994b, 1999b) and for *E. editha* (Harrison et al. 1988, Harrison 1989). Topographic heterogeneity within habitat patches has been shown to be another key element in determining extinctions in the Bay area *E. editha* (chapter 3). Other researchers on other butterfly species have arrived at similar conclusions (C. D. Thomas et al. 1992).

More recent butterfly studies, however, have challenged the emphasis on patch areas and connectivities and instead suggested that patch quality is a key determinant of patch occupancy (J. A. Thomas et al. 2001b, Fleishman et al. 2002b). There cannot be any disagreement about the significance of patch quality because delimitation of habitat patches must always be based on patch quality. We believe that different researchers have arrived at different conclusions for two reasons. First, patch area and quality together determine the expected size of local populations (the carrying capacity of the patch), which is expected to significantly affect, for example, the risk of extinction. However, habitat quality may be much more variable across space and time for some species of butterflies than for others, resulting in the finding that area is not significant when quality is included in the analysis for species with more diverse patch qualities. It may be that there is less spatial variation in habitat quality in our study systems than in some other systems, though we emphasize that the influence of habitat quality on patch occupancy, colonizations, and extinctions has also been demonstrated for *M. cinxia* (Hanski 1999b) and that habitat heterogeneity decreases population fluctuations and thereby reduces extinction risk for *E. editha* (chapter 3). Second, species mobility relative to the scale of the metapopulation study will affect the detection of any influence of patch isolation. *Speyeria nokomis*, studied by Fleishman et al. (2002b) in the Great Basin in the western United States, did not show an effect of patch isolation. This species is somewhat more dispersive than the checkerspots, with a mean dispersal distance of 1 km and a maximum observed of 4.5 km in a landscape where the most isolated patch was 3 km from its nearest neighbor.

11.5 Beyond Butterflies

How widely can the results we have obtained for *Euphydryas editha* and *Melitaea cinxia* be generalized beyond butterflies? We focus here on two particular results for our focal species, the role of habitat heterogeneity in buffering the dynamics of local population dynamics and the occurrence of species in a balance between local extinctions and recolonizations in highly fragmented landscapes.

One of the most important results from research on *Euphydryas editha bayensis* is that topographic heterogeneity within a patch interacts with macroclimate to produce favorable microclimatic conditions for larval development (chapter 3). Thus heterogeneity is a critical factor in habitat quality for *E. editha bayensis* (Weiss et al. 1988) and may well be equally important for other insects and small-bodied endotherms (e.g., Ehrlich and Murphy 1987a). Consequently, availability of heterogeneous thermal environments (or other microhabitats) is likely to be important to insect conservation (Weiss et al. 1993). There is a vast theoretical literature on various consequences of habitat heterogeneity, starting from Levins (1968), but empirical data on heterogeneity effects of the sort seen in *E. editha* are surprisingly scarce. Some of the most convincing data come from Kindvall's (1995, 1996) work on the bush cricket *Metrioptera bicolor* in southern Sweden. In this case it is not slope differences that influence survival under different macroclimatic regimes but variation in soil composition and vegetation type. In dry years, tall, dense vegetation is most favorable to survival; in wet years, low grassland is most favorable. Habitat heterogeneity increases the chances that the crickets will survive weather extremes, and Kindvall supports the conservation conclusions of the work on *E. editha*: "it is a better strategy to give priority to patches with a high diversity of habitats, rather than to preserve as large patches as possible of the habitat that has proven to be most productive in normal years" (Kindvall 1996, p. 213).

It was shown long ago that slope aspect controls time of emergence from hibernation in ground squirrels (*Spermophilus columbianus*), causing differences of as much as 10 days within the same population (Shaw 1925). The population dynamic consequences of these differences have not been elucidated, but one can imagine that they are significant. Ground squirrels in environments with short growing seasons are in a race to put on fat in order to survive hibernation, which would be aided by early emergence; however, early emerging individuals might be more vulnerable to starvation due to slow development of vegetation, late-season storms, or predation (K. O'Keefe pers. comm.). Similarly, small birds have been shown to take advantage of habitat heterogeneity (microclimate and vegetation gradients) by shifting nest positions from year to year in response to changing precipitation, and such microhabitat selection affects fitness (Martin 1998, 2001).

We still lack sufficient sample systems to allow us to judge precisely how broadly these conclusions from the checkerspot system apply to small vertebrates. For example, the pest murid rodent *Mastomys natalensis* shows the influence of both endogenous and exogenous factors in its dynamics, as does *E. editha* (Leirs et al. 1997), but we do not yet know what influence habitat heterogeneity may have in influencing the numbers of these multimammate rats. One can imagine, considering that habitat specialization is common in the genus (Nowak 1991), that heterogeneity may well be important.

In plants, buffering effects of habitat patch heterogeneity have been suggested by an experimental system with the annual grass *Stipa capensis*, a common species in North Africa and the Middle East (Kadmon 1993). The potential for such effects being common is clearly suggested by many experiments, such as that of Wedin and Tilman (1993), demonstrating how soil gradients that are common in nature influence competitive abilities of plants.

The importance of habitat heterogeneity on larger spatial scales is better documented. On the island of Hirta off Scotland, feral sheep have a structured population whose units have somewhat different dynamics, apparently related to areal variation in forage quality (and perhaps parasite load; Coulsen et al. 1999). Serengeti mongoose dynamics appear to be governed by factors such as grass depth, availability of sheltering holes, and the density of termitaries (mongooses eat termites), but whether heterogeneity buffers dynamics is not yet known (Waser et al. 1995). Snail kites provide another suggestive example; they use special emergency habitats in times of drought, when homogeneous normal habitat is not satisfactory (Takekawa and Beissinger 1989). Migratory herbivores, especially wildebeeste, take advantage of rainfall, vegetation, and soil differences to maintain large populations in the Serengeti (Sinclair and Norton-Griffiths 1979, Fryxall 1995), and many tropical birds migrate altitudinally to take advantage of different conditions at different seasons (or at different stages of their life cycles, Diamond 1972, Hilty 1994). A striking example of this is the resplendent quetzal (*Pharomachrus mocinno*) in Costa Rica (Stiles and Skutch 1989). The quetzal breeds in cloud forests in the dry season and then descends into the wet forests to live in warm environments, where foraging is superior (Hilty 1994). One might even think of long-distance migratory birds as occupying two spatially widely separated parts of one very heterogeneous "patch." But few of us would consider patch heterogeneity when thinking of the biology of the arctic tern (*Sterna paradisaea*), which breeds in the arctic summer and spends the antarctic summer at the other end of their "patch" in the southern oceans.

Turning to the question of the prevalence of classic metapopulation dynamics, table 11.2 lists examples from the recent literature covering taxa from rust fungus to insects, frogs, and mammals. In this sample of studies, patch occupancy levels vary greatly, but in all studies occupancy is at most 80%, that is, a considerable fraction of habitat patches have been empty at the studied point in time. The effects of habitat patch area and connectivity on occupancy were observed in every study, which underscores the generality of these effects. Admittedly, one cannot assume that the sample of studies in table 11.2 represents a random sample of natural metapopulations; researchers interested in metapopulation dynamics naturally tend to work on systems that appear to exhibit the expected structure. Furthermore, whether one will detect the effects of patch area and connectivity on patch occupancy depends on the spatial scale of the study. Nonetheless, we can minimally conclude that it is not difficult to demonstrate these effects in a wide range of taxa living in fragmented landscapes. The likely importance of spatial dynamics in producing these patterns is further strengthened by the frequently observed effects of patch area on extinction and of connectivity on colonization (table 11.2; Hanski 1999b; Hanski 2001). We do not thus entirely agree with Harrison (Harrison 1991, 1994), who has concluded that classic metapopulations are rare. A truly objective assessment of how common metapopulation structures actually are would require a study of a substantial sample of species living within some sufficiently large area—an undertaking that is practically impossible with the resources available to ecologists. Even without such a survey, one is tempted to guess that metapopulation structures are prevalent in many modern landscapes in densely populated areas, such as southern England, the Netherlands, and lowland China, where most natural habitats are exceedingly sparse and fragmented. In these cases the structure clearly is largely anthropogenically imposed. One might assume that metapopulation structures are less common in, for instance, natural boreal forests because they are seemingly homogeneous, but even in this case most species actually live in patchily distributed microhabitats and may exhibit a metapopulation structure (Hanski and Hammond 1995).

11. 6 Coevolution of Plants and Butterflies

The species richness of phytophagous (plant-eating) insects is phenomenal: an estimated 35% of all animal species are such insects (Schoonhoven et al. 1998), and butterflies, with about 17,000 species (Ackery et al. 1999) account for a modest proportion. The checkerspots, like most other butterflies and many other phytophagous insects, are highly specialized, using often just one or a few host plant species. These patterns have long intrigued researchers (Brues 1920). Why is phytophagy such a successful way of living, and what are the advantages of host plant specialization? The answers to these questions are not yet all in, and, as we discuss below, controversy still remains about the fundamental patterns and processes in the evolution of host plant use in insects.

In an influential paper on insect–plant interactions, Ehrlich and Raven (1964) suggested that plant secondary chemical compounds have evolved as defenses and that a novel toxic compound could allow the plant to escape from herbivory. This could set the stage for an adaptive radiation of plants that had evolved the compound. Any insect evolving resistance to that secondary compound would have, in turn, an abundance of potential host plants, and another radiation could take place in the insects. The hypothesis of Ehrlich and Raven (1964) proposes that there is coevolution between insects and their host plants; in other words, both groups affect each other in an evolutionary context. Through the process of coevolution, also called an "evolutionary arms race" between insects and plants, insects have become highly specialized, with most species using only one or a few plant species.

In the years since Ehrlich and Raven (1964) first

Table 11.2. Large-scale metapopulation studies that have examined the effects of habitat patch area and connectivity on patch occupancy, extinction, and colonization (based on Hanski 2001).

Species	No. Patches	Patch Occupancy Level (%)	Area Effect on Occupancy	Connectivity Effect on Occupancy	Area Effect on Extinction	Connectivity Effect on Colonization	Comments	References[a]
Rust fungus *Uromyces valerianae* infecting hosts	30	43–73	Yes	Yes	Yes	Not studied	Data from 13 years	1
Plants in urban vegetation fragments	423	3–74	Yes	Yes	Not studied	Not studied	22 spp.; fragment age and habitat quality effects	2
Grasshopper *Oedipoda caerulescens*	312	28/48	Yes	Yes	Not studied	Yes	Occupancy from two years; three extinctions observed	3
Froghopper *Neophilaenus albipennis*	506	18	Yes	Yes	Yes	(Yes)	Three colonizations observed	4
Butterfly *Coenonympha tullia*	148/33	82/70	Yes	Yes	Not studied	Not studied	Habitat quality effect; different patches each year	5
Butterfly *Hipparchia semele*	77	64	Yes	Yes	Not studied	Not studied	Mainland–island system; occupancies from different years	6
Butterfly *Lopinga achine*	168	47	Yes	Yes	Not studied	Not studied		7
Tree frog *Hyla arborea*	378	48–73	Yes	Yes	Yes	No	Data from four years	18, 9
Hazel grouse *Bonasa bonasia*	38	Not reported	Yes	Yes	Yes	(No)	Matrix quality effect; only relatively high connectivities	10
Marsh rabbit *Sylvilagus palustris hefneri*	59	(71)	Yes	Yes	Not studied	Not studied	Occupancy includes all patches occupied in three years	11
Tree hyrax *Dendrohyrax arboreus*	199	18	Yes	Yes	Not studied	Not studied	Mainland–island system; habitat quality effects	12
Blue duiker *Philantomba monticola*	199	18	Yes	Yes	Not studied	Not studied	Mainland–island system; habitat quality effects	12
Samango monkey *Cercopithecus mitis labiatus*	199	7	Yes	Yes	Not studied	Not studied	Mainland–island system	12

[a]References: 1, Ericson et al. (1999); 2, Bastin and Thomas (1999); 3, Appelt and Poethke (1997); 4, Biedermann (2000); 5, Dennis and Eales (1999); 6, Dennis et al. (1998); 7, Bergman and Landin (2001); 8, Carlson and Edenhamn (2000); 9, Edenhamn (1996); 10, Saari et al. (1998); 11, Forys and Humphrey (1999); 12, Lawes et al. (2000).

examined butterfly–plant coevolution in response to a question about checkerspot diets (chapter 14), coevolution has developed into a vast and varied field of research. It has transformed the study of plant ecology and evolution, which in the 1950s paid almost no attention to the role of herbivores in shaping the morphology, phenology, and chemistry of plants. The 1956 edition of Oosting's standard plant ecology contains only five pages on "animals as factors," including pollinators, seed dispersers, and soil tillers, as well as some cursory descriptions of the economic damage plant pests may cause. The transformation of botany traces primarily to one of Ehrlich and Raven's basic arguments, building on earlier suggestions by others (e.g., Fraenkel 1959) that most plant secondary compounds evolved to play a defensive role, rather than functioning as excretory products (e.g., Muller 1969). This argument is now widely accepted (Rhoades and Cates 1976, Brattsten et al. 1977, Berenbaum 1983, 1995b, Krischik and Denno 1983, Berenbaum and Zangerl 1988, Brattsten 1988, Marquis 1992, Thompson 1994, Futuyma 1998, Cornell and Hawkins 2003), although we now realize that the relationships among plants, secondary compounds and plant enemies can be dauntingly complex (for wonderful butterfly-based examples, see Berenbaum and Zangerl 1988, Chew 1988).

Ehrlich and Raven also emphasized the evolutionary pressures placed on plants by herbivores. This is clear without reference to secondary compounds, as illustrated by the many specific structures that have evolved in plants in response to the attack by herbivores. The best examples involving butterflies are perhaps the special hooked structures on the leaves and the egg mimics that have evolved in tropical *Passiflora* in apparent response to selection pressures from *Heliconius* (Gilbert 1971, 1983, Benson et al. 1975, K. S. Williams and Gilbert 1981) and a similar egg-load assessment and egg mimicry system in some temperate pierids (Shapiro 1981). Nonetheless, some researchers (Jermy 1976, 1984, 1993, Schoonhoven et al. 1998) continue to maintain that "plants exert strong selection pressure on insects, whereas insects exert selective pressure on plants only in rare cases and even then weakly" (Schoonhoven et al. 1998). We do not agree with this for several reasons. First, extremely heavy selection pressures can be surmised from the recorded dramatic impacts of herbivores on plant survival and reproduction. Second, selection coefficients that might be described as "weak" (say, 0.05 or 0.02) are very strong over just hundreds or thousands of generations (e.g., Hartl and Clark 1997), but virtually undetectable under most field conditions. As Gilbert (1983) put it, "the arguments that butterflies are minor components of plant faunas and therefore are not likely to select for plant defensive traits are similar to arguments against the role of birds as selective agents in butterfly mimicry (on the grounds that most mortality occurs to eggs and larvae)" (p. 279).

The view that butterflies have little effect on plants perpetuates Jermy's (1976) earlier argument, derived from a statement Bates (1958) made in another context: "the great bulk of plant material never goes through the animal part of the energy-food cycle so neatly diagrammed in the textbooks" (p. 212). Jermy suggested that the energy-rich compounds plants manufacture to protect themselves are selectively "neutral" and that insects serve as mutualists for plants by thinning their populations (1976). By the same logic, one might claim that overdominance for fitness has not produced the sickle-cell polymorphism in *Homo sapiens* (Allison 1954, 1964, Hartl and Clark 1997) because "the great bulk of human material never goes through the *Plasmodium* part of the energy-food cycle" and that malarial parasites were mutualists of *Homo sapiens*, favored by selection because they reduced dangerous overpopulation (which would amount to a naïve group selection argument). That plants manage to defend their otherwise edible tissues so effectively is exactly because of hundreds of millions of years of evolution of defenses—and largely chemical defenses at that.

Research on butterflies has played a major role in demonstrating the impact of herbivores on plants. Examples on checkerspots include heavy defoliation of *Aster* plants by *Chlosyne harrisii* recorded by Dethier (1959) and the cases discussed in chapters 3 and 4. White (1974) found that in many years populations of *E. editha* destroyed 30–70% of their food plants—a stunning result in the context of findings showing that reduction of seedling leaf area due to herbivory by *E. editha* by as little as 5% could significantly reduce fecundity in the host plant *Collinsia torreyi* (Parmesan 2000). Selection pressure in the form of destruction of flowers and seeds is even easier to see. An early example was the demonstration that predation on inflorescences of *Lupinus* and *Thermopsis* (Fabaceae) by tiny lycaenid butterflies (*Glaucopsyche lygdamus*) could reduce potential seed set by much more than 50% (Breedlove and Ehrlich 1968, 1972, Dolinger et al. 1973). The rates of attack on these plants were shown to be related to the quantities and mixes of defensive alkaloids they contain (Dolinger et al. 1973). The existence of spines, thorns, sticky gums and resins, and many other physical obstacles to ingestion and digestion of leaves makes the evolutionary response of plants to folivores as selective agents rather obvious. In short, the idea that by devouring foliage herbivores do not put much selection pressure on plants is persistent (Raffa 1991) but falsified by the available evidence.

Much research on butterflies has supported the conclusion that plant chemistry is under selection by plant enemies (e.g., Mauricio and Rausher 1997). Just consider the elegant studies on swallowtails by Berenbaum (e.g., Berenbaum and Feeny 1981, Berenbaum and Zangerl 1988), Feeny (1991, Feeny et al. 1985), Scriber and colleagues (e.g., Scriber et al. 1989, 1991, 1995) and Thompson (e.g., 1988, 1993, 1995, 1996). An extensive body of work on plant–herbivore chemical ecology has been produced by Bowers and co-workers on iridoid glycosides and checkerspots (Bowers 1981, 1983a, 1983b, 1986, 1988, Belofsky et al. 1989). Of course, although we are butterfly chauvinists, we admit that other insects are being used to examine plant–herbivore interactions (e.g., de Jong and Nielsen 1998). Nonetheless, below we outline another checkerspot example, involving the relationship between *Euphydryas chalcedona* at Jasper Ridge and its major host plant at that site, the shrub *Diplacus aurantiacus,* suggesting a close correspondence between the chemistry of the plant and predation pressure by the butterfly.

The phenologies of *Euphydryas chalcedona* and *Diplacus aurantiacus* at Jasper Ridge are highly synchronized (Mooney et al. 1981; plate VIII.4). Postdiapause larvae become active as the plant begins to grow after the winter rains have reduced the plants' drought stress, usually in January or February. Adults fly in May and June, and prediapause larvae feed in June and July until about the time plant growth ceases. *Diplacus* sheds most of its leaves during California's summer drought, but those leaves that persist have a high content of water, protein, nonstructural carbohydrate (sugars), and a high per unit-area concentration of a leaf-surface resin (Mooney et al. 1980). The resin consists of monomers containing a flavonoid nucleus with several phenolic groups, and it apparently reduces protein availability and serves as a feeding deterrent, leading to reduced larval growth rate and size as well as reduced survival (Lincoln 1980, 1985, Lincoln et al. 1982). There is less resin in wet-season leaves, which are fed upon by *E. chalcedona* larvae. In this respect, *Diplacus* shows an antiherbivore strategy reminiscent of that in the desert shrub *Larrea tridentata* (creosote bush; Rhoades and Cates 1976). Both plants maintain a few terminal leaves during drought periods and heavily defend these "apparent" resources (Feeny 1976).

In *Diplacus* temporal distribution of the feeding-deterrent resin leads to maximal protection of the plant's carbon-producing capacity (Mooney et al. 1981). *Euphydryas chalcedona* larvae feeding on *Diplacus* recognize the quality of the leaves and preferentially feed on those with the highest nitrogen/resin ratio. In the absence of high-quality food, the larvae may choose not to eat, feed on low-quality leaves, migrate to other *Diplacus* individuals, or return to diapause and "try again next year" (K. S. Williams et al. 1983a). This complex relationship is maintained in an area where a more nutritious but less accessible larval food plant is also utilized. *Euphydryas chalcedona*, like *E. editha,* has complex population-by-population adaptations to host plants, influenced both by the traits of the plants and their availabilities. Different populations have evolved diverse strategies which, among other things, may influence their ability to colonize new habitats with novel hosts (e.g., Bowers 1986; chapter 9).

In conclusion, research on butterflies has produced much evidence showing that powerful ecological and evolutionary forces shape plant–herbivore relationships and that these forces may involve much more than simple interplay between plant secondary chemicals and herbivore strategies to avoid or detoxify them (Gilbert and Singer 1975, Bernays and Graham 1988). It is true that we need more well-worked-out examples of coevolution where selection is measured directly (Rausher 1988, Marquis 1992), but this is true of selection in nature in general (Endler 1986).

11.7 Evolution of Host Plant Use in Checkerspots

The coevolutionary model outlined above posits that speciation in phytophagous insects is influenced by properties of their host plants and that speciation in plants is reciprocally influenced by the impact of phytophagous insects. Those who assume that insects rarely exert strong selective pressures on plants (Jermy 1976, 1984, 1993, Schoonhoven et al. 1998) favor the hypothesis of sequential evolution, in which insects primarily track the evolution of plant secondary compounds without having an influence on that evolution. In other words, speciation in phytophagous insects can be brought about by plants, but speciation in plants is not caused by insects feeding on them. A rigorous test of these hypotheses requires that patterns of host plant use first be placed in an evolutionary perspective.

The relationships between checkerspots and their larval host plants have been much studied since Singer (1971b, 1983, White and Singer 1974) first discovered that different populations of *Euphydryas editha* preferred different host plant species. Bowers (1981, 1983b) showed that the host plants of *Euphydryas* species in North America all contained iridoids, a class of plant secondary chemicals. Bowers (1980, 1981) was intrigued by the fact that *Euphydryas* larvae and adults are aposematically colored, and she found that they were unpalatable and even emetic to vertebrate predators. The reason for their unpalatability turned out to be the ability of larvae to sequester iridoids (Bowers and Puttick 1986). Iridoids are bitter-tasting compounds; they have even been used as insecticides against generalist insect herbivores (Seigler 1998). Iridoids were found to be important feeding stimulants for *E. chalcedona* larvae (Bowers 1983b). The larvae refused to feed on pure artificial diet, but when catalpol (an iridoid glycoside) was added to the diet, the larvae fed actively (Bowers 1983b).

The ability of checkerspot larvae to sequester iridoids has been studied closely in several species. Iridoids are a diverse group of more than 1000 chemical compounds, which can be divided into two major groups, iridoid glycosides and seco-iridoids (Jensen 1991, Seigler 1998). So far, all iridoids that have been found to be sequesterable by checkerspots are iridoid glycosides (Bowers and Puttick 1986, Stermitz et al. 1986, 1994, Franke et al. 1987, Belofsky et al. 1989, L'Empereur and Stermitz 1990a, 1990b, Mead et al. 1993, Bowers and Williams 1995), with two compounds that are sequestered in large quantities, catalpol and aucubin. The ability to sequester iridoid glycosides has been recorded in five species of *Euphydryas* (Bowers and Puttick 1986, Stermitz et al. 1986, 1994, Franke et al. 1987, Gardner and Stermitz 1988, Belofsky et al. 1989, L'Empereur and Stermitz 1990a), two species of *Chlosyne* (Mead et al. 1993, Stermitz et al. 1994), *Poladryas minuta* (L'Empereur and Stermitz 1990b), and *Melitaea cinxia* (Suomi et al. 2001). Bowers and Williams (1995) reported that though the larvae of *Euphydryas gillettii* are found mainly on plants containing seco-iridoids, they were unable to sequester these compounds, but they were able to sequester iridoid glycosides from other plants that post-diapause larvae occasionally feed on. More generally, the host plants of checkerspots naturally contain a large range of compounds other than iridoid glycosides. How sensitive checkerspots are to these other secondary compounds has not been studied in detail. Stermitz et al. (1989) reported that quinolizidine alkaloids that the hemiparasitic host plants of *E. editha* obtain from other plants do not affect most larvae, but some larvae were affected.

The finding that *E. chalcedona* larvae refused to feed on artificial diet lacking catalpol led Bowers (1983b) to postulate that the ability to utilize iridoids by *Euphydryas* species has enabled them to colonize a variety of plant families containing these chemical compounds. Recent studies by Wahlberg (2001b), based on the phylogenetic hypothesis of Wahlberg and Zimmermann (2000), allows a comprehensive assessment of this hypothesis for all checkerspot species. Most checkerspot species (or populations) are oligophagous or even monophagous on plant species belonging to 16 plant families. Fourteen families belong to two distinct clades of the subclass Asteridae (Olmstead et al. 1993, 2001, Angiosperm Phylogeny Group 1998). The families Scrophulariaceae, Lamiaceae, Plantaginaceae, Oleaceae, Acanthaceae, Verbenaceae, Gentianaceae, Orobanchaceae, and Convolvulaceae belong to the asterid clade I, whereas Asteraceae, Adoxaceae, Caprifoliaceae, Valerianaceae, and Dipsacaceae belong to the asterid clade II. The remaining two families are entirely unrelated to the previous families: Urticaceae, belonging to the subclass Rosidae, and Amaranthaceae, belonging to the subclass Caryophyllidae. Twelve of these families are united by the presence of iridoids (Jensen et al. 1975, Jensen 1991). Four families, Asteraceae, Convolvulaceae, Urticaceae, and Amaranthaceae, do not contain iridoids.

In this perspective, it is apparent that, as originally suggested by Bowers (1983b), iridoid glycosides have had a substantial impact on the evolution of host plant use in checkerspot butterflies (Wahlberg 2001b). Mapping the presence of iridoid glycosides in the host plants of the extant checkerspot species onto the phylogeny of the butterflies shows that this trait is very conservative; that is, there is not much switching back and forth between character states (figure 11.7). In contrast, when the use of host plant families is mapped onto the phylogeny, the pattern is much more dynamic (figure 11.8). The evolutionary dynamics of host plant use are evident as the widening of the host plant range in clades using plants with iridoids and as host shifts to chemically dissimilar plants.

From Female Preference to Host Plant Use in Clades of Species

In checkerspots, host plant use is primarily controlled by the ovipositing female, as newly hatched larvae are not able to disperse over distances greater than a few centimeters (Moore 1989a; chapter 7). All checkerspot species that have been studied show similar oviposition behavior to that observed in *Euphydryas editha* (Singer 1994, C. D. Thomas and Singer 1998) and *Melitaea cinxia* (Kuussaari et al. 2000) (for instance, for *Euphydryas maturna,* see Wahlberg [1998] and for *Melitaea diamina,* see Wahlberg [1997a]). Detailed studies of female hostplant preference have furthermore shown that females choose individual plants rather than plant species on which to oviposit (Ng 1988, Singer and Lee 2000; chapter 6). This means that an individual of one plant species may be preferred over an individual of another plant species, which in turn may be preferred more than another individual of the first species. These findings suggest that females choose to oviposit on individual plants based on their biochemical profile.

The choices made by females may, of course, also relate to factors other than plant secondary chemicals. For instance, *Euphydryas editha* in Colorado oviposits preferentially, and perhaps exclusively, on *Castilleja linariifolia*, although two other potential host species in the area, *Castilleja chromosa* and *Penstemon strictus,* proved equally or more satisfactory to the larvae in feeding trials (Holdren and Ehrlich 1982). Apparently, a combination of high plant density, year-to-year persistence, phenology, and relative resistance to mammalian herbivory makes *C. linariifolia* a more satisfactory host. Similarly, in the case of *Euphydryas chalcedona* at Jasper Ridge, even though *Scrophularia californica* is nutritionally superior and larvae show a strong preference for feeding on *Scrophularia*, females oviposit preferentially on *Diplacus aurantiacus* (K. S. Williams 1983, K. S. Williams et al. 1983b). *Diplacus* is the spatially and temporally more abundant and persistent larval resource in the chaparral community occupied by *E. chalcedona* and more often grows in sunny and thermally superior environments for larval development than does *Scrophularia* (K. S. Williams 1983).

Within a given population, most females prefer to oviposit on one species of host plant (Singer et al. 1994, Kuussaari et al. 2000), indicating that biochemical profiles tend to be specific to a plant species, though exceptions do occur (Dolinger et al. 1973, Singer and Lee 2000). An environmental perturbation may introduce a new plant species, with a biochemical profile that is more acceptable to some ovipositing females than that of the original host species (Singer et al. 1993, C. D. Thomas and Singer 1998). If the new host supports a high rate of larval survival, female preference in the population may evolve rapidly (Singer et al. 1993; for an example, see figure 6.9). Major environmental perturbations have been common, especially for Nearctic and Palaearctic species over the past 5 million years (My), because of repeated glaciations.

Turning from the level of populations to the level of species, one finds that different populations have become specialized on different host plant species, but usually in the same plant family. This pattern is clear in many Holarctic species, whereas the apparently monophagous Neotropical species may just reflect a dearth of information from this region. In *Euphydryas editha*, which has been studied in great detail, the evolution of host plant use appears to have been very dynamic, with several host plant genera being lost and recolonized several times by different populations of the butterfly (Radtkey and Singer 1995).

As one moves up still further to the level of clades of species, the inferred dynamics in the pattern of the use of host plant genera is repeated in the use of host plant families by related butterfly species (Wahlberg 2001b). It is at this level that one should see signs of coevolution over longer periods of time. But, in fact, no such signs are apparent. Most important, the coevolutionary scenario would suggest that the ages of the insect and host plant clades should be similar (Mitter and Farrell 1991), whereas in this case the host plant families are much older than the age of the tribe Melitaeini, about 5 My, inferred from low sequence divergence and the biogeography of the tribe (Wahlberg and Zimmermann 2000). For instance, the ages of Acanthaceae, Asteraceae, and Caprifoliaceae are all between 20 and 50 My (Eriksson and Bremer 1992).

Figure 11.7. Optimization of the presence of two classes of iridoids in the host plants of checkerspots. Fast optimization was used. The two classes were optimized independently but are shown together to save space.

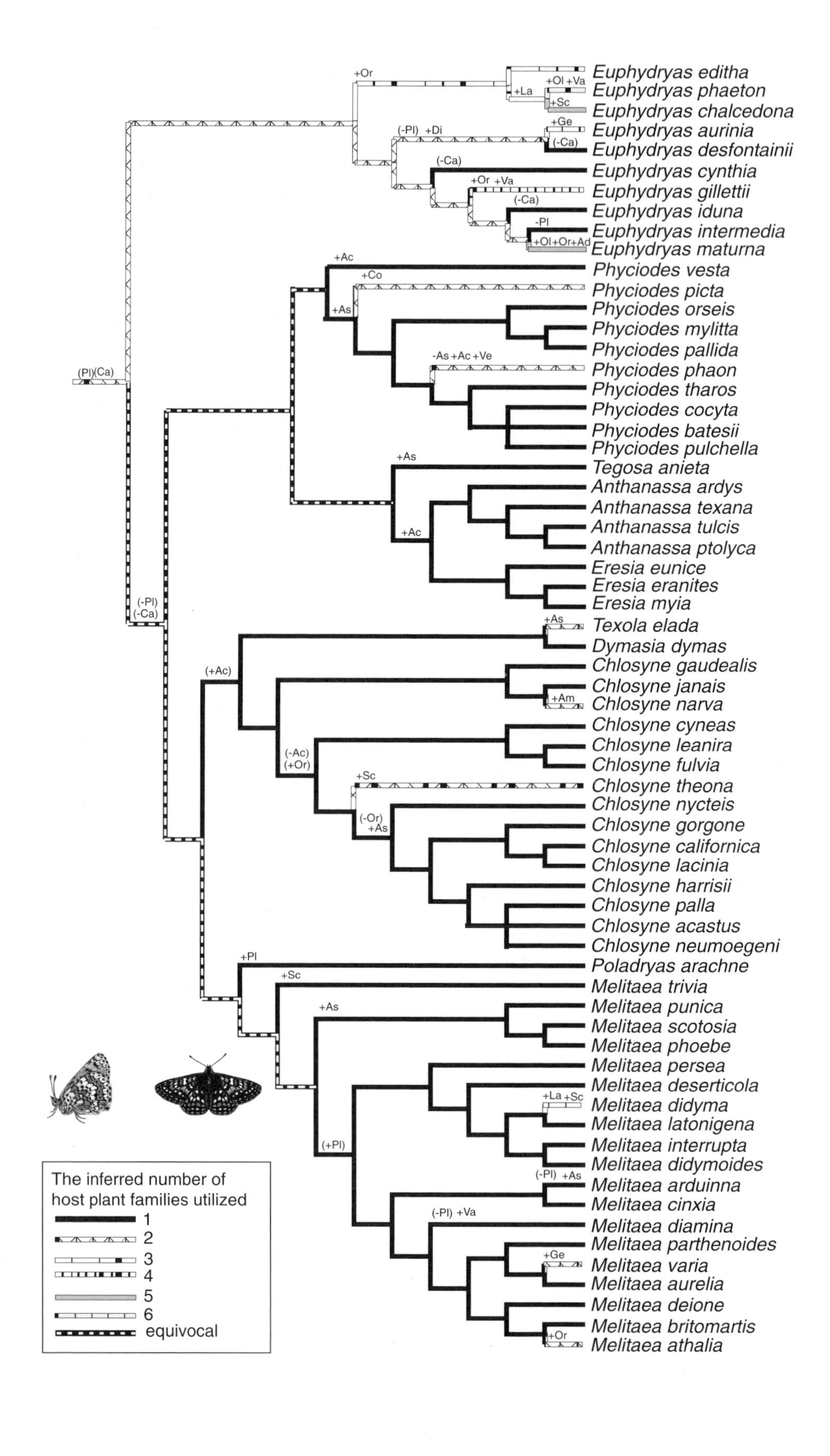

+Or
Euphydryas editha
+La
+Ol +Va
Euphydryas phaeton
+Sc
Euphydryas chalcedona
+Ge
Euphydryas aurinia
(-Pl) +Di
(-Ca)
Euphydryas desfontainii
(-Ca)
Euphydryas cynthia
+Or +Va
Euphydryas gillettii
(-Ca)
Euphydryas iduna
-Pl
Euphydryas intermedia
+Ol+Or+Ad
Euphydryas maturna
+Ac
Phyciodes vesta
+Co
Phyciodes picta
+As
Phyciodes orseis
Phyciodes mylitta
Phyciodes pallida
-As +Ac +Ve
Phyciodes phaon
(Pl)(Ca)
Phyciodes tharos
Phyciodes cocyta
Phyciodes batesii
Phyciodes pulchella
+As
Tegosa anieta
Anthanassa ardys
Anthanassa texana
Anthanassa tulcis
+Ac
Anthanassa ptolyca
Eresia eunice
Eresia eranites
(-Pl)
(-Ca)
Eresia myia
+As
Texola elada
Dymasia dymas
(+Ac)
Chlosyne gaudealis
Chlosyne janais
+Am
Chlosyne narva
Chlosyne cyneas
Chlosyne leanira
(-Ac)
(+Or)
Chlosyne fulvia
+Sc
Chlosyne theona
Chlosyne nycteis
(-Or)
+As
Chlosyne gorgone
Chlosyne californica
Chlosyne lacinia
Chlosyne harrisii
Chlosyne palla
Chlosyne acastus
Chlosyne neumoegeni
+Pl
Poladryas arachne
+Sc
Melitaea trivia
+As
Melitaea punica
Melitaea scotosia
Melitaea phoebe
Melitaea persea
Melitaea deserticola
+La +Sc
Melitaea didyma
Melitaea latonigena
(+Pl)
Melitaea interrupta
Melitaea didymoides
(-Pl) +As
Melitaea arduinna
Melitaea cinxia
(-Pl) +Va
Melitaea diamina
Melitaea parthenoides
+Ge
Melitaea varia
Melitaea aurelia
Melitaea deione
Melitaea britomartis
+Or
Melitaea athalia
The inferred number of
host plant families utilized
1
2
3
4
5
6
equivocal

The most likely explanation for the pattern of checkerspot host plant use is that the butterflies have colonized an already diverse assemblage of plants. Checkerspots have not coevolved with iridoid-containing plants, but rather have been able to circumvent the plants' defenses and are now able to exploit the plants. Whether insect herbivory was one driving force in the evolution of iridoids has not been answered, but it appears clear that if such herbivory-driven evolution has occurred, the checkerspots have not been instrumental. Some ancestral checkerspot lineages have specialized on plants containing seco-iridoids in addition to those containing iridoid glycosides. One lineage has speciated to form the *Euphydryas* group, but two lineages, consisting of *Melitaea diamina* and *M. varia,* have not speciated so far. *Melitaea diamina* is widespread and phenotypically variable, suggesting that the ancestral populations have evolved a way to circumvent the negative effects of seco-iridoids, and thus the butterflies have merely followed the plants, rather than caused the evolution of a more potent plant defense.

There are three groups of checkerspot species that depart from the general pattern of using host plants containing iridoids, and all three groups use host plants in the Acanthaceae or Asteraceae (Wahlberg 2001b), which are devoid of iridoids (Jensen et al. 1975, Jensen 1991). The three groups of species are in the subtribes Phyciodina and Melitaeina, with several species in the genus *Melitaea* and many species in the genera *Chlosyne, Texola, Microtia,* and *Dymasia.* Whether the host plants in these two families confer some sort of chemical protection to the larvae feeding on them is still an open question. Adult *Chlosyne harrisii* were palatable to a bird predator (Bowers 1983a), but larvae of *C. lacinia* were unpalatable to amphibian predators (Clark and Faeth 1997a). Both species feed on plants belonging to Asteraceae. In the case of the *Chlosyne* group, species ancestral to the *C. nycteis* clade used host plants in Orobanchaceae, which are hemiparasitic plant species. These plants are able to take up plant secondary compounds from their hosts (Stermitz et al. 1989). If the hosts of the ancestral Orobanchaceae fed upon by the *Chlosyne* species belonged to Asteraceae, the butterfly would be exposed to the secondary compounds of Asteraceae, which would have facilitated the colonization of Asteraceae. In the other two groups, the host plant affiliations of ancestral species are unclear.

In conclusion, though there is little doubt that phytophagous insects including checkerspots influence the fitness of their host plants, there is no sign of the checkerspots having coevolved, in the sense of interacting adaptive radiations, with their host plants. Iridoids in the host plants of checkerspots may well have evolved in response to herbivory, but the task of tracing the long evolutionary process with the information that is available seems utterly impractical. In contrast, it is evident that the presence of iridoid glycosides in the plants have significantly influenced the radiation of checkerspot butterflies.

11.8 Checkerspots and Generalities

As biologists, we search for generalities that could explain a broad range of phenomena we see in unrelated species in nature. Since Darwin's work became fully and widely understood, all population biologists have had natural selection in the back of their mind while attempting to understand a structure or a behavior. And since Nicholson (1933) and Andrewartha and Birch (1954), no ecologist interested in population dynamics should fail to consider both exogenous and endogenous factors influencing changes in numbers.

So what have the checkerspots told us that we did not know before studying them and that would be of wide significance for population biology? One general lesson relates to the spatial structure of populations. Checkerspots have helped us understand the patchy nature of populations and have

Figure 11.8. Optimization of host plant families used by checkerspot butterflies. Gains (+) and losses (–) are shown above the relevant branches. The plant families have been optimized as independent binary characters. Abbreviations for plant families are as follows: Ac = Acanthaceae, Ad = Adoxaceae, Am = Amaranthaceae, As = Asteraceae, Ca = Caprifoliaceae, Co = Convolvulaceae, Di = Dipsacaceae, Ge = Gentianaceae, La = Lamiaceae, Ol = Oleaceae, Or = Orobanchaceae, Pl = Plantaginaceae, Sc = Scrophulariaceae, Va = Valerianaceae, and Ve = Verbenaceae. Parentheses around the abbreviation denote ambiguities resolved using fast optimization. Branch shading indicates the inferred number of plant families used by ancestral populations.

brought to the forefront the frequency of population extinctions (e.g., Ehrlich et al. 1980) and, through that, the magnitude and conservation importance of those extinctions (Ehrlich and Daily 1993, Hughes et al. 1997, Ceballos and Ehrlich 2002). Through checkerspots, we also have gained a much better understanding of how assemblages of local populations interact with each other and thus influence the persistence of a species in heterogeneous landscapes. Indeed, metapopulations have been called to the attention of ecologists in no small degree because of work on the checkerspot system, and the concept is now applied widely in population biology and in conservation biology (Hanski and Simberloff 1997).

Checkerspots were instrumental in starting an entire subdiscipline in evolutionary ecology, that of the study of coevolution (Ehrlich and Raven 1964). Although checkerspots have not played a major role in developing this field further, information on checkerspot host plant use has contributed greatly to understanding the interactions between plants and insects (Singer 1971b, 1983, Bowers 1983b, Singer et al. 1993, Parmesan 2000, Wahlberg 2001b).

Finally, a strength of the checkerspot system at this point is the vast and diverse array of data that have been accumulated, which can be placed in the context of that available for other groups. As a result, other aspects of checkerspot biology are now primed for study using the comparative approach at taxonomic levels ranging from genera to families: the evolution of wing patterns, the possible correlation between average egg batch size and degree of larval gregariousness (see table 11.1), and the evolution of mate-searching behavior are just three examples. Armed with this array of data and a phylogenetic hypothesis, the path is open to uncover further broad evolutionary patterns. Such work provides the context for checkerspots as a model system, as outlined in the next chapter, facilitating further advances in population biology.

12

Checkerspots as a Model System in Population Biology

ILKKA HANSKI, JESSICA J. HELLMANN, CAROL L. BOGGS, AND JOHN F. McLAUGHLIN

12.1 Model Systems in Population Biology

Model systems such as *Drosophila*, *Arabidopsis*, *Mus*, *Caenorhabditis*, and the squid giant axon are familiar to geneticists, developmental biologists, neurobiologists, and other laboratory biologists. A chief characteristic of these systems is that numerous researchers have used them over a long period of time to study a range of research questions, often greatly benefiting from the diverse knowledge accumulated by previous investigators. Successful model systems possess some definite advantages for research. For example, it is critical for laboratory model systems that the organisms can be cultured in large numbers under controlled conditions and have short generation times.

A classic population biological model system is the Galapagos finches, a group of 14 currently recognized species (Petren et al. 1999) on the 13 oceanic Galapagos Islands. Charles Darwin came across the Galapagos in 1835 during his round-the-world voyage. Darwin did not select the finches to be a model system; he just happened to sample them on several islands. To his later frustration, he even failed to keep all the specimens from different islands separate (Darwin 1842). The Galapagos finches have gradually become a prime model system for studies of speciation, adaptive radiation, and character displacement because of the initial knowledge gained by Darwin and because of the remarkable diversification of the finches. David Lack (1947), Robert Bowman (1961), and Peter and Rosemary Grant (Abbott et al. 1977, P. Grant 1986) and co-workers have made lasting contributions to the conceptual framework of population biology using the finches. Like other successful model systems, the Galapagos finches allow studies that would be difficult to accomplish with other combinations of species and environments (Schluter 2000).

The sample of population biological model systems in table 12.1 shows that such systems have been developed to study most major ecological and evolutionary phenomena, including behavior, population dynamics, species interactions, community assembly, and ecosystem structure and function, as well as adaptation and speciation. Classic model systems in biology in general are single species, but in ecology and evolution, population interactions are often the focal issue, leading to a broadening of the model system concept to include communities and even ecosystems (e.g., Vitousek 2003). Likewise, variation across habitats and environments, or variation due to phylogenetic history, is important for the development of a conceptual framework in ecology and evolution, in which case guilds or other groups of species have attained the status of model systems. Clusters of model systems exist for some questions of particular interest to population biologists, such as the fitness concept and interspecific competition (table 12.1). Some model systems

Table 12.1. A small sample of model systems in population and community ecology.

Concept or Phenomenon	Study System	Major Focus of Research	Selected References
Adaptation/fitness	Galapagos finches	Adaptation of morphology, foraging, and life-history strategies to resource availability	Grant (1999)
	Great and blue tits	Evolution of life history, mating systems, and foraging strategies	Both et al. (1998), Naef-Daenzer (2000)
	Scrub jays	Social systems influence individuals' fitness	Koenig et al. (1992), Woolfenden and Fitzpatrick (1984)
	Red deer	Influence of evolved life histories on population dynamics	Clutton-Brock and Albon (1979), Clutton-Brock et al. (1982), Benton et al. (1995)
	Sulfur butterflies	Adaptation of metabolic systems to environmental conditions to determine individual fitness	Watt (2003)
Speciation	Galapagos finches	Competition drives divergence in morphology and resource use	Darwin (1859), Schluter et al. (1985), Grant and Grant (1997)
	Stickleback fish	Parallel speciation by niche separation	Schluter (1994), Rundle et al. (2000)
	Rhagoletis flies	Sympatric speciation	Bush (1969), Filchak et al. (2000)
	Lake cichlids	Rampant speciation	Barlow (2002), Kornfield and Smith (2000)
	Anolis lizards	Interspecific competition determines species composition; coexisting species partition resources	Roughgarden (1995), Losos et al. (1998)
Competition	Desert rodents and ants	Ecologically similar species compete for access to food resources	Brown and Lieberman (1973), Brown and Davidson (1977), Brown et al. (2001), Heske et al. (1994)
	Prairie plants	Competition for nitrogen leads to trade-off between competitive and dispersal abilities	Tilman (1982, 1988, 1994), Rees et al. (2001)
	Intertidal invertebrates	Competition for space structures intertidal communities and creates patterns of zonation; keystone predator maintains species diversity by opening space	Connell (1983), Paine (1974), Wooton (1997)
Predation/trophic structure	Lynx-hare	Interaction with both food and predators drives cyclic dynamics of hares Interaction with hares drives lynx cycles	Elton (1942), Krebs et al. (1995)
	Rodents	Specialist predators drive complex dynamics in rodent populations	Krebs and Myers (1974), Finerty (1980), Hanski et al. (2001)

	Orcas, otters, urchins, kelp	Species abundances regulated by predation from higher trophic level	Estes and Duggins (1995), Estes et al. (1998), Lubchenco and Menge (1978)
	Intertidal invertebrates	Bottom-up and top-down effects can regulate species abundances	Menge (2000)
Host-pathogen dynamics	*Silene-Ustilago*	Migration between host populations affects local infection; regional dynamic in patterns of disease	Antonovics et al. (1994), Alexander et al. (1996)
Mutualism	Yucca-yucca moth	Plant prevents overexploitation by controlling pollinator access to resources	Pellmyr and Huth (1994), Addicott (1996)
		Coevolution can generate obligate mutualisms.	
Lepidopteran Model Systems	Gypsy moth	Cyclic dynamics driven by complex web of top-down and bottom-up trophic interactions	Hunter (1993), Jones et al. (1998)
		Invasive species dramatically alters forest ecosystem	
	Peppered moth	Predation can drive rapid evolution in an altered environment	Kettlewell (1955), Brakefield and Liebert (2000), Majerus (1998)
	Epirrita autumnata	Food quality influences forest insect dynamics, induced plant defenses	Haukioja (1980), Haukioja et al. (1983)
	Tent caterpillars	Habitat heterogeneity and virus mediate population outbreaks	Myers (1993, 2000), Roland (1993)
	Winter moth	Predators stabilize population dynamics with aid of parasitoids	Varley et al. (1973a), Roland (1994), Liebhold et al. (2000)
	Bicyclus spp.	Evolutionary development of wing patterns	Brakefield et al. (1996), Brakefield and Monteiro (2003)
	Papilionidae	Dynamics of hybrid zones; coevolutionary patterns of host plant use; speciation processes	Thompson (1994), Scriber (2003), Sperling (2003)
	Heliconius spp.	Mimicry; host plant use; mating strategies; male nuptial gifts	Gilbert (1971), Boggs and Gilbert (1979), Deinert et al. (1994), Mallet and Gilbert (1995)
	Pieridae	Thermoregulatory strategies; see also "adaptation"	Kingsolver and Watt (1983), Kingsolver (1985)
	Precis spp.	Genetics and ontogeny of butterfly color patterns	Nijhout (1991, 1994)

are exceptionally versatile. Grassland plants might be considered to compose one grand model system, covering 20% of the earth. They have been used to study competition (table 12.1; Pacala and Silander 1990), spatial pattern formation (Herben et al. 2000), the relationship between diversity and stability (Pfisterer and Schmid 2002), habitat fragmentation (Jackel and Poschlod 1996), and climate change (Chiariello and Field 1996, Buckland et al. 2001). The Rothamsted Park Grass Experiment, begun in 1856 and still running, is a model system par excellence (Dodd et al. 1995). Biologists have even used biogeographical regions as model systems, such as the Caribbean island system (e.g., Schoener and Schoener 1983, Roughgarden 1995, Losos 1998) or the Hawaiian island system (e.g., Vitousek 2003), for understanding phenomena operating on long time scales (Barnosky et al. 2001). Characteristically, many of the model systems in table 12.1 have been studied for decades, with early research suggesting new questions for subsequent investigation.

Some model systems in table 12.1, such as the cichlid fish in the large lakes in East Africa, were not selected for research because they represent many other species in other environments, but rather because some particular phenomenon or process is strikingly evident in that system (rampant speciation in the case of lake cichlids in East Africa). Such model systems can be especially powerful because their dynamics may be dominated by a particular process, thereby bringing them closer to a controlled laboratory system. Even though such model systems are not good models for any other species or phenomena, their intensive study can make a significant contribution toward a better understanding of a particular phenomenon or process. Other model systems in table 12.1 were originally selected on other bases, but they turned out to be convenient model systems for the purpose mentioned in table 12.1, or they came to be good model systems by much effort and by the amount of information that has been accumulated.

Some laboratory systems used in population biology, such as *Lucilia* blowflies (Nicholson 1954, 1957), *Tribolium* flour beetles (King and Dawson 1972, Constantino and Desharnais 1991), and *Drosophila* fruit flies (Mueller 1985, 1997), have become important model systems because of the large amount of biological information available for them. Existing genetic information is particularly useful in the case of *Drosophila*. Mueller and Joshi (2000) emphasize the value of carefully controlling environmental conditions, which is only possible with laboratory systems. This strength in a laboratory model system is, however, a key weakness for the model's use in population biology: the environmental effects are not just a nuisance to be controlled but are an essential part of most phenomena worth studying. Some of the most promising ecological models are those that allow the effective study of both laboratory and field populations. For example, one of us has worked on *Lucilia* and other blowflies using laboratory and field experiments (Hanski 1976, 1987, Hanski and Kuusela 1977, Kouki and Hanski 1995, Prinkkilä and Hanski 1995) and field observations (Hanski and Kuusela 1980). This combination of empirical work and associated theory (Atkinson and Shorrocks 1981, Hanski 1981) has contributed to understanding the conditions that allow ecologically similar species to coexist on patchy, ephemeral resources (Hanski 1990).

So far we have not attempted to define what makes a model system in population biology. The most successful model systems in biology in general are systems that are repeatedly used for multiple studies of related phenomena by several groups of investigators. This is not necessarily so in population biology. Table 12.1 makes the point that ecologists and evolutionists have not concentrated their research efforts around a small number of much-studied model systems in the same way as biologists working in the laboratory have done. Instead, we appear to have a multitude of "minor" model systems. Dugatkin (2001) presented a fascinating overview of 25 model systems in behavioral ecology alone. What might be the reasons for lack of major model systems in population biology? One reason, surely, is our fascination with diversity itself—how attractive it is to study a species, an environment, or perhaps an interaction that no one else has examined before. Such studies may evolve into minor model systems if focused on an outstanding issue, and such a model system study may contribute to further development of the relevant conceptual framework and theory. In the best cases, represented by the Galapagos finches among others, the conceptual, theoretical, and empirical elements are all integrated into a common undertaking in the context of the model system, which leads to hand-in-hand development and testing of new ideas. Though this sort of model system is different from the classic model systems in laboratory-based biology, it may nonetheless represent a substantial concentration of research effort and may advance science much more than short-term, isolated case studies. We should also recognize that turning any one of these minor model systems into a major population biological model system would not be entirely satisfactory. The natural world is widely diverse both in terms of the environmental conditions and the species and their interactions. It would be practically impossible to select and define just one system for comprehensive studies of all questions of interest in ecology and evolution. The middle ground of multiple minor model systems serves population biology well. What is needed is more focus on a set of well-selected minor model systems.

Butterflies as Model Systems

Butterflies have been used widely as model systems for the study of the ecology and evolution of natural populations (Boggs et al. 2003, Watt and Boggs 2003). Basic questions about coevolution between insects and plants and the evolution of host specialization have been studied in a number of butterfly taxa, including the papilionids and nymphalids (table 12.1). The entire field of coevolution began through investigation of interactions between butterflies and their host plants (chapter 11). Butterfly projects have addressed the evolution of mimicry, developmental biology of wing patterns, and metabolic strategies. Butterflies are useful model systems in population ecology because of their sensitivity to environmental changes, short generation times, and our exceptionally good knowledge of their natural history.

An important practical advantage of butterflies for field studies is that individuals can be easily marked and recaptured in the field. In fact, as argued in chapter 9, butterflies are close to an optimal size in that they are sufficiently large to be individually marked but at the same time sufficiently small to be cheaply raised in large numbers in captivity. Another great advantage for field studies is that the larvae of many butterflies feed on just a single host plant or a few plant species. These characteristics imply that in many landscapes the species have a well-defined spatial population structure and that the researcher can determine relatively easily what that structure is. This is a key point for our own research, making it possible to study large systems of small or relatively small local populations with rapid turnover. Many other species may have equally distinct spatial population structures, but because the individuals are cryptic or otherwise difficult to find, we may never know exactly what that structure actually is.

The two long-term checkerspot research projects described in this volume have functioned as model systems for the study of population and metapopulation biology. Over the past 40 years, the *Euphydryas editha* project at Stanford, and its offspring at Texas, have contributed to understanding the local population concept, population regulation and dynamics, local adaptation, evolution of host plant use and life-history strategies, and even the biological species concept. Research on *E. editha* has also made major contributions to metapopulation biology, which has been the particular focus of the *Melitaea cinxia* project in Finland. This latter project played a key role in the development of the notion and theory of metapopulation biology over the past decade, including the construction of modeling approaches that can be applied more generally to metapopulations living in fragmented landscapes. The two butterfly projects complement each other in several ways, and together they have contributed much more to population biology than either would have done alone. In the following sections, we first review selected aspects of each model system. We then consider some lessons for population biology that stem from our experiences.

12.2 *Euphydryas editha*: A Model System for the Modern Population Concept

Populations are the basic functional units for ecological and evolutionary processes. Local populations are sets of individuals that share the same environment, including both biotic and abiotic components of that environment, and which mix freely in reproduction. Although populations are not always easy to distinguish in practice, the population concept is so fundamental to modern biology that we can scarcely discuss our science without using the term population. Knowledge of populations provides a basis for understanding natural processes ranging from the interactions of individual organisms to the composition of communities. The science of population biology seeks to explain how and why populations vary in size and in demographic and genetic structure over time and across space. This goal often is achieved by studying factors affecting individuals and then ascending to the level of populations.

The Bay checkerspot evolved into a model system both because of its characteristic population structure and because Ehrlich developed a research program to systematically study the full range of population processes in *Euphydryas editha*. The project began by delimiting checkerspot populations both ecologically and genetically (figure 1.2). The predominant view in population ecology in the early 1960s considered population extinctions unlikely in the presence of effective regulation, wide dispersal, and generally large population sizes. Spatial population structure was given only limited attention (Allee et al. 1949). Nicholson (1954, 1957), the principal architect of the population regulation paradigm, actually did recognize the possibility of spatially structured populations and extinctions of small local populations, but only in the case of host–parasitoid dynamics with strong density dependence leading to oscillations with increasing amplitude (Nicholson 1933). In this context, the early work on *Euphydryas editha*, showing apparently independent dynamics of similar populations over short distances in the absence of obvious density dependence, was quite disturbing to the conventional wisdom (Ehrlich 1961b, 1965, Singer, 1972). Nevertheless, these results were consistent with the minority view in population ecology, largely due to the Australian ecologists Andrewartha and Birch (1954), who were not impressed by the effectiveness of population regulation in preventing extinctions, and who developed proto-metapopulation ideas of large-scale persistence of species with ephemeral local populations (see discussion in Hanski 1999b).

Euphydryas editha qualifies as a model system because work on its dynamics and ecology did not stop here. Instead, it expanded to a multitude of processes: from individual behavior of larvae and adults to life histories and mating strategies, from population regulation to genetic structuring of populations at multiple scales, and from the evolutionary dynamics of host plant use in individual populations to the species concept. There are many other research projects on other species that have made lasting contributions in these areas, but an exceptional feature of the checkerspot system is the range of investigations that has been undertaken and the comprehensive understanding of the population biology of the butterflies that has followed.

The power of the model system approach is well exemplified by work on population responses of *E. editha* to environmental change. In 1996, Parmesan documented patterns of population losses in *E. editha* in California that may be attributable to climate change, a pioneering study of its kind for any organism. Thanks to the knowledge and modeling results of the effects of climate variables on population dynamics (Hellmann 2002a, McLaughlin et al. 2002a, Hellmann et al. 2004; chapter 3), the potential for climate to drive population losses is now understood in a mechanistic and predictive way. A wealth of information now exists for the influence of other global changes, such as habitat loss, invasion of non-native plants that compete with native hosts, and habitat alteration due to atmospheric pollution and nitrogen deposition (these processes are discussed in chapters 3 and 13). Each can contribute to the vulnerability of *E. editha* to climate change (McLaughlin et al. 2002a).

Local Population Structure and Dynamics

By applying unique marks to adult butterflies at Jasper Ridge in 1960 (figure 2), Ehrlich found that butterflies in the seemingly continuous grassland habitat were divided into three "demographic units" or local populations (figure 1.2; see also figure 12.1, plate V) with relatively independent population dynamics (Harrison et al. 1988, Hellmann et al. 2003; figure 3.6). Long-term mark–recapture data showed that migration between local populations averaged less than 2% of recaptured individuals (Ehrlich et al. 1975). This discovery set the scene for the study of the ecological and evolutionary dynamics within local populations and for the subsequent work that has contributed to the development of the metapopulation concept. A novel finding at the time was that local populations can be small and spatially limited, yet be located close to each other. Other general conclusions followed, as discussed below.

Understanding Local Spatial and Temporal Dynamics Requires Understanding Processes Affecting Individuals Singer (1972) first hypothesized that population fluctuations in *E. editha* were primarily due to survival at the larval stage, mediated by a phenological race between larval development and host plant senescence (chapter 3). This work identified a mechanism explaining a key mortality factor. Subsequent work has demonstrated the subtlety of the interactions between individual larvae and adult butterflies with their host plants and the often drastic consequences that these interactions have at the population level (e.g., Hellmann

Figure 12.1. Model systems often develop around study sites that are the focus of research for long periods of time. The Jasper Ridge Biological Preserve is one such example, where a variety of basic population studies have been pursued since 1960 (Hellmann et al. 2003). See also figure 1.2 and plate V.

2002c). One example of the importance of host resources to population dynamics occurs in the Sierra Nevada mountains. Here, a novel host plant for *E. editha* in man-made habitats turned out to be superior to the traditional host in the native habitat, but the novel host was far more susceptible to extreme weather events. In one well-studied example, this led to a catastrophic extinction of an entire set of high-density populations in disturbed habitats that had been acting as source populations for those in undisturbed habitats in the regional metapopulation (Singer and Thomas 1996, C. D. Thomas et al. 1996). This example shows that human intervention may set up source–sink relationships in which sources are more extinction prone than sinks, thereby causing source–sink relationships to be misleading cues to conservation biologists as they attempt to identify which populations to preserve. More generally, emphasis on details of individuals' biology has been a hallmark of research on *E. editha*. This approach contrasts with the traditional life table and key factor approaches to population dynamics (Morris 1959, Varley and Gradwell 1960, Stiling 1988), which focus on counting numbers and ascribing mortality to a few general factors (such as natural enemies and "unknown factors"). Although the traditional approach can yield models predicting changes in population size, the lack of detailed understanding of the biology risks faulty predictions under changed environmental conditions (Royama 1996).

Migration Does Not Equal Gene Flow It is nowadays taken for granted that migration is not the same as gene flow (Ehrlich et al. 1975), but this was not the case in the 1960s. Detailed research on *E. editha* showed why the observed migration rate often greatly overestimates gene flow (Harrison 1989; chapter 9). At Jasper Ridge, larvae emerging from eggs laid after the first week of the flight season, which is typical for the offspring of migrating females, have low probability of surviving to diapause (Cushman et al. 1994; see chapter 9). In populations in the Sierra Nevada, gene flow is curtailed because of differences in phenology among nearby populations using different host plants (Boughton 2000). In the case of males, migration does not easily lead to gene flow because females mate only once, or rarely twice, and typically soon after eclosion (chapter 5). These processes reduce the likelihood that migration will cause gene flow and thereby increase the roles of local adaptation and genetic drift in the evolutionary dynamics of local populations. It may not be a coincidence that local host plant adaptation is widespread in the more stable checkerspot populations (chapter 6). In contrast, in metapopulations consisting of many unstable local populations, the relative strengths of migration and gene flow may be reversed, as shown by recent findings for *Melitaea cinxia* (Saccheri et al. 1998) and other taxa (Ebert et al. 2002). An immigrant may have an enhanced reproductive success when it arrives to a population suffering from inbreeding depression (Ingvarsson and Whitlock 2000, Saccheri and Brakefield 2002). In this case a direct estimate of migration rate gives an underestimate of gene flow.

Life-History Strategies Influence Population Dynamics Life-history strategies, including age-specific reproductive and survival schedules as well as mating habits, form part of the mechanistic basis for a population's ecological dynamics and interact with its evolutionary dynamics. Labine's (1964, 1966a) work on fecundity, mating strategies, and sperm precedence in *E. editha* was an important early contribution to the study of sexual selection, apart from establishing the basic understanding of the reproductive biology of *E. editha* (chapter 5). This early work allowed proper interpretation of the relationship between migration and gene flow, as outlined above. More recent work (Boggs 1990, 1997a, 1997b, O'Brien et al. 2004) has been aimed at understanding patterns of resource allocation to connect adult feeding, paternal investment, and ovarian dynamics first to life history strategies and thence to population dynamics. Such work underscores once again that fundamental understanding of population processes requires a proper understanding of processes at the level of individuals.

Both Exogenous and Endogenous Factors Influence Temporal Population Dynamics, and Their Roles May Vary Even in Adjacent Local Populations One benefit of combining short-term mechanistic studies and long-term observational studies is the opportunity to determine the relative roles of different processes in the dynamics of populations. The debate over density-dependent versus density-independent population dynamics has been one of the most contentious issues in population ecology (Sinclair 1989, Hanski 1990, Turchin 1995, 2002). A resolution to this debate has been elusive because both long-term data and mechanistic understanding of the individual processes for select taxa have generally been lacking. McLaughlin et al. (2002b) used results of field studies on larval survival and plant phenology to predict the dynamics of the two more persistent populations of *E. editha* at Jasper Ridge. They then tested the predictions using nonlinear modeling of data on adult abundances. Models included endogenous factors, exogenous weather effects, or both. Results showed that the two populations, though located close to each other, differed in their responses to the two kinds of factors (McLaughlin et al. 2002b). The population in the more homogeneous habitat patch varied more widely, fluctuated more severely with climate, and went extinct first. The dynamics of the population occupying the topographically diverse habitat patch were more complex, containing damped oscillations and weaker influence of weather (figure 12.2; chapter 3).

Evolutionary Dynamics

One of the most fundamental evolutionary questions concerns the degree of ecological specialization, an area of research in which studies on herbivorous insects have played a key role (Futuyma and Moreno 1988, Jaenike 1990, Thompson 1994, Wahlberg 2001b). Among herbivorous insects, butterflies are particularly useful for experimental studies of the fitness consequences of host choice because, in contrast to many beetles, aphids, and *Drosophila*, adult butterflies do not usually feed on the larval resources. Female butterflies choose oviposition plants exclusively for their offspring to eat, thereby simplifying decision making both for themselves and for the biologists attempting to dissect the adaptive significance of their behavior. These tasks are further simplified in checkerspots, where natural selection on host use can be measured directly by inducing butterflies to oviposit in the field on experimentally chosen plants, including plants that they would not normally accept (chapter 6). The discrete population structure of checkerspots facilitates adaptation to local plant communities, generating intricate spatial patterns of diet (figure 12.3). This apparent complexity actually simplifies our attempts to elucidate the evolution of specialization. Dispersive species with diffuse population structure are under natural selection to be adapted to some unspecified and often unspecifiable set of hosts and habitats that they might possibly encoun-

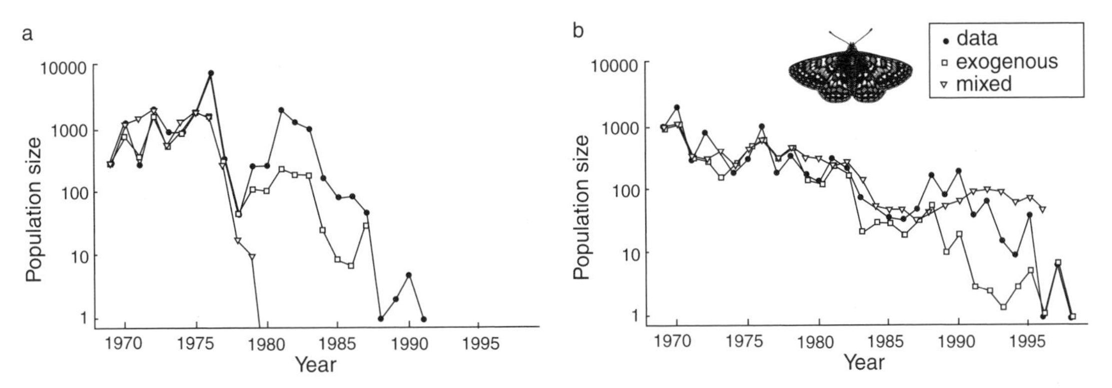

Figure 12.2. Observed and simulated trajectories of numbers of males in populations of *Euphydryas editha* at Jasper Ridge (McLaughlin et al. 2002b). The statistical models on which the simulations were based are described in chapter 3 (box 3.1; figure 3.15). Observed abundances are plotted as filled circles, simulated values are plotted as open squares (exogenous models) and triangles (mixed endogenous–exogenous models). Results for (a) population C (based on models in figure 3.15a, c) and (b) population H (figure 3.15b, d).

ter. The nature and extent of such adaptation is, in practice, untestable. In contrast, we can understand the adaptation (or maladaptation) conferred by host specialization of a checkerspot population in terms of a combination of selection acting locally and selection acting in some nearby source populations that can be found and examined (plus, of course, the usual suspects, drift and historical constraints; Singer et al. 1992a, 1993, 1994, Singer and Thomas 1996).

Diet Evolution in Single Populations Anthropogenic changes in butterfly habitat can trigger rapid evolution of host preference. Changes in the beak size of Galapagos finches in response to weather events (Grant and Grant 1993) is a classic example of rapid evolution of resource utilization in the field. An equally impressive example is provided by changes in the oviposition preference of *E. editha* in response to anthropogenic habitat alteration (Singer et al. 1993, Singer and Thomas 1996, C. D. Thomas et al. 1996). Field estimates of selection on diet and heritability of oviposition preference predicted that rapid evolution of checkerspot diet should occur—and, indeed it does (figure 6.9). In two well-studied examples, humans have triggered such bouts of diet evolution, in one case by introducing an exotic host and in another by "improving" a native plant from the insects' perspective (chapter 6). In each case a dramatic increase in the strength of preference for the novel host took place over < 10 insect generations. Genetic changes leading to dissimilar preference occurred in the absence of mating barriers between butterflies using different hosts, and the preference changes did not appear to constitute incipient speciation. In fact, analyses of mtDNA sequences suggest that no host shifts accompanied the speciation event between *E. editha* and *E. chalcedona*, whereas a plethora of host shifts have occurred within *E. editha* itself, including repeated reversals of preference (Radtkey and Singer 1995). Evidently, checkerspots are not en route to rival *Enchenopa* (Wood and Tilmon 1999), *Rhagoletis* (Feder 1998), pea aphids (Via 1999, Hawthorne and Via 2001), or larch budmoths (Emelianov et al. 2001) as models for speciation triggered by host shifts. In contrast, they present the best-known example to date of a system in which host shifts occur readily with minimal effects on gene flow and in which the anatomy of these host shifts is available for contemporary observation and experiment, rather than indirectly inducible from phylogenetic analysis.

Diet Evolution in Multiple Populations: Roles of Spatial Scale and Metapopulation Structure Natural selection on host use varies among local populations in metapopulations as well as among isolated populations. Spatial genetic variation in oviposition preference in *E. editha* has at least three causes (Singer and Thomas 1996). Preference variation among two habitat patch types with different host plants arises because butterfly movements among the patches are influenced by their genetically determined preference (chapter 9). Second, variation in preference

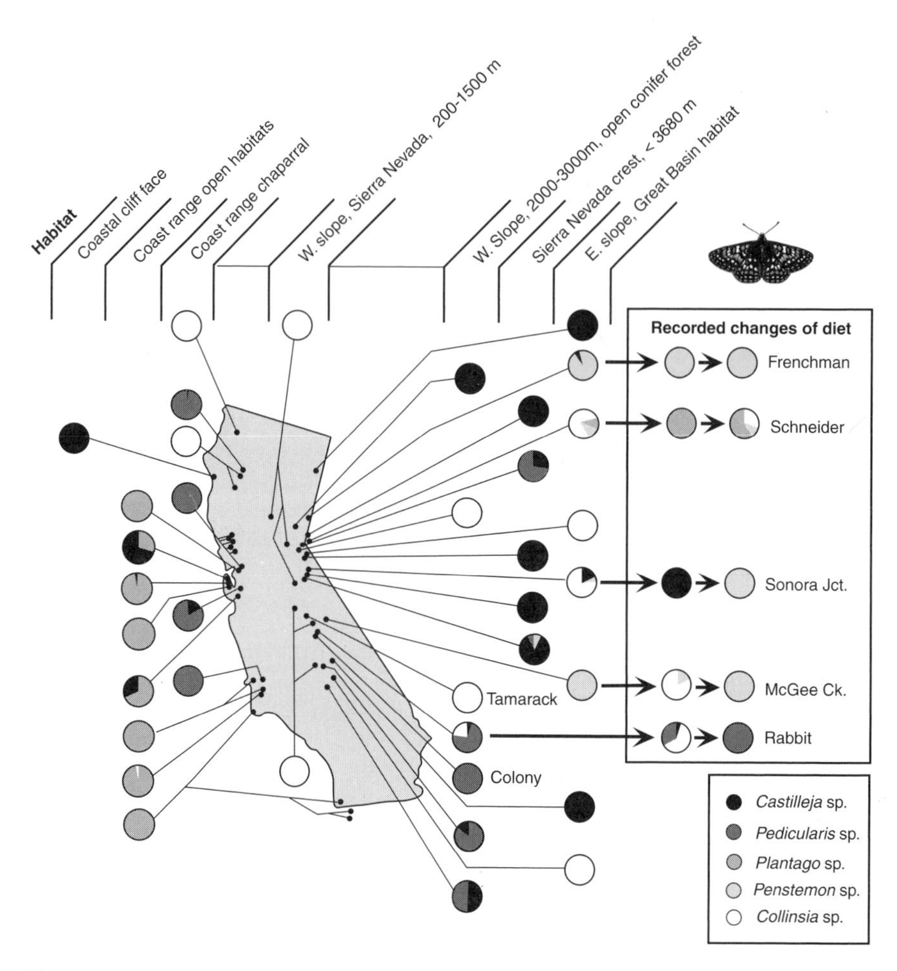

Figure 12.3. Proportions of eggs and/or prediapause larvae found on each host genus at different sites in California. Arrows connect observations made in different years at the same site. For the names of the sites, see figure 3.1.

among populations with the same host plants may be caused by dissimilar connectivity to large source populations in which another host plant is used. In this case, gene flow among populations may drive evolution against the direction of local natural selection. Third, a preference difference between (meta)populations located only a few kilometers apart may be driven by a difference in natural selection. Still greater complexity existed in a metapopulation with two patch types that demonstrated a reversible source–sink relationship with ecological and evolutionary dynamics intertwined (chapter 9).

Genetic Differentiation across Populations, Ecotypes, and Species The findings concerning host plant adaptations in checkerspots combined with other ecological, phenotypic and genetic studies demonstrate that variability in many characters crosses taxonomic and population boundaries (Ehrlich 1961a, McKechnie et al. 1975; chapter 10); genetic differentiation is continuous. The demonstration that some structural gene loci in *Euphydryas editha* are under selection while some are neutral (McKechnie et al. 1975, Mueller et al. 1985, Baughman et al. 1990a) helped clarify the relative roles of natural selection versus genetic drift in shaping the allele frequencies in natural populations. As detailed in chapter 10, the selectionist/neutralist debate has been resistant to resolution, but progress is now being made in developing theory to predict situations and loci for which selection or neutrality should dominate (Watt 1992, 1994, Mitton 1997, Watt and Dean 2000).

12.3 *Melitaea cinxia*: A Model System for the Classic Metapopulation Concept

Suitable habitat for *E. editha* and other checkerspots often occurs in discrete patches, which means that local populations are easy to delimit (Ehrlich 1961b). The *E. editha* study system stimulated the early development of the metapopulation concept, though because it occupied a relatively small set of habitat patches, often isolated by considerable distance (figure 3.2), it was of limited use for the study of metapopulation biology. In contrast, the *Melitaea cinxia* metapopulation in the Åland Islands in southwestern Finland occupies a huge network of thousands of small meadows (figure 4.3). This system fits the classic metapopulation concept very closely (Hanski et al. 1995b, Hanski 1999b). Indeed, the project was started in 1991 for the very purpose of testing the predictions of metapopulation models (Gyllenberg and Hanski 1992, Hanski and Gyllenberg 1993). The initial aim was to find a study system with multiple metapopulations living in multiple patch networks to investigate the possibility of alternative stable states in metapopulation dynamics. In hindsight, this was perhaps an overly ambitious aim for a new project, but after several years of work the conclusion was reached that the dynamics of *M. cinxia* metapopulations do indeed exhibit alternative stable states (Hanski et al. 1995b).

The large *Melitaea cinxia* metapopulation in the Åland Islands functions as an effective model system because several features in the landscape and in the natural history of the butterfly greatly facilitate empirical study. In terms of the landscape, the suitable habitat mostly occurs as discrete, small, dry meadows; there is easy access to most of the some 4000 meadows; and the meadows are naturally clustered into "semi-independent patch networks" around small villages (plate XII; Hanski et al. 1996c), which vary in their properties: number of patches, distribution of patch sizes, and connectivities (chapter 4). The butterfly has gregarious larvae that spin a conspicuous web (plate IX), which greatly helps in the annual survey of the presence or absence and the sizes of local populations. As with other butterflies, it is easy to mark *M. cinxia* and, luckily, its rather slow flight makes it fairly easy to recapture. Only the short-lasting pupal stage is cryptic and difficult to find in the field.

An essential element of the metapopulation concept is that spatially localized interactions of individuals lead to more or less asynchronous local dynamics. More specifically, the classic metapopulation consists of discrete local populations, all of which have a substantial and independent risk of local extinction. The long-term persistence of a species with such a spatial population structure is possible only at the regional level, in a balance between local extinctions and establishment of new local populations. The basic concept was established by Levins (1969, 1970) and formalized in his well-known metapopulation model, which assumes an infinitely large network of identical habitat patches, all of which are equally connected to existing local populations. This model is clearly greatly simplified—real patch networks have a finite number of dissimilar patches. The evident conflict between the overly simple Levins model and reality has led to doubts about the significance of the classic metapopulation concept for ecology and conservation (Harrison 1991, 1994). Perhaps the most significant contribution that the *M. cinxia* project has made to population biology is to facilitate the development of metapopulation theory that can readily be applied to real metapopulations. The long-term empirical study has both stimulated new modeling ideas and constantly provided opportunities to test model assumptions and predictions with abundant and biologically rich data.

Spatially Realistic Metapopulation Theory

The purpose of the spatially realistic metapopulation theory (SMT; Hanski 1998b, 1999b, 2001, Ovaskainen and Hanski 2001, Hanski and Ovaskainen 2003) is to add a realistic description of landscape structure to the classic model. Thus, while the Levins model and related models assume an infinitely large network of identical patches, SMT assumes a finite network of dissimilar patches (box 12.1). The simplifying assumption of the Levins model that is still retained is the presence/absence description of local populations (hence the name "patch occupancy models"). SMT contributes to the unification of research in population ecology in several ways (Hanski 2001, Hanski and Ovaskainen 2003). First, it may be viewed as representing a synthesis of Levins's metapopulation concept and MacArthur and Wilson's (1963, 1967) dynamic theory of island biogeography (Hanski 2001). Second, the theory helps bring

BOX 12.1 Spatially Realistic Metapopulation Ttheory

Consider a network of n habitat patches with areas A_i and spatial locations x_i. Let $p_i(t)$ denote the probability that patch i is occupied at time t. The rate of change in $p_i(t)$ is given by (Hanski and Ovaskainen 2000, Ovaskainen and Hanski 2001):

$$dp_i/dt = (\text{colonization rate})_i\,(1 - p_i) - (\text{extinction rate})_i\,p_i\,. \quad (12.1)$$

This model consists of n equations, which are dynamically coupled, as the colonization rate of patch i depends on the occupancy states (here probabilities) of the other patches. Equation 12.1 represents a skeleton to which one may add specific assumptions about patch-specific colonization and extinction rates and how they depend on landscape structure to derive a spatially realistic metapopulation model.

A wide range of assumptions about the colonization and extinction rates are possible and would be appropriate for different biological situations. Hanski and Ovaskainen (2000) use the following simple assumptions. First, $(\text{extinction rate})_i = e/A_i$, where e is a parameter. This assumption is justified by large patches tending to have large local populations, and the extinction risk roughly scaling as the inverse of expected population size in taxa affected by moderate environmental stochasticity (Lande 1993). Second, $(\text{colonization rate})_i = c\Sigma_{j \neq i} \exp(-\alpha d_{ij})\,A_j\,p_j\,(t)$, where $1/\alpha$ is the average migration distance, d_{ij} is the distance between patches i and j, and c is a parameter. The colonization rate is thus a sum of contributions from all patches surrounding the focal patch, weighted by their areas (surrogates for population sizes), distances from the focal patch, and their probabilities of being occupied. More refined biology can be captured, at the cost of a few extra parameters, by assuming that extinction, immigration, and emigration rates scale by patch areas as power functions (Ovaskainen and Hanski 2001, Ovaskainen 2002b).

The fundamental idea of this model is that the key processes of classic metapopulation dynamics, colonization and extinction, are related to the key structural features of fragmented landscapes, the areas, qualities, and spatial locations of the habitat patches. The theory thus retains patch-specific information about the landscape and makes patch-specific predictions about the occurrence of the species. Although the model is complex in the sense that there are n coupled equations for a patch network with n patches, several useful results can be obtained through mathematical analysis (Ovaskainen and Hanski 2001, 2002, 2003). Most important, the condition for metapopulation persistence in the habitat patch network is given by

$$\lambda_M > \delta \quad (12.2)$$

where λ_M is the leading eigenvalue of a 'landscape' matrix, which is constructed with assumptions about how habitat patch areas and connectivities influence extinction and colonization rates (Hanski and Ovaskainen 2000, Ovaskainen and Hanski 2001), and δ is defined as the ratio of the extinction and colonization rate parameters, $\delta = e/c$. Note that this condition shows explicitly how the ability of a species to persist in a fragmented landscape depends on both the properties of the species (δ) and the properties of the landscape (λ_M). Hanski and Ovaskainen (2000) coined the term "metapopulation capacity" for λ_M, which is a measure of landscape structure summarizing the influence of both the amount and spatial configuration of habitat on metapopulation dynamics, including, as expression 12.2 shows, the condition for metapopulation persistence.

The essential dynamic behavior of the n-dimensional model is well described by a 1-dimensional approximation, which in the case of the specific assumptions about

(*continued*)

extinction and colonization rates described above leads to (Ovaskainen and Hanski 2002b):

$$dp_\lambda/dt = \tilde{c}p_\lambda\,(1-p_\lambda) - \tilde{e}p_\lambda \qquad (12.3)$$

which is the familiar Levins model with the colonization and extinction parameters interpreted as $\tilde{c} = c\lambda_M/\omega$ and $\tilde{e} = e/\omega$. ω is defined as the weighed average of patch areas, the weights being given by patch values, which are obtained as elements of the leading eigenvectors of the landscape matrix (Ovaskainen and Hanski 2001). Metapopulation size p_λ is measured by the weighted average of the patch-specific probabilities of occupancy, the weights again being the patch values. The equilibrium metapopulation size as solved from equation (12.3) is

$$p_\lambda{}^* = 1 - \delta/\lambda_M. \qquad (12.4)$$

A mathematically rigorous and general formulation of both the deterministic (as outlined here) and stochastic theory and their ramifications is presented in Ovaskainen and Hanski (2001, 2002, 2003) and Ovaskainen (2002a, 2002b). For an ecologically oriented discussion, see Hanski (2001).

spatial ecology and landscape ecology closer together (Hanski and Ovaskainen 2003). Population ecologists interested in spatial dynamics have often ignored the landscape structure to facilitate development of population theory (e.g., Dieckmann et al. 2000), whereas landscape ecologists have sacrificed population theory for detailed description of landscape structure (Turner et al. 2001). SMT strikes a compromise between these conflicting demands (Hanski 1998b). Third, SMT is closely related mathematically to matrix population models (Caswell 2000), the metapopulation being structured by the distribution of individuals among the finite set of habitat patches. Finally, SMT has highlighted new similarities between metapopulation models and epidemiological models (Ovaskainen and Grenfell 2003).

The foundation of the theory is the assumption that the processes of classic metapopulation dynamics, local extinction and recolonization of currently empty patches, can be related to the area, quality, and spatial location of the habitat patches. This assumption is amply demonstrated by results on *M. cinxia* (chapter 4) and many other species (Hanski 1994a, 1999b). Rather than modeling the proportion of occupied patches as in the Levins model, the dynamic variable in SMT is a vector of patch occupancy probabilities (in stochastic models, the vector consists of actual occupancy states of the patches). Initially, modeling was based on numerical simulations, like many other approaches to metapopulation modeling, but with the advantage that model parameters could be estimated statistically from empirical data (Hanski 1994b). Subsequently, the methods of parameter estimation have been greatly improved (Moilanen and Hanski 1998, ter Braak et al. 1998, Moilanen 1999, 2000, O'Hara et al. 2002), and a rigorous mathematical theory has been developed (Hanski and Ovaskainen 2000, 2003, Ovaskainen and Hanski 2001, 2002, 2003, Ovaskainen 2002a). Combining modeling with large-scale empirical work in the model system context has helped clarify many fundamental issues about classic metapopulation dynamics.

Fundamental Metapopulation Processes

Below we highlight five key issues in classic metapopulation dynamics and the contributions that research on *M. cinxia* has made toward a better understanding of these issues.

Metapopulations Persist in Fragmented Landscapes in Spite of Unstable Local Dynamics The large habitat patch network in the Åland Islands consists of about 4000 meadows, about 10% of which are occupied by *M. cinxia* in any one year. The extinction rate of local populations is so high that of the 500 local populations recorded in 1993, only 36 survived until 2001 (chapter 4; figure 4.10). There is reason to believe that all the populations that survived for the longest time did so just by chance (M. Nieminen, pers. comm.). In contrast, of the 72 patch networks that were occupied in 1993, half were continuously occupied until 2001 (figure 4.10), demonstrating the much greater stability of occurrence at the patch network level than at the patch level. A successful experimental establishment of a *M. cinxia* metapopulation in the island of Sottunga (figure 2.3) with some 20 small habitat patches represents a clear-cut example of metapopulation-level persistence (see figure 13.7). This experimental metapopulation has persisted until present, though local extinction rate is so high that the entire system has turned over more than once in 12 years. In short, research on *M. cinxia* has provided a prime empirical example of regional persistence of a species as a classic metapopulation through the balance between extinction and recolonization.

An Extinction Threshold Exists for Metapopulation Persistence For a metapopulation to persist, a single local population in an otherwise empty patch network has to establish at least one other local population during its lifetime; otherwise the metapopulation will decline to extinction. A similar threshold condition, typically expressed as the requirement that the basic reproductive ratio, R_0, has to be > 1, is fundamental in population dynamics in general. The particular feature of the spatially realistic metapopulation theory is that the threshold condition is explicitly expressed in terms of the properties of the landscape as well as the properties of the species (box 12.1). Thus, the threshold condition requires that the habitat patches are "sufficiently" large and that the density of patches is "sufficiently" high, because otherwise extinction rate (in small patches) would be "too" high, and the recolonization rate (of isolated patches) would be "too" low to allow long-term persistence. The theory quantifies in a single number, called the metapopulation capacity of the fragmented landscape, what is "sufficient" and what is "too" high or low (box 12.1). The metapopulation capacity increases with the amount of habitat in the landscape, but it also takes into account the spatial configuration of the habitat on metapopulation processes. The metapopulation capacity thus measures how favorable a fragmented landscape is for the focal species. Figure 12.4 shows the relationship between metapopulation size and metapopulation capacity in *M. cinxia* metapopulations in the Åland Islands, with clear evidence for an extinction threshold. Note that each data point in figure 12.4 refers to one habitat patch network (chapter 4). This example also illustrates the practical value of the theory: the different fragmented landscapes (networks) that are compared in figure 12.4 might as well be different managed landscapes, and the theory allows one to quantify the population dynamic consequences of habitat loss, habitat restoration, or any other form of management that would alter the landscape structure for a focal species.

Emigration and Losses Due to Mortality in Population Dynamics Must Be Distinguished Migration and gene flow are key processes in metapopulation dynamics, but unfortunately their empirical study is complicated by many difficulties. For instance, it is generally difficult to distinguish between emigration

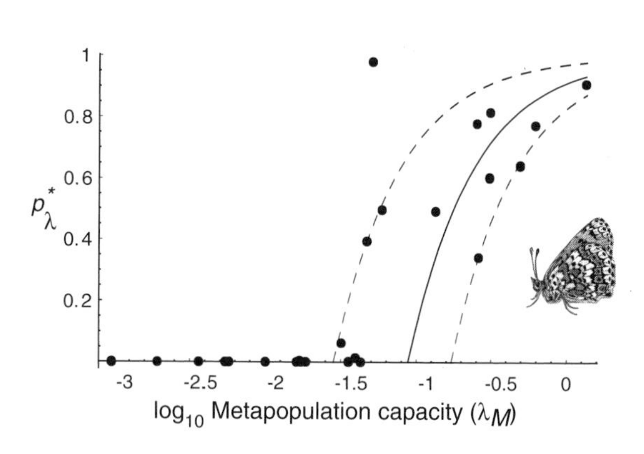

Figure 12.4. Plot of metapopulation size (p_λ*; see box 12.1) against the logarithm of metapopulation capacity (λ_M) in 25 real habitat patch networks that are potentially occupied by *Melitaea cinxia* in the Åland Islands. The value of p_λ^* was calculated based on patch areas, spatial locations, and the occurrence of the butterfly in the patches in 1993. For each network with $p_\lambda^* > 0.3$, the threshold value for persistence was calculated using the formula $\delta = \lambda_M (1 - p_\lambda^*)$. The continuous line is based on the average of the estimated Δ values, the broken lines give the minimum and maximum estimates omitting the two networks yielding the most extreme value. From Hanski and Ovaskainen (2000).

and mortality, to estimate the level of mortality during migration, and to estimate the demographic and genetic contributions that immigrants make to local dynamics (the rescue effect). Emulating the success of metapopulation models based on the effects of patch area and connectivity on local extinction and colonization, Hanski et al. (2000) developed an analogous statistical model for individual movements in metapopulations. In this model, emigration and immigration rates are assumed to be power functions of patch area, and while survival within habitat patches is assumed to be constant, survival during migration is assumed to be an increasing function of the connectivity of the source patch to other habitat patches in the network. With the latter assumption, one may tease apart mortality during migration and mortality within habitat. One may also calculate quantities such as the numbers of individual-days spent by migrants from patch i in patch j, which is potentially closely related to gene flow (though see the caveats in section 12.2). The model has been applied to mark–release–recapture data from studies on checkerspot butterflies (Hanski et al. 2000, Petit et al. 2001, Wahlberg et al. 2002b; chapter 11). It is worth stressing that the idea for this model of individual movement behavior in fragmented landscapes would not have emerged without the combination of our experience with mark–release–recapture work on checkerspot butterflies and building spatially realistic metapopulation models.

Many Causes Contribute to Extinctions of Local Populations in Metapopulations As one could have expected, the most consistent predictor of local extinction risk of *M. cinxia* populations has been the size of the population in the previous year (Hanski 1999b; chapter 4): small populations have a high risk of extinction. But the long-term project has produced evidence for a multitude of mechanisms influencing the risk of extinction. Both demographic and environmental stochasticity increase the risk of extinction, immigration reduces it (the rescue effect), and inversely density-dependent emigration (chapters 4 and 9) and inbreeding depression (chapter 10) and parasitism (chapter 8) increase the extinction risk. Additionally, extinction risk is related to regional changes in population density, which are driven by correlated environmental effects (chapter 4). The list of causes and mechanisms of local extinction detected in *M. cinxia* practically exhausts the plausible processes that could cause population extinction (Caughley 1994). Examining any one of these mechanisms in isolation would run the risk of misleading conclusions, and it is apparent that the model system approach has produced extra value to the study of local extinction. It may also be added that recurrent extinctions in metapopulations allow a wider range of processes to influence extinction than could be expected in isolated populations, including emigration increasing and immigration decreasing extinction risk and specialized natural enemies causing local prey extinctions (Hanski 1998b).

Interspecific Interactions Influence Metapopulation Dynamics Much current metapopulation biology is focused on single species, whereas all species interact with other species. These interactions may influence the metapopulation dynamics of the focal species and, vice versa, the metapopulation dynamics of interacting species may significantly influence the outcome of their interaction at a large spatial scale. *Melitaea cinxia* interacts with several other species in the Åland Islands, including two larval host plants (chapter 4) and two primary larval parasitoids (chapter 8). The genetically determined host–plant oviposition preference varies at the scale of tens of square kilometers with the relative abundances of the two host plant species, *Veronica spicata* and *Plantago lanceolata* (Kuussaari et al. 2000). Spatial variation in this key life-history trait could have been demonstrated even in a short-term study, but having the long-term and large-scale project in place has allowed much more detailed analyses (chapter 9). Of the two primary parasitoids, the dynamics of the braconid *Cotesia melitaearum* are strongly coupled with the host butterfly, to the extent that the parasitoid occasionally increases the risk of extinction of host populations (Lei and Hanski 1997). On the other hand, this parasitoid has such a limited colonization ability (van Nouhuys and Hanski 2002a; chapter 8) that it is liable to go regionally extinct where the host density is low. Another interesting feature of the system is that the extinction–colonization dynamics of the parasitoid are influenced by the plant use of its host (van Nouhuys and Hanski 2000). These large-scale interactions of *M. cinxia* and its host plants and natural enemies and the interplay between local and regional processes are described in greater detail in other chapters in this volume. Here we recapitulate the point that the large amount of detailed spatial information about the study system has allowed a more comprehensive understanding of the interspecific interactions than would have been possible to achieve in a short-term and small-scale study.

Integrating Ecological, Genetic, and Evolutionary Processes

The counterpart of Levins's metapopulation model in population genetic theory is the island model of Wright (1931). One could argue that including elements of landscape structure in the genetic theory of metapopulations would bring advantages comparable to those obtained from including landscape structure into the ecological theory. Research in this direction is only beginning to appear (Whitlock and Barton 1997). The *M. cinxia* model system has provided three significant examples of the genetic and evolutionary consequences of highly fragmented landscape structure and of the value of incorporating in the analyses information about the spatial configuration of the habitat. These studies have started to integrate ecological, genetic, and evolutionary questions in the metapopulation context.

Inbreeding Depression May Increase Local Extinction Risk Empirical evidence for inbreeding depression in captive animal populations is widespread (Ralls et al. 1988, Thornhill 1993), and studies on plant populations have frequently reported reduced fitness in small populations (Ellstrand and Elam 1993, Ouborg and Vantreuren 1994). However, only a few field studies have attempted to test directly the hypothesis that inbreeding depression increases the risk of extinction of natural populations (Jiménez et al. 1994, Keller et al. 1994, Newman and Pilson 1997, Haag et al. 2002), and one of the most comprehensive examples comes from *M. cinxia* in the Åland Islands. Laboratory results have demonstrated that just a single round of brother–sister matings, which occur commonly in newly established and small local populations in the field, have significant consequences for several fitness components, including egg-hatching rate (Haikola et al. 2001). Observational (Saccheri et al. 1998) and experimental (Nieminen et al. 2001) field studies have shown that inbreeding depression increases the risk of extinction of small local populations. The extensive database on population sizes and other ecological factors in several hundred local populations for many years allowed Saccheri et al. (1998) to control for the influence of these ecological factors on extinction rate. Once again, the fact that the inbreeding studies have been conducted in the model system context has allowed a much more penetrating analysis than would have been possible otherwise (chapter 10).

Metapopulation Dynamics Influence the Evolution of Host Plant Use Classic metapopulations are characterized by a high rate of population turnover, which opens up the possibility that extinctions and colonizations are influenced by some genetically determined traits and lead to natural selection at the level of local populations. One life-history trait of particular significance in the case of herbivorous insects is the host-plant oviposition preference. Studies on *M. cinxia* have shown that there is spatial variation among populations and metapopulations in the oviposition preference (Kuussaari et al. 2000). Hanski and Singer (2001) showed that the colonization rate of currently empty patches is influenced by the match between the oviposition preference of immigrant females and the host plant composition of the focal patches. Given such biased colonizations, one might expect that extinction–colonization dynamics would play some role in the evolution of oviposition preference in metapopulations. This indeed appears to be the case (Hanski and Heino 2003), as a model of the evolution of oviposition preference involving the biased colonizations explains the observed spatial variation in host use better than the availability of host plants alone. This is an exciting finding because it indicates that the ecological extinction–colonization dynamics influence the evolutionary dynamics of oviposition preference and vice versa. Another implication is that landscape structure affecting spatial dynamics may influence the evolution of life-history traits, underscoring the value of the spatially realistic approach to metapopulation biology more generally.

Evolution of Migration Rate in Metapopulations The evolution of migration rate is a classic topic in evolutionary ecology (van Valen 1971, Hamilton and May 1977, Comins et al. 1980, Hastings 1983; for a review, see Clobert et al. 2001). Research on *Melitaea cinxia* has contributed much empirical information (chapter 9) and also stimulated novel modeling work. Heino and Hanski (2001) constructed an individual-based model of the evolution of migration rate in metapopulations, with the checkerspot system as their primary example. Two essential building blocks in their model are a spatially realistic metapopulation model, which drives the metapopulation-level dynamics, and the statistical model of individual movement behavior in metapopulations (Hanski et al. 2000) discussed above. Thereby the evolutionary model is also "spatially realistic"; that is, it allows the investigation of how the spatial configuration of the landscape influences the evolution of migration rate. Indeed, a generic prediction of the model is that the evolutionarily stable migration rate is likely to vary both in space and in time in response to the structure of the landscape. The model was parameterized for *Melitaea diamina*, and it predicted the observed migration rate of this species in its naturally fragmented landscape quite well (Heino and Hanski 2001). The theoretical work has in turn stimulated new field experiments on *M. cinxia*. One such experiment showed that females originating from newly established local populations are more dispersive than females from old populations (Hanski et al. 2002), supporting the theoretical expectation (Olivieri and Gouyon 1997) that colonizations select for increased migration rate in metapopulations with fast population turnover.

12.4 Integration of the Two Butterfly Model Systems

The two long-term butterfly projects have efficiently though unintentionally complemented each other, to a large extent because of the phylogenetic relatedness of the species, effectively making the checkerspots a population biological model system. Research on *E. editha* was originally initiated to elucidate both the concept and the dynamics of local populations as distinct biological entities, whereas the *M. cinxia* study system was selected for intensive research as a tractable example of a classic metapopulation, an assemblage of many local populations.

Local population and metapopulation are twin concepts; one does not exist without the other. The difference in the spatial structure of the two model systems is quantitative rather than qualitative. The Jasper Ridge *E. editha* system consisted of three local populations that used to be quite large and appeared to have a long time to extinction. They are now all extinct, following a long period of high-amplitude fluctuations and steadily declining average numbers, apparently caused by climate change (McLaughlin et al. 2002a; chapter 3). The *M. cinxia* metapopulation in the Åland Islands consists of several hundred local populations, which are all so small that they have only a short expected lifetime, but the metapopulation as a whole has remained relatively stable (chapter 4). The common feature in the two systems is the primary role that the landscape structure plays in shaping population structure and dynamics.

The two species are also biologically similar though by no means identical. The subspecies *E. editha bayensis* is particularly sedentary, and its biology further constrains gene flow among populations (Ehrlich et al. 1975). *Melitaea cinxia* is more mobile than *E. editha bayensis*, though recent modeling (Heino and Hanski 2001) and empirical results (Hanski et al. 2002) suggest that selection has modified migration rate in *M. cinxia* among landscapes with a different degree of fragmentation. Comparing and contrasting results for these two species is an effective way of testing and reinforcing conclusions reached for each.

One of the key strengths of the checkerspot systems for our research is the often discrete nature of the habitat patches occupied by local butterfly populations. The ease with which habitat patches can be delimited in the field led the patch area to play a key role in our analyses (Hanski 1994b, 1999b, Hanski et al. 2000, Moilanen and Cabeza 2002, Ovaskainen 2002b). There is a potential danger here, however. The impression may be created that the habitat patches are more discrete and well delimited than they actually are. Boundaries are truly evident in some cases, such as in *E. editha* at Jasper Ridge where habitat suitability is largely determined by abrupt changes in soil type (chapter 3). In Åland, many suitable meadows for *M. cinxia* border on forests or cultivated fields (plate XII). But in other cases the definition of the exact patch boundary is a matter of judgment, and the boundaries may occasionally change from one year to the next (chapters 3 and 4). Other butterfly biologists have reacted to the strong focus on the spatial configuration of habitat by emphasizing spatially varying habitat quality (J. A. Thomas et al. 2001, Fleishman et al. 2002b). The tension between the need to simplify for the purpose of modeling and analysis and the wish to keep the full "reality" in the picture is inevitable and is part of healthy progress in ecology. Incidentally, much of our own research on checkerspots is based on quantifying spatial variation in habitat quality, including long-term research on *E. editha* in the large Morgan Hill population near Stanford (chapter 3) and the studies of its source–sink dynamics in the Sierra Nevada mountains in California (C. D. Thomas et al. 1996, Boughton 1999, 2000; chapter 9).

One important similarity in the two systems that should be emphasized is the role of host plant preference and use in the spatial dynamics of the butterfly. Both butterfly species are locally specialized to use only one or a few host plant species, though they use a larger number of plant species across their geographic ranges (White and Singer 1974). In the Sierra Nevada, numerous host colonizations and host switches have occurred due to plant invasions and habitat changes (Singer et al. 1993, C. D. Thomas et al. 1996; section 12.2). In Finland, the spatial dynamics of *M. cinxia* are influenced by the correspondence between the oviposition preference of migrant females and the local host plant availabilities (Hanski and Singer 2001; chapter 9). The facility of local adaptation, the dynamic coupling between host plant use, spatial dynamics, and local population size, and the fact that mating in checkerspot species does not typically occur near the larval host plants (chapter 5) probably all help explain why local adaptation in host plant use and preference has not played an important role in speciation in this group of insects (chapter 2).

There are also differences in the structure and dynamics of the two butterfly systems, though the differences tend to be quantitative rather than qualitative. One such difference relates to the generally more marked multi-annual climatic fluctuations in California than in Finland. Generally, colonizations and extinctions of *E. editha* populations in California occur randomly in time but with high correlation in space. This is because the dynamics are largely driven by weather conditions that affect many populations simultaneously at a large spatial scale. Thus, in the Morgan Hill system studied by Harrison et al. (1988), the 1976–77 drought caused the extinction of surrounding small populations, and widespread extinctions occurred in other California populations during the same period (Ehrlich et al. 1980; figure 3.11). Colonization events are also likely to be highly synchronous because colonizations often arise from large, common source populations at times when weather conditions are most favorable for migration and population establishment. Recolonization of the surrounding habitat patches in the Morgan Hill system occurred when the Kirby Canyon "mainland" population was particularly large (Harrison et al. 1988). The situation is not fundamentally different in the *M. cinxia* metapopulation in Finland, with large-scale correlation in year-to-year population changes being driven by the prevailing weather conditions (chapter 4).

12.5 Lessons for Population Biological Research

Populations interact along three dimensions. First, the size and the structure of populations fluctuate and evolve through time according to the familiar mechanisms of population dynamics and natural selection. Second, local populations interact via migration and gene flow through space. And third, populations interact with populations of other species in webs of direct and indirect interactions. It would be foolish to try to analyze everything together in a single short-term project, yet the biology of most natural populations is influenced by all these interactions all the time. Here is the dilemma: How do we avoid reaching oversimplified or even misleading conclusions in studies focused on a particular dimension in the three-dimensional space of population biology? Our solution, the model system approach, allows all three dimensions to be explored around one focal taxon. This approach raises several fundamental questions about how population biological research is conducted in practice.

First, the large amount of diverse information available for a model system provides a helpful context for a particular study; that is, it provides a picture of the species in the three-dimensional space. As an example, Saccheri et al. (1998), in their study of inbreeding and local extinction in *M. cinxia*, were able to take into account the ecological factors influencing extinction risk (chapter 4), thereby making the demonstration of the inbreeding effect more convincing (chapter 10). Similarly, Kuussaari et al. (2000) were able to demonstrate that the use of the two host plant species in local populations of *M. cinxia* is influenced by the regional adaptation in terms of host plant preference of the butterfly populations. In *E. editha*, Hellmann et al. (2003) and McLaughlin et al. (2002b) were able to examine how changes in weather conditions affect populations thanks to existing long-term data.

Second, meaningful integration of ecological and genetic research on natural populations is difficult to achieve in short-term projects because of the breadth of information and expertise required. The long-term study of the white campion plant, *Silene alba*, conducted in Virginia by Antonovics and colleagues (1994, 1997, Thrall and Antonovics 1995) is a good example of the benefits of using a system that is well-known ecologically to conduct population genetic research (McCauley 1993, McCauley et al. 1995, C. M. Richards et al. 1999). The checkerspot system described here is similarly positioned. Abundant ecological information is available to support genetic and evolutionary studies.

Third, ecological phenomena are variable in space and time, and arguably it is essential to know the details of this variability for proper interpretation of many specific results. Information about large-scale and long-term variability for many of the best-known taxa is being accumulated in national and international monitoring programs. The new atlas of British butterflies is a prime example of the results of such programs (Asher et al. 2001). Valuable as this information is, there is still a need for more detailed studies at the landscape scale as opposed to the national scale and a need for studies that complement distributional data with other relevant population biological information. In other words, there is a need for model systems. Both the *E. editha* and *M. cinxia* projects have strived to achieve this aim.

Fourth, as it is not possible or efficient for every student and researcher to establish a new model system, there is a need for researchers and funding agencies to find other ways of conducting meaningful research in the model system context. The network of Long-Term Ecological Research sites in the United States is one example, though in this case the focus is on a study area rather than on a focal species. Likewise, the U.S. National Science Foundation has specific mechanisms to support the continuation of long-term data collection on specific populations or communities. Model system studies such as those highlighted in table 12.1 have typically been created and maintained by individual researchers. An interesting question is whether such ecological and evolutionary model systems would benefit from coordination among multiple research teams, as is often the case for laboratory-based models like *Drosophila* and *E. coli*.

Our final comment is about the contrast between laboratory and field model systems in population biology. Mueller and Joshi (2000) presented a spirited argument for the study of laboratory model systems such as their three best examples, *Lucilia* blowflies, *Tribolium* flour beetles, and *Drosophila* fruit flies. Given all the complexities of the natural populations, Mueller and Joshi (2000) ask, "why should we bother to study natural populations at all?" The answer, of course, is that we are ultimately interested in the ecology and evolution of natural populations. Though we admit that there are great advantages of laboratory model systems, including the comprehensive knowledge of the genetic architecture of fitness components in *Drosophila*, we can also list a number of drawbacks. First, the dynamics of laboratory populations are typically greatly influenced by the artificial maintenance regime, and it is not easy to tease apart such effects from the effects of the biological traits of the study organisms. Second, the greatest strength of laboratory systems, the ability to control everything of significance, may also be considered a weakness: the researcher turns the laboratory model system to a biological computer. We could just as well ask why bother; why not use the real computer? Third and most important, given that everyone is ultimately interested in contributing to the knowledge of the biology of organisms in their natural environments, the question arises to what extent the laboratory model systems can be helpful in this respect. Chapter 7 on natural populations in Mueller and Joshi's (2000) otherwise admirable book is not entirely convincing.

The truth, of course, is that we need both laboratory and field model systems. What we especially lack in population biology are model systems that can be studied both in the laboratory and in the field. Unfortunately, checkerspot butterflies do not make ideal laboratory organisms, though much more could be achieved in this respect if sufficient resources were available. Until such a paragon of tractability is found, checkerspots display great strength as one of the more prominent model systems in population biology.

13

Checkerspots and Conservation Biology

ILKKA HANSKI, PAUL R. EHRLICH, MARKO NIEMINEN,
DENNIS D. MURPHY, JESSICA J. HELLMANN, CAROL L. BOGGS,
AND JOHN F. McLAUGHLIN

13.1 Introduction

The days are long gone when we could be happy in simply satisfying our curiosity about the ecology, evolution, and behavior of butterflies. In our view, biologists today have an obligation to determine how their work can contribute to solving the human predicament of declining environmental quality and increasing economic inequity. This obligation is strongest for population biologists. Populations are the basis for species diversity (Ehrlich and Daily 1993), and they provide the ecosystem goods and services that support our civilization (Daily 1997, Hughes et al. 1997). The role of most butterflies in the ecosystems that support them is not such that their extinction would lead to immediate and obvious repercussions. Nonetheless, butterflies do matter because occasionally they may play important roles, for example, as pollinators, and the disruption of those roles might cause long-term cascading effects. Equally important is that many people consider butterflies symbolic of healthy landscapes, and thus the disappearance of butterflies can be readily understood by the public as a signal of more serious environmental degradation.

Local disappearance of many butterfly species in the latter part of the 20th century called attention to population extinctions of butterflies in Europe (e.g., Heath 1981, J. A. Thomas 1984, van Swaay 1990). For example, 19 of the 64 indigenous species in Flanders (northern Belgium) have already gone extinct, and half of the remaining species are threatened (Maes and Van Dyck 2001). The exceptionally informative postage stamp shown in figure 13.1 details the level of threat to butterflies in the Netherlands. At present, 20% of European butterfly species are threatened or near threatened, and perhaps 15% of those in the United States and Canada are similarly imperiled in all or parts of their ranges (section 13.2). It is evident that if current extinction trends continue into the near future, many regions will be left with highly impoverished butterfly communities, which will primarily consist of ubiquitous generalist species that thrive in highly modified landscapes. Checkerspots have not been immune to the processes that have led to widespread erosion of population diversity of butterflies (e.g., Lavery 1993, Warren 1994) and many other taxa.

The disappearance of the large blue butterfly *Maculinea arion* from the United Kingdom in 1979 (J. A. Thomas 1980) vividly illustrates the race against time that can occur when ecologists and conservationists try to learn more about the biology of species before it is too late to respond. Local populations of *M. arion* had been disappearing from seemingly suitable locations since the 19th century. In the last phase of its extinction, *M. arion* declined from some 30 populations and an estimated 100,000 individuals in the mid-1950s to just 1 population of some 250 adults in the early 1970s.

Maculinea arion plummeted to extinction in spite of substantial conservation efforts, largely because

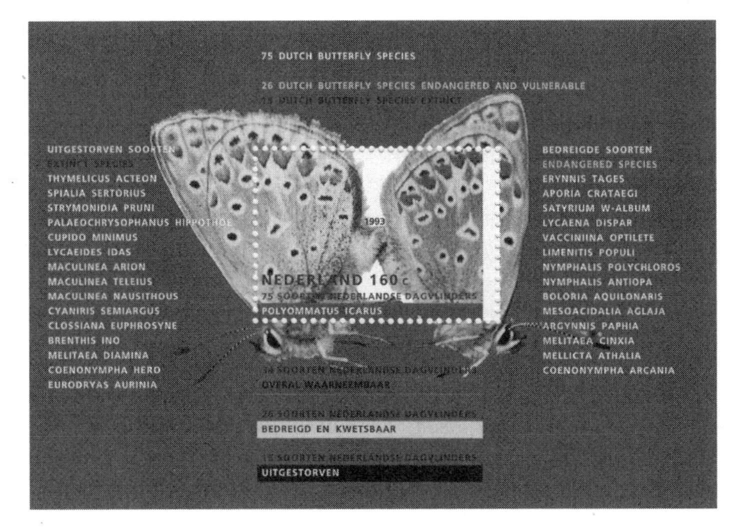

Figure 13.1. A Dutch stamp indicating the level of threat to butterflies in the Netherlands.

for 50 years conservation projects in Britain were based on erroneous information about the ecology of the species. By the time Jeremy Thomas (1980, 1995b, and references therein) determined the causes of most extinctions, it was too late; a series of four years of exceptionally unfavorable weather finished off that last population. Thomas found that the larvae of *M. arion* were obligatory predators on the larvae of just one species of ant (*Myrmica sabuleti*), a species that is largely restricted to warm, south-facing slopes with turf height <3 cm. Additionally, Thomas showed that for larvae to enter ant nests during the latter stage of their development, flowering *Thymus* host plants of *Maculinea* had to be within 2 m of *M. sabuleti* nests. Research conducted on the dwindling populations of *M. arion* in Britain increased awareness of the highly specialized habitat requirements of many butterflies.

In North America, studies of butterflies have documented declining abundances and shrinking distributions due to destruction, alteration, and fragmentation of butterfly habitats, especially near human population centers (for a general discussion of the conservation implications of habitat destruction and fragmentation, see Harrison and Bruna 1999). In contrast, extinction of the Jasper Ridge colony of *Euphydryas editha*, studied intensively by Ehrlich and his students, was unexpected. Ehrlich originally assumed that the 400-ha nature reserve on which the butterfly lived was adequate to support *E. editha* in perpetuity. As studies progressed from the early 1960s, it was discovered that the colony consisted of three populations that were much more ephemeral than originally thought. This discovery helped shift the focus of butterfly research in North America toward studies of population vulnerability and the reasons that populations go extinct (Gall 1984, Schultz 1998, Ries and Debinski 2001). Nevertheless, with the exception of *E. editha* and a few other species, such as, *Plebejus icarioides fenderi* (Schultz and Dlugosch 1999), several prairie butterflies (Swengel 1996, Swengel and Swengel 2001), and *Speyeria* species (Hammond and McCorkle 1983, Debinksi and Kelly 1998, Fleishman et al. 2002b, Boggs 2003), North American butterfly biologists lag far behind their European colleagues in the effort to document changes in the distribution and abundance of butterflies, in the study of the dynamics of their populations, and in the understanding of the causes of their vulnerability. This is probably because much of North America has more intact landscapes than Europe, and the latter has a much higher density of both amateur and professional students of butterflies.

The information on the population biology of checkerspots and other butterflies that has been compiled over the past two centuries forms an excellent basis for developing sound conservation principles for invertebrates and other small-bodied animals. The detailed data on the distribution of butterflies that are available for much of Europe have allowed exceptionally reliable conservation assessments of particular species; the new *Millennium Atlas of Butterflies in Britain and Ireland* (Asher et al. 2001) is truly outstanding in this respect (see figure 1.1). Such data also have played a key role in demonstrating the extent of population extinctions (Ehrlich 1995). The availability of such information is one of the reasons butterflies have functioned as a model group in the development of conservation principles. In general, butterfly studies have confirmed that the species in the greatest trouble are those with specialized ecological requirements, because these species are most likely to suffer from changes in land use practices, which are the predominant cause of declining abundances and extinctions.

The purpose of this chapter is to discuss the implications of our work on checkerspots to the preservation of populations and to place that work in a broader conservation context. We start with a brief overview of the conservation status of butterflies in Europe and North America (section 13.2). Next, we outline a series of conservation lessons that can be learned from our work on the biology and dynamics of populations and metapopulations of checkerspots. We then broaden our discussion to a set of topical issues to which these lessons apply, including the importance of conserving networks of sites, the potential role for reintroductions in butterfly conservation, the use of butterflies as indicators and umbrella species in conservation, and the responses of butterflies to global environmental changes. We conclude by turning to some of the policy issues that surround the conservation of butterflies in Europe and North America and to possible actions that can be taken to slow the global loss of butterflies and the loss of biodiversity in general.

13.2 Causes of Threat and Extinction in Europe and North America

The broad patterns in Finland, California, Europe, and North America (table 13.1) speak for themselves. The comparisons have to be interpreted cautiously, however, as the assessment of threat at different spatial scales tends to be based on somewhat different criteria (Gardenfors 2001, Daily et al.

Table 13.1. Species richness, number of threatened species, primary habitats, and causes of threat to butterflies in Finland (Rassi et al. 2001), Europe (van Swaay and Warren 1999), California (California Insect Survey 2002), and North America (Scott 1986, Opler et al. 1995) where data are available.

	Finland	Europe	California	North America
Total number of species	115	576	193	679
Threatened and extinct species	18	71	16	111
Near threatened species	11	43	30[a]	4[a]
Main habitats (%)				
Grasslands	—	39		
Cultural habitats	50	5		
Mountain grasslands	11	10		
Cliffs, rocky habitats	9	3		
Forests	15	18		
Scrubs	—	9		
Bogs, mires etc.	3	11		
Water-fringe habitats	12	5		
Main causes of threat (%)				
Changes in agricultural land	41	33		
Changes in forests and woodlands	17	13		
Construction, development	10	20		
Habitat deterioration, pollution	2	15		
Limited amount of habitat	15	—		
Natural population fluctuations	15	5		
Collecting	—	7		
Climate change	—	7		

[a]Proposed for listing (California and United States) or designated species of special concern (Canada).

unpubl. ms.), and because the natural environment and human pressures are substantially different in Finland and California.

Finland and Europe

The Finnish butterfly fauna includes 115 species, of which 2 are thought to be extinct in the country, 16 species are threatened, and an additional 11 species or subspecies are near threatened, as assessed by applying the current International Union for the Conservation of Nature red list criteria and categories at the national level (Rassi et al. 2001). In Europe as a whole, 576 species have been recorded, of which 71 are threatened and 43 near threatened (van Swaay and Warren 1999). The former figure includes 19 European endemics, mostly occurring on isolated islands such as Madeira and the Azores. Given the very different spatial scales, the percentages of threatened species are surprisingly similar in Finland (16%) and Europe as a whole (12%). More than half of the threatened butterflies both in Finland and Europe inhabit various types of grasslands. (There are hardly any natural grasslands in Finland, but the cultural habitats in table 13.1 consist of pastures, waste lands, and other open habitats created by traditional agriculture and other forms of land use.) Not surprisingly, then, the main threats to European butterflies relate to agriculture, including the abandonment of traditional agricultural practices and intensification of land use. In general terms, the primary threats can be lumped under the broad category of habitat loss and fragmentation. The situation is worse in the western part of Europe, though prospects for conservation are also rapidly deteriorating elsewhere. In particular, the countries in eastern Europe that have applied for a membership in the European Union are rapidly intensifying their agriculture, which is bad news for butterflies and innumerable other taxa of open habitats (for the impact on bird populations, see Donald et al. 2001).

The magnitude of the decline of butterfly populations in Europe has been assessed with butterfly atlases compiled at different times (Warren et al. 2001). However, the spatial resolution of such atlases is relatively crude (e.g., 10 × 10 km grids), and therefore the decline in the number of occupied grid cells may underestimate the true rate of decline of common butterflies (Ehrlich 1995, Daily et al. unpubl. ms). León-Cortés et al. (1999) provide an example. Using transect data, they assessed changes in the occurrence of the common blue butterfly *Polyommatus icarus* within an area of 35 km^2 in Wales. No decline during the past century could be inferred at the scale used in most common "square-bashing" surveys (using 1 km^2 grid squares), and even at higher resolution (500 × 500 m grid squares), the decline would only be estimated at about 7%. But based on the occurrence of the species in much higher resolution surveys of adults and changes in habitat, the true decline in the area occupied by populations was as great as 75%. The lesson here is that the current large-scale estimates of changes in butterfly populations are likely to underestimate, probably severely, the rate of loss of local populations and decline in local population sizes. Incidentally, the same observation has been made about global loss of mammals (Ceballos and Ehrlich 2002) and is likely to apply to virtually all organisms whose distributions are mapped using standard methods. More detailed surveys of a diverse sample of taxonomic groups and habitat types is required to get a better general picture of the decline of animals and plants in general.

California and North America

Although no extinction of a butterfly species has been recorded in North America so far, unknown numbers of populations have been lost, and several subspecies have disappeared. Most North American information comes from the United States, where several dozen subspecies have been recognized as candidates for protection under the federal Endangered Species Act (ESA). To date 18 butterfly taxa, all subspecies, have been conferred protection in the United States. Thirteen protected subspecies are from California, which amounts to about 3% of the named butterfly subspecies in the state, and the 12 species from which they are drawn represent 5% of California's 240 resident butterfly species. Thus, 1 out of every 20 California butterfly species is imperiled in at least a part of its distribution.

The above statistics may underestimate the current risk to North American butterflies, however. While butterflies are well represented on the endangered species list compared to other insects, only 44 of the 514 animal species on the list are insects. Considering that insects make up more than 70% of the species richness of animals in the United States, terrestrial invertebrates seem to be grossly under-represented by current protection efforts. To some extent this reflects shortcomings in the statute itself, as well as lacunae in the knowledge of most invertebrate species. Although distinct population segments of vertebrates (well-recognized populations, suites of populations, "subspecies," etc.) at risk can be protected under the ESA, invertebrates can be protected only when the entire subspecies or species is endangered. Therefore, butterflies must be much further down the track toward regional extirpation than vertebrates before they can receive federal protection. And, indeed, they often are.

Telling numbers can be gleaned from the city and county of San Francisco. Of 46 species recorded there before the 20th century, 21 have been lost. Yet only one of those that have disappeared has protected status elsewhere in its now diminished distribution—the Bay checkerspot butterfly. In the adjacent San Francisco Bay area and Central Valley (an area of around 25,000 km^2), less than 5% of native grasslands, coastal strand, riparian oak woodland, and vernal pools survive intact, and although many butterflies have proven remarkably resistant to sustained losses of their habitat, for many species remnant habitats and diminished populations seem poised to vanish in the face of changing environmental conditions.

13.3 Lessons for the Conservation of Populations

The challenge of conserving biodiversity amounts, in practice, to the challenge of arresting population declines before the extinction of populations and, eventually, of species. Many lessons for meeting this challenge have emerged from or have been reinforced by our studies of *Euphydryas editha* and *Melitaea cinxia* and other checkerspot species, by analyses of extended time series of their dynamics, and by testing population dynamic models with empirical data. These lessons are particularly applicable to invertebrates and small-bodied specialist vertebrates living in fragmented natural and managed landscapes, which are now characteristic of much of Earth's land surface.

1. Understanding the Spatial Structure of Populations Is Crucial to Preserving Patchily Distributed Species Ehrlich's early assumption that Jasper Ridge might contain not one but several largely independent demographic units (local populations) of *Euphydryas editha* was quickly validated by mark–release–recapture studies (Ehrlich 1961b; figure 3.6). Had the relatively independent dynamics of those units not been detected, the real fate of the butterflies on Jasper Ridge, from the early extirpation of the population in the small habitat patch G to the unexpected disappearance of butterflies from the largest patch, C, would have been obscured (figure 3.7). Not only would independent extinction events have been overlooked, but the ability of *E. editha* on Jasper Ridge to persist in the face of stochastic events would have been even more seriously overestimated than it actually was.

Although *Euphydryas editha* populations on Jasper Ridge and other coastal sites in northern California inhabit discrete patches of serpentine soil-based grassland that support their resources, the extent and boundaries of most other areas occupied by *Euphydryas* species are much less clear. Spatial distribution of allozyme frequencies in Great Basin populations of the *Euphydryas anicia-chalcedona-colon* complex suggests that populations extend across many dozens of square kilometers of habitat of indeterminate quality (Brussard et al. 1989). Alternatively, the pattern in the allozyme markers might be the result of spatially correlated selection in relatively discrete populations, but, unfortunately, it is not known whether the loci that have been studied are under selection. In Colorado, sex ratios vary spatially within a population of *E. editha* across ridgelines, hilltops, and benches (flat areas on slopes), as well as temporally with population size, suggesting that details of habitat structure are important in determining population structure as well as its size and persistence. In southern California, *E. editha* survive in areas of inferior habitat adjacent to previously high-quality habitats that have lost *E. editha* populations due to disturbance or fragmentation (D. Murphy, unpubl. obs.). This finding suggests that undocumented habitat refugia commonly support more extensive populations. The primary lesson is that superficial inventory or even population censuses will rarely document the entire area occupied by individual populations or metapopulations, uncover population subdivision that may affect its persistence, or identify habitat refugia that may be critical to sustaining species at risk.

2. Topographic Heterogeneity and Other Aspects of Habitat Diversity May Critically Influence the Capacity of Habitat Patches to Maintain Viable Populations Field studies indicate that weather and topography have interacted to produce the population fluctuations observed in the *E. editha* populations at Jasper Ridge (e.g., Singer 1972, Ehrlich et al. 1975, 1980; chapter 3). Long-term census data suggest that since the 1976–77 drought, the standard deviation of the population growth rate observed in the topographically homogeneous patch C has been at least 50% greater than in the more heterogeneous patch H (Hellmann et al. 2003). The population in the more homogeneous area also went extinct first. Thus, at least in this system, topographic heterogeneity appears to reduce variance in population growth rate and to extend population persistence time.

Intensive studies conducted at the large area of suitable habitat at Morgan Hill have shown how landscape heterogeneity helps buffer populations against environmental extremes (Weiss and Murphy 1988a, 1988b, 1990, Weiss et al. 1988). As discussed in chapter 3, a complex phase relationship exists between the timing of the development of the butterfly and that of its host plants. Seasonal patterns of rainfall and temperature are the primary determinants of larval survival, but the spatial arrangement of slope exposures can significantly ameliorate the population-level consequences of year-to-year fluctuations in weather. The availability of moist and cool slopes reduces larval mortality by providing favorable microclimates during dry and warm years. Similarly, survival of larvae on warm slopes during cool and wet years facilitates population increase, which in turn contributes to large population sizes that buffer populations against declines in the inevitable drought years. While conservation plans for invertebrates, other small-bodied animals, and plants should clearly minimize habitat loss and fragmentation, it is equally important to give priority to habitats within which population fluctuations remain constrained in spite of the exogenous factors that tend to increase population variability.

3. Variable Weather Conditions Are the Most Important Factor in Population Extinctions Aside from Anthropogenic Habitat Loss and Alteration A wide array of factors that can lead to population extinction have been observed or inferred to operate in checkerspot populations, from inbreeding and parasitism to emigration from natal sites (chapter 4). The multiplicity of potential causes of extinction becomes especially apparent in large metapopulations consisting of many small local populations with fast local population turnover (section 13.4, lesson #15). In the case of larger populations not threatened immediately by habitat loss and alteration, events related to weather (an example of environmental stochasticity in the parlance of theoretical ecologists; May 1973) are almost always the triggering mechanism for extinction. These events reduce numbers to the point where populations become vulnerable to many other mechanisms that alone or interacting can cause their ultimate demise. Quite often, weather events alone suffice to cause extinction. Ehrlich et al. (1980) found that a severe two-year drought caused the extinction of several *E. editha* populations throughout central California, presumably because the drought disconnected the critical phenological links between checkerspot butterflies and their larval resources (chapters 3 and 7). The effects of climate and weather on extinctions are also reflected in host plant choice itself. In Gunnison County, Colorado, populations of *E. editha* scattered through sagebrush meadows feed on drought-resistant *Castilleja linariifolia*, although two other less-resistant Scrophulariaceae, *Castilleja chromosa* and *Penstemon strictus*, are equally nutritious for the larvae (Holdren and Ehrlich 1982). In this case, extinction would likely occur in severe drought years if the populations specialized on these latter potential host plants. The most suitable conservation response to weather-related complexity is to design and manage reserves to maintain habitat diversity, including (where possible) topographic heterogeneity, a breadth of resources, and refugia that can sustain some individuals through adverse periods. The flexibility thus provided could be especially important in the face of global climate change.

4. Population Variability Provides an Effective Measure of Extinction Risk The rationale for using habitat quality to assess the conservation value of a site assumes that population density is proportional to habitat quality (Anderson and Gutzwiller 1996), an assumption that may often be incorrect (e.g., Van Horne 1983). More robust definitions of quality would take into account the potential for habitats to maintain populations in the long run despite fluctuations in abundance. The capacity of a habitat to facilitate population recovery from low numbers may provide a better indication of long-term quality than the potential of the habitat to support large populations in years with optimal conditions. Data from the Jasper Ridge *Euphydryas editha* populations reinforce this point, which has been recognized by many others (Schoener and Spiller 1992; for general theory of population extinction due to environmental stochasticity, see Lande 1993, Foley 1994, Middleton et al. 1995, Ludwig 1996, Belovsky et al. 1999, McLaughlin et al. 2002a; see also Hellmann et al. 2003). At Jasper Ridge the topographically diverse habitat patch H was of higher quality than the other areas, despite being smaller in extent and supporting a smaller average population size during much of the study period. Note, however, that the rank order of sites in terms of the variability of their populations may vary with the time period considered because population variance tends to increase over time (Pimm and Redfearn 1988, Ariño and Pimm 1995), and the rate of such increase may be different in different habitats.

5. The Ability to Use More Than One Larval Host Plant Decreases the Likelihood of Population Extinction Early studies on *Euphydryas editha* suggested that most populations were monophagous. More detailed studies have shown that many checkerspot populations in western North America use two or more larval hosts (figure 12.3). Use of more than one resource is predicted to increase overall stability, and in the Rabbit Meadow metapopulation of *E. editha,* this was shown to be the case (Boughton and Singer, 2004). Singer's and Hellmann's work at Jasper Ridge and Morgan Hill (Singer 1972, Hellmann 2002a) strongly suggested that population resistance to extinction depends on the availability of a plant that remains edible long enough during the spring, effectively mitigating against frequent impacts from drought and high temperatures.

When a habitat contains more than one host species, the proportion of eggs laid on each host may change dramatically from year to year in association with changes in plant quality or quantity (chapter 6; figure 12.3). Sometimes these changes are purely ecological; sometimes they entail rapid evolution of butterfly preference. Such evolution may occasionally imperil the butterfly. For example, individuals in one *E. editha* population living in a grazed meadow evolved a preference for ovipositing on an exotic weed from Europe, *Plantago lanceolata* (Singer 1983). Evolving dependence on this host renders the butterflies gradually more vulnerable to changes in human land management, just like European checkerspots that feed on *P. lanceolata*. Many European checkerspot populations have become extinct after traditional techniques of haymaking or light grazing were abandoned and *P. lanceolata* became scarce or unsuitable. This is one of causes of local extinction of *Melitaea cinxia* in the Åland Islands (chapter 4).

6. Resources at Any One Life-History Stage May Be Limiting While larval resources clearly are a primary determinant of habitat suitability and directly affect population structure and dynamics in many butterflies and other herbivorous insects, adult resources may also be limiting (Boggs 2003, O'Brien et al. 2004). Checkerspots may visit a wide variety of spring-blooming wildflowers that offer a landing platform and accessible nectar (figure 13.2), but they also exhibit clear preferences for certain species and demonstrate constancy of utilization of preferred sources (Murphy 1984). And while it appears that nectar use may enhance fitness of individuals only in some years (Murphy et al. 1983), populations are only infrequently found where preferred nectar sources are sparse, and checkerspots are not known to occur where nectar is entirely lacking. At the individual level, we know that emigration rate is elevated from areas with fewer nectar flowers (Kuussaari et al. 1996). The availability of other resources such as standing water has been

Figure 13.2. Adult male *Euphydryas chalcedona* nectaring on golden yarrow, *Eriophyllum confertiflorum*, at Jasper Ridge. Photo by Irene Brown.

observed to affect population structure in *Euphydryas* (Ehrlich and White 1980, Launer et al. 1993 [1996]) and *Chlosyne* (Schrier et al. 1976). Both male and female checkerspots have been found to commute up to hundreds of meters from the parts of habitat that contain larval resources to find adult resources.

Considering the role of larval resources in limiting population size in checkerspots, it is important to keep in mind that larvae of many species live in large groups and especially that small larvae have limited powers of movement (chapter 7). Thus larvae of *M. cinxia* often defoliate all available host plants within small areas, which can lead to starvation even though host plants are plentiful in other parts of the habitat patch (Kuussaari 1998). Large-scale defoliation of larval host plants affects the dynamics of local populations of *Euphydryas editha* (White 1974).

7. *Invasion of Non-Native Plant Species May Pose a Serious Threat to Butterfly Populations* Many natural plant communities are being replaced by non-native species, and, with that replacement, butterfly host plants and nectar species are often lost (although new nectar sources may be gained). In many parts of the world, the landscape is becoming less and less natural, challenging conservation planners to develop reserve designs and management strategies in response to new threats to the survival of native species. *Euphydryas editha bayensis* has probably been relegated to serpentine soil-based grasslands by the virtually complete conversion of California grasslands on other substrates to non-native vegetation. If so, that will have reduced the butterfly's distribution to just a small fraction of its presettlement extent. Thus it may be struggling to maintain populations in habitat patches just marginally suitable to meet its needs for larval and adult resources (chapter 3). Such exclusion from more productive areas is a common situation for many threatened species, which persist as relict populations in relatively low-quality habitats (Perry 1995, Root 1998).

Weedy alien plant species are now invading even serpentine soils, some apparently evolving serpentine tolerance (Harrison et al. 2001, Williamson and Harrison 2002), and some aided by nitrogen deposition (Weiss 1999). Cattle grazing appears to be a useful tool to manage against this risk factor for *Euphydryas editha* (figure 13.3). Thus alien grazers become the management response to threats from alien plants (Weiss 1999). Such possible fixes may be limited, however. In southern California, the invasion of alien plants is even more advanced than in the San Francisco area. In southern California, the host plants of *E. editha* survive in different edaphic circumstances, on clay lenses, where compact soil and cryptogamic soil crusts resist the establishment of weedy species (Mattoni et al. 1997). In these situations cattle grazing disturbs soil-surface integrity and actually encourages invasion by weeds. As yet unidentified mechanical, chemical, and fire treatments will be necessary to save the vanishing remnants of southern California's native grassland ecosystem and the checkerspots it supports.

13.4 Lessons for the Conservation of Metapopulations

Long-term research on *Euphydryas editha* in California has provided many insights and lessons for the conservation of individual populations. These studies also opened the door to the investigation and conservation of metapopulations, assemblages of local populations inhabiting networks of habitat patches (chapter 12). The metapopulation scenario has been the primary focus of the research on *Melitaea cinxia* conducted in the Åland Islands in Finland since 1991, and this research has yielded a wealth of information that is relevant to conservation. Empirical studies on *M. cinxia* have further stimulated development of metapopulation theory and provided valuable data to test model predictions. Our first lessons for the conservation of metapopulations stem from this mutual development of field studies on *M. cinxia* and new theory.

8. *Metapopulation Theory Can Be Used to Predict How the Structure of a Fragmented Landscape Influences Metapopulation Size and Persistence* In classic metapopulations, all local populations have a substantial risk of extinction. Therefore, over the long term, species that exhibit classic metapopulation dynamics can only persist regionally in a stochastic balance between local extinctions and establishment of new populations (Hanski 1999b). A principal conservation lesson that can be drawn from classic metapopulation models and supporting empirical studies is that a species will go regionally extinct when the amount of suitable habitat in the fragmented landscape falls below a threshold value (Levins 1969, Lande 1987, Hanski 1999b, Hanski and Ovaskainen 2000). Below the threshold, frag-

Figure 13.3. Combined effects of cattle grazing and nitrogen deposition on *Euphydryas* habitat on serpentine soil in the south San Francisco Bay area. To the left of the fence line, non-native vegetation dominates in the absence of cattle grazing. To the right of the fence line, cattle grazing has favored native vegetation used by the butterfly. Photo by Stuart Weiss.

ments of suitable habitat are so few or so small that the rate of extinction will exceed the rate of reestablishment even when most habitat fragments are unoccupied and there seems to be potential for metapopulation growth. Theory predicts that, as habitat is lost and becomes increasingly fragmented, not only does the total area occupied decrease, but also the proportion of occupied habitat declines. The fundamental processes leading to this prediction are the extinction proneness of small populations in small and poor-quality habitat fragments, and the reduced rate of recolonization of isolated, unoccupied fragments (Hanski 1994a).

Early analyses of the occurrence of *Melitaea cinxia* in the Åland Islands involved comparisons among regions with large versus small average patch areas and among regions with high versus low density of habitat patches. The empirical observations supported the model predictions of reduced metapopulation size (fraction of habitat occupied) with increasing fragmentation, in the sense of small average patch area and low patch density (Hanski et al. 1995b). These results also gave credence to the conclusion that *M. cinxia* had gone extinct in many parts of northern Europe (Hanski and Kuussaari 1995) because the amount of habitat had fallen below the extinction threshold (Hanski 1999b). More recently, mathematical theory was developed that allows a more efficient description of exactly how the structure of a fragmented landscape will influence metapopulation size and persistence (Hanski and Ovaskainen 2000, Ovaskainen and Hanski 2001, 2002b; box 12.1). This theory involves a measure of landscape structure, "metapopulation capacity" (Hanski and Ovaskainen 2000), which represents in a single real number the capacity of a fragmented landscape to support a viable metapopulation. Metapopulation persistence requires that the metapopulation capacity of the landscape exceeds a threshold value set by the properties of the species (box 12.1). The example on *M. cinxia* described in chapter 12 shows the fit of the model to data from multiple patch networks and provides strong evidence for the existence of a threshold value for metapopulation persistence (figure 12.3).

One apparent limitation of metapopulation theory as depicted above and described in detail by Hanski and Ovaskainen (2000) and Ovaskainen and Hanski (2001) is that it is focused on the effects of habitat fragment size and connectivity, while often other environmental attributes such as the topographic heterogeneity (lesson #2) and the quality of the intervening matrix (lesson #23 below) also influence metapopulation dynamics (e.g., Roland et al. 1999, Ricketts 2001, J. A. Thomas et al. 2001). Fortunately, such factors can be incorporated into metapopulation models by replacing real patch areas and pairwise distances by measures corrected for the effects of habitat quality on population dynamics and movements. It would also be possible but more difficult to incorporate factors related to interactions between life-history traits and the environment into the models (lessons #4 to #6). In most cases, however, the real limiting factor is likely to be availability of empirical information on the effects of these factors on individual behavior and population processes. If adequate information is not available, no predictive models involving these factors can be constructed.

Another limitation of much of the metapopulation theory is that it is deterministic. This is not a major problem for gaining general insights, but a stochastic theory should provide superior quantitative predictions (Ovaskainen 2002a). A stochastic model can be applied via simulation, with parameters estimated for a particular metapopulation (Moilanen et al. 1998, Hanski 1999b, Moilanen 1999, 2000, O'Hara et al. 2002). The incidence function model (IFM), originally developed by Hanski (1994b) in the context of the *Melitaea cinxia* project, is a prime example of a stochastic patch occupancy model. The IFM has been successfully applied to more than a dozen real metapopulations of butterflies and other species (for reviews, see Hanski 1999b, 2001), as exemplified in the lessons below.

9. *Models Developed for Well-studied Common Species May Provide Insight to the Dynamics of Ecologically Related Species That Are Rare and Threatened* Wahlberg and co-workers (1996) applied the incidence function model (IFM) developed for *M. cinxia* (Hanski et al. 1996c) to a close relative, *Melitaea diamina*, an endangered species restricted to two small regions in Finland (Wahlberg 1997a, Rassi et al. 2001). Using the parameter values estimated for *M. cinxia*, Wahlberg et al. (1996) predicted the probability of occurrence of *M. diamina* in its network. These predicted occurrence probabilities (incidences) were found to be in good agreement with the observed distribution of *M. diamina* (figure 13.4). This example suggests that one may use information on a related species to construct a model for an endangered species, for which sufficient information might not be available. This approach has to be used with caution,

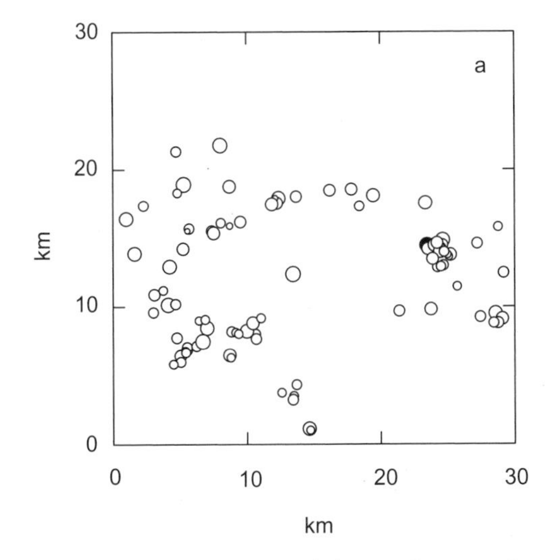

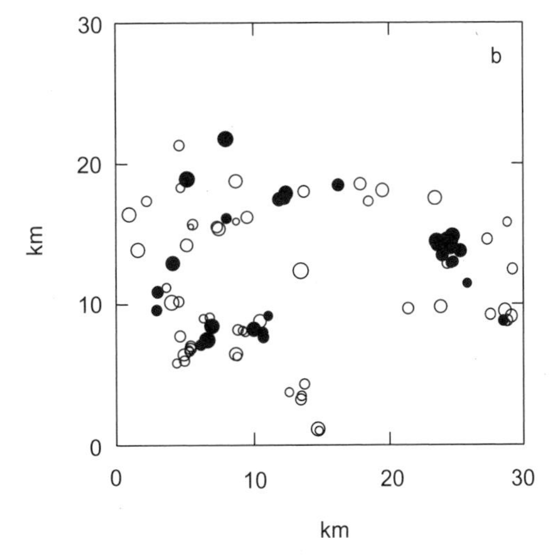

Figure 13.4. (a) A map of the patch network occupied by *Melitaea diamina* showing the relative patch areas, their spatial locations, and the model-predicted probability of occupancy (incidence) based on parameter values estimated for *Melitaea cinxia*. Greater probabilities are shown by darker shading. (b) A snapshot of patch occupancy in 1995. Compare with the predicted occupancies in (a). From Wahlberg et al. (1996).

however, because one can never be certain that there are no significant biological differences between two closely related species.

In conservation, it is generally more helpful to compare the population dynamic consequences of alternative management scenarios than to predict the absolute risk of population and metapopulation extinction (e.g., Hanski 1999b, Possingham et al. 2002). The IFM is particularly well suited for comparing different landscape (management) scenarios, because the model predictions are based on an explicit description of landscape structure. Drechsler et al. (2003) present an example, in which they incorporate formal decision analysis in the comparison of several management alternatives. The following lesson illustrates how the IFM can be coupled with and used to add spatial dynamics to reserve-site selection procedures.

10. Metapopulation Models Combined with Reserve-Site Selection Algorithms Provide a Robust Framework for Conservation of Metapopulations Over the past 15 years, conservation biologists interested in the protection of biodiversity over large areas have developed two distinct approaches: reserve-site selection algorithms (Pressey and Nicholls 1989, Margules et al. 1994, P. Williams 2000) and spatially extended population viability analysis (Soulé 1987, Possingham et al. 2000). Reserve-site selection algorithms are used to choose a set of sites that includes a large fraction of the species (or other forms of biodiversity) to be conserved. The selection of sites is required to be cost effective in the sense that a required minimum level of representation of species is achieved with minimum cost or area of habitat. However, what has been missing from the site selection algorithms is parameterization of the spatial dynamics of the species (Cabeza and Moilanen 2001); the sites are selected based on the occurrence of the species without any consideration for their chances of survival in the future. The smaller the reserves in consideration, the more important it is to incorporate spatial dynamics in the analysis.

Using the IFM parameterized for *Melitaea diamina*, Moilanen and Cabeza (2002) applied an optimization method to determine which subset of the sites in the network of 151 *M. diamina* patches should be selected to maximize the long-term persistence of the species, given that protecting each site has an associated cost and given that the amount of money to acquire sites for conservation is limited. The value of each candidate selection of sites for promoting persistence was evaluated using the IFM. This was not a trivial task given the astronomical number of possible combinations of sites. Moilanen and Cabeza's (2002) work demonstrated that a rigorous optimization can be conducted, effectively combining ecology, spatial dynamics, and economics in the context of reserve site selection.

Several key results of this work deserve to be highlighted. First, the sites in the optimal set were clustered, which facilitated recolonization after extinction of local populations. However, depending on the quantity of resources available for conservation, the optimal set included, in the example studied, either one, two, or three separate clusters of sites (figure 13.5). Second, a particular cluster of sites may or may not be included in the optimal set under different funding scenarios (figure 13.5; look at the central cluster of selected sites). In practical applications, other considerations such as preservation of genetic diversity may further increase the benefits of protecting several relatively independent clusters of sites (Moilanen and Cabeza 2002). Third, there were decreasing conservation returns with increasing land allocation for conservation.

11. An Intermediate Level of Connectivity Is Generally Most Beneficial for Long-Term Persistence The metapopulation theory discussed in lesson #8 allows one to assess the relative merits of changes in the spatial configuration of habitat patches. One of the general conclusions that emerge relates to the spatial scale of population dynamics and population extinction. Typically, in real metapopulations, both changes in the sizes of local populations and population extinctions are to some extent spatially correlated (Hanski 1999b, chapter 4). As a consequence, optimal spacing of protected sites is often likely to be a compromise between increased connectivity to maximize the probability of recolonization events and increased isolation to avoid simultaneous extinction events. Unfortunately, because the strength and spatial scale of correlation varies among metapopulations in different landscapes, and even among years (chapter 4), it is not easy to arrive at general recommendations about the optimal degree of patch clustering within networks. Perhaps the only general recommendation that we can make, when considering the conservation of a single species with metapopulation structure, is that the patches should be located no farther apart than

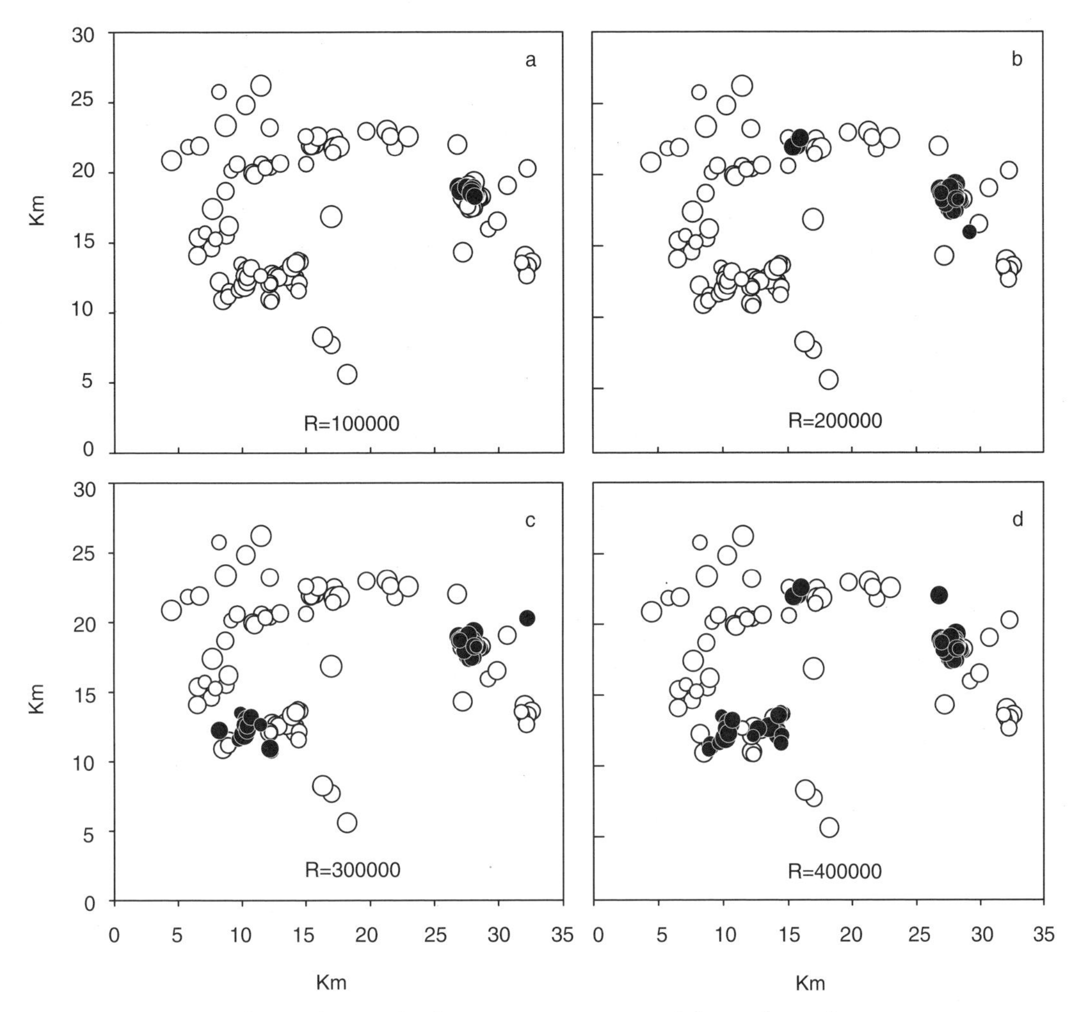

Figure 13.5. Optimal site selection for long-term persistence of the *Melitaea diamina* metapopulation (figure 13.4) at different levels of resource availability (R, money) for the acquisition of habitat for conservation. For further explanation, see text. From Moilanen and Cabeza (2002).

would allow for recolonization within a few years after extinction events.

Within-patch topographical heterogeneity and other forms of habitat diversity are likely to reduce the risk of extinction of local populations (lesson #2). Topographical heterogeneity tends to buffer populations against environmental stochasticity by reducing correlation in the survival and reproduction of individuals. The same mechanism operates at the network level, making it potentially advantageous to have habitat patches of dissimilar quality in a network. Such among-patch variation in quality would also allow increased clustering of sites without greatly increasing the risk of metapopulation extinction due to regional stochasticity (spatially correlated environmental stochasticity).

12. Species Inhabiting Successional Habitats Often Exhibit Metapopulation Dynamics Many butterflies live in early successional habitats, and many such species occur as metapopulations in their naturally dynamic landscape. For example, *Euphydryas maturna* prefers forest edges, which are suitable for this species only briefly after the opening of closed forest cover. Warren's (1987c) detailed studies of *Melitaea* (*Mellicta*) *athalia* in coppiced woodlands in England and Wahlberg et al.'s (2002b) modeling study of *Euphydryas aurinia* living in clear-cuts in Finland demonstrate the critical role of connectivity for species dependent on successional habitats. Species can only persist in landscapes in which distances between the currently occupied patches and the new successional patches are not too great. Apart from the difference in the process generating unoccupied but suitable patches, the dynamics of species living in successional habitats are not fundamentally different from the dynamics of species living in more stable habitats (lesson #15). In both cases, whatever the intrinsic cause of local population extinction, anthropogenic habitat alteration and loss are currently the primary cause of regional extinction of butterflies (section 13.2) and of most other taxa (Caughley 1994, Harrison 1994, Simberloff 1994, Wilcove et al. 1998).

13. Single Large, High-Quality Patches May Play an Important Role in Maintaining Regional Persistence It appears that regional persistence of the Bay area ecotype of *Euphydryas editha* is due to the presence of large "mainland" populations, such as that at Morgan Hill (figure 3.2). The relatively large habitat patches have substantial spatial heterogeneity in slope aspect and resource quality and thus spatial variation in larval survival. As a matter of fact, the Morgan Hill population may be viewed as a sort of metapopulation, although the landscape does not consist of physically well-delimited patches as does the landscape of *M. cinxia* in the Åland Islands (chapters 3 and 4). Such mainland populations periodically supply migrants that can reestablish local populations in the surrounding smaller habitat patches, where local populations have gone extinct (Harrison et al. 1988, Harrison 1989). When connectivity to the mainland population is lost, as happened with the Jasper Ridge populations of *E. editha*, such rescue becomes improbable or impossible.

The theory discussed in lesson #8 and in box 12.1 allows one to calculate the contribution that a particular habitat patch makes to the metapopulation capacity of the network as a whole. Ovaskainen and Hanski (2003) have investigated possible measures of patch "value." Apart from permitting the evaluation of existing habitat patches, the theory also allows one to calculate the potential value of conserving an additional patch of certain size at a certain location in the landscape (for an example, see section 13.5). Such measures have obvious applications in conservation and landscape management. In real situations, other factors apart from the ones included in the metapopulation theory have to be considered, however, and can be potentially important. These factors include genetics and interspecific interactions. It is nonetheless helpful to quantify the patch-specific effects related to the spatial configuration of the fragmented landscape.

14. The Presence of a Metapopulation in the Current Landscape May Be Deceptive Because Metapopulation Dynamics Track Environmental Changes with a Substantial Delay Conservation applications of the metapopulation theory would be greatly enhanced if one could make the assumption that metapopulations generally occur in a stochastic balance between local extinctions and recolonizations. Unfortunately, that assumption is not always warranted, partly because metapopulations are constantly being perturbed by environmental changes, but also because the landscapes that support them are subject to change. Species commonly become endangered because of habitat loss and fragmentation, and for some time after environmental change the metapopulation is likely to be out of stochastic equilibrium, with the dynamics lagging behind the change. In other cases, a landscape may become more favorable for a species due to improvement in habitat quality (Hanski and Thomas 1994) or climate change (C. D. Thomas et al. 2001, Warren et al. 2001), and, again, it takes time before the environmental changes are fully reflected in the size of the metapopulation. The duration of the transient dynamics may be predicted with metapopulation models, provided one has independent estimates of model parameters, which were obtained, for example, from an environment that has remained relatively stable. Hanski et al. (1996b) presented an example on *Melitaea cinxia* in one of its semi-independent patch networks in the Åland Islands (see figure 4.8). Unfortunately, it is difficult or impossible to estimate the values of the model parameters if the information comes from the changing landscape itself (see Wahlberg et al. 2002a for an example on *Euphydryas aurinia*).

An important aspect of transient dynamics in a changing environment is that the time it takes for the metapopulation to move to a new equilibrium after perturbation is longest for species that are close to their extinction threshold (Hanski and Ovaskainen 2002, Ovaskainen and Hanski 2002b). This general result is evident in the example in figure 4.8.

Because so many landscapes have experienced recent anthropogenic habitat loss and fragmentation, it is expected that many species are currently "too abundant" with respect to the present landscape structure because they are still in the transient state after environmental change. In this situation, the probability of eventual disappearance may be severely underestimated (Hanski and Ovaskainen 2002).

15. Many Causes of Local Extinction May Operate in a Large Metapopulation Individual populations in the large *M. cinxia* metapopulation in the Åland Islands tend to be so small, often consisting of just one or a few larval groups (chapter 4), that one might expect demographic stochasticity to be the primary cause of local extinctions. Surprisingly, however, research has demonstrated that numerous mechanisms of extinction have operated at least in some years and at least in some parts of the metapopulation (Hanski 1998b). These mechanisms include habitat loss, environmental stochasticity (which is to some extent spatially correlated; chapter 4), migration (emigration increasing, immigration decreasing extinction risk; chapter 9), parasitism (chapter 8), and inbreeding (chapter 10). Although the primary correlate of extinction risk is small size and poor quality of habitat patches (chapter 4), at the mechanistic level many different processes are involved. A similar situation is likely to occur in other species living in highly fragmented or naturally patchy landscapes, in which processes that lead to extinction that are unlikely to be important in isolated single populations can play a major role. Processes that are likely to influence local extinction in metapopulations but not in more isolated populations include migration, inbreeding, and specialist natural enemies (Hanski 1998b).

16. Inbreeding Depression May Be an Important Cause of Local Extinction in Species with a Fragmented Population Structure Observations and experimental field studies have demonstrated that inbreeding depression affecting several fitness components increases the risk of extinction of small local populations of *M. cinxia* (Saccheri et al. 1998, Haikola et al. 2001, Nieminen et al. 2001; chapter 10). The strong effect of inbreeding on population dynamics reflects the life history of *M. cinxia* and many other checkerspot butterflies—larvae live in large groups of full sibs, which typically greatly reduces effective population size and increases the chance of breeding among relatives (chapter 10). Although local populations repeatedly pass through severe bottlenecks in population size, the fragmented structure of checkerspot populations with frequent turnover and distance-dependent migration does not seem to effectively purge the slightly deleterious recessive alleles that apparently cause inbreeding depression in *M. cinxia* (Saccheri et al. 1998, Haikola et al. 2001). Nor is there any evidence for avoidance of mating with close kin in this species (Haikola et al. 2004), perhaps because the cost of reducing inbreeding by this means would be too great in small populations with limited mating opportunities (Kuussaari et al. 1998).

17. Habitat Fragmentation May Select for Increased Migration Rate There is currently much interest in the evolution of migration rate in response to environmental changes (C. D. Thomas et al. 1998, Parvinen et al. 2000, Ronce et al. 2000, Heino and Hanski 2001). Modeling results for *M. cinxia* indicate that habitat fragmentation may select for higher migration rate at the metapopulation level (Heino and Hanski 2001), primarily because habitat fragmentation increases the extinction rate of increasingly small local populations, thereby providing increased opportunities for the migrants to establish new populations (Leimar and Norberg 1997). In an empirical study, Hanski et al. (2002) showed that female *M. cinxia* from newly established populations have higher migration rates than females from older populations. The likely explanation is that new populations are typically established by females with greater than average tendency to migrate, a trait that may be inherited by offspring. Selection operating at the metapopulation level may thus tend to increase migration rate (Olivieri and Gouyon 1997).

From the conservation viewpoint, the important question is whether the evolutionary changes in life-history characteristics, such as migration tendency, can make a difference to the long-term survival of species in increasingly fragmented landscapes. The answer will depend on the rate of environmental change and on the rate and magnitude of evolutionary response by the species. Heino and Hanski's (2001) modeling of the population dynamics of *M. cinxia* and *M. diamina* suggested that though an "evolutionary rescue" is a possibility, it is unlikely to be of practical significance in the real metapopulations of these species. Certainly, we cannot assume that evolution will solve the extinction crisis created by habitat loss and fragmentation.

18. Local Adaptation May Influence the Persistence of Populations in a Metapopulation Host-plant oviposition preference in checkerspots provides many instructive examples of how populations may be locally adapted at a surprisingly small spatial scale (chapter 6). Local adaptation is an important consideration for possible introduction of populations to new environments (section 13.5). We know, for instance, that female checkerspots have a higher emigration rate from habitat patches that lack their preferred host plant (Hanski and Singer 2001, Hanski et al. 2002; chapter 9). Therefore, if a population is introduced at a site with nonpreferred host plants, the performance of the introduced population may be compromised by a high rate of emigration.

13.5 Broader Conservation Implications of Checkerspot Studies

The lessons for the conservation of single populations and metapopulations discussed in the previous sections relate to a suite of broader topics in conservation biology. Among these issues, studies on checkerspots and other butterfly species have produced results that elucidate the contrast between preserving individual habitat patches versus networks of multiple patches, the potential role for reintroductions to reduce biodiversity loss, and the biological consequences of regional and global environmental changes.

Protecting Single Sites versus Networks of Sites

Views about local versus regional conservation have changed over the years and will probably continue to change in the future. In the 1980s, the contrast was encapsulated in the SLOSS (single large or several small) debate (Gilpin and Diamond 1980, Diamond and May 1981, Simberloff 1988, Meffe et al. 1997): would one large reserve better meet conservation goals than a series of small reserves of the same total area? With the necessary caveat that other physical and biotic features had to be equal, the choice was a red herring. All else is never equal in nature, and all else to a large extent determines the fates of populations, metapopulations, and ultimately of species. The argument, however, did set the stage for an important challenge: To what extent should one emphasize the protection of individual local populations via habitat management versus the conservation of networks of sites?

In butterfly biology, the debate about protecting single populations versus metapopulations relates to the question of what actually constitutes a population. Differences in the mobility of butterfly species are so striking—from the famous long-distance migrants such as the monarch butterfly (*Danaus plexippus*) in North America and the red admiral (*Vanessa atalanta*) in Europe to extremely sedentary species such as the silver-studded blue (*Plebejus argus*; C. D. Thomas and Harrison 1992, Lewis et al. 1997)—that butterflies have been classified as having either "open" or "closed" population structures (section 11.5). The notion of closed populations may be traced back to the early studies by Ford and colleagues (Ford and Ford 1930, Ford 1945) and to the work of Ehrlich and colleagues on *E. editha* and other checkerspots (Ehrlich et al. 1975, Singer and Ehrlich 1979, Ehrlich 1984a). J. A. Thomas (1984) more formally established the contrast between open and closed population structures and argued that in the case of closed populations, species conservation efforts should be directed toward active management of habitat quality.

Over the past 10 years, the evidence on which the distinction between closed and open population structures in butterflies was established has to some extent been eroded. Though there is no doubt that many butterflies have well-delimited local populations (lesson #1), extensive mark–release–recapture studies conducted simultaneously in several populations have revealed more movements among nearby populations than was previously assumed (Hanski et al. 1994, Hanski and Kuussaari 1995, Nève et al. 1996, C. D. Thomas and Hanski 1997, Roland et al. 1999, Petit et al. 2001, Fleishman et al. 2002b, Wahlberg et al. 2002b). Lifetime emigration rates of specialist butterflies from small habitat patches (0.01–1 ha) are often 10–30% or greater (C. D. Thomas and Hanski 1997, Hanski et al. 2000, Wahlberg et al. 2002b), and though most migrants travel only hundreds of meters, some individuals reach patches several kilometers away from the natal patch (Hanski 1999b, Kuussaari and Nieminen 2004). Given the new evidence of short-range mobility of many "sedentary" species and an improved understanding of the more or less ephemeral nature of their local populations (lesson #8), it is logical that conservation efforts at the habitat patch network level have been called for (C. D. Thomas and Hanski 1997). The debate has not disappeared, however, and recently the pendulum appears to be swinging back again. Several researchers have failed to find clear evidence for the dominant role of patch area and connectivity in determining the occurrence of species in fragmented or naturally patchy landscapes (C. D. Thomas et al. 2001, Fleishman et al. 2002b). Instead, these studies have found that some measures of habitat quality are more important determinants of occupancy, from which the conclusion has been drawn that patch area and isolation are less important for long-term persistence than previously thought.

The critical issue is not whether habitat quality makes a difference. As the lessons learned from checkerspot studies clearly illustrate, the quality of habitat patches often makes a big difference (lessons #2 and #5), and so does the quality and permeability of the matrix (Ricketts 2001, Ries and Debinski 2001, Keyghobadi et al. 2003). The more important question is whether, in a particular landscape and for a particular species, variation in habitat quality is great enough, and indeed whether necessary empirical information exists, to warrant considering spatial variation in habitat quality in developing conservation strategies (for an analysis of *Melitaea cinxia*, see Moilanen and Hanski 1998). The critical issue is whether population connectivity makes a difference in the spatial dynamics of the species. It is unfortunate that most empirical studies continue to use a simplistic measure of connectivity, distance to the nearest population (or even distance to the nearest habitat patch), which is known to yield unreliable results (Moilanen and Nieminen 2002). Instead, one should use a connectivity measure that takes into account the distances to and the sizes of all neighboring populations (Hanski 1999b, Moilanen and Nieminen 2002; see box 12.1). It is also clear that many species have such great mobility that there are no local breeding populations within the relatively small areas typically studied, and hence no role for classic metapopulation dynamics at this spatial scale.

Given the history of opinions that we have briefly sketched above and the slow pace of improving understanding of species' spatial population structures and dynamics, it would be foolish to suggest that there is universal answer to the questions of which landscape parameters best explain the distribution and abundance of species and which elements of the landscape should be the primary focus of conservation; the answers will depend both on the landscape and on the species. With further work, it may become possible to determine classes of metapopulations and landscapes for which particular parameters are likely to be most important. In the meantime, we emphasize that if metapopulation dynamics are relevant in a particular case, the models discussed in lessons #8 through #10 above allow the significance of habitat quality to be assessed, as the following lesson illustrates.

19. The Relative Merits of Improving Habitat Quality at Single Sites versus Conserving Networks of Sites Can Be Assessed with Metapopulation Theory Figure 13.6 depicts a hypothetical habitat patch network. In this figure, the value of metapopulation capacity (box 12.1) is 10.3. Let us now consider an example in which the quality of patch 1 in figure 13.6 is increased fivefold through very successful habitat management, as indicated by a fivefold increase in population size. This change will increase the metapopulation capacity of the network by 14%. A much greater increase (77%) is obtained by increasing the area of the same patch fivefold (but retaining the original quality). The model assumes that immigration and emigration rates scale by patch area (box 12.1); hence the latter change is more helpful for metapopulation persistence than increasing population density within a given area. The reason is that increased area often increases immigration rate to and reduces emigration rate from the patch. Another point often favoring larger areas, though not considered in this model, is that increased area probably increases topographic heterogeneity, which is likely to reduce population variability and thereby lower its risk of extinction; lessons #2 and #4. Metapopulation capacity is increased by only 5% if the area of patch 2 instead of patch 1 (figure 13.6) is increased fivefold, because patch 2 is poorly connected to the remaining network and hence plays only a minor role in the dynamics of the metapopulation as a whole. We can also calculate the benefits of adding a new patch to the network. The metapopulation capacity of the original network is increased by 18% if the patch shown by an open circle is added to the network (figure 13.6). These examples illustrate how metapopulation theory can be used to answer some of the specific questions that might arise in an active management context.

One limitation of the above model predictions is that they relate to long-term metapopulation be-

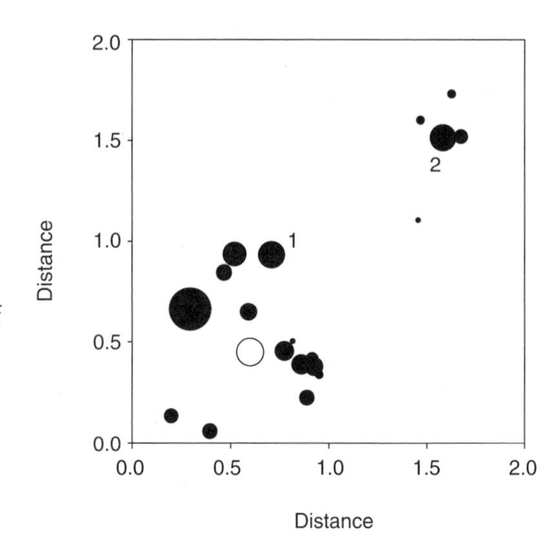

Figure 13.6. A hypothetical habitat patch network, which is used to compute the metapopulation capacity (box 12.1) for alternative management scenarios (details in the text). The patches have log-normally distributed areas (the size of the circle is proportional to the logarithm of area). The results are calculated for the spatially realistic Levins model (Ovaskainen and Hanski 2001) with the following parameter values: $\zeta_{im} = \zeta_{em} = \zeta_{ex} = 0.5$, $\alpha = 2$ and $\delta = 1$. For the significance of the numbered patches, see text; for further explanation, see Ovaskainen and Hanski (2001) and Ovaskainen (2002b, 2002c).

havior, whereas often the management goal is to help species survive in a catastrophically bad situation for the next few years. In the latter situation, intensive management of particular populations and sites may be the only meaningful option available, and if any model predictions are needed, they are best produced via simulation (lesson #10). The most useful applications of the metapopulation theory can probably be found in the management of "ordinary" species in "ordinary" fragmented landscapes, with the aim of preventing these species from becoming the next critically endangered species (Hanski 1998b).

Reintroductions

Whether habitat loss occurs locally via successional change in habitat quality (lesson #12) or via outright habitat destruction, species may persist as metapopulations as long as connectivity to currently suitable habitat is sufficiently high to allow establishment of new local populations (Hanski 1999a). When the necessary level of connectivity is lost, while some isolated fragments of habitat still remain, a possible conservation measure is to introduce species to sites that are too distant for natural recolonization. There have been thousands of attempts to introduce butterflies, for various purposes, to unoccupied sites where they have or have not been recorded previously. The most comprehensive compilation of butterfly introductions is Oates and Warren's (1990) analysis of 323 introductions in the United Kingdom. Based on their results and on other studies, including ours on checkerspots, a number of conclusions with conservation implications can be drawn, as discussed in the lessons below.

20. Most Introduced Populations Survive for a Short Time Only Oates and Warren (1990) found that 26% of the introductions survived for at least 3 years, but only 20 introduced populations (6% of all) survived for more than 10 years. Though many introductions failed because of unsuitable habitat, high extinction rates characterize small butterfly populations even in satisfactory habitats (Warren 1992), and introduced populations are typically small. One should not assume, however, that every small population will go quickly extinct, and from a conservation viewpoint there is always some hope as long as even a minimal population exists. Holdren and Ehrlich's (1981) 1977 introduction of *Euphydryas gillettii* many hundreds of kilometers south of its known range yielded a population that survived at an astonishingly small size. During most years over two decades, adult population size was so small that the presence of *E. gillettii* could only be detected by intensive search for egg masses on the butterfly's larval host plant, *Lonicera involucrata,* and the adult population probably never exceeded 100 individuals. Then, after 25 years, the population exploded and spread, possibly resulting from a series of winters with low snow fall and early snow melt (Boggs et al. unpubl. ms).

There are classic avian examples of similarly persistent small invader populations in North America, including the Eurasian tree sparrow (*Passer mon-*

tanus) and the European skylark (*Alauda arvensis*). The first was introduced to St. Louis in 1870 and has spread only as far as east-central Missouri, west-central Illinois, and the southeastern corner of Iowa, occuring in an area of about 200 × 400 km (Ehrlich et al. 1988, Barlow and Leckie 2000). The skylark was brought to the Saanich Peninsula of southern Vancouver Island a century ago and has only spread as far as a small portion of San Juan Island in Washington, 18 km east (Ehrlich et al. 1988). The Saanich area occupied has fluctuated and by the 1980s extended only to about 30–40 km^2 (Campbell et al. 1997). In both of these cases the reasons for the failure to spread widely are not understood. The tree sparrow's congener, the house sparrow (*Passer domesticus*), has spread over virtually all of the North American continent south of the Arctic as well as invading many other areas (Long 1981, Lowther and Cink 1992). The skylark has also been a successful invader elsewhere, including Hawaii, New Zealand, and southeastern Australia, but without the general success of the house sparrow (Long 1981).

21. Long-Term Persistence of Introduced Populations Requires Either Large Tracts of Habitat or the Introduction of the Species into a Network of Sites That introduction to networks of sites may lead to long-term persistence is demonstrated by the results on *Melitaea cinxia* in Finland. All populations introduced to single sites vanished in a few years (some 10 introductions; M. Kuussaari and I. Hanski, unpub. data), whereas the only successful introduction, a metapopulation that has survived since 1991, took place on the island of Sottunga east of the main Åland Island (figure 2.3), which has a network of more than 20 small habitat patches (figure 13.7). The introduced metapopulation has survived, though not a single local population has persisted for the period of 12 years. The metapopulation capacity (box 12.1) of the Sottunga network is 0.82. Compared with networks in the main Åland Island, theory predicts that Sottunga should indeed support a viable metapopulation (see figure 12.4; note that the logarithm of 0.82 is -0.09). Incidentally, *M. cinxia* was introduced to Sottunga as larvae, which resulted in the simultaneous introduction of the primary parasitoid *Hyposoter horticola* (chapter 6).

22. There is Often No Relationship between the Number of Introduced Individuals and the Success of Butterfly Introductions Oates and Warren (1990) found no relationship between number of introduced butterflies and their success in their general survey of British butterfly introductions. These results are contrary to many other studies, which have reported a propagule size effect for introductions of birds and small mammals (Sheppe 1965, Crowell 1973, Ebenhard 1987, 1991, Veltman et al. 1996). Lack of significant effect of the number of introduced individuals in butterfly studies most likely relates to the great significance of habitat quality at the site of introduction (lesson #2), and possibly to the importance of prevailing weather conditions during and immediately after introduction (lesson #3), which may be assumed to be less significant factors for birds and other vertebrates.

Although some butterfly introductions with few individuals have succeeded, in such introductions the deleterious consequences of inbreeding may be significant. Using captive-bred individuals as the source of introduction would reduce the need to remove individuals from natural populations, but captive-bred individuals may have reduced fitness in the field (Lewis and Thomas 2001, Nieminen et al. 2001), further reducing the chance of successful establishment of introduced populations. Though there is no hard evidence on this, we speculate that introductions should include at least 20 mated females originating from a large source population or populations.

The potential harm that can be caused by translocations leading to mixing of locally adapted genotypes or introducing genotypes into environments in which they are poorly adapted is an argument against introductions. Conservation priority should indeed be given to protecting existing populations and metapopulations rather than to expending limited resources in the often risky attempt to create new populations. However, under favorable conditions—a species with well-known biology, an unoccupied landscape that clearly appears to satisfy the criteria for long-term persistence—translocations may provide a cost-effective way to promote butterfly conservation (J. A. Thomas 1989, Pullin et al. 1995). It will be interesting to see whether the long-distance translocation of *Euphydras gillettii* (Holdren and Ehrlich 1981, Boggs et al. unpubl. ms.) will eventually result in this species colonizing the extensive areas of apparently suitable habitat in central Colorado. Reintroduction opportunities are now available for *E. editha* at Jasper Ridge and

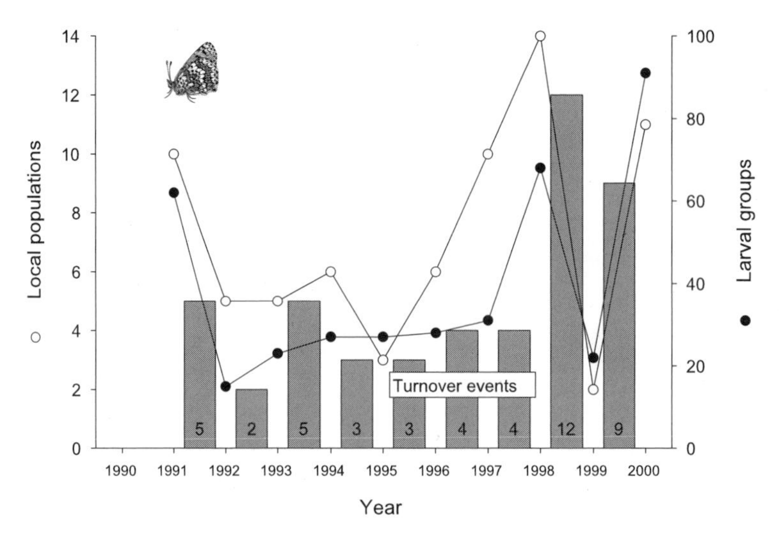

Figure 13.7. The island of Sottunga, which is 4 km long and 2 km wide, did not have *Melitaea cinxia* in June 1991, though there were some 20 small meadows apparently suitable for the butterfly (for the location of this island, see figure 2.3). In August 1991, 72 larval groups from the main Åland Island were translocated to 10 meadows in Sottunga. This metapopulation has persisted for the past 11 years, though it went through a bottleneck of just three occupied meadows in 1999. None of the original populations that were established in 1991 has survived until present, and hence this metapopulation has survived in a stochastic balance between local extinctions and recolonizations of empty meadows, as predicted by the theory. The figure shows the observed changes in metapopulation size in terms of the number of larval groups and the number of occupied meadows, as well as the number of turnover events (extinctions and colonizations) between consecutive years.

along the northern portion of the species' range in British Columbia (Guppy and Shepard 2001).

Landscape-Level and Global Changes

In section 13.4 we discussed the significance of classical metapopulation dynamics for the occurrence of butterflies in fragmented landscapes and the conservation lessons gleaned from our research on that topic. However, the metapopulation framework is only one perspective worthy of attention when considering large-scale spatial issues. For instance, landscape ecologists have examined how much species richness of butterflies is increased by landscape heterogeneity (Weibull et al. 2000) and to what extent species richness depends on particular types of land cover, such as old, fallow fields (Balmer and Erhardt 2000). Some human-dominated landscapes offer opportunities for habitat restoration without significant loss of other land uses. Examples include restoration of prairie habitat on roadsides in the United States (Ries et al. 2001) and conversion of the land below power lines into early successional habitats that satisfy the requirements of endangered butterfly species (M. Kuussaari, pers. comm.).

23. *The Quality of the Matrix Landscape and Corridors Is Expected to Influence Movements* Several studies have examined the significance of corridors in facilitating movements of butterflies in fragmented landscapes. In some cases, apparent corridors, such as hedgerows in Britain, may simply represent linear habitat patches for butterflies (Dover and Sparks 2000). More generally, the structure of the matrix separating habitat patches can be expected to influence movements of butterflies among the patches. Haddad (1999a, see also Haddad 1999b, Haddad and Baum 1999, Tewksbury et al. 2002)

conducted a large-scale replicated experiment in which fragments of early successional habitat were either connected or not by similar habitat across the matrix of pine forest. Corridors facilitated movements by three species of butterflies, *Junonia coenia, Euptoieta claudia,* and *Phoebis sennae*. Keyghobadi et al. (2003), working in Alberta, Canada, found that *Parnassius smintheus* moved more readily between meadows separated by open vegetation types than between meadows separated by forest. Another study by Ricketts (2001) demonstrated similar effects for 21 butterfly species in Colorado. Given that most butterflies inhabiting open habitats rarely fly in closed forests, it is not surprising that an effect of matrix type can be detected in carefully selected comparisons. To what extent the quality of the matrix influences spatial dynamics in heterogeneous landscapes with a fine-scale mixture of different kinds of land cover is a somewhat different issue. A modeling study by Moilanen and Hanski (1998) on *Melitaea cinxia* suggested that in a heterogeneous landscape, the overall effect of the matrix quality was limited, apparently because in a fine-scale landscape mosaic the effects of different matrix types become averaged.

At an even larger scale, studies of checkerspots and other butterflies continue to play a significant role in demonstrating the biological consequences of global climate change (Hellmann 2002b), as well as pointing to genetic mechanisms that can influence the responses of populations to environmental change (e.g., Watt 1992). Parmesan (1996) showed that southern and lower elevation populations of *Euphydryas editha* in California had significantly higher extinction rates than northern and higher elevation populations during the latter portion of the past century (figure 13.8). One causal mechanism for the extinctions may have been increased population variability induced by increasing variance in precipitation (McLaughlin et al. 2002a; see figure 3.16b). Southern populations of *E. editha* are likely to be even more vulnerable than the well-studied Jasper Ridge populations (chapter 3) to increased variance in precipitation because phenological constraints are more severe in warmer climates. The link between population variability and extinction risk is well known (Pollard and Yates 1992, Lande 1993, Foley 1994, Ariño and Pimm 1995, Ludwig 1996, Belovsky et al. 1999).

Changes in geographical ranges similar to those observed in *E. editha* in California have occurred in a large number of butterfly species in Europe (Hill et al. 1999b, 2001, 2002, 2003, Parmesan et al. 1999, Hughes 2000, Roy and Sparks 2000, Warren et al. 2001). European studies have demonstrated that the impact of climate change will interact with other factors in determining the ultimate consequences for the abundance and distribution of species. In areas close to the northern range limit of species, climate change may broaden the range of microhabitats available for colonization for some species, thereby increasing the density of suitable habitat patches in the landscape and thus potentially facilitating the spread of the species (C. D. Thomas et al. 2001). But this opportunity for enhanced colonization cannot be realized if the landscape has already become so severely fragmented that the newly colonizable sites remain too isolated. Thus, in an analysis of the 46 nonmigratory butterfly species reaching their northern range limits in Britain, half of the mobile habitat generalist species have increased their distribution over the past decades, apparently in response to climate change. But the distribution of 89% of the habitat specialists has instead decreased (Warren et al. 2001), primarily because of loss of suitable habitat and connectivity. Even when range shifts do occur, as exemplified by *Pararge aegeria,* the current distribution of the species appears to lag behind the occupancy pattern predicted by current climate and habitat availability (Hill et al. 2003). Such a delay presumably reflects the general delay in the response of species to environmental change.

24. Climate Change Related to Global Warming Does Not Always Improve Conditions for Southerly Species at Their Northern Range Limit In the Åland Islands in Finland, several summers in the 1990s included exceptionally dry periods in July or August, which caused widespread withering of larval host plants and, subsequently, larval mortality of *M. cinxia*. In contrast, the populations have done well in relatively cool and wet summers (figure 4.7). The explanation for this result is that much of the *M. cinxia* habitat occurs on rocky outcrops, which are especially susceptible to drying during even relatively short periods of drought. Similarly, recent reports suggest that northern populations of *E. editha* have suffered declines and possibly extinctions in the last several years due to drought (Guppy and Fischer 2001, Guppy and Shepard 2001). If climate change comes in the form of increased variability, as observed at Jasper Ridge (McLaughlin et al. 2002a), populations throughout the range of *E. editha*, in-

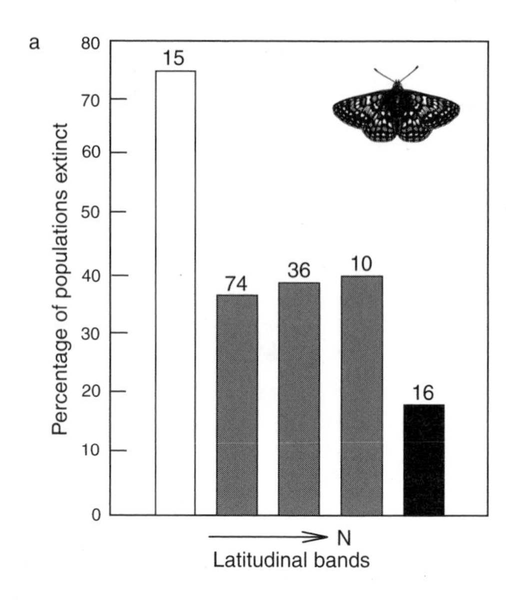

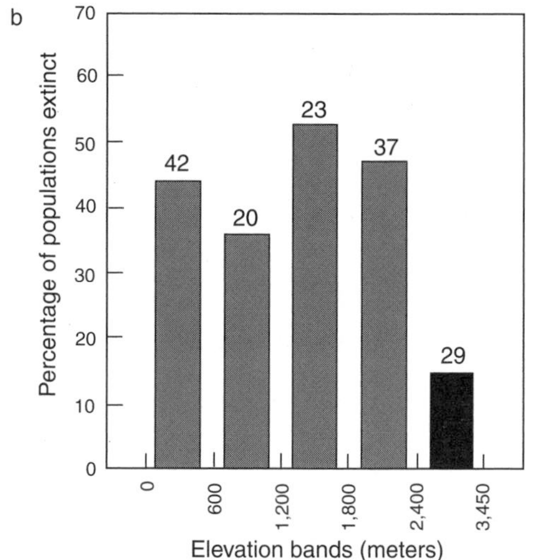

Figure 13.8. Population extinctions of *Euphydryas editha* on west coast of North America, as recorded by Parmesan (1996). (a) Observed extinction rate with latitude, showing increased extinctions in southern part of range. (b) Observed extinction rate with elevation, showing increased extinctions at lower elevations. In (a) and (b), latitudinal and elevational bands of different shades are significantly different at $p \leq .05$. For further details, see Parmesan (1966).

cluding those in the north, may be affected. This may be especially true in the case of butterfly populations that, like many *E. editha* populations, are adapted to local host species. *Euphydryas editha* populations in submontane Colorado appear to feed on the one host plant (of several that are nutritionally satisfactory) that is relatively drought resistant and phenologically suitable (Holdren and Ehrlich 1982). Even small changes in climate, such as an increasing frequency or severity of droughts, might easily lead to population extinctions.

In the face of global change, conservation biologists are often forced to rely on "quick and dirty" ways to evaluate the impact of habitat alteration. One approach they have taken is to search for focal species. Those are species thought to be representative of the species richness, occurrence, or status of other, less easily studied taxa (indicator species); species whose protection could help conserve other taxa (umbrella species); or species that might have a disproportionately large impact in the ecosystem, relative to their numbers or biomass (keystone species) (Fleishman et al. 2000a, 2000c). Butterflies have been suggested for the first two purposes (e.g., Prendergast et al. 1993, Kremen 1994, Daily and Ehrlich 1995, Holl 1995, Prendergast 1997). However, the practical utility of some focal species concepts has been called into question (e.g., Daily and Ehrlich 1996c, Andelman and Fagan 2000, Rubinoff 2001).

25. Checkerspots and Other Butterflies Are Useful for Evaluating the Focal Species Concept in Conservation Planning Studies by Carroll and Pearson (1998), Blair (1999), and Fleishman et al. (2001, 2002a, 2003) have shown that some butterfly species may serve as effective umbrella species. *Euphydryas editha* has been shown to have potential value for protecting the suite of organisms endemic to the increasingly rare serpentine grasslands of the San Francisco Bay area of California (Murphy and Ehrlich 1989) and for predicting butterfly species richness in the mountains of the Great Basin (interior drainage) of the western United States (Mac Nally and Fleishman 2002).

13.6 Conservation of Butterflies: Policy and Action

What steps should be taken to slow down and ultimately halt the disappearance of butterflies? The main step, of course, is the same that should be taken to halt the decimation of biodiversity in general: we must reduce the scale of the human enterprise (Ehrlich 1997, Ehrlich and Ehrlich 2004). The basic problem is summarized in the well-known I = PAT formula (e.g., Ehrlich and Holdren 1971, Holdren and Ehrlich 1974, Ehrlich and Ehrlich 1991): environmental impact (I) is a product of the number of people (P), their level of affluence or per capita consumption (A), and a measure of the environmental impact of the technologies and socioeconomic-political arrangements used to service the consumption (T). As long as the human population continues to increase, consumption per capita rises, and people refuse to adopt environmentally benign technologies, checkerspots, other butterflies, and the rest of biodiversity will continue to disappear (e.g., Ehrlich and Ehrlich 1981, Ehrlich 1995). But beyond the critical long-term actions of moving toward a sustainable human population size of perhaps 1–2 billion people (Daily et al. 1994), finding ways to curb overconsumption (Ehrlich and Ehrlich 2004), and facilitating the transition to more environmentally friendly technologies, other actions can help protect butterflies in the short term. Specific measures have been discussed earlier in the chapter; here we turn to more general issues.

One of the main conclusions we can draw from our work on checkerspots is that informed conservation decisions often can be made without waiting for results of exhaustive research. Most conservation decisions are necessarily implemented with incomplete knowledge about the species of concern (Ludwig et al. 1993), and one of the most vexing dilemmas is determining how much information is necessary to conserve a target population or species. Lessons from the checkerspot systems, and in particular from *Euphydryas editha*, help address this dilemma (Ehrlich 1992). Using data from several decades of study of *E. editha* in the Bay area, we were able to develop models that accurately predicted fluctuations of adjacent populations at Jasper Ridge (McLaughlin et al. 2002a; figure 12.2). However, the models are site specific and therefore population specific—each performs poorly if applied to the other population. Fortunately, that work showed that intensive research is not required before initiating conservation actions. Relative extinction risk can be determined without detailed study with the use of metrics such as population variability and its surrogate, topographical and other within-site habitat diversity. Less time may be needed to reach conservation decisions about many plants and invertebrates now that climate-induced variability is understood for the Bay area *E. editha* and other intensively studied checkerspots.

In the case of *Melitaea cinxia* in the Åland Islands in Finland, long-term research has allowed us to parameterize metapopulation models with substantial predictive power (Hanski 1994b, Hanski et al. 1996c, Wahlberg et al. 1996, 1999, Hanski and Ovaskainen 2000). Repeating the same research exercise with many other species would be very expensive and time consuming. Fortunately, research on *M. cinxia* has produced results and stimulated theory indicating that research on relatively common, well-known species sometimes can be used to draw inferences regarding related species for which much fewer data exist (lesson #9). Just calculating the metapopulation capacity for fragmented landscapes is a potentially powerful tool for comparing alternative management scenarios in patchy or highly fragmented landscapes (Hanski and Ovaskainen 2000; see examples in section 13.5).

Control of human population size and human consumption of resources is most essential in rich nations, since the actions of rich nations drive extinction events far beyond their borders (Ehrlich and Ehrlich 1989, 2004). As exemplified by conservation efforts in the United Kingdom, rich nations often have sufficient resources to take relatively expensive steps to delay butterfly extinctions (Pullin 1995, 1996). But those resources are sometimes applied without considering the existing scientific knowledge. For instance, 13 butterfly species and subspecies have been legally protected in Finland since 1997, but this protection principally means only that collecting them is forbidden. This kind of protection is clearly not efficient, as collecting adult butterflies is rarely a major threat to population persistence. Ten butterfly species currently have specially protected status, which means that their populations are protected against any actions that might adversely affect them. Unfortunately, the legislation of 1997 was partly based on inaccurate information. Therefore, the 10 specially protected species represent a mixed bag: some are not currently considered to be threatened (e.g., *Parnassius apollo*), some are increasing their ranges (e.g., *Lycaena dispar*), some apparently have erratic long-term dynamics (e.g., *Lopinga achine*), and some have severely declined and are critically endangered (e.g., *Maculinea arion* and *Pseudophilotes baton*). This situation exemplifies the challenges of translating biological information, which is continuously being updated, into conservation action and legislation.

Prospects appear better for butterfly conservation in North America (north of Mexico) than in much of Europe. North America has a relatively rich butterfly fauna, with 10 times more species than Britain. North America is topographically and ecologically diverse, and has 80 times the area and only a fifteenth of Britain's human population density (the United States alone has about an eighth). Nonetheless, even in North America the butterfly fauna is in widespread decline through loss of populations. The United States has a pioneering federal Endangered Species Act (ESA), which has been helpful in protecting species, especially prominent vertebrates. The ESA can protect any species and any distinct population segment of a vertebrate, but populations of invertebrates are eligible for protection only if they are designated as subspecies. Unfortunately, subspecific differentiation based on morphological characters is only weakly concordant with patterns of larval host plant use and other key ecological characteristics of populations (chapter 2). Thus ESA listings may poorly reflect important patterns of genetic and ecological diversity. But although implementation of the ESA can be inefficient and subject to time-consuming and expensive legal challenges, it is much better than nothing. It is the main legal tool citizens of the United States are likely to have until and unless the public at large insists more be done to protect biodiversity. The legal battles could be made much worse by the needed alteration of the ESA into a device for protecting populations and ecosystems, which is unlikely in the short term but essential in the medium term. Emerging interest in Canada for legal protection of biodiversity may lead to new beneficial conservation laws in that country as well (dubbed the Species at Risk Act as of 2002). Eventually the burden of proof must be shifted to those who believe that sprawling development and spewing of greenhouse gases into the environment have benefits that outweigh their heavy costs.

In the European Union, the most powerful conservation legislation concerns those habitats and species that are included in the annexes of the Habitat Directive. The EU-wide Natura 2000 reserve network is based on these habitats. The annexes list many crucial butterfly habitats, such as different types of managed and unmanaged grasslands, as well as 26 species of butterflies. The species list was much criticized in a recent conference on the directive species, however, as many species and subspecies that are clearly endangered in the EU (according to van Swaay and Warren 1999) have been left off the list. Apart from the European legislation, most European countries have their own nationally protected butterfly species and reserves, as we have described for Finland.

Much of the action to preserve biodiversity in general and butterflies in particular will need to occur in the tropical centers of butterfly diversity, where forests and other natural habitats are disappearing rapidly. Unfortunately, not much of what we have learned from the checkerspots and other temperate model systems is likely to be applicable in tropical nations. The future of tropical biodiversity depends largely on limiting human population growth and using ecologically sustainable development to bring poor people out of desperate poverty, so that they will have the education and opportunity to take a long-range view rather than continually scrabbling to keep their families fed, housed, and healthy. Only then would they be in a position to value butterflies both for their role in ecosystems and their beauty and have the necessary resources to make substantial investments in conservation. The chances for sensible development in the developing tropical countries naturally relate to the way international companies operate in these countries and how much pressure can be taken off the tropics by reducing population and restraining wasteful consumption in already-rich temperate-zone nations. Currently much of the destructive land use practices in the tropics is the responsibility of North American, European, and Japanese companies and the trade agreements favoring the rich nations.

In some very limited situations the esthetic appreciation of butterflies may actually help their preservation in the tropics by creating a flow of hard currency to tropical butterfly farming enterprises. Perhaps the best example is the combined conservation and economic success of raising rare birdwing butterflies (Ornithoptera; for an excellent discussion, see Parsons 1999, chapter 6). In terms of research, neither the funds nor the trained personnel are available to conduct even the most basic investigations in most of the tropical centers of biodiversity, and the political structures are often not in place to implement conservation recommendations for charismatic megavertebrates, let alone for insects. A hope for tropical biodiversity in most areas may reside in the development of countryside biogeography, which is seeking principles that will guide actions to increase the capacity of human-disturbed landscapes to maintain populations of endangered organisms (Daily 2001, Daily et al. 2001). Butterflies have been a useful model system in the development of this new field (e.g., Horner-Devine et al. 2002), and because most butterflies are denizens of open and successional habitats, they are especially likely to benefit from the findings of countryside biogeography (e.g., Carreck and Williams 2002).

While efforts of many kinds are needed, there is one contribution that we can all make to help conserve butterflies. That is to greatly expand the public's familiarity with the beauty and intricate biology of butterflies—to bring to others the joy that we have had in our long-term association with these fascinating insects. Then they, like us, will press decision makers to maintain butterflies as a valued component of the human environment.

14

What Have We Learned?

PAUL R. EHRLICH, ILKKA HANSKI, AND CAROL L. BOGGS

14.1 Introduction

In this chapter we recap the major lessons to be distilled from the work on checkerspots. Some of the lessons are novel; some reinforce previous conclusions by other researchers. The first, and perhaps the most significant lesson relates to the model system approach that we have taken in our research. The diversity of natural populations is so great that model systems may never play quite the same role in population biology as *Drosophila* in genetics or *Caenorhabditis elegans* in developmental biology. Nonetheless, the checkerspot system has demonstrated that concentrating research efforts on particular systems will greatly contribute to better understanding of many fundamental issues in ecology, evolution, behavior, and conservation biology. The knowledge gained from the model systems ultimately can be applied to answering questions about less intensively studied species. For instance, if we were now asked to evaluate the conservation status of a little-known endangered insect species, we would first look at its probable metapopulation structure, the topographic heterogeneity of its habitat, and how these will interact with climatic variability. None of this may prove critical in a particular system, but the work on checkerspots tells us that here would be a good place to start.

Ehrlich and his research group began work on the population dynamics and evolution of *Euphydryas editha* at Jasper Ridge as a single point in what Ehrlich hoped would become a taxonomically and geographically stratified sample of model systems investigated by population biologists. But even that single point did not remain as such for very long, as Ehrlich and others became interested in other populations of *E. editha*, other species of *Euphydryas*, other checkerspots, and other butterflies. Research on *Euphydryas* helped Hanski decide to start a large-scale project on the closely related *Melitaea cinxia*, which in turn led to research by his group on other European checkerspots and on the set of species that interact with *M. cinxia,* the focal model species. Such an expansion to related species may represent the natural evolution of single-population model systems in population biology, needed to overcome the inevitable limitations set by particular environmental and biological conditions of any one population and landscape. This expansion has been aided by the numerous students who have been trained in the Ehrlich and Hanski groups, many of whom have established their own research groups, often working on butterflies though not just on checkerspots. These scientists and many others, some tracing their contacts to the work of E.B. Ford at Oxford and Charles Remington at Yale, have helped research on butterflies as a group to begin to merge into one large model system (Ehrlich 2001, Boggs et al. 2003, Watt and Boggs 2003).

Although it is appropriate to view our research in this broader taxonomically defined context, putting the research in the conceptual framework of population and metapopulation biology is more important. We next highlight what we consider to be the main contributions of this research to population biology, systematics, evolutionary studies, and conservation biology.

14.2 Contributions of Checkerspot Research to Population Biology

Population Dynamics

One of the most basic conclusions from our work is that in order to understand the dynamics of populations, it is essential to understand their spatial structure (box 14.1). When Ehrlich first arrived at Stanford in the fall of 1959, a butterfly-collecting colleague, Elton Sette, told him there was "a population" of *Euphydryas editha* on Jasper Ridge. Suppose that Ehrlich had simply assumed that the colony on the ridge, occupying a 1500 m strip of habitat, was a single population and had not subdivided the ridge so that he could keep track of where each individual in his study was marked and released. He would not have discovered that there were three local populations (demographic units) on Jasper Ridge. That, in turn, would have produced a misleading impression of the factors controlling population size, since the extinction of one local population would not have been detected, and averaging would have concealed the somewhat different dynamics of the other populations. Considering that the earlier literature on population dynamics (e.g., Huntsman 1937, MacLulich 1937, Schwerdtfeger 1941, Elton 1942, Baltensweiler 1964) focused on economically important organisms and paid little attention to spatial structure, it seems probable that the development of metapopulation theory would have been retarded without the early emphasis on local population structure in *E. editha*. The realization that population dynamics can be properly understood only in the context of their spatial structure has been an important outcome of our work and is a conclusion with consequences for everything from strategies for harvesting Peruvian anchovetas (Boerema and Bulland 1973) to controlling Mediterranean fruit flies (W. Watt, pers. comm.).

We often speak casually of "the ecology of" or "population regulation in" species X (Barney and Anson 1920, Cullenward et al. 1979). One of the lessons learned from the work on checkerspots is that ecological relationships and the mechanisms of population regulation may vary greatly among populations within the same species. This is shown dramatically by the ecological differences (host plant use, timing of adult flight, vulnerability to population extinction, etc.) among populations of *E. editha*. We have shown that the local populations C and H on Jasper Ridge exhibited significant differences in the mixture of endogenous and exogenous factors influencing their dynamics (McLaughlin et al. 2002a). It is hardly surprising that more distant populations of *E. editha*, such as those at Del Puerto Canyon in the Inner Coast Range, Ebbett's Pass in the Sierra

---

BOX 14.1 Key Lessons: Population Ecology

The more than four decades of research on *Euphydryas editha* have elucidated many facets of the dynamics of populations:

1. Identifying demographic units is critical for a proper understanding of population dynamics.
2. The dynamics and life histories of populations can vary dramatically among populations of the same species (chapter 3).
3. Topographic heterogeneity is often a critical factor influencing the dynamics of populations because it reduces the extent to which the impact of temporal variation in environmental conditions (environmental stochasticity) is correlated among individuals (chapter 13).
4. The relative importance of endogenous and exogenous factors in population dynamics varies from one population to another (chapter 3).
5. Individual movement is not a reliable measure of gene flow (chapter 10).

Nevada, and Lower Otay near the Mexican border have very different ecological characteristics (Ehrlich et al. 1975), leading to differences in population regulation.

One of the main lessons from research on *M. cinxia* is that, in the large metapopulation in the Åland Islands, practically every mechanism that has been proposed to play any role in population extinction (e.g., Caughley 1994) has, in fact, been shown to make a difference (Hanski 1998b, 1999b). Of equal importance is the realization that not all mechanisms have been biologically significant in all the populations all the time. As a matter of fact, very different conclusions would have been reached had the research been restricted to some particular local population or a small assemblage of such populations, studied in some years only. Ecologists must be alert to the fact that regulation of different populations of the same species, or of the same population in different years, may be very different.

Some may read this great spatial and temporal variation in the processes of population regulation as a warning against any generalizations. This is not our intention, however. There are, of course, factors that are inherently difficult or impossible to predict, such as year-to-year variation in weather conditions, but the effects of climatic conditions on individuals and populations can be investigated to yield useful generalizations. For example, we have reached a good mechanistic understanding of the interaction between temporal variation in environmental conditions and spatial variation in habitat quality for *E. editha* (chapters 3 and 13). Based on this information, we expect this interaction to be of fundamental importance in the dynamics of a great many populations of checkerspots, other butterflies, other insects, and yet other small organisms. Another example is the influence of landscape structure on the dynamics and persistence of metapopulations. It would be impossible to predict accurately the dynamics or even the presence of a species in each particular part of the landscape, but it may be possible make robust predictions about persistence and dynamics at the landscape level (chapter 12), regardless of the small-scale variability generated by the multitude of population processes.

One particular feature of the life history of checkerspots, as well as of many other insects, is that females lay eggs in smaller or larger clutches. Due to correlated survival within clutches (chapter 7), ecological risks are not as well spread (den Boer 1968) as they would be if eggs were laid singly. Combined with the fact that most females are singly mated (chapter 5), clutch laying and correlated survival of individuals among clutches strongly reduces the genetic effective size of populations, thus enhancing the loss of genetic variability from small populations (chapter 10). Clutch-laying can also be expected to increase variability in population size (Hanski 1987), which will further increase the risk of extinction of small populations. In *M. cinxia*, overwinter survival strongly depends on the size of the larval group (chapter 7), apparently because small groups are unable to construct a high-quality winter nest within which the larvae diapause. In general, highly gregarious larval behavior as in *M. cinxia* brings a special twist to questions about density dependence of population dynamics because one has to recognize potentially conflicting effects of density at different spatial scales: positive effects at the level of larval groups, negative effects at larger scales within populations (chapter 7). In *E. editha*, variation in clutch size among populations is at least partly explained by the size and species of the larval host plant (chapter 7). Clutch laying and gregarious larval behavior are particularly fascinating areas for research because they integrate many behavioral, ecological, genetic, and evolutionary issues. Such integration, of which there are more examples below, is one facet of the checkerspot research that has benefited from the long-term, concerted research effort.

A major finding from the checkerspot research that will not surprise anybody is that *Homo sapiens* is presently a force shaping the dynamics and future of many, if not most, populations, including the checkerspot populations that we have studied in California and in Finland. Land clearing and development, agricultural practices, nitrogen deposition, and climate change all influence checkerspots in dramatic ways, and sadly these influences appear likely only to strengthen in the decades ahead.

Metapopulation Dynamics

Metapopulation structure may be the rule rather than the exception for most small animals, especially for many insects that are habitat specialists and relatively sedentary, as they are likely to experience most landscapes as fragmented at the population scale. There has been a tendency to classify metapopulations by type of spatial population structure. However, just naming populations and metapopulations as having this or that structure is unlikely to be helpful (Hanski 1999b). It is evident that real landscapes and metapopulations cover a continuum of structures, and the real challenge is to understand the processes that are significant under different conditions and that may in turn mold these structures. Checkerspots have proved to be excellent systems for developing this sort of understanding. The group provides well-studied examples of relatively continuous populations living in extensive tracts of suitable habitat but with important spatial variation in habitat quality (e.g., *E. editha* at Morgan Hill; Harrison et al. 1988, chapter 3), examples of mainland–island metapopulations (Morgan Hill and the outlying small patches of habitat; Harrison et al. 1988), examples of classic metapopulations without any large, persistent local populations (*M. cinxia* in the Åland Islands; Hanski et al. 1995a, Hanski 1999b; chapter 4), and examples of source–sink metapopulations (*E. editha* in the Sierra Nevada; Boughton 1999, unpubl. ms; chapter 9).

Research on *M. cinxia* has focused attention on the biology of species living in highly fragmented landscapes—that is, in landscapes where suitable habitat for the focal species comprises only a small fraction of the total landscape and occurs as discrete patches (box 14.2). This checkerspot system has functioned as a model system for metapopulation biology, in which theory, modeling, and field studies have been developed hand in hand (chapter 12). Empirical research on checkerspots, as well as on a large number of other organisms, has provided ample evidence for the great influence of habitat fragment size, quality, and spatial connectivity on the key processes of classic metapopulation dynamics, local extinction and establishment of new populations. Research on checkerspots has stimulated the development of the spatially realistic metapopulation theory (Hanski 2001, Hanski and Ovaskainen 2004; chapter 12), which provides a framework for modeling the influence of landscape structure on population processes, thus building a bridge between metapopulation ecology and landscape ecology. This theory also allows one to analyze spatial structures ranging from the mainland–island scenario to the Levins-type metapopulation structures. Indeed, the theory represents a synthesis of the dynamic theory of island biogeography of MacArthur and Wilson (1967) and the classic metapopopulation model of Levins (1969).

We have learned that habitat fragment area and isolation effects can be dissected not only at the level of populations but also at the level of individuals. The timing and rates of emigration and the range of migration vary with environmental conditions, including the sizes and the spatial positions of the habitat fragments. Typically, emigration rate increases and immigration rate decreases with decreasing area of the habitat fragment, and the rate of

BOX 14.2 Key Lessons: Metapopulation Ecology

Long-term research in the Åland Islands has covered the entire range of *Melitaea cinxia* in Finland, an area of 50 × 70 km. The suitable habitat is fragmented into more than 4000 distinct meadows that vary in their area (though all are small or relatively small) and degree of connectivity to other meadows. Key lessons learned include:

1. The spatial configuration of suitable habitat together with variation in habitat quality determines the probability of a species persisting in the landscape (chapter 4). There is a threshold value of landscape structure below which the species will go extinct.
2. Theory incorporating the spatial configuration of the suitable habitat leads to spatially structured models that are akin to matrix population models (chapter 12). Ecologists are able to make quantitative predictions about the population dynamic consequences of habitat loss and fragmentation.
3. Local populations often exhibit significant variation in genetic composition and life-history traits due to both spatial variation in the environment and spatial population processes (founder effects, migration). Such variation may influence local extinctions (chapter 10) and recolonizations (chapter 9).

successful migration decreases with increasing isolation. Though these effects are not unexpected, conscious attention to habitat fragment area and isolation effects suggests a modeling approach with which useful insights can be gained about migration of individuals in highly fragmented landscapes (Hanski et al. 2000). Migration in heterogeneous landscapes also raises questions about individual variation in traits that influence the perception of habitat quality. Not only is there variation in the factors that influence the performance of individuals in space and time, but different individuals have different perceptions about variation in such factors, and these differences among individuals can be large enough to influence population and metapopulation dynamics. A striking example is the influence of oviposition host-plant preference by *M. cinxia* on the rate of migration among populations (Kuussaari et al. 2000) and the colonization of currently empty habitat patches (Hanski and Singer 2001; chapter 9).

Several chapters in this volume have commented on the sedentary behavior of checkerspots, which facilitates the delimitation of local populations (chapters 3 and 4) and frequent local adaptation in host plant use (chapters 6 and 9). Checkerspots are "sedentary" compared with many other organisms, though many other butterflies are equally sedentary (chapter 11) and have clear-cut local populations, often largely constrained by the patchy distribution of the host plants. Among checkerspot species (chapter 11), and even among populations (chapter 9), there are also differences in mobility, which undoubtedly reflect differences in the spatial distribution of larval and adult resources, the degree of habitat fragmentation, the likelihood of inbreeding and sib competition, and probably also variation in correlated life-history traits such as readiness to reproduce soon after eclosion (chapter 5). Intriguing research questions remain about how fast migration rate evolves in response to environmental changes and about the actual mechanisms on which such changes might be based, whether behavioral, physiological, or morphological.

Trophic Interactions and Coevolution

Herbivores are located "between the devil and the deep blue sea" as Lawton and McNeill (1979) put it. The devil is the set of natural enemies attacking the herbivores, and the deep blue sea is the vegetation that seems so benign but may in reality be anything but benign. The checkerspot system has taught us a great deal about the factors controlling herbivore populations from both above and below (box 14.3).

From "below" in food chains, host plants have a profound influence on the ecology and evolution of checkerspot populations. One of the first general rewards of working on the checkerspot system came in the early 1960s. It followed from Ehrlich commenting to his plant-evolutionist colleague Peter Raven about the host plant choices of Jasper Ridge *Euphydryas*. Wasn't it strange, he asked, for *E. editha* to feed on *Plantago erecta* (Plantaginaceae) and *E. chalcedona* to feed on *Diplacus aurantiacus* (Scrophulariaceae)? Raven responded that it wasn't

BOX. 14.3 Key Lessons: Trophic Interactions

The checkerspot system has contributed to our understanding of trophic interactions in several major ways:

1. Research on checkerspots sparked the field of plant–herbivore coevolution (chapter 11).
2. Checkerspots have been used to investigate the details of the relationships between herbivores and plant populations, including spatial and temporal patterns in host plant use, evolution of specific insect–plant interactions, and the basis for host shifts (chapter 6).
3. Checkerspot research has elucidated the interplay between tri-trophic interactions and spatial dynamics. Parasitoid attack may depend on the host plant(s) utilized by the host larvae in a particular population, with significant metapopulation-level consequences to the parasitoid (chapter 8).

strange at all: "don't you know that Plantaginaceae are just wind-pollinated Scrophulariaceae?" (Ehrlich 1984b). That initial discussion between Ehrlich and Raven led to the first major spin-off from research on checkerspots. Their discussion of the patterns of host plant utilization by butterfly larvae caused them to look for more general patterns, and they ransacked a large literature produced largely by amateur lepidopterists. Collectors often published records of host plant use, and by around 1960 host plants had been reported for about half of all the genera of true butterflies (Papilionoidea). Butterflies had become the largest group of herbivores (about 15,000 species) whose food choices were quite well documented.

Analyzing that database from the perspective of both the butterflies and their host plants, Ehrlich and Raven (1964) concluded that "the evolution of secondary plant substances and the stepwise evolutionary responses to these by phytophagous organisms have clearly been the dominant factors in the evolution of butterflies and other phytophagous groups." Right or wrong on that point (chapter 11), a conversation about checkerspots, stimulated by some of the first research done on the system, launched the field of coevolution (e.g., Ehrlich and Raven 1964, Berenbaum 1983, Futuyma 1991, Farrell et al. 1992, Becerra 1997, Farrell 1998, Janz and Nylin 1998). As we have seen in the earlier chapters of this volume, checkerspots have contributed substantially to our understanding of plant–herbivore coevolution, especially in the case of oviposition plant preference (chapter 6), and relationships of adult butterflies and larvae with the larval host plants remain a major focus of checkerspot research. One of the outstanding contributions of the checkerspot system to evolutionary ecology was the discovery of the unexpected and extraordinary complexity that can underlie the decision of a female insect about whether to lay its eggs on a given plant. This work, largely due to Singer and colleagues, comprises one of the most thorough and ingenious investigations ever carried out on the behavioral and evolutionary relationships of an herbivore and its host plants (chapter 6).

Many checkerspot populations are attacked from "above" in their food chains by specialized parasitoids. The sequestered host-plant secondary compounds that appear to protect larvae against many other natural enemies are ineffective against the specialist parasitoids (chapter 8). Much more research has been done on the parasitoids of checkerspots in Europe than in North America, and hence the latter communities are not yet well known, but our impression is that the role of parasitoids is smaller in the dynamics of the well-studied *Euphydryas editha* populations in California than in the dynamics of *Melitaea cinxia* in northern Europe. One possible reason for this difference is the ephemeral nature of many host plants in California's Mediterranean climate, as compared to host plants of *M. cinxia* in northern Europe. One consequence is that *E. editha* larvae disperse at a much earlier stage than the larvae of *M. cinxia*; temporally extended gregarious behavior in *M. cinxia* may make them especially vulnerable to parasitism.

Trophic interactions also include indirect connections between the species below and above the focal checkerspot butterflies, in the form of the influence of butterfly host-plant species on parasitoid dynamics. Such tri-trophic interactions have been much studied by ecologists and entomologists (for a review, see Tscharntke and Hawkins 2002), but generally without consideration for the spatial dimension of the interaction (but see van Nouhuys and Hanski 2002b). Research on *M. cinxia* has placed the study of such interactions squarely in a spatial context, considering hundreds of dynamically coupled local populations not only of the butterfly but also of its specialist natural enemies, as well as population-to-population variation in the relative abundances of the two host plant species. One striking result is that the large-scale spatial dynamics of one of the parasitoid species strongly depend on which of the two possible host plants the butterfly larvae feed on (van Nouhuys and Hanski 1999). These results highlight the possibility that large-scale spatial dynamics can be critically influenced by individual behavioral decisions, which in this case underlie the tri-trophic interactions locally.

One of the most venerable generalizations in ecology is the species–area relationship (Rosenzweig 1995), which may be driven by the size of the area that influences metacommunity dynamics (Holt and Gaines 1992, Holt 1997). The food chains associated with *M. cinxia* in Åland nicely illustrate these principles. The main Åland Island, with several thousand habitat patches (figure 2.3), has the full food chain consisting of host plants (*Plantago lanceolota* and *Veronica spicata*), the butterfly (*M. cinxia*), the primary parasitoid *Hyposoter horticola*, and the specialist hyperparasitoid *Mesochorus stigmaticus* (chapter 8). In contrast, the nearby island of Kumlinge which have an order of magnitude smaller habitat patch network, lacks the hyperparasitoid, apparently because this specialist does not have a viable metapopulation within a network of < 100 small habitat fragments (for more examples and discussion, see Hanski 1999b).

Behavior

One of the strengths of the checkerspot system is the extent to which researchers have been able to explore the behavioral and physiological mechanisms that underlie the ecological and evolutionary dynamics of the butterflies, thereby providing a solid basis for understanding those dynamics (box 14.4). This work began with the pioneering studies of Labine on *E. editha*. She documented the mating patterns of females, describing the way in which mating plugs prevent female remating and suggesting that a stretch receptor in the bursa could induce behavioral nonreceptivity (Labine 1964). Labine (1966a) also did one of the first studies of sperm competition in insects, again using *E. editha*. These analyses gave us a basis for understanding the genetic effects of female migration (chapter 10). Labine's work also broadened the research arena on the evolution of mating systems (chapter 5).

More recent work on mating systems in checkerspots has focused on inbreeding in the context of metapopulations. Work on *M. cinxia* has shown that inbreeding in small local populations can be sufficiently strong to have significant population dynamic consequences: inbred populations have an elevated risk of local extinction (Saccheri et al. 1998, Nieminen et al. 2001). This being the case, one might expect that behaviors would evolve to minimize the incidence of inbreeding. The inversely density-dependent emigration rate found in *M. cinxia* (Kuussaari et al. 1996) and other checkerspots (Gilbert and Singer 1973, Brown and Ehrlich 1980) may be interpreted as a mechanism to reduce inbreeding, though emigration by males from low-density populations might also simply reflect limited mating opportunities. Inbreeding avoidance through evolution of female discrimination against close kin might also be expected, but two experiments found no evidence for such discrimination (Haikola et al. 2004). Perhaps the cost of inbreeding avoidance (reduced chance of mating by discriminating females) in small populations in a metapopulation is so high that it outweighs the potential genetic benefit. It is apparent that the consequences of individual behaviors have to be assessed in a proper population dynamic context.

Physiological Ecology

In butterflies and many other taxa, both thermal ecology and resource allocation significantly affect individual behaviors and life histories. Checkerspot research has contributed to our understanding in both areas, including the connections upward to the population level. With respect to thermal ecology, interactions between slope heterogeneity and larval thermal preferences and tolerances help shape the dynamics of Bay area populations of *E. editha*, as described in chapters 3 and 13 (Dobkin et al. 1987,

BOX 14.4 Key Lessons: Organismal Biology

Research on checkerspots has helped establish a mechanistic foundation for the study of the evolution of mating systems and life histories (chapter 5). Key lessons include:

1. Female reluctance to remate can be controlled by multiple mechanisms.
2. Inbreeding avoidance has not evolved in small ephemeral populations in a metapopulation, perhaps because the cost of such behavior would be too great.
3. Mating success can vary dramatically across space and time, influencing effective population size and population structure.
4. The combination of ovarian dynamics and adult feeding habits affects the role of both male- and female-derived adult nutrients in reproduction, influences age-specific fecundity and survival patterns, and modifies the impact of variation in food supply.

Weiss et al. 1987, Weiss and Murphy 1988b). The influences of the adults' thermal ecology on their activity patterns, oviposition site preference, and dispersal are less well known, although initial studies suggest that they are not dissimilar to those affecting the few other well-studied butterflies (Boggs et al. unpubl. data for *E. editha*; see also for other species, e.g., Watt 1968, Roland 1982, Kingsolver 1983, van Dyck and Matthysen 1999).

Studies of checkerspot adult life histories and feeding habits have examined how resource allocation determines life-history patterns. In combination with work on other taxa, these data provide the empirical base for the beginnings of a resource-based understanding of the evolution of, and constraints on, life-history strategies (Boggs 2003; chapter 5). The work on *Euphydryas* indicates that adult feeding makes a relatively small contribution to successful reproduction, although the contribution increases somewhat with female age (Murphy et al. 1983, Boggs 1997c, O'Brien et al. 2004). The importance of adult feeding becomes even less significant when one considers the short life span of most females and the constraints on larval survival due to plant phenology (Singer and Ehrlich 1979, Cushman et al. 1994). Likewise, nutrients donated by males at mating do not influence egg numbers or hatchability in *E. editha* (Jones et al. 1986), in contrast to what has been found in many other insects (Boggs 1995). One implication of these findings is that *Euphydryas* and other organisms with similar ovarian dynamics and feeding habits have life histories that are not particularly sensitive to variation in the adult nutritional environment. The caveat has to be added, however, that sugars in nectar commonly power flight and survival in other butterflies (e.g., Watt and Boggs 1987), and thus nectar feeding could be important to survival and dispersal of checkerspots even if it plays a minor role in reproduction.

14.3 Contributions of Checkerspot Research to Systematics and Evolution

The brilliant system of taxonomic nomenclature developed by the Swedish botanist Carolus Linnaeus (1758; for details see Mayr et al. 1953), has served biologists well for almost 250 years, but it has not proven perfect (no taxonomic system could be). The system must communicate about multidimensional sets of morphological, behavioral, and genetic relationships, a process that always involves cutting continua of differentiation. This has attracted much attention from philosophers (e.g., Ghiselin 1975, Hull 1976, Ruse 1987), and one result has been a gigantic, expanding, and often confused literature trying to solve the "species problem" that was definitively recognized as insoluble (or, rather, badly posed) 40 years ago.

Research on checkerspots played a major role in demonstrating that the "species problem" was, in fact, not a problem, at least from the viewpoint of evolutionists and ecologists (box 14.5). A phenetic analysis of *E. editha* and *E. chalcedona* specimens (Ehrlich 1961a) showed that individuals have different degrees of similarity and relatedness. This analysis also made it clear that the genus *Euphydryas*, as well as many of other Nearctic butterflies, did not divide neatly into distinct species (Ehrlich 1961a). Then came the discovery of the diversity of habitats, flight periods, oviposition plants, secondary host plants, nectar sources, and life histories among populations of *E. editha*, demonstrating that species were not necessarily monolithic ecological units (Ehrlich et al. 1975). And, finally, research on *E. editha* showed that the supposed coherence of species could not be credited to sharing a single gene pool (Ehrlich and White 1980). The fundamental reason that checkerspots and the rest of the natural world are not divided into easily identifiable, unambiguous units is that evolution creates continua of differentiation, which is displayed throughout the checkerspots. Phil DeVries (1987), in his outstanding treatise on Costa Rican butterflies, comments extensively on the difficulties of determining what should and should not be considered species in the Phycioditi (*Eresia, Anthanassa, Phyciodes*, etc.), and the problem carries over into the Nearctic realm (e.g., Scott 1986).

Early results showed those continua of differentiation through phenetics; more recent research has shown that the same applies to the underlying genetic differentiation. Checkerspot populations exhibit nearly continuous degrees of genetic divergence, regardless of which particular genes and populations are examined or what method of analysis is used (Brussard et al. 1985, 1989, Baughman et al. 1990a, Wahlberg and Zimmermann 2000). With respect to sexually reproducing organisms, most evolutionists are satisfied to follow Ernst Mayr's (1942) lead and consider as distinct species those populations of interbreeding individuals that are sympatric without showing abundant signs of hybridization, be they populations of hippos and

BOX 14.5 Key Lessons: Species and Speciation

Work on checkerspots has played a key role in highlighting the species pseudo-problem and in helping us understand the mechanisms of geographic speciation:

1. Geographic variation is ubiquitous and complex in checkerspots (Brussard et al. 1989), suggesting the presence of virtually every stage of speciation (e.g., Wahlberg and Zimmermann 2000).
2. Although it is usually simple in practice to apply the biological species concept to sympatric populations, it is impossible in practice and in theory to apply it definitively to allopatric populations (Ehrlich 1961a). Checkerspots support the idea that allopatric speciation is a major (probably the major) mode of differentiation of animal populations.
3. Work on *Euphydryas editha,* among other studies, led Ehrlich and Raven (1969) (chapter 10) to question the orthodoxy that barriers to gene flow are the prime mover in speciation, rather than similar selective pressures that keep species cohesive. The issue of the degree to which reduction in gene flow as opposed to different selection pressures are responsible for speciation remains an important question today (Templeton 1989, Wilson 1989, Ridley 1996), along with questions about the importance of sympatric speciation (Futuyma 1998, Johannesson 2001, Schluter 2001, Via 2001).

lions or of *Euphydryas aurinia* and *E. maturna.* This viewpoint sensibly recognizes the evolutionary importance of discontinuities in gene pools, even though technically determining the amount of interbreeding and hybridization can be complex (Sokal and Crovello 1970).

But with respect to allopatric populations, using the "potential interbreeding" criterion of the "biological species concept" is nearly impossible in all interesting cases (Ehrlich 1961a, Ehrlich and Holm 1962). And when asexual organisms are considered, it becomes obvious that taxonomists have no one clear way of determining how to define species, and that the best "definition" is simply "kind," or some arbitrary level of phenetic or genetic divergence. That definition, in fact, has proven quite useful in practice; evolutionists, ecologists, and behaviorists usually know what sort of entity is being discussed, most biologists never ponder the "problem" of "what a species is," and amateur birdwatchers or butterfly collectors ordinarily don't care. For a sample of the literature on the species non-problem, see Mischler and Donoghue (1982), Cracraft (1983), Patterson (1985), De Queiroz and Donoghue (1988), Nixon and Wheeler (1990), Vrana and Wheeler (1992) Vogler and DeSalle (1994), Mallet (1995), Davis (1996), Avise and Wollenberg (1997), De Queiroz (1998), R. G. Harrison (1998), Sterelny and Griffiths (1999), and Hey (2001).

A related question of interest is whether to recognize taxonomic entities below the level of species. Wilson and Brown (1953) pointed out 50 years ago that subspecies are not usually evolutionary units because of discordant character variation, and studies of checkerspots have supported this view (e.g., Brussard et al. 1989, Baughman et al. 1990a). There is, of course, considerable interest in studying patterns of genetic variation within species (Avise 1994), and subspecies names can sometimes make biogeographic discussions smoother. Another reason to name and describe new subspecies relates to conservation. In some circumstances in the United States, populations in danger of extinction can be protected legally if they are designated subspecies (chapter 13). Unfortunately, rather than just making a subspecific designation when it appears worth the effort, some taxonomists have engaged in an intellectually empty debate about what constitutes an "evolutionary significant unit" within a species (Ryder 1986, Moritz 1994). As Tom Brooks cogently put it, at the global scale this amounts to "fiddling while Rome burns" (quoted in DeWeerdt 2002).

Work on the checkerspot system has also illustrated the futility of a notion that is persistent in the taxonomic community, a notion that promotes an uneconomical way of sampling nature. This was recently expressed by one of the best taxonomists: "Our task is to chart the diversity of life, in its entirety, from the tiniest tips of the Tree to every one of its branches" (Donoghue 2001, p. 755). The "tiniest tips" are individuals that die without offspring, but knowing the context, we can assume that the statement refers to species (however defined) as "tips," and "chart" as producing superficial species descriptions and some sort of estimate of times of occurrence of the myriad crotches of the cladistic tree. Even trying to sample just an individual or two from every species would make that quite an ambitious project, and in our view a poor allocation of the efforts of scientists. Remember, the checkerspot system has shown in great detail that species are often not unitary "tips" (ecological or genetic units), but rather complexes of interrelated entities in constant flux as environments change, as demographic units go extinct and are reestablished, and as populations evolve in response to varying selection and migration pressures and genetic drift. Thus the nature of the twigs of the tree, even considering the twigs to be species, is not understood now, even for intensively studied model systems. Our understanding of how the world works would be much more enhanced by in-depth studies of a smaller sample of twigs and branches than by attempting the total "charting" (even if it was purely cladistic and ignored the evolutionarily critical phenetic dimension). It has been almost four decades since a simple analysis showed the futility of trying to "completely" chart life—even before its complexity was understood as it is today (Ehrlich 1964). In the light of what has been learned from the checkerspot system, and considering the accelerating extinction crisis, that goal now looks particularly preposterous.

Finally, on the taxonomic front, we believe that this multiauthored book demonstrates the basic utility of the Linnaean nomenclature system for communication among scientists. We have had no difficulty letting each other know exactly which organisms we were discussing and have not been hampered by the inevitable ambiguities of imposing a discontinuous system of names on a natural system permeated with various forms of continua. Indeed, if those interested in biology had stuck to the Linnaean system (originally designed for international communication) rather than also introducing a plethora of common names, and treated the expanded Linnaean hierarchy conservatively (Ehrlich and Murphy 1981a), our job would have been even easier. The multiplicity of "scientific" and "common" names (Murphy and Ehrlich 1983) and the instability of both is a major reason that systematics (taxonomy) is an unappreciated discipline. Most scientists and nature lovers are only exposed to it in the form of frequent, annoying, and often senseless changing of names. Following its traditional death wish, systematics is now debating changing all names, including all species names, to ones based on the uncertain and (for communication) relatively uninformative estimates of recency of common ancestry (Donoghue 2001, Pennisi 2001). Most people, including most biologists, are not mesmerized by cladistics. They care little about the time at which birds, crocodiles, and lizards split from one another (the basis of cladistic relationships); they are much more concerned with communicating about the phenetic relationships of those organisms, the relationships that are based on time and rate of evolution, and thus describe how different or similar those organisms are. Told that a "Birdcroc" (a member of a cladistic taxon) is sneaking up on her, a naturalist wouldn't know whether it would sing her a song or drag her into a lake and eat her.

14.4 Lessons for Developing Model Systems in Population Biology and Conservation Biology

High-Profile Model Systems

Large-scale and long-term research projects attract attention through the publications they generate and through the students and researchers who become associated with those projects in one way or another. Such research projects may be conceived with model systems in mind, and they will start to function as symbols for the field of research they represent. Examples are numerous (chapter 12), ranging from model systems for a narrow discipline of science to model systems that epitomize a critical area of research of great and broad importance. At this stage, a particular concept or phenomenon immediately brings to mind a particular model system, and the model system is featured as a "classic" example in textbooks. For instance, what does sympatric speciation due to host shift in phytophagous insects bring to mind? Probably the tephritid fly *Rhagoletis pomonella*, a pest of apples and the subject of a long series of investigations by Guy Bush and colleagues (Bush 1969, Feder et al. 1988, 1990, 1993).

Work on *Euphydryas editha* began with the explicit goal of developing a model system for population biology at a time when few people thought in terms of developing such systems in ecology and evolution, and long-term studies of populations in the field were rare. *Melitaea cinxia* has to some extent become the icon of classic metapopulation dynamics, through the conclusive empirical demonstrations that the metapopulation in the Åland Islands persists in a stochastic balance between ongoing local extinctions and recolonizations (Hanski et al. 1995a, Hanski 1999b) and the successful verification of related theoretical predictions. The latter include the presence of a distinct threshold value for metapopulation persistence (Hanski and Ovaskainen 2000) and immigration reducing the risk of local extinction, which leads to an Allee effect in local dynamics (Kuussaari et al. 1998) and alternative stable states in metapopulation dynamics (Hanski et al. 1995b). Many of the chapters in this volume elaborate on additional ecological, genetic, and evolutionary questions that have a special significance in the population and metapopulation context and that have been addressed using checkerspots.

What are the advantages and the disadvantages for scientists associating a particular research topic with a particular model system? It is evident that for a researcher or a research group in the media-driven world, the extra publicity is beneficial. But what about the issue of advancing science and of making its results available to society? A well-known model system often helps provide a baseline to which a phenomenon found in another system can be compared. Where would genetics be without *Drosophila* and *Escherichia coli*, or great ape behavior without Jane Goodall's (1986) Gombe chimps? A potential problem arises if the model system is perceived as a special case, something of a curiosity that has somehow been tailored to fit the conditions under which a particular phenomenon can be expected to occur. This has, to some extent, happened with the *M. cinxia* project, though in our opinion this judgment is unwarranted. There are many reasons to believe that the *M. cinxia* metapopulation in the Åland Islands is representative of a vast array of ecologically specialized species living in highly fragmented landscapes, and the many similarities between the *M. cinxia* and *E. editha* systems support this conclusion. What makes *M. cinxia* a little unusual are the features that make field work especially feasible (chapter 4), and, of course, that many people have invested a lot of time in studying this system and developing the theory and modeling approaches inspired by it (chapter 12).

One of Ehrlich's concerns in starting long-term work on the checkerspot system was that results would accrue slowly, which would negatively affect his ability to get tenure, grants, and graduate students. His fears proved groundless. Interesting results appeared almost immediately, such as the very sedentary behavior of Jasper Ridge *E. editha* and the existence of independent dynamics of the three demographic units there. There was no difficulty in finding good projects for doctoral candidates, some of whom are co-authors in this volume, and results were abundant enough to continue to attract support from the National Science Foundation. Indeed, the National Science Foundation eventually cited the checkerspot project as an excellent example of the benefits from long-term research. Hanski has been equally well supported by the Finnish National Science Foundation (the Academy of Finland) because novel results have been produced in abundance by the combination of long-term monitoring and specific research projects conducted by particular individuals or groups of individuals. It has become clear that the combination of long-term research and more specific projects may work well and the former may greatly facilitate the latter.

Conservation Lessons from Populations to Global Change

The current extinction crisis—the ongoing sixth global mass extinction—is generally viewed at the level of species. Species extinctions, however, are preceded and greatly outnumbered by extinctions of populations. There are several reasons to be gravely concerned about population extinctions. First, populations are often unique biological entities, as is amply demonstrated by the results on *Euphydryas editha* and many other species. Second, for species with a metapopulation structure, a critical density of suitable sites for local populations is needed for the long-term persistence of the metapopulation and ultimately of the species as a whole. The results for *Melitaea cinxia* in the Åland Islands provide conclusive evidence of this. Third, and perhaps most important, what matters locally and regionally for the conservation of biodiversity and ecosystem services is population diversity—the populations of species that are found, or that could potentially be found, in the focal area. As discussed in chapter 13, work on *Euphydryas* led Ehrlich's research group to the realization that population diversity is a critical element in conservation (Ehrlich 1992) and especially in maintaining ecosystem services (Daily 1997). This caused them to examine the general problem of population extinctions (Ehrlich and Daily 1993) and to make the first quantitative estimates of the global extent of the problem (Hughes et al. 1997, Ceballos and Ehrlich 2002).

Moving from local populations to the other extreme of global change and its biological consequences, which are becoming increasingly well documented (Parmesan and Yohe 2003, Root et al. 2003), work by Parmesan (1996) evaluating the response of *E. editha* populations to global warming has pioneered the study of long-term influence of climate change on the ranges of organisms (box 14.6). This work led to a thorough investigation of climate-induced range shifts in European butterflies (Parmesan et al. 1999) and the impact of climate change on other organisms (Parmesan et al. 2000).

Research on model systems has the potential to inform conservation decisions and politics on many fronts. However, such applications can encounter difficulties that stem from lack of appreciation by policy makers of the connection between specific studies and general concepts and thus the value of model systems. Misunderstanding and possibly intentional misrepresentation arose when Hanski actively advanced the view that the current level of forest protection in southern Finland, currently a mere 1% of the forested land, is not enough to prevent a wave of extinctions from taking place. He used the butterfly study to elucidate some of the key biological processes pertaining to species living in highly fragmented landscapes, such as the threshold for persistence (Hanski and Ovaskainen 2000) and the delayed response of metapopulations to habitat loss and fragmentation (Hanski et al. 1996b, Ovaskainen and Hanski 2002b). Since it is scientifically sound and the biology is well understood, the butterfly work was used as an example in the hope that it would strengthen nonscientists' understanding of the general issues. The counter-argument has been made, however, that results from a butterfly living in meadows surely cannot be applied to forest-dwelling species. Thus, the model system approach may not be helpful in conservation unless we learn how to convince nonscientific audiences that there is value in using general concepts and theories when examining particular cases, and value in sampling nature as the basis for the framework within which particular new questions and situations can be examined.

14.5 Details and Generalities

When we became scientists, we chose the profession with the hope of discovering general principles—the population biological equivalents of Boyle's Law or Newton's Law of Universal Gravitation. Are we then to be disappointed that, in our examination of the

BOX 14.6 Key Lessons: Global Change

Checkerspot studies have demonstrated the often unexpected impact of several drivers of global change (chapter 13):

1. Among populations of *Euphydryas editha* in California, Oregon, Washington, and British Columbia, those in the south have exhibited a greater rate of extinction than northern populations, leading to a northward shift of geographic range (Parmesan 1996). The mechanism underlying these changes is likely to relate to the frequent dependence of checkerspot population dynamics on weather conditions.
2. Land-use changes have contributed to the rate of population extinction. In metapopulations, elimination of previous source populations, or just any populations, has resulted in eventual disappearance of populations whose habitat was not directly affected.
3. Nitrogen deposition leads to changes in vegetation that may gradually alter habitat quality to the detriment of populations.

population biology of a suite of closely related organisms, we seem to have discovered no equivalent of the gravitational constant? We think not, for the following basic reason. The physical laws just mentioned are generalizations of what happens under idealized conditions. Newton's law is of little help in determining whether a certain soccer player's kick will send the ball into a goal. Too much depends on the characteristics of the player, his motion, the motion of the ball, the quality of the opposition, and so forth. They are details in Newton's scheme, but that sort of detail must be understood if our scientific goal is predicting the future of populations in the real world. The devil is in the details. One of our contributions has been to provide insight into the great complexities of population ecology and evolution in the real world.

That is not all, however. Population biology has its own laws, though these are not derived from first principles but have rather emerged during the long course of research. These "laws," often of the type indicated in the boxes, should be viewed more as general statements of how things usually are than as mathematical formulas to be trusted under all circumstances. The application of these laws in both research and management is complicated by the need to pay attention to the details and by the need to know exactly which details to include and which to ignore. In spite of these difficulties, population biology is gradually turning to a predictive science. Our goal in conducting research on checkerspots has been to make a contribution to that development.

15

A Look to the Future

ILKKA HANSKI AND PAUL R. EHRLICH

The diversity of the living world is so huge that "knowledge of everything" will not be attained. We might develop a theory of everything, in some sense that we do not attempt to define here, but complete knowledge of the evolving living world is not an option. The reasons are simple. First, human beings have not been, are not, and will not be observing, recording, and analyzing everything of significance in every part of the globe. Even if we narrowed our focus to just a single species, or even a single small population, the goal would not be attainable. Second, the living world is in a state of perpetual ecological and evolutionary change. That not everything worth knowing for population biology, and for humanity, will ever be known keeps scientists busy—the excitement of making new discoveries will always be there. At the same time, though knowing more of checkerspot butterflies and the other parts of the living world always raises new questions, one hopes that the broad picture of how the world works will become increasingly sharp and accurate. We have every reason to believe that this will occur.

In the case of checkerspots, our knowledge of many aspects of their taxonomy, life histories, and population and metapopulation biology is already comprehensive enough to allow us to make useful predictions for science and conservation. Many lessons have been learned that, combined with those from studies of other organisms, should help us and others forge answers to much broader issues, ultimately to questions about humanity's relationship with the rest of the world. At the moment, however, we are satisfied just contemplating the exciting challenges for future research on checkerspots. Our research in North America and Europe has taken somewhat different paths, partly because of differences in the ecological conditions under which the species live, and partly also due to differences in research traditions. The great diversity of landscapes, habitats, and potential host plants that *Euphydryas editha* encounters in western North America is not matched by the landscapes of northern Europe, where the bulk of research on *Melitaea cinxia* has taken place. But there are populations of *M. cinxia* in the Alps and other mountains in southern and southeastern Europe, Asia (figure 15.1), and North Africa, about which little is known and which are obvious targets for future investigations. There may still be significant broad differences in the geographical ecology of *E. editha* and *M. cinxia*, and, in fact, the likely European analogue of *E. editha* is *Euphydryas aurinia*, the species that was the study subject of E.B. Ford more than half a century ago (Ford and Ford 1930).

Even more interesting would be studies on populations in the steppe and forest steppe regions in Russia and Asia, not least because this is probably the habitat in which *M. cinxia* and many other Old World checkerspots have spent much of their evolutionary past. The steppes cover huge areas, often providing large stretches of seemingly unfragmented

Figure 15.1. Tian Shan mountains in northwestern China, where Hanski and his group have started to do research with Guangchun Lei and his students on *Melitaea cinxia* populations feeding on *Veronica spicata*, the same host plant as used in Finland. Photo by Ilkka Hanski.

habitat for the species. Will the populations in the steppes nonetheless be spottily distributed with rather sedentary individual behavior, or might they show spatial structures more reminiscent of the well-studied *Erebia epipsodea* in Colorado (Brussard and Ehrlich 1970c)? Will the spatial population structures in more diffusely patchy habitat than the Åland landscape in Finland stimulate some novel theory of spatial dynamics? Will geographical patterns of host plant choice in *M. cinxia* prove to be as variable as those in *E. editha* in California, or are Asian *M. cinxia* primarily associated with just a single common steppe species, *Veronica incana* (a close relative of *V. spicata*)?

More questions await us. It remains to be seen whether many *E. editha* populations have equally complex interactions with their natural enemies as do the more thoroughly investigated *M. cinxia* populations. Another complexity that awaits study is voltinism, the number of generations per year, which is more variable in *M. cinxia* across its geographic range than in *E. editha*. The phylogeographic patterns of *M. cinxia* in Europe show a strong signature of glacial history (chapter 10), but little is known about patterns of population differentiation and how they might differ between the distinct landscape structures of northern and southern Europe. One difference that we know about is the apparently greater genetic load of deleterious recessive alleles in *M. cinxia* populations in southern France with greater expanses of suitable habitat than in Åland with a highly fragmented landscape (Haikola et al. 2001).

These questions obviously can and should be extended to comparisons of checkerspots in general. Knowledge of the population biology of most checkerspot species is fragmentary at best. Perhaps the largest gap is the lack of knowledge of the tropical Phyciodini. This group is still poorly known taxonomically (DeVries 1987) and should be targeted for a modern cladistic analysis (Wahlberg and Zimmermann 2000). In general, tropical Phyciodini should be undergoing rapid evolution because they are denizens of open habitats (figure 15.2). In the American tropics their habitats have been greatly expanded by deforestation in response to demand (much of it generated in the United States and Europe) for cattle, timber, bananas, coffee, and other tropical products (Tucker 2000). The opportunities

Figure 15.2. Landscape with low-intensity agriculture near San Vito, in southern Costa Rica, where the Ehrlich group plans to do research on the countryside biogeography of tropical melitaeines. Photo by Paul R. Ehrlich.

for research there seem virtually endless, and the time to do it most effectively, while the transition is still going on, is depressingly short.

Finally, as knowledge of checkerspot larval behavior and morphology builds, the checkerspot system could provide an opportunity to answer at least one fundamental question of evolution and development: whether the elements of the genome that interact with the environment to produce the larval phenome are partly independent of those that produce the adult butterfly. Note, for instance, how different the larval phenotypes can be in California populations of *Euphydryas editha* with relatively similar adults (plate VIII.3). How much, we should ask, can selection alter larval color or ability to detoxify host-plant defensive chemicals without seriously affecting adult color or metabolic capabilities? Given the obvious significance of iridoid glycosides in the larval biology of checkerspots (chapter 7), this immediately leads to some more specific questions. Investigations of these questions would expand the use of the checkerspot model system into a new area and complement other studies of the same topic that seem certain to be undertaken with laboratory-friendly organisms such as *Drosophila*. Checkerspots have the advantage of possessing well-studied differences of adult and larval ecology under natural conditions, and there is at least a start toward understanding their genetics. More modestly but equally importantly, the well-known ecology of checkerspots is already facilitating the study of intriguing evolutionary issues, including the long-term research into the evolution of host-plant oviposition preference (chapter 6) but also the evolution of migration rate (chapter 9). Extending research in these areas to the molecular level is an exciting challenge for future work.

When we mentioned challenges for future research and opportunities for investigators at the beginning of this chapter, we meant both professional and amateur scientists. It is all too easy in this era of high-energy physics and molecular biology to forget that much of the foundations of modern science was built by amateurs; in our own area of interest, we need only think of Charles Darwin. Today new opportunities for amateur participation have arisen. For example, checkerspots provide an opportunity for documenting large-scale and fine-scale patterns of population extinction and refining knowledge of the causal factors—habitat destruction, invasive plants, climate change, and so forth. We need to recruit more people into the ranks of butterfly enthusiasts and organize their efforts so that they can contribute to developing large databases on population changes. A starting point could be organizing "square bashing" efforts of the sort that have produced relatively fine-scale understanding of the British butterfly fauna (Heath et al. 1984, J. Thomas and Webb 1984, Dennis and Shreeve 1996, Asher et al. 2001) on *Euphydryas editha* in North America and *Melitaea cinxia* in Eurasia. For example, the geographic and phenological scope of the Fourth of July butterfly counts in North America could be expanded and the procedures made more rigorous. Organizations of butterfly enthusiasts might eventually conduct systematic transect counts of checkerspot populations or entire butterfly communities. There is a long-standing tradition of such work in the United Kingdom (Pollard and Yates 1993) and Belgium (Maes and Van Dyck 2001). More recently, systematic butterfly transect counts have been started in Finland (Kuussaari et al. 2001), The Netherlands (van Swaay et al. 1997), and Catalonia in northern Spain (Stefanescu 2000). Such censuses are invaluable in detecting trends in distribution and abundance and for giving insight into mechanisms, as Root's (1988) brilliant analysis of the data gathered by amateur ornithologists in Christmas censuses of birds in the United States demonstrates.

A splendid example of the value of censuses is the work of Warren et al. (2001), showing that over the past few decades, several butterfly species have expanded their ranges in the United Kingdom, most probably in response to climate change. But, and this is significant, the species that have expanded their ranges are habitat generalists, whereas the distributions of habitat specialists such as checkerspots have shrunk. As observed by Warren et al. (2001), the most probable explanation for the contrasting patterns in the generalist and specialist species relates to the other major global change, habitat loss and fragmentation. The specialist species have lost so much of their habitat (which is also the cause of their decline) that their populations cannot benefit from the ameliorating climatic conditions. Even though the environmental conditions might otherwise be more favorable, most of the specialists cannot spread in their excessively fragmented landscapes in the United Kingdom and much of the rest of the western Europe. Though the message in this case is clear enough, many more efforts such as Warren et al.'s are called for.

We trust that more amateur lepidopterists, be they butterfly watchers or collectors, can be recruited to scientific efforts, as long as the goals and value of the project are made clear. We tend to prefer observation to collecting because it fits better in with today's increasing environmental consciousness. But the removal of small numbers of adults from most butterfly populations for the pleasure of some amateur butterfly collectors can promote the development of important natural history information. Scientists already owe a debt to collectors for the very things that have helped make butterflies a key model group: relatively detailed information on distributions, larval host plant use, and general habitat requirements. And it is an error to blame the widespread decline of butterflies on collectors, who may put pressure on a few rare species or (as commercial collectors) cause problems for local populations in the tropics. Habitat alteration and destruction is the primary culprit. With proper education and encouragement, butterfly watchers and collectors can contribute to preserving humanity's biological capital and to making not just checkerspots, but butterflies as a whole, a model group that can serve as a sentinel providing warnings of impending losses of that capital.

What are the prospects for long-term conservation of checkerspots? Like most butterflies, they are open-country organisms—and humanity has been a potent force converting forests into open country. Unhappily, though, for most if not all checkerspots, the kind of open country that humanity often creates—pesticide-treated farm fields, overgrazed pastures, golf courses, and treeless subdivisions—are not suitable habitats. In Europe, checkerspots and many other butterflies used to abound in the traditional less intensively cultivated landscapes, which are now rapidly disappearing and causing the decline and extinction of populations and species. The extinction of *Melitaea cinxia* from mainland Finland in the 1970s is a case in point (chapter 4). In western North America, by contrast, many *E. editha* populations occur in montane and arid areas where human population densities remain low and grazing pressure is relatively light. All in all, we think that *E. editha* is in no danger of species extinction, even if the next decades show substantial climate change. In contrast, it seems certain that population extinctions will outnumber the establishment of new populations in *E. editha*, and hence the species will continue to decline. The situation is much worse for *M. cinxia* and most other checkerspots in Europe, such as *Euphydryas aurinia* in the United Kingdom (Asher et al. 2001), where the primary cause of decline has been and continues to be habitat loss. It remains to be seen whether anything good for checkerspots and other species of open landscapes will come from the changes in the agricultural policies that are currently being debated in the European Union.

Tropical checkerspots in areas where agriculture and grazing are not too intensive should, overall, benefit from the rape of tropical landscapes. The weedy open country that usually replaces tropical forests tends to be hospitable to the checkerspots, as it is to neotropical nymphaline weeds such as *Anartia fatima* and *A. jatrophae*. So with any luck the checkerspots should, in large part, weather the extinction crisis, and perhaps radiate into empty niches, replacing other groups of herbivorous insects that are not so well adapted to human-caused landscape changes.

There is thus a little bit of good news for checkerspots on which to end this book. For tropical forest butterflies and many species elsewhere that are losing populations steadily to the expansion of the human enterprise, the news is not so good. And to the degree that the butterfly model system serves as a "miner's canary" for the state of the human environment, the news seems rather grim for *Homo sapiens* as well. It remains to be seen whether there will be scientists around to study the evolutionary future of the checkerspots, and whether scientists in that future world will be interested in the biology of these insects we find so fascinating. But scientists today are, and all of us who contributed to *On the Wings of Checkerspots* have enjoyed the field work, laboratory experiments, and theoretical analyses immensely. The two of us intend to continue to develop our model system further, and we suspect our collaborators will also. We hope some of our readers will join in the effort and share the joy.

References

Abbott, I., L. K. Abbott, and P. R. Grant. 1977. Comparative ecology of Galapagos Ground Finches (*Geospiza* Gould): Evaluation of the importance of floristic diversity and interspecific competition. *Ecological Monographs* 47: 151–184.

Abrahamson, W. G., and A. E. Weis. 1997. *Evolutionary Ecology across Three Trophic Levels: Goldenrods, Gallmakers and Natural Enemies*. Princeton University Press, Princeton, NJ.

Ackery, P. R. 1984. Systematics and faunistic studies on butterflies. Pages 9–21 in R. I. Van-Wright and P. R. Ackery, eds. *The Biology of Butterflies*. Academic Press, London.

Ackery, P. R., R. de Jong, and R. I. Vane-Wright. 1999. The butterflies: Hedyloidea, Hesperoidea and Papilionoidea. Pages 263–300 in N. P. Kristensen, ed. *Lepidoptera, Moths and Butterflies. 1. Evolution, Systematics and Biogeography. Handbook of Zoology*, Vol. 4, *Lepidoptera*. de Gruyter, Berlin.

Addicott, J. F. 1996. Cheaters in yucca/moth mutualism. *Nature* 380: 114–115.

Agnew, K., and M. C. Singer. 2000. Does fecundity constrain the evolution of insect diet? *Oikos* 88: 533–538.

Agrawal, A. 2000. Plant defense: signals in insect eggs. *Trends in Ecology and Evolution* 15: 357.

Ahnesjö, I., C. Kvarnemo, and S. Merilaita. 2001. Using potential reproductive rates to predict mating competition among individuals qualified to mate. *Behavioral Ecology* 12: 397–401.

Alcock, J. 1987. Leks and hilltopping in insects. *Journal of Natural History* 21: 319–328.

———. 1994. Alternative mate-locating tactics in *Chlosyne californica* (Lepidoptera, Nymphalidae). *Ethology* 97: 103–118.

Alexander, H. M., P. H. Thrall, J. Antonovics, A. M. Jarosz, and P. V. Oudemans. 1996. Population dynamics of plant diseases: A case study of anther-smut disease of *Silene alba* caused by the fungus *Ustilago violancea*. *Ecology* 77: 990–996.

Allee, W. C., O. Park, A. E. Emerson, T. Park, and K. P. Schmidt. 1949. *Principles of Animal Ecology*. W. B. Saunders Co., Philadelphia.

Allison, A. C. 1954. The distribution of the sickle-cell trait in East Africa and elsewhere, and its apparent relationship to the incidence of subtertian malaria. *Transactions of the Royal Society of Tropical Medicine and Hygiene* 48: 312–318.

———. 1964. Polymorphism and natural selection in human populations. *Cold Spring Harbor Symposium for Quantitative Biology* 29: 137–149.

Alonso-Mejia, A., E. Montesinos-Patino, E. Rendon-Salinas, L. P. Brower, and K. Oyama. 1998. Influence of forest canopy closure on rates of bird preation on overwintering monarch butterflies *Danaus plexippus* L. *Biological Conservation* 85: 151–159.

Amarasekare, P. 2000. Coexistence of competing parasitoids on a patchily distributed host: Local vs. spatial mechanisms. *Ecology* 81: 1286–1296.

Andelman, S. J., and W. F. Fagan. 2000. Umbrellas and flagships: Efficient conservation surrogates, or expensive mistakes? *Proceedings of the National Academy of Sciences USA* 97: 5954–5959.

Anderson, R. M., and R. M. May. 1981. The population dynamics of micro parasites and their invertebrate hosts. *Philosophical Transactions of the Royal Society of London* 291: 451–524.

Anderson, S. H., and K. J. Gutzwiller. 1996. Habitat evaluation methods. Pages 592–606 in T. A. Bookhout, ed. *Research and Management Techniques for Wildlife and Habitats*. The Wildlife Society, Betheda, MD.

Andersson, J., A. K. Borg Karlson, and C. Wiklund. 2000. Sexual cooperation and conflict in butterflies: A male-transferred anti-aphrodisiac reduces harassment of recently mated females. *Proceedings of the Royal Society of London B*. 267: 1271–1275.

Andrewartha, H. G., and L. C. Birch. 1954. *The Distribution and Abundance of Animals*. University of Chicago Press, Chicago.

Angiosperm Phylogeny Group. 1998. An ordinal classification for the families of flowering plants. *Annals of the Missouri Botanical Garden* 85: 531–553.

Antolin, M. F. 1999. A genetic perspective on mating systems and sex ratios of parasitoid wasps. *Researches in Population Ecology* 41: 29–37.

Antolin, M. F., and D. R. Strong. 1987. Long distance dispersal by a parasitoid (*Anagrus delicatus*, Mymaridae) and its host. *Oecologia* 73: 288–292.

Antonovics, J., P. H. Thrall, and A. M. Jarosz. 1997. Genetics and the spatial ecology of species interaction. Pages 158–183 in D. Tilman and P. Kareiva, eds. *Spatial Ecology*. Princeton University Press, Princeton, NJ.

Antonovics, J., P. H. Thrall, A. M. Jarosz, and D. Stratton. 1994. Ecological genetics of metatpopulations: The *Silene-Ustilago* plant-pathogen system. Pages 146–170 in L. Real, ed. *Ecological Genetics*. Princeton University Press, Princeton, NJ.

Appelt, M., and H. J. Poethke. 1997. Metapopulation dynamics in a regional population of the blue-winged grasshopper (*Oedipoda caerulescens*; Linnaeus, 1758). *Journal of Insect Conservation* 1: 205–214.

Ariño, A., and S. L. Pimm. 1995. On the nature of population extremes. *Evolutionary Ecology* 9: 429–443.

Arms, K., P. Feeny, and R. C. Lederhouse. 1974. Sodium: Stimulus for the puddling behavior by tiger swallowtail butterflies, *Papilio glaucus*. *Science* 185: 372–374.

Arnason, A. N. 1973. The estimating of population size, migration rates and survival in a stratified population. *Researches on Population Ecology* 15: 1–8.

Asher, J., M. Warren, R. Fox, P. Harding, G. Jeffcoate, and S. Jeffcoate. 2001. *The Millennium Atlas of Butterflies in Britain and Ireland*. Oxford University Press, Oxford.

Askew, R. R., and M. R. Shaw. 1997. *Pteromalus apum* (Retzuis) and other pteromalid (Hym.) primary parasitoids of butterfly pupae in western Europe, with a key. *Entomologist's Monthly Magazine* 133: 67–72.

Atkinson, W. D., and B. Shorrocks. 1981. Competition on a divided and ephemeral resource: A simulation model. *Journal of Animal Ecology* 50: 461–471.

Aubert, J., L. Legal, H. Descimon, and F. Michel. 1999. Molecular phylogeny of swallowtail butterflies of the tribe Papilionini (Papilionidae, Lepidoptera). *Molecular Phylogenetics and Evolution* 12: 156–167.

Austin, G. T., and D. D. Murphy. 1998. Patterns of phenotypic variation in the *Euphydryas chalcedona* complex (Lepidoptera: Nymphalidae) of the southern intermountain region. Pages 419–432 in T. C. Emmel, ed. *Systematics of Western North American Butterflies*. Mariposa Press, Gainesville, FL.

Avise, J. 1994. *Molecular Markers, Natural History, and Evolution*. Chapman & Hall, New York.

Avise, J. C. 1998. The history and purview of phylogeography: A personal reflection. *Molecular Ecology* 7: 371–379.

Avise, J. C., J. Arnold, R. M. Ball, E. Bermingham, T. Lamb, J. E. Neigel, C. A. Reeb, and N. C. Saunders. 1987. Intraspecific phylogeography: The mitochondrial DNA bridge between population genetics and systematics. *Annual Review of Ecology and Systematics* 18: 489–522.

Avise, J. C., and J. L. Hamrick, eds. 1996. *Conservation Genetics: Case Histories from Nature*. Chapman & Hall, New York.

Avise, J. C., and K. Wollenberg. 1997. Phylogenetics and the origin of species. *Proceedings of the National Academy of Sciences USA* 94: 7748–7755.

Baguette, M., and G. Nève. 1994. Adult movements between populations in the specialist butterfly *Proclossiana eunomia* (Lepidoptera, Nymphalidae). *Ecological Entomology* 19: 1–5.

Baguette, M., S. Petit, and F. Queva. 2000. Population spatial structure and migration of three butterfly species within the same habitat network: Consequences for conservation. *Journal of Applied Ecology* 37: 100–108.

Baguette, M., C. Vansteenwegen, I. Convi, and G. Neve. 1998. Sex-biased density-dependent migration in a metapopulation of the butterfly *Proclossiana eunomia. Acta Oecologica* 19: 17–24.

Baker, R. R. 1968. Sun orientation during migration in some British butterflies. *Proceedings of the Royal Entomological Society of London A* 143: 89–95.

Baker, R. R., and M. A. Bellis. 1995. *Human Sperm Competition: Copulation, Masturbation and Infidelity.* Chapman & Hall, London.

Balmer, O., and A. Erhardt. 2000. Consequences of succession on extensively grazed grasslands for central European butterfly communities: Rethinking conservation practices. *Conservation Biology* 14: 746–757.

Baltensweiler, W. 1964. The case of *Zeiraphera griseana* (Hb. (= *diniana* Gn.) (Lep., Tortricidae) in the European Alps. A contribution to the problem of cycles. *The Canadian Entomologist* 96: 790–800.

Baptista, L. F. 1975. Song dialects and demes in sedentary populations of the White-crowned Sparrow. *University of California Publications in Zoology* 105: 1–53.

Baptista, L. F., and L. Petrinovich. 1986. Song development in the White-crowned Sparrow: Social factors and sex differences. *Animal Behaviour* 35: 1359–1371.

Barbosa, P., ed. 1998. *Conservation Biological Control.* Academic Press, San Diego, CA.

Barbosa, P., J. A. Saunders, J. Kemper, R. Trumble, J. Olechno, and P. Martinat. 1986. Plant allelochemicals and insect parasitoids: Effects of nicotine on *Cotesia congregata* (Say) (Hymenoptera: Braconidae) and *Hyposoter annulipes* (Cressen) (Hymenoptera: Ichneumonidae). *Journal of Chemical Ecology* 12: 1319–1328.

Barlow, G. W. 2002. *The Cichlid Fishes: Nature's Grand Experiment in Evolution.* HarperCollins, Scranton, PA.

Barlow, J. C., and S. N. Leckie. 2000. Eurasian tree sparrow. *The Birds of North America* 560: 1–20.

Barney, R. L., and B. J. Anson. 1920. Life history and ecology of the pygmy sunfish *Elassoma zonatum. Ecology* 1: 241–256.

Barnosky, A. D., E. A. Hadly, B. A. Maurer, and M. I. Christie. 2001. Temperate terrestrial vertebrate faunas in North and South America: Interplay of ecology, evolution, and geography with biodiversity. *Conservation Biology* 15: 658–674.

Barrett, S. C. H., and D. Charlesworth. 1991. Effects of a change in the level of inbreeding on the genetic load. *Nature* 352: 522–524.

Barton, N. H. 1987. The probability of establishment of an advantageous mutation in a subdivided population. *Genetical Research* 50: 35–40.

Barton, N. H., and M. C. Whitlock. 1997. The evolution of metapopulations. Pages 183–214 in I. Hanski and M. E. Gilpin, ed. *Metapopulation Biology.* Academic Press, San Diego, CA.

Bastin, L., and C. D. Thomas. 1999. The distribution of plant species in urban vegetation fragments. *Landscape Ecology* 14: 493–507.

Bates, M. 1958. Food-getting behavior. Pages 206–223 in A. Roe and G. G. Simpson, eds. *Behavior and Evolution.* Yale University Press, New Haven, CT.

Baughman, J. F. 1991. Do protandrous males have increased mating success? The case of *Euphydryas editha. The American Naturalist* 138: 536–542.

Baughman, J. F., P. F. Brussard, P. R. Ehrlich, and D. D. Murphy. 1990a. History, selection, drift, and gene flow: Complex differentiation in checkerspot butterflies. *Canadian Journal of Zoology* 68: 1967–1975.

Baughman, J. F., D. D. Murphy, and P. R. Ehrlich. 1988a. Emergence patterns in male checkerspot butterflies: Testing theory in the field. *Theoretical Population Biology* 33: 102–112.

———. 1988b. Population structure in a hilltopping butterfly. *Oecologia* 75: 593–600.

———. 1990b. A reexamination of hilltopping in *Euphydryas editha. Oecologia* 83: 259–260.

Beattie, A. J. 1985. *The Evolutionary Ecology of Ant–Plant Mutualisms.* Cambridge University Press, Cambridge.

Becerra, J. X. 1997. Insects on plants: Chemical trends in host use. *Science* 276: 253–256.

Beck, J., E. Mühlenberg, and K. Fiedler. 1999. Mud-puddling behavior in tropical butterflies: In search of proteins or minerals? *Oecologia* 119: 140–148.

Begon, M., and G. A. Parker. 1986. Should egg size decrease with age? *Oikos* 47: 293–302.

Bellows Jr., T. S., R. G. van Driesche, and J. R. Elkinton. 1992. Life table construction and analysis in the evaluation of natural enemies. *Annual Review of Entomology* 37: 587–614.

Belofsky, G., M. D. Bowers, S. Janzen, and F. Stermitz. 1989. Iridoid glycosides of *Aureolaria flava* and their sequestration by

Euphydryas phaeton butterflies. *Phytochemistry* 28: 1601–1604.

Belovsky, G. E., C. Mellison, C. Larson, and P. A. van Zandt. 1999. Experimental studies of extinction dynamics. *Science* 286: 1175–1177.

Belshaw, R. 1993. Tachinid flies Diptera: Tachinidae. Pages 1–169. *Handbook for the Identification of British Insects*. Royal Entomological Society of London.

Benrey, B., R. F. Denno, and L. Kaiser. 1997. The influence of plant species on attraction and host acceptance in *Cotesia glomerata* (Hymenoptera: Braconidae). *Journal of Insect Behaviour* 10: 619–630.

Benson, W. W., K. S. Brown, and L. E. Gilbert. 1975. Coevolution of plants and herbivores: Passion flower butterflies. *Evolution* 29: 659–680.

Benton, T. G., A. Grant, and T. H. Clutton-Brock. 1995. Does environmental stochasticity matter: Analysis of red deer life histories on Rum. *Evolutionary Ecology* 9: 559–574.

Berenbaum, M. R. 1983. Coumarins and caterpillars: A case for coevolution. *Evolution* 37: 163–169.

———. 1995a. *Bugs in the System*. Addison-Wesley, Reading, MA.

———. 1995b. The chemistry of defense: Theory and practice. *Proceedings of the National Academy of Sciences USA* 92: 2–8.

Berenbaum, M. R., C. Favret, and M. A. Schuler. 1996. On defining "key innovations" in an adaptive radiation: Cytochrome P450S and Papilionidae. *The American Naturalist* 148: S139–S155.

Berenbaum, M. R., and P. Feeny. 1981. Toxicity of angular furanocoumarins to swallowtails: Escalation in the coevolutionary arms race. *Science* 212: 927–929.

Berenbaum, M. R., and A. B. Zangerl. 1988. Stalemates in the coevolutionary arms race: Syntheses, synergisms, and sundry other sins. Pages 113–132 in K. C. Spencer, ed. *Chemical Mediation of Coevolution*. Academic Press, New York.

Bergman, K. O., and J. Landin. 2001. Distribution of occupied and vacant sites and migration of *Lopinga achine* (Nymphalidae: Satyrinae) in a fragmented landscape. *Biological Conservation* 102: 183–190.

Bernays, E. A. 1988. Host specificity in phytophagous insects: Selection pressure from generalist predators. *Entomologia Experimentalis et Applicata* 49: 1153–1160.

Bernays, E. A., and M. L. Cornelius. 1988. Generalist caterpillar prey are more palatable than specialists for the generalist predator *Iridomyrmex humilis*. *Oecologia* 79: 427–430.

Bernays, E. A., and C. DeLuca. 1981. Insect antifeedant properties of an iridoid glycoside: Ipolamiide. *Experimentia* 37: 1289–1290.

Bernays, E., and M. Graham. 1988. On the evolution of host specificity in phytophagous arthropods. *Ecology* 69: 886–892.

Bernstein, C. 2000. Host-parasitoid models: The story of a successful failure. Pages 41–57 in M. E. Hochberg and A. R. Ives, eds. *Parasitoid Population Biology*. Princeton University Press, Princeton, NJ.

Biedermann, R. 2000. Metapopulation dynamics of the froghopper *Neophilaenus albipennis* (F., 1798) (Homoptera, Cercopidae): What is the minimum viable metapopulation size? *Journal of Insect Conservation* 4: 99–107.

Bijlsma, R., J. Bundgaard, and W. F. van Putten. 1999. Environmental dependence of inbreeding depression and purging in *Drosophila melanogaster*. *Journal of Evolutionary Biology* 12: 1125–1137.

Bissoondath, C. J., and C. Wiklund. 1997. Effect of male body size on sperm precedence in the polyandrous butterfly *Pieris napi* L. (Lepidoptera: Pieridae). *Behavioral Ecology* 8: 518–523.

Blair, R. B. 1999. Birds and butterflies along an urban gradient: Surrogate taxa for assessing biodiversity. *Ecological Applications* 9: 164–170.

Boerema, L. K., and J. A. Bulland. 1973. Stock assessment of the Peruvian anchovy (*Engraulis ringens*) and management of the fishery. *Journals of Fishery Research Board, Canada* 30: 2226–2235.

Boggs, C. L. 1979. Resource allocation and reproductive strategies in several heliconiine butterflies. PhD dissertation, University of Texas at Austin.

———. 1981. Selection pressures affecting male nutrient investment at mating in heliconiine butterflies. *Evolution* 35: 931–940.

———. 1986. Reproductive strategies of female butterflies: Variation in and constraints on fecundity. *Ecological Entomology* 11: 7–15.

———. 1987a. Ecology of nectar and pollen feeding in Lepidoptera. Pages 369–391 in F. Slansky Jr. and J. G. Rodriguez, eds. *Nutritional Ecology of Insects, Mites and Spiders and Related Invertebrates*. John Wiley & Sons, New York.

———. 1987b. Within population variation in the demography of *Speyeria mormonia* (Lepidoptera: Nymphalidae). *Holarctic Ecology* 10: 175–184.

———. 1990. A general model of the role of male-donated nutrients in female insects'

reproduction. *The American Naturalist* 136: 598–617.
———. 1992. Resource allocation: Exploring connections between foraging and life history strategies. *Functional Ecology* 6: 508–518.
———. 1995. Male nuptial gifts: Phenotypic consequences and evolutionary implications. Pages 215–242 in S. R. Leather and J. Hardie, eds. *Insect Reproduction.* CRC Press, Boca Raton, FL.
———. 1997a. Dynamics of reproductive allocation from juvenile and adult feeding: Radiotracer studies. *Ecology* 78: 192–202.
———. 1997b. Reproductive allocation from reserves and income in butterfly species with different adult diets. *Ecology* 78: 181–191.
———. 1997c. Resource allocation in variable environments: Comparing insects and plants. Pages 73–92 in F. Bazzaz and J. Grace, eds. *Plant Resource Allocation.* Academic Press, San Diego, CA.
———. 2003. Environmental variation, life histories and allocation. Pages 185–206 in C. L. Boggs, W. B. Watt, and P. R. Ehrlich, eds. *Butteflies: Ecology and Evolution Taking Flight.* University of Chicago Press, Chicago.
Boggs, C. L., and B. Dau. ND. Resource specialization in puddling Lepidoptera. Manuscript.
Boggs, C. L., P. R. Ehrlich, G. C. Daily, C. E. Holdren, and I. Kulahci. ND. Population explosion of the transplanted checkerspot *Euphydryas gillettii*: A response to climate change? Manuscript.
Boggs, C. L., and L. E. Gilbert. 1979. Male contribution to egg production in butterflies: Evidence for transfer of nutrients at mating. *Science* 206: 83–84.
Boggs, C. L., W. B. Watt, and P. R. Ehrlich, eds. 2003. *Butterflies: Ecology and Evolution taking Flight.* University of Chicago Press, Chicago.
Bonsall, M. B., and M. P. Hassell. 1998. Population dynamics of apparent competition in a host-parasitoid assemblage. *Journal of Animal Ecology* 67: 918–929.
Bossart, J. L., and J. M. Scriber. 1995. Maintenance of ecologically significant genetic variation in the tiger swallowtail butterfly through differential selection and gene flow. *Evolution* 49: 1163–1171.
Both, C., J. M. Tinbergen, and A. J. van Noordwijk. 1998. Offspring fitness and individual optimization of clutch size. *Proceedings of the Royal Society of London, Series B* 265: 2303–2307.
Boughton, D. A. 1998. Ecological and Behavioral Mechanisms of Colonization in a Metapopulation of the Butterfly *Euphydryas editha.* PhD dissertation, University of Texas, Austin.
———. N.D. Habitat selection in a source-sink metapopulation of butterfly *Euphydryas editha*: Population patterns. Manuscript.
———. 1999. Empirical evidence for complex source-sink dynamics with alternative states in a butterfly metapopulation. *Ecology* 80: 2727–2739.
———. 2000. The dispersal of a butterfly: A test of source-sink theory suggests the intermediate-scale hypothesis. *The American Naturalist* 156: 131–144.
Boughton, D. A., and M. C. Singer. 2004. Reproductive investment portfolios in a butterfly. Manuscript.
Bourn, N. A. D., and M. S. Warren. 1997. *Species Action Plan. Glanville Fritillary* Melitaea cinxia. Butterfly Conservation Society, Colchester, Essex, UK.
Bowers, M. D. 1978. Over-wintering behavior in *Euphydryas phaeton* (Nymphalidae). *Journal of the Lepidopterists' Society* 32: 282–288.
———. 1980. Unpalatability as a defense strategy of *Euphydryas phaeton* (Lepidoptera: Nymphalidae). *Evolution* 34: 586–600.
———. 1981. Unpalatability as a defense strategy of western checkerspot butterflies (*Euphydryas scudder*, Nymphalidae). *Evolution* 35: 367–375.
———. 1983a. Mimicry in North America checkerspot butterflies: *Euphydryas phaeton* and *Chlosyne harrisii* (Nymphalidae). *Ecological Entomology* 8: 1–8.
———. 1983b. The role of iridoid glycosides in host-plant specificity of checkerspot butterflies. *Journal of Chemical Ecology* 9: 475–493.
———. 1986. Population differences in larval hostplant use in the checkerspot butterfly, *Euphydryas chalcedona. Experimental and Applied Entomology* 40: 61–69.
———. 1988. Chemistry and coevolution: Iridoid glycosides, plants, and herbivorous insects. Pages 133–165 in K. C. Spencer, ed. *Chemical Mediation of Coevolution.* Academic Press, San Diego, CA.
———. 1990. Recycling plant natural products for insect defense. Pages 353–386 in D. L. Evans and J. O. Schmidt, eds. *Insect Defenses: Adaptive Mechanisms and Strategies of Prey and Predators.* State University of New York Press, Albany.
———. 1991. Iridoid glycosides. Pages 297–325 in G. A. Rosenthal and M. R. Berenbaum, eds. *Herbivores: Their Interactions with Secondary Plant Metabolites.* Vol. 1. Academic Press, San Diego, CA.

———. 1993. Aposematic caterpillars: Life-styles of the warningly coloured and unpalatable. Pages 331–371 in N. E. Stamp and T. M. Casey, eds. *Caterpillars: Ecological and Evolutionary Constraints on Foraging.* Chapman & Hall, New York.

Bowers, M. D., I. L. Brown, and D. Wheye. 1985. Bird predation as a selective agent in a butterfly population. *Evolution* 39: 93–103.

Bowers, M. D., and S. Farley. 1990. The behaviour of grey jays, *Perisoreus canadensis*, towards palatable and unpalatable Lepidoptera. *Animal Behaviour* 39: 699–705.

Bowers, M. D., and G. M. Puttick. 1986. Fate of ingested iridoid glycosides in Lepidopteran herbivores. *Journal of Chemical Ecology* 12: 169–178.

———. 1988. Response of generalist and specialist insects to qualitative allelochemical variation. *Journal of Chemical Ecology* 14: 319–334.

Bowers, M. D., and N. E. Stamp. 1993. Effects of plant age, genotype, and herbivory on *Plantago* performance and chemistry. *Ecology* 74: 1778–1791.

Bowers, M. D., N. E. Stamp, and S. K. Collinge. 1992. Early stage of host-range expansion by a specialist herbivore, *Euphydryas phaeton* (Nymphalidae). *Ecology* 73: 526–536.

Bowers, M. D., and E. H. Williams. 1995. Variable chemical defence in the checkerspot butterfly Euphydryas gillettii (Lepidoptera: Nymphalidae). *Ecological Entomology* 20: 208–212.

Bowman, R. I. 1961. Morphological differentiation and adaptation in the Galapagos finches. *University of California Publications in Zoology* 58: 1–302.

Box, G. E. P., and N. R. Draper. 1987. *Empirical Model-building and Response Surfaces.* Wiley, New York.

Bradbury, J. W. 1985. Contrasts between insects and vertebrates in the evolution of male display, female choice, and lek mating. Pages 273–289 in B. Hölldobler and M. Lindauer, eds. *Experimental Behavioral Ecology and Sociobiology*. Fisher, New York.

Brakefield, P. M., E. El Filali, R. Van der Laan, C. J. Breuker, I. J. Saccheri, and B. Zwaan. 2001. Effective population size, reproductive success and sperm precedence in the butterfly, *Bicyclus anynana*, in captivity. *Journal of Evolutionary Biology* 14: 148–156.

Brakefield, P. M., J. Gates, D. Keys, F. Kesbeke, P. J. Wijngaarden, A. Monteiro, V. French, and S. B. Carroll. 1996. Development, plasticity and evolution of butterfly eyespot patterns. *Nature* 384: 236–242.

Brakefield, P. M., F. Kesbeke, and P. B. Koch. 1998. The regulation of phenotypic plasticity of eyespots in the butterfly *Bicyclus anynana. The American Naturalist* 152: 853–860.

Brakefield, P. M., and T. G. Liebert. 2000. Evolutionary dynamics of declining melanism in the peppered moth in The Netherlands. *Proceedings of the Royal Society of London B* 267: 1953–1957.

Brakefield, P. M., and A. Monteiro. 2003. The evolution of butterfly eyespot patterns. Pages 243–258 in C. L. Boggs, W. B. Watt, and P. R. Ehrlich, eds. *Butterflies: Ecology and Evolution Taking Flight*. University of Chicago Press, Chicago.

Brakefield, P. M., and I. J. Saccheri. 1994. Guidelines in conservation genetics and the use of population cage experiments with butterflies to investigate the effects of genetic drift and inbreeding. Pages 165–179 in V. Loeschcke, J. Tomiuk, and S. K. Jain, eds. *Conservation Genetics*. Birkhäuser Verlag, Basel, Switzerland.

Brattsten, L. B. 1988. Enzymatic adaptations in leaf-feeding insects to host-plant allelochemicals. *Journal of Chemical Ecology* 14: 1919–1939.

Brattsten, L. B., C. F. Wilkinson, and T. Eisner. 1977. Herbivore-plant interactions: Mixed-function oxidases and secondary plant substances. *Science* 196: 1349–1352.

Breedlove, D. E., and P. R. Ehrlich. 1968. Plant-herbivore coevolution: Lupines and lycaenids. *Science* 162: 671–672.

———. 1972. Coevolution: Patterns of legume predation by a Lycaenid butterfly. *Oecologia* 10: 99–104.

Brewer, B. A., R. C. Lacy, M. L. Foster, and G. Alaks. 1990. Inbreeding depression in insular and central populations of *Peromyscus* mice. *Journal of Heredity* 81: 257–266.

Britten, H. B., P. F. Brussard, D. D. Murphy, and P. R. Ehrlich. 1995. A test for isolation-by-distance in central Rocky Mountain and Great Basin populations of Edith's checkerspot butterfly (*Euphydryas editha*). *Journal of Heredity* 86: 204–210.

Brodeur, J. 2000. Host specificity and trophic relationships of hyperparasitoids. Pages 163–183 in M. E. Hochberg and A. R. Ives, eds. *Parasitoid Population Biology*. Princeton University Press, Princeton, NJ.

Brower, A. V. Z. 2000. Phylogenetic relationships among the Nymphalidae (Lepidoptera) infered from partial sequences of the *wingless* gene. *Proceedings of the Royal Society of London B* 267: 1201–1211.

Brower, A. V. Z., and M. G. Egan. 1997. Cladistic analysis of *Heliconius* butterflies and relatives (Nymphalidae: Heliconiiti): A revised phylogenetic position for *Eueides* based on sequences from mtDNA and a nuclear gene. *Proceedings of the Royal Society of London Series B Biological Sciences* 264: 969–977.

Brown, I., and P. R. Ehrlich. 1980. Population biology of the checkerspot butterfly, *Euphydryas chalcedona*: Structure of the Jasper Ridge colony. *Oecologia* 47: 239–251.

Brown, J. H., and D. W. Davidson. 1977. Competition between seed-eating rodents and ants in desert ecosystems. *Science* 196: 880–896.

Brown, J. H., and A. Kodric-Brown. 1977. Turnover rates in insular biogeography: Effect of immigration and extinction. *Ecology* 58: 445–449.

Brown, J. H., and G. A. Lieberman. 1973. Resource utilization and coexistence of seed-eating rodents in sand dune habitats. *Ecology* 54: 788–797.

Brown, J. H., T. G. Whitham, S. K. M. Ernest, and C. A. Gehring. 2001. Complex species interations and the dynamics of ecological systems: Long-term experiments. *Science* 293: 643–650.

Brues, C. T. 1920. The selection of food-plants by insects, with special reference to lepidopterous larvae. *The American Naturalist* 54: 313–332.

Brunton, C. F. A., and G. D. D. Hurst. 1998. Mitochondrial DNA phylogeny of brimstone butterflies (genus *Gonepteryx*) from the Canary Islands and Madeira. *Biological Journal of the Linnean Society* 63: 69–79.

Brunton, C. F. A., and M. E. N. Majerus. 1995. Ultraviolet colors in butterflies: Intraspecific or interspecific communication. *Proceedings of the Royal Society of London Series B Biological Sciences* 260: 199–204.

Brussard, P. F., J. F. Baughman, D. D. Murphy, and P. R. Ehrlich. 1989. Complex population differentiation in checkerspot butterflies (*Euphydryas* spp.). *Canadian Journal of Zoology* 67: 330–335.

Brussard, P. F., and P. R. Ehrlich. 1970a. Adult behavior and population structure in *Erebia epipsodea* (Lepidoptera: Satyrinae). *Ecology* 51: 880–885.

———. 1970b. Contrasting population biology of two species of butterfly. *Nature* 227: 91–92.

———. 1970c. The population structure of *Erebia epipsodea* (Lepidoptera: Satyrinae). *Ecology* 51: 119–129.

Brussard, P. F., P. R. Ehrlich, D. D. Murphy, B. A. Wilcox, and J. Wright. 1985. Genetic distances and the taxonomy of checkerspot butterflies (Nymphalidae: Nymphalinae). *Journal of the Kansas Entomological Society* 58: 403–412.

Brussard, P. F., P. R. Ehrlich, and M. C. Singer. 1974. Adult movements and population structure in *Euphydryas editha*. *Evolution* 28: 408–415.

Brussard, P. F., and A. T. Vawter. 1975. Population structure, gene flow and natural selection in populations of *Euphydryas phaeton*. *Heredity* 34: 407–415.

Bryant, S. R., C. D. Thomas, and J. S. Bale. 2000. Thermal ecology of gregarious and solitary nettle-feeding nymphalid butterfly larvae. *Oecologia* 122: 1–10.

Buckland, S. M., K. Thompson, J. G. Hodgson, and J. P. Grime. 2001. Grassland invasions: Effects of manipulations of climate and management. *Journal of Applied Ecology* 38: 301–309.

Bulmer, M. G. 1983. Models for the evolution of protandry in insects. *Theoretical Population Biology* 23: 314–322.

Bush, G. L. 1969. Sympatric host race formation and speciation in frugivorous flies of the genus *Rhagoletis* (Diptera, Tephritidae). *Evolution* 23: 237–251.

Cabeza, M., and A. Moilanen. 2001. Design of reserve networks and the persistence of biodiversity. *Trends in Ecology and Evolution* 16: 242–248.

Cain, A. J., and P. M. Sheppard. 1954. Natural selection in Cepaea. *Genetics* 39: 89–116.

Caldwell, R. L., and J. P. Hegmann. 1969. Heritability of flight duration in the milkweed bug *Lygaeus kalmii*. *Nature* 223: 91–92.

California Insect Survey. 2002. *California's Endangered Insects*. Essig Museum of Entomology, University of California at Berkeley.

Calvert, W. 1974. The external morphology of foretarsal receptors involved with host descrimination by the nymphalid butterfly, *Chlosyne lacinia*. *Annals of the Entomological Society of America* 67: 853–856.

Camara, M. D. 1997a. Physiological mechanisms underlying costs of chemical defense in *Junonia coenia* Hubner (Nymphalidae): A gravimetric and quantitative genetic. *Evolutionary Ecology* 11: 451–469.

———. 1997b. Predator responses to sequestered plant toxins in buckeye caterpillars: Are tritrophic interactions locally variable? *Journal of Chemical Ecology* 23: 2093–2106.

———. 1997c. A recent host range expansion in *Junonia coenia* (Nymphalidae). Oviposition preference, survival, growth and chemical defense. *Evolution* 51: 873–884.

Camin, J. H., and P. R. Ehrlich. 1958. Natural selection in water snakes (*Natrix sipedon* L.) on islands in Lake Erie. *Evolution* 12: 504–511.

Campbell, B. C., and S. F. Duffy. 1979. Tomatine and parasitic wasps: Potential incompatibility of plant antibiosis with biological control. *Science* 205: 700–702.

Campbell, D. L., A. V. Z. Brower, and N. E. Pierce. 2000. Molecular evolution of the *wingless* gene and its implications for the phylogenetic placement of the butterfly family Riodinidae (Lepidoptera: Papilionoidea). *Molecular Biology and Evolution* 17: 684–696.

Campbell, I. M. 1962. Reproductive capacity in the genus *Choristoneura* Led. (Lepidoptera: Tortricidae). I. Quantitative inheritance and genes as controllers of rates. *Canadian Journal Genetics and Cytology* 4: 272–288.

Campbell, R. W., L. M. Van Damme, and S. R. Johnson. 1997. Sky lark. *The Birds of North America* 286: 1–19.

Cappuccino, N., and P. W. Price, eds. 1995. *Population Dynamics: New Approaches and Synthesis*. Academic Press, San Diego, CA.

Carlson, A., and P. Edenhamn. 2000. Extinction dynamics and the regional persistence of a tree frog metapopulation. *Proceedings of the Royal Society Biological Sciences Series B* 267: 1311–1313.

Carlson, R. W. 1979. Ichneumonidae. Pages 315–740 in K. V. Krombein, P. D. Hurd, D. R. Smith, and B. D. Burk, eds. *Catalog of Hymenoptera in America North of Mexico*, vol. 1. *Symphyta and Apocrita (Parasitica)*. Smithsonian Institute, Washington, DC.

Carreck, N. L., and I. H. Williams. 2002. Food for insect pollinators on farmland: Insect visits to flowers of annual seed mixtures. *Journal of Insect Conservation* 6: 13–23.

Carroll, S. S., and D. L. Pearson. 1998. Spatial modeling of butterfly species richness using tiger beetles as a bioindicator taxon. *Ecological Applications* 8: 531–543.

Carson, R. 1962. *Silent Spring*. Houghton Mifflin, Boston.

Casey, T. M. 1993. Effects of temperature on foraging of caterpillars. Pages 5–28 in N. E. Stamp and T. M. Casey, eds. *Caterpillars: Ecological and Evolutionary Constraints on Foraging*. Chapman & Hall, New York.

Casey, T. M., B. Joos, T. D. Fitzgerald, M. Yurlona, and P. Young. 1988. Synchronized group feeding, thermoregulation, and growth of eastern tent caterpillars in relation to microclimate. *Physiological Zoology* 61: 372–377.

Cassel, A., J. Windig, S. Nylin, and C. Wiklund. 2001. Effects of population size and food stress on fitness-related characters in the scarce heath, a rare butterfly in western Europe. *Conservation Biology* 15: 1667–1673.

Caswell, H. 2000. *Matrix Population Models: Construction, Analysis and Interpretation.* Sinauer Associates, Sunderland, MA.

Caterino, M. S., R. D. Reed, M. M. Kuo, and F. A. H. Sperling. 2001. A partitioned likelihood analysis of swallowtail butterfly phylogeny (Lepidoptera: Papilionidae). *Systematic Biology* 50: 106–127.

Caughley, G. 1994. Directions in conservation biology. *Journal of Animal Ecology* 63: 215–244.

Ceballos, G., and P. R. Ehrlich. 2002. Mammal population losses and the extinction crisis. *Science* 296: 904–907.

Chapman, T., L. F. Liddle, J. M. Kalb, M. F. Wolfner, and P. L. 1995. Cost of mating in *Drosophila melanogaster* females is mediated by male accessory gland products. *Nature* 373: 241–244.

Charlesworth, B., and D. Charlesworth. 1999. The genetic basis of inbreeding depression. *Genetical Research* 74: 329–340.

Charlesworth, D., and B. Charlesworth. 1987. Inbreeding depression and its evolutionary consequences. *Annual Review of Ecology and Systematics* 18: 237–268.

Chew, F. S. 1977. Coevolution of pierid butterflies and their cruciferous foodplants. II. The distribution of eggs on potential foodplants. *Evolution* 31: 568–579.

———. 1988. Searching for defensive chemistry in the Cruciferae, or, do glucosinolates always control interactions of Crucifera with their potential herbivores and symbionts? No! Pages 81–112 in K. C. Spencer, ed. *Chemical Mediation of Coevolution.* Academic Press, New York.

Chew, F. S., and R. K. Robbins. 1984. Egg-laying in butterflies. Pages 65–79 in R. Vane-Wright and P. R. Ackery, eds. *The Biology of Butterflies*. Academic Press, New York.

Chiariello, N. R., and C. B. Field. 1996. Annual grassland responses to elevated CO_2 in multiyear community microcosms. Pages 139–157 in X. F. A. Bazzaz, ed. *Carbon Dioxide, Populations and Communities.* Academic Press, New York.

Christiansen, F. B., and O. Frydenberg. 1974. Geographical patterns of four polymorphisms in *Zoarces viviparus* as evidence for selection. *Genetics* 77: 765–770.

Clark, B. R., and S. H. Faeth. 1997. The consequences of larval aggregation in the butterfly *Chlosyne lacinia*. *Ecological Entomology* 22: 408–415.

———. 1998. The evolution of egg clustering in butterflies: A test of the egg desiccation hypothesis. *Evolutionary Ecology* 12: 543–552.

Clausen, C. P., D. W. Clancy, and Q. C. Chock. 1965. Biological control of the oriental fruitfly (*Dacus soralis* Hendel) and other fruit flies in Hawaii. Pages 1–102. U.S. Department of Agriculture, Washington, DC.

Clobert, J., E. Danchin, A. A. Dhondt, and J. D. Nichols, eds. 2001. *Dispersal*. Oxford University Press, Oxford.

Clutton-Brock, T. H., and S. D. Albon. 1979. The roaring of red deer and the evolution of the honest advertisement. *Behavior* 69: 145–170.

Clutton-Brock, T. H., F. E. Guinness, and S. D. Albon. 1982. *Red Deer: Behavioral Ecology of Two Sexes*. University of Chicago Press, Chicago.

Colfer, R. G., and J. A. Rosenheim. 2001. Predation on immature parasitoids and its impact on aphid suppression. *Oecologia* 126: 292–304.

Comins, H. N., W. D. Hamilton, and R. M. May. 1980. Evolutionary stable dispersal strategies. *Journal of Theoretical Biology* 82: 205–230.

Connell, J. H. 1983. On the prevalence and relative importance of interspecific competition: Evidence from field experiments. *The American Naturalist* 122: 661–696.

Conradt, L., E. J. Bodsworth, T. Roper, and C. D. Thomas. 2000. Non-random dispersal in the butterfly *Maniola jurtina*: implications for metapopulation models. *Proceedings of the Royal Society B* 267: 1505–1510.

Constantino, R. F., and R. A. Desharnais. 1991. *Population Dynamics and the Tribolium Model: Genetics and Demography*. Springer-Verlag, New York.

Cook, S. E., J. G. Vernon, M. Bateson, and T. Guilford. 1994. Mate choice in the polymorphic African swallowtail butterfly, *Papilio dardanus*: Male-like females may avoid sexual harrassment. *Animal Behaviour* 47: 389–397.

Cornell, H. V., and B. A. Hawkins. 2003. Herbivore responses to plant secondary compounds. *American Naturalist* 161: 507–522.

Coulsen, T., S. Albon, J. Pilkington, and T. Clutton-Brock. 1999. Small-scale spatial dynamics in a fluctuating ungulate population. *Journal of Animal Ecology* 68: 658–671.

Courtney, S. P. 1982. Coevolution of pierid butterflies and their cruciferous foodplants. IV. Crucifer apparency and *Anthocharis cardamines* oviposition. *Oecologia* 52: 258–265.

———. 1984. The evolution of egg clustering by butterflies and other insects. *The American Naturalist* 123: 276–281.

Courtney, S. P., G. K. Chen, and A. Gardner. 1989. A general model for individual host selection. *Oikos* 55: 55–65.

Courtney, S. P., and J. Hard. 1990. Host acceptance and life-history traits in *Drosophila busckii*: tests of the hierarchy-threshold model. *Heredity* 64: 371–375.

Cracraft, J. 1983. Species concepts and speciation analysis. *Current Ornithology* 1: 159–187.

Crawford, H. S., and D. T. Jennings. 1989. Predation by birds on spruce budworm *Choristoneura fumiferana*: Functional, numerical, and total responses. *Ecology* 70: 152–163.

Crawley, M. J. 1984. *Herbivory*. Blackwell, Oxford.

Crnokrak, P., and D. A. Roff. 1999. Inbreeding depression in the wild. *Heredity* 83: 260–270.

Crow, J. F. 1948. Alternative hypotheses of hybrid vigour. *Genetics* 33: 477–487.

———. 1952. Dominance and overdominance. Pages 282–297 in J. W. Gowen, ed. *Heterosis*. Iowa State College Press, Ames.

———. 1993. Mutation, mean fitness, and genetic load. *Oxford Surveys in Evolutionary Biology* 9: 3–42.

Crowell, K. L. 1973. Experimental zoogeography: Introduction of mice to small islands. *The American Naturalist* 107: 535–558.

Crozier, R. H. 1977. Evolutonary genetics of the Hymenoptera. *Annual Review of Entomology* 22: 263–288.

Cullenward, M. J., P. R. Ehrlich, R. R. White, and C. E. Holdren. 1979. The ecology and population genetics of an alpine checkerspot butterfly, *Euphydryas anicia*. *Oecologia* 38: 1–12.

Cushman, J. H., C. L. Boggs, S. B. Weiss, D. D. Murphy, A. W. Haney, and P. R. Ehrlich. 1994. Estimating female reproductive success of a threatened butterfly: Influence of emergence time and hostplant phenology. *Oecologia* 99: 194–200.

Daily, G. C., ed. 1997. *Nature's Services*. Island Press, Washington, DC.

———. 2001. Ecological forecasts. *Nature* 411: 245.

Daily, G. C., A. H. Ehrlich, and P. R. Ehrlich. 1994. Optimum human population size. *Population and Environment* 15: 469–475.

Daily, G. C., and P. R. Ehrlich. 1995. Preservation of biodiversity in small rainforest patches: Rapid evaluation using butterfly trapping. *Biodiversity and Conservation* 4: 35–55.

———. 1996a. Global change and human susceptibilty to disease. *Annual Review of Energy and the Environment* 21: 125–144.

———. 1996b. Impacts of development and global change on the epidemiological environment. *Environment and Development Economics* 1: 309–344.

———. 1996c. Nocturnality and species survival. *Proceedings of the National Academy of Sciences USA* 93: 11709–11712.

Daily, G. C., P. R. Ehrlich, D. Goehring, and S. Dasgupta. ND. Range occupancy and endangerment: A test with a butterfly community. *Manuscript.*

Daily, G. C., P. R. Ehrlich, and A. Sanchez-Azofeifa. 2001. Countryside biogeography: Utilization of human-dominated habitats by the avifauna of southern Costa Rica. *Ecological Applications* 11: 1–13.

Daily, G. C., P. R. Ehrlich, and D. E. Wheye. 1991. Determinants of spatial distribution in a population of the subalpine butterfly *Oeneis chryxus. Oecologia* 88: 587–596.

Darrow, K., and M. D. Bowers. 1999. Effects of herbivore damage and nutrient level on induction of iridoid glycosides in *Plantago lanceolata. Journal of Chemical Ecology* 25: 1427–1440.

Darwin, C. 1842. *Journal of Researches.* Henry Colborn, London.

———. 1859. *On the Origin of Species.* John Murray, London.

Davis, J. I. 1996. Phylogenetics, molecular variation, and species concepts. *BioScience* 46: 502–511.

Day, J. R., and H. P. Possingham. 1995. A stochastic metapopulation model with variability in patch size and position. *Theoretical Population Biology* 48: 333–360.

Debinski, D. M. 1994. Genetic diversity assessment in a metapopulation of the butterfly *Euphydryas gillettii. Biological Conservation* 70: 25–31.

Debinksi, D. M., and L. Kelly. 1998. Decline of Iowa populations of the regal fritillary (*Speyeria idalia* Drury). *Journal of the Iowa Academy of Sciences* 105: 16–22.

de Bono, M., and C. I. Bargmann. 1998. Natural variation in a neuropeptide Y receptor homolog modifies social behavior and food response in *C. elegans. Cell* 94: 679–689.

Deinert, E. I., J. T. Longino, and L. E. Gilbert. 1994. Mate competition in butterflies. *Nature* 370: 23–24.

de Jong, P. W., and J. K. Nielsen. 1998. Polymorphism in a flea beetle for the ability to use and atypical host plant. *Proceedings of the Royal Society of London B* 266: 103–111.

de Jong, R., R. I. Vane-Wright, and P. R. Ackery. 1996. The higher classification of butterflies (Lepidoptera): Problems and prospects. *Entomologica scandinavica* 27: 65–101.

De Moraes, C. M., W. J. Lewis, P. W. Paré, H. T. Albarn, and J. H. Tumlinson. 1998. Herbivore infested plants selectively attract parasitoids. *Nature* 393: 570–573.

Dempster, J. P. 1983. The natural control of populations of butterflies and moths. *Biological Review* 58: 461–481.

———. 1984. The natural enemies of butterflies. Pages 97–104. *The Biology of Butterflies.* Academic Press, London.

den Boer, P. J. 1968. Spreading of risk and stabilization of animal numbers. *Acta Biotheoretica* 18: 165–194.

Dennis, R. L. H. 1977. *The British Butterflies: Their Origin and Establishment.* E.W. Classey, Faringdon, UK.

Dennis, R. L. H., and H. T. Eales. 1999. Probability of site occupancy in the large health butterfly *Coenonympha tullia* determined from geographical and ecological data. *Biological Conservation* 87: 295–301.

Dennis, R. L. H., and T. G. Shreeve. 1996. *Butterflies on British and Irish Offshore Islands.* Gem, Wallingford, UK.

Dennis, R. L. H., T. H. Sparks, and T. G. Shreeve. 1998. Geographical factors influencing the probability of *Hipparchia semele* (L.) (Lepidoptera: Satyrinae) occurring on British and Irish off-shore islands. *Global Ecology and Biogeography Letters* 7: 205–214.

Denno, R. F., and B. Benrey. 1997. Aggregation facilitates larval growth in the neotropical nymphalid butterfly *Chlosyne janais. Ecological Entomology* 22: 133–141.

De Queiroz, K. 1998. The general lineage concept of species, species criteria, and the process of speciation. Pages 57–75 in D. J. Howard and S. H. Berlocher, eds. *Endless Forms: Species and Speciation.* Oxford University Press, Oxford.

De Queiroz, K., and M. J. Donoghue. 1988. Phylogenetic systematics and the species problem. *Cladistics* 4: 317–338.

Dethier, V. G. 1959. Food-plant distribution and density and larval dispersal as factors affecting insect population. *The Canadian Entomologist* 41: 581–596.

Dethier, V. G., and R. H. MacArthur. 1964. A field's capacity to support a butterfly population. *Nature* 201: 728–729.

Devine, M. C. 1975. Copulatory plugs in snakes: Enforced chastity. *Science* 187: 844–845.

———. 1977. Copulatory plugs, restricted mating opportunities and reproductive competition among male garter snakes. *Nature* 267: 345–346.

DeVries, P. J. 1987. *The Butterflies of Costa Rica and Their Natural History: Papilionidae, Pieridae, Nymphalidae*. Princeton University Press, Princeton, NJ.

DeVries, P. J., I. J. Kitching, and R. I. Vane-Wright. 1985. The systematic position of *Antirrhea* and *Caerois*, with comments on the classification of the Nymphalidae (Lepidoptera). *Systematic Entomology* 10: 11–32.

DeWeerdt, S. 2002. What *really* is an evolutionarily significant unit? The debate over integrating genetics and ecology in conservation biology. *Conservation Biology in Practice* 3: 10–17.

Dewsbury, D. A. 1984. Sperm competition in muroid rodents. Pages 547–571 in R. L. Smith, ed. *Sperm Competition and the Evolution of Animal Mating Systems*. Academic Press, Orlando, FL.

Diamond, J. M. 1972. *Avifauna of the Eastern Highlands of New Guinea*. Nuttall Ornithological Club, Cambridge, MA.

Diamond, J. M., and R. M. May. 1981. Island biogeography and the design of natural reserves. Pages 228–252 in R. M. May, ed. *Theoretical Ecology*. Blackwell, Oxford.

Dias, P. C., and J. Blondel. 1996. Local specialization and maladaptation in Mediterranean blue tits *Parus caeruleus*. *Oecologia* 107: 79–86.

Dickinson, J. L., and R. L. Rutowski. 1989. The function of the mating plug in the chalcedon checkerspot butterfly. *Animal Behaviour* 38: 154–162.

Dieckmann, U., R. Law, and J. A. J. Metz. 2000. *The Geometry of Ecological Interaction: Simplifying Spatial Complexity*. Cambridge University Press, Cambridge.

Diggle, P. J., and A. G. Chetwynd. 1991. Second-order analysis of spatial clustering for inhomogeneous populations. *Biometrics* 47: 1155–1163.

Dingle, H. 1968. The influence of environment and heredity on flight activity in the large Milkweed bug, *Oncopeltus*. *Journal of Experimental Biology* 48: 175–184.

———. 1991. Evolutionary genetics and animal migration. *American Zoologist* 31: 253–264.

Dingle, H., N. R. Blackley, and E. R. Miller. 1980. Variation in body size and flight performance in milkweed bugs (Oncopeltus). *Evolution* 34: 371–385.

Doak, P. 2000. The effects of plant dispersion and prey density on parasitism rates in a naturally patchy habitat. *Oecologia* 122: 556–567.

Dobkin, D. S., I. Olivieri, and P. R. Ehrlich. 1987. Rainfall and the interaction of microclimate with larval resources in the population dynamics of checkerspot butterflies (*Euphydryas editha*) inhabiting serpentine grassland. *Oecologia* 71: 161–166.

Dobson, A., and J. Poole. 1998. Conspecific aggregation and conservation biology. Pages 93–208 in T. Caro, ed. *Behavioral Ecology and Conservation Biology*. Oxford University Press, New York.

Dobson, F. S. 1982. Competition for mates and predominant male dispersal in mammals. *Animal Behaviour* 30: 1183–1192.

Dobzhansky, T. 1947. Genetics of natural populations. XIV. A response of certain gene arrangements in the third chromosome of *Drosophila pseudoobscura* to natural selection. *Genetics* 32: 142–160.

———. 1950. Mendelian populations and their evolution. *The American Naturalist* 84: 401–418.

———. 1955. A review of some fundamental concepts and problems of population genetics. *Cold Spring Harbor Symposia on Quantitative Biology* 20: 1–15.

Dodd, M., J. Silvertown, K. Mcconway, J. Potts, and M. Crawley. 1995. Community stability: A 60-year record of trends and outbreaks in the occurrence of species in the Park Grass Experiment. *Journal of Ecology* 83: 277–285.

Dolinger, P. M., P. R. Ehrlich, W. L. Fitch, and D. E. Breedlove. 1973. Alkaloid and predation patterns in Colorado lupine populations. *Oecologia* 13: 191–204.

Donald, P. F., R. E. Green, and M. F. Heath. 2001. Agricultural intensification and the collapse of Europe's farmland bird populations. *Proceedings of the Royal Society of London B* 268: 25–29.

Donoghue, M. J. 2001. A wish list for systematic zoology. *Systematic Biology* 50: 755–757.

Dover, J. W., S. A. Clarke, and L. Rew. 1992. Habitats and movement patterns of satyrid butterflies (Lepidoptera: Satyridae) on arable farmland. *Entomologist's Gazette* 43: 29–44.

Dover, J., and T. Sparks. 2000. A review of the ecology of butterflies in British hedgerows. *Journal of Environmental Management* 60: 51–63.

Dowdeswell, W. H. 1981. *The Life of the Meadow Brown*. Heinemann, London.

Dowdeswell, W. H., R. A. Fisher, and E. B. Ford. 1949. The quantitative study of populations in the Lepidoptera. 2. *Maniola jurtina* L. *Heredity* 3: 67–84.

Drechsler, M., K. Frank, I. Hanski, R. B. O'Hara, and C. Wissel. 2003. Ranking metapopulation extinction risk for conservation: From patterns in data to management decisions. *Ecological Applications* in press.

Drummond, B. A. I. 1984. Multiple mating and sperm competition in the Lepidoptera. Pages 291–370 in R. L. Smith, ed. *Sperm Competition and the Evolution of Animal Mating Systems*. Academic Press, New York.

Drummond, B. A. I., G. L. Bush, and T. C. Emmel. 1970. The biology and laboratory culture of *Chlosyne lacinia* Geyer (Nymphalidae). *Journal of the Lepidopterists' Society* 24: 135–142.

Dugatkin, L. A., ed. 2001. *Model Systems in Behavioral Ecology*. Princeton University Press, Princeton, NJ.

Dunlap-Pianka, H., C. L. Boggs, and L. E. Gilbert. 1977. Ovarian dynamics in heliconiine butterflies: Programmed senescence versus eternal youth. *Science* 197: 487–490.

Dwyer, G. 1994. Density dependence and spatial structure in the dynamics of insect pathogens. *The American Naturalist* 143: 533–562.

Dwyer, G., J. Dushoff, J. R. Elkinton, and S. A. Levin. 2000. Pathogen-driven outbreaks in forest defoliators revisited: Building models from experimental data. *The American Naturalist* 156: 105–120.

Dyer, L. A. 1995. Tasty generalist and nasty specialist? Antipredator mechanisms in tropical lepidopteran larvae. *Ecology* 76: 1483–1496.

Dyer, L. A., and M. D. Bowers. 1996. The importance of sequestered iridoid glycosides as a defence against an ant predator. *Journal of Chemical Ecology* 22: 1527–1539.

Ebenhard, T. 1987. An experimental test of the island colonization survival model: Bank vole (*Clethrionomys glareolus*) populations with different demographic parameter values. *Journal of Biogeography* 14: 213–223.

———. 1991. Colonization in metapopulations: A review of theory and observations. *Biological Journal of the Linnean Society* 42: 105–121.

Ebert, D., C. Haag, M. Kirkpatrick, M. Riek, J. W. Hottinger, and V. I. Pajunen. 2002. A selective advantage to immigrant genes in a *Daphnia* metapopulation. *Science* 295: 485–488.

Edenhamn, P. 1996. Spatial dynamics of the European tree frog (*Hyla arborea* L.) in a heterogeneous landscape. Swedish University of Agricultural Sciences, Uppsala.

Efron, B., and R. J. Tibshirani. 1993. *An Introduction to the Bootstrap*. Chapman & Hall, Boca Raton, FL.

Ehrlich, A. H., and P. R. Ehrlich. 1978. Reproductive strategies in the butterflies: I. Mating frequency, plugging and egg number. *Journal of the Kansas Entomological Society* 51: 666–697.

Ehrlich, P. R. 1958a. The comparative morphology, phylogeny and higher classification of the butterflies (Lepidoptera: Papilionoidea). *The University of Kansas Science Bulletin* 39: 305–370.

———. 1958b. The integumental anatomy of the monarch butterfly *Danaus plexippus* L. (Lepidoptera: Danaiidae). *The University of Kansas Science Bulletin* 38: 1315–1349.

———. 1961a. Has the biological species concept outlived its usefulness? *Systematic Zoology* 10: 167–176.

———. 1961b. Intrinsic barriers to dispersal in checkerspot butterfly. *Science* 134: 108–109.

———. 1964. Some axioms of taxonomy. *Systematic Zoology* 13: 109–123.

———. 1965. The population biology of the butterfly *Euphydryas editha*. II. The structure of the Jasper Ridge colony. *Evolution* 19: 327–336.

———. 1983. Butterfly nomenclature, stability, and the rule of obligatory categories. *Systematic Zoology* 32: 451–453.

———. 1984a. The structure and dynamics of butterfly populations. Pages 25–40 in R. I. Van-Wright and P. R. Ackery, eds. *The Biology of Butterflies*. Academic Press, London.

———. 1984b. This week's citation classic. *Current Contents* no. 37, 10 September, p. 16.

———. 1992. Population biology of checkerspot butterflies and the preservation of global biodiversity. *Oikos* 63: 6–12.

———. 1995. The scale of the human enterprise and biodiversity loss. Pages 214–226 in J. H. Lawton and R. M. May, eds. *Extinction Rates*. Oxford University Press, Oxford.

———. 1997. *A World of Wounds: Ecologists and the Human Dilemma*. Ecology Institute, Oldendorf/Lune.

———. 2000. *Human Natures: Genes, Cultures, and the Human Prospect*. Island Press, Washington, DC.

———. 2001. Tropical butterflies: A key model group that can be "completed." *Lepidoptera News* 2 (1): 10–12.

———. 2002. Human natures, nature conservation, and environmental ethics. *BioScience* 52: 31–43.

Ehrlich, P. R., and L. C. Birch. 1967. The "balance of nature" and "population control". *The American Naturalist* 101: 97–107.

Ehrlich, P. R., and J. H. Camin. 1960. Natural selection in Middle Island water snakes (*Natrix sipedon* L.). *Evolution* 14: 136.

Ehrlich, P. R., and G. C. Daily. 1993. Population extinction and saving biodiversity. *Ambio* 22: 64–68.

Ehrlich, P. R., and S. E. Davidson. 1960. Techniques for capture-recapture studies of Lepidoptera populations. *Journal of the Lepidopterists' Society* 14: 227–229.

Ehrlich, P. R., D. S. Dobkin, and D. Wheye. 1988. *The Birder's Handbook: A Field Guide to the Natural History of North American Birds*. Simon and Schuster, New York.

———. 1992. *Birds in Jeopardy*. Stanford University Press, Stanford, CA.

Ehrlich, P. R., and A. H. Ehrlich. 1981. *Extinction: The Causes and Consequences of the Disappearance of Species*. Random House, New York.

———. 1989. Too many rich folks. *Populi* 16: 21–29.

———. 1991. *Healing the Planet*. Addison-Wesley, Reading, MA.

———. 2004. *One with Nineveh: Politics, Consumption, and the Human Future*. Island Press, Washington, DC.

Ehrlich, P. R., and L. E. Gilbert. 1973. Population structure and dynamics of the tropical butterfly *Heliconius ethilla*. *Biotropica* 5: 69–82.

Ehrlich, P. R., and J. Holdren. 1971. Impact of population growth. *Science* 171: 1212–1217.

Ehrlich, P. R., and R. W. Holm. 1962. Patterns and populations. *Science* 137: 652–657.

———. 1963. *The Process of Evolution*. McGraw-Hill, New York.

Ehrlich, P. R., A. E. Launer, and D. D. Murphy. 1984. Can sex ratio be defined or determined? The case of a population of checkerspot butterflies. *The American Naturalist* 124: 527–539.

Ehrlich, P. R., and L. G. Mason. 1966. The population biology of the butterfly *Euphydryas editha*. III. Selection and the phenetics of the Jasper Ridge colony. *Evolution* 20: 165–173.

Ehrlich, P. R., and D. D. Murphy. 1981a. Butterfly nomenclature: A critique. *Journal of Research on the Lepidoptera* 20: 1–11.

———. 1981b. The population biology of checkerspot butterflies (*Euphydryas*). *Biologisches Zentralblatt* 100: 613–629.

———. 1987a. Conservation lessons from long-term studies of checkerspot butterflies. *Conservation Biology* 1: 122–131.

———. 1987b. Monitoring populations on remnants of native vegetation. Pages 201–210 in D. A. Saunders, G. W. Arnold, A. A. Burbidge, and A. J. M. Hopkins, eds. *Nature Conservation: The Role of Remnants of Native Vegetation*. Surrey Beatty and Sons, Chipping Norton, NSW.

Ehrlich, P. R., D. D. Murphy, M. C. Singer, and C. B. Sherwood. 1980. Extinction, reduction, stability and increase: The responses of checkerspot butterfly (*Euphydryas*) populations to the California drought. *Oecologia* 46: 101–105.

Ehrlich, P. R., and P. H. Raven. 1964. Butterflies and plants: A study in coevolution. *Evolution* 18: 586–608.

———. 1969. Differentiation of populations. *Science* 65: 1228–1232.

Ehrlich, P. R., and J. Roughgarden. 1987. *The Science of Ecology*. Macmillan, New York.

Ehrlich, P. R., and D. Wheye. 1984. Some observations on spatial distribution in a montane population of *Euphydryas editha*. *Journal of Research on the Lepidoptera* 23: 143–152.

———. 1986. "Nonadaptive" hilltopping behavior in male checkerspot butterflies (*Euphydryas editha*). *The American Naturalist* 127: 477–483.

———. 1988. Hilltopping checkerspot butterflies revisited. *The American Naturalist* 132: 460–461.

Ehrlich, P. R., and R. R. White. 1980. Colorado checkerspot butterflies: Isolation, neutrality, and the biospecies. *The American Naturalist* 115: 328–341.

Ehrlich, P. R., R. R. White, M. C. Singer, S. W. McKechnie, and L. E. Gilbert. 1975. Checkerspot butterflies: A historical perspective. *Science* 188: 221–228.

Ehrlich, P. R., and E. O. Wilson. 1991. Biodiversity studies: Science and policy. *Science* 253: 758–762.

Eliasson, C. 1991. Studier av boknätfjärilens *Euphydryas maturna* (Lepidoptera: Nymphalidae) förekomst och biologi i Västmanland. *Entomologisk Tidskrift* 112: 113–124.

Elkinton, J. S., W. M. Healy, J. P. Buonaccorsi, G. H. Boettner, A. M. Hazzard, H. R. Smith, and A. M. Liebhold. 1996. Interactions among gypsy moths, white-footed mice, and acorns. *Ecology* 77: 2332–2342.

Elkinton, J. S., and A. M. Liebhold. 1990. Population dynamics of the Gypsy moth in North America. *Annual Review of Entomology* 35: 571–596.

Ellers, J., and C. L. Boggs. 2003. The evolution of wing color: Male mate choice opposes adaptive wing color divergence in *Colias* butterflies. *Evolution* 57: 1100–1106.

Ellers, J., and J. J. M. van Alphen. 1997. Life history evolution in *Asobara tabida*: Plasticity in allocation of fat reserves to survival and reproduction. *Journal of Evolutionary Biology* 10: 771–785.

Ellner, S., A. R. Gallant, D. McCaffrey, and D. Nychka. 1991. Convergence rates and data requirements for Jacobian-based estimates of Lyapunov exponents from data. *Physics Letters A* 153: 357–363.

Ellstrand, L. C., and D. R. Elam. 1993. Population genetic consequences of small population size: Implications for plant conservation. *Annual Review of Ecology and Systematics* 24: 217–242.

Elton, C. 1924. Periodic fluctuations in the numbers of animals: Their causes and effects. *British Journal of Experimental Biology* 2: 119–163.

———. 1942. *Voles, Mice, and Lemmings. Problems in Population Dynamics.* Clarendon Press, Oxford.

———. 1949. Population interspersion: An essay on animal community patterns. *Journal of Ecology* 37: 1–23.

Emelianov, I., M. Dres, W. Baltensweiler, and J. Mallet. 2001. Host-induced assortative mating in host races of the larch budmoth. *Evolution* 55: 2002–2010.

Emlen, J. M. 1984. *Population Biology: The Coevolution of Population Dynamics and Behavior*. Macmillan, New York.

Emlen, S. T., and L. W. Oring. 1977. Ecology, sexual selection and the evolution of mating systems. *Science* 197: 215–223.

Endler, J. A. 1986. *Natural Selection in the Wild.* Princeton University Press, Princeton.

Ericson, L., J. J. Burdon, and W. J. Müller. 1999. Spatial and temporal dynamics of epidemics of the rust fungus *Uromyces valarinae* on populations of its host *Valeriana salina. Journal of Ecology* 87: 649–658.

Eriksson, O., and B. Bremer. 1992. Pollination systems, dispersal modes, life forms, and diversification rates in angiosperm families. *Evolution* 46: 258–266.

Estes, J., and D. Duggins. 1995. Sea otters and kelp forest in Alaska. *Ecological Monographs* 65: 75–100.

Estes, J. A., M. T. Tinker, T. M. Williams, and D. F. Doak. 1998. Killer whale predation on sea otters linking oceanic and nearshore ecosystems. *Science* 282: 473–476.

Etienne, R. S. 2002. Striking the metapopulation balance. PhD dissertation, University of Wageningen, Netherlands.

Excoffier, L. 2001. Analysis of population subdivision. Pages 271–307 in D. J. Balding, M. Bishop, and C. Cannings, eds. *Handbook of Statistical Genetics*. John Wiley & Sons, Chichester, UK.

Farrell, B. D. 1998. "Inordinate fondness": Explained: Why are there so many beetles? *Science* 281: 555–559.

Farrell, B. D., C. Mitter, and D. J. Futuyma. 1992. Diversification at the insect-plant interface: Insights from phylogenetics. *BioScience* 42: 34–42.

Fauvergue, X., F. Fleury, C. Lemaitre, and R. Allemand. 1999. Parasitoid mating structures when hosts are patchily distributed: Field and laboratory experiments with *Leptopilina boulardi* and *L. heterotoma. Oikos* 86: 344–356.

Feder, J. L. 1998. The apple maggot fly, *Rhagoletis pomonella*: Flies in the face of conventional wisdom about speciation? Pages 130–144 in D. J. Howard and S. H. Berlocher, eds. *Endless Forms: Species and Speciation.* Oxford University Press, Oxford.

Feder, J. L., C. A. Chilcote, and G. L. Bush. 1988. Genetic differentiation between sympatric host races of the apple maggot fly *Rhagoletis pomonella. Nature* 336: 61–64.

———. 1990. Geographic pattern of genetic differentiation between host-associated populations of *Rhagoletis pomonella* (Diptera: Tephritidae) in the eastern United States and Canada. *Evolution* 44: 570–594.

Feder, J. L., T. A. Hunt, and G. L. Bush. 1993. The effects of climate, host plant phenology and host fidelity on the genetics of apple and hawthorn infesting races of *Rhagoletis pomonella. Entomologia Experimentalis et Applicata* 69: 117–135.

Feeny, P. P. 1976. Plant apparency and chemical defense. Pages 1–40 in J. W. Wallace and R. L. Mansell, eds. *Recent Advances in Phytochemistry: Biochemical Interaction between Plants and Insects.* Plenum Press, New York.

———. 1991. Chemical constraints on the evolution of swallowtail butterflies. Pages 315–340 in P. W. Price, T. M. Lewinsohn, G. W. Fernandes, and W. W. Benson, eds. *Plant-animal Interactions: Evolutionary Ecology in Tropical and Temperate Regions.* John Wiley & Sons, New York.

Feeny, P., W. S. Blau, and P. Kareiva. 1985. Larval growth and survivorship of the black swallowtail butterfly in central New York. *Ecological Monographs* 55: 167–187.

Feeny, P. P., E. Stadler, I. Ahman, and M. Carter. 1989. Effects of plant odor on oviposition by the Black Swallowtail butterfly, *Papilio polyxenes* (Lepidoptera: Papilionidae). *Journal of Insect Behavior* 2: 803–827.

Feldman, T. S., and W. A. Haber. 1998. Oviposition behavior, host plant use and diet breadth of *Anathassa* butterflies (Lepidoptera: Nymphalidae) using plants in the Acanthaceae in a Costa Rican community. *Florida Entomologist* 81: 396–406.

Ferguson, C. S., J. S. Elkinton, J. R. Gould, and W. E. Wallner. 1994. Population regulation of gypsy moth (Lepidoptera: Lymantriidae) by parasitoids: Does spatial density dependence lead to temporal density dependence? *Environmental Entomology* 23: 1155–1164.

Filchak, K. E., J. B. Roethele, and J. L. Feder. 2000. Natural selection and sympatric divergence in the apple maggot *Rhagoletis pomonella*. *Nature* 407: 739–742.

Finerty, J. P. 1980. *The Population Ecology of Cycles in Small Mammals*. Yale University Press, New Haven, CT.

Fisher, R. A. 1930. *The Genetic Theory of Natural Selection*. Clarendon, Oxford.

Fisher, R. A., and E. B. Ford. 1947. The spread of a gene in natural conditions in a colony of the moth *Panaxia dominula* L. *Heredity* 1: 143–174.

Fitzgerald, T. D. 1993. Sociality in caterpillars. Pages 372–403 in N. E. Stamp and T. M. Casey, eds. *Caterpillars: Ecological and Evolutionary Constraints on Foraging*. Chapman & Hall, New York.

Fleishman, E., J. F. Baughman, A. E. Launer, and P. R. Ehrlich. 1993. The effect of fluorescent pigments on butterfly copulation. *Ecological Entomology* 18: 165–167.

Fleishman, E., C. J. Betrus, R. B. Blair, R. Mac Nally, and D. D. Murphy. 2002a. Nestedness analysis and conservation planning: The importance of place, environment, and life history across taxonomic groups. *Oecologia* 113: 78–89.

Fleishman, E., R. B. Blair, and D. D. Murphy. 2001. Empirical validation of a method for umbrella species selection. *Ecological Applications* 11: 1489–1501.

Fleishman, E., B. G. Jonsson, and P. Sjögren-Gulve. 2000a. Focal species modeling for biodiversity conservation. *Ecological Bulletins* 48: 85–99.

Fleishman, E., A. E. Launer, S. B. Weiss, D. D. Murphy, J. M. Reed, and P. R. Ehrlich. 2000b. Effects of microclimate and oviposition timing on prediapause larval survival of the Bay checkerspot butterfly, *Euphydryas editha bayensis*. *Journal of Research on the Lepidoptera* 36: 31–44.

Fleishman, E., R. Mac Nally, and J. P. Fay. 2003. Validation tests of predictive models of butterfly occurence using environmental variables. *Conservation Biology* 17: 806–817.

Fleishman, E., D. D. Murphy, and P. F. Brussard. 2000c. A new method for selection of umbrella species for conservation planning. *Ecological Applications* 10: 569–579.

Fleishman, E., C. Ray, P. Sjögren-Gulve, C. L. Boggs, and D. D. Murphy. 2002b. Assessing the relative roles of patch quality, area, and isolation in predicting metapopulation dynamics. *Conservation Biology* 16: 706–716.

Fleury, F., R. Allemand, P. Fouillet, and M. Bouletreau. 1995. Genetic variation in locomotor activity rhythm among populations of *Leptopilina heteroma* (Hymenoptera: Eucoilidae), a larval parasitoid of *Drosophila* species. *Behavior Genetics* 25: 81–89.

Foley, P. 1994. Predictng extinction times from environmental stochasticity and carrying capacity. *Conservation Biology* 8: 124–137.

Ford, E. B. 1945. *Butterflies*. Collins, London.

———. 1975. *Ecological Genetics*. 4th ed. Chapman & Hall, London.

Ford, E. B., and P. M. Sheppard. 1969. The *Medionigra* polymorphism of *Panaxia dominula*. *Heredity* 24: 561–569.

Ford, H. D., and E. B. Ford. 1930. Fluctuation in numbers and its influence on variation in *Melitaea aurinia*, Rott. (Lepidoptera). *Transactions of the Entomological Society of London* 78: 345–351.

Ford, T. H., M. R. Shaw, and D. M. Robertson. 2000. Further host records of some west Paleartic tachinidae (Diptera). *Entomologist's Record and Journal of Variation* 112: 25–36.

Forman, R. T. T., and M. Godron. 1986. *Landscape Ecology*. John Wiley & Sons, New York.

Forys, E., and S. R. Humphrey. 1999. The importance of patch attributes and context to the management and recovery of an endangered lagomorph. *Landscape Ecology* 14: 177–185.

Fowler, K., and L. Partridge. 1989. A cost of mating in female fruitflies. *Nature* 338: 760–761.

Fox, C. W. 1993. The influence of maternal age and mating frequency on egg size and offspring performance in *Callosobruchus maculates* (Coleoptera, Bruchidae). *Oecologia* 96: 139–146.

Fraenkel, G. S. 1959. The raison d'etre of secondary plant substances. *Science* 129: 1466–1470.

Franke, A., H. Rimpler, and D. Schneider. 1987. Iridoid glycosides in the butterfly *Euphydryas cynthia* (Lepidoptera, Nymphalidae). *Phytochemistry* 26: 103–106.

Frankel, O. H., and M. E. Soulé. 1981. *Conservation and Evolution*. Cambridge University Press, Cambridge.

Frankham, R. 1995. Effective population size/adult population size ratios in wildlife: A review. *Genetical Research* 66: 95–107.

Frankham, R., and K. Ralls. 1998. Inbreeding leads to extinction. *Nature* 392: 441–442.

Fraser, F. C. 1954. High mortality of larvae of *Euphydryas aurinia*(Rott.) (Lep., Nymphalidae) from parasitisation by *Apanteles bignelli* Marsh (Hym., Braconidae). *The Entomologists Monthly Magazine* XC: 253.

Fryxall, J. M. 1995. Aggregation and migration by grazing ungulates in relation to resources and predators. Pages 257–273 in A. R. E. Sinclair and P. Arcese, eds. *Serengeti II. Dynamics, Management, and Conservation of an Ecosystem*. University of Chicago Press, Chicago.

Futuyma, D. J. 1986. *Evolutionary Biology*, 2nd ed. Sinauer Associates, Sunderland, MA.

———. 1991. Evolution of host specificity in herbivorous insects: Genetic, ecological, and phylogenetic aspects. Pages 431–454 in P. W. Price, T. M. Lewinsohn, G. W. Fernandes, and W. W. Benson, eds. *Plant-Animal Interactions: Evolutionary Ecology in Tropical and Temperate Regions*. John Wiley & Sons, New York.

———. 1998. *Evolutionary Biology*, 3rd ed. Sinauer Associates, Sunderland, MA.

Futuyma, D. J., and G. Moreno. 1988. The evolution of ecological specialization. *Annual Review of Ecology & Systematics* 19: 203–233.

Gadgil, M. 1971. Dispersal: Population consequences and evolution. *Ecology* 52: 253–261.

Gall, L. F. 1984. Population structure and recommendations for conservation of the narrowly endemic alpine butterfly, *Boloria acrocnema* (Lepidoptera: Nymphalidae). *Biological Conservation* 28: 111–138.

Gardener, M. C., and M. P. Gillman. 2001. The effects of soil fertilizer on amino acids in the floral nectar of corncockle, *Agrostemma githago* (Caryophyllaceae). *Oikos* 92: 101–106.

Gardenfors, U. 2001. Classifying threatened species at national versus global levels. *Trends in Ecology and Evolution* 16: 511–516.

Gardner, D. R., and F. R. Stermitz. 1988. Host plant utilization and iridoid glycoside sequestration by *Euphydryas anicia* (Lepidoptera: Nymphalidae). *Journal of Chemical Ecology* 14: 2147–2168.

Gaston, K. J. 1994. *Rarity*. Chapman & Hall, London.

Gause, G. F. 1934. *The Struggle for Existence*. Williams & Wilkins, Baltimore, MD.

Geiger, R. 1965. *The Climate Near the Ground*. Harvard University Press, Cambridge, MA.

Ghiselin, M. T. 1975. A radical solution to the species problem. *Systematic Zoology* 23: 536–544.

Giebultowicz, J. M., A. K. Raina, E. C. Uebel, and R. L. Ridgway. 1991. 2-Step regulation of sex-pheromone decline in mated gypsy moth females. *Archives of Insect Biochemistry and Physiology* 16: 95–105.

Gilbert, L. E. 1971. Butterfly-plant coevolution: Has *Passiflora adenopoda* won the selectional race with heliconiine butterflies? *Science* 172: 585–586.

———. 1973. Pollen feeding and reproductive biology of *Heliconius* butterflies. *Proceedings of the National Academy of Sciences USA* 69: 1403–1407.

———. 1976. Postmating female odor in *Heliconius* butterflies: Male-contributed antiaphrodisiac. *Science* 193: 419–420.

———. 1983. Coevolution and mimicry. Pages 263–281 in D. J. Futyuma and M. Slatkin, eds. *Coevolution*. Sinauer Associates, Sunderland, MA.

Gilbert, L. E., and M. C. Singer. 1973. Dispersal and gene flow in a butterfly species. *The American Naturalist* 107: 58–72.

———. 1975. Butterfly Ecology. *Annual Review of Ecology and Systematics* 6: 365–397.

Gillespie, J. H. 1991. *The Causes of Molecular Evolution*. Oxford University Press, New York.

Gillett, J. D. 1972. *The Mosquito: Its Life, Activities, and Impact on Human Affairs*. Doubleday, Garden City, NY.

Gillham, N. W. 1965. Geographic variation and the subspecies concept in butterflies. *Systematic Zoology* 5: 100–120.

Gilligan, D. M., L. M. Woodworth, M. E. Montgomery, D. A. Briscoe, and R. Frankham. 1997. Is mutation accumulation a threat to the survival of endangered populations? *Conservation Biology* 11: 1235–1241.

Gilpin, M. E., and J. M. Diamond. 1980. Subdivision of nature reserves and the maintenance of species diversity. *Nature* 285: 567–568.

Godfray, H. C. J. 1994. *Parasitoids: Behavioural and Evolutionary Ecology*. Princeton University Press, Princeton, NJ.

———. 2000. Host resistance, parasitoid virulence, and population dynamics. Pages 121–138 in M. E. Hochberg and A. R. Ives, eds. *Parasitoid Population Biology*. Princeton University Press, Princeton, NJ.

Goodall, J. 1986. *The Chimpanzees of Gombe: Patterns of Behavior*. Harvard University Press, Cambridge, MA.

Goodall, J., A. Bandora, E. Bergman, C. Busse, H. Matma, E. Mpongo, A. Pierce, and D. Riss. 1979. Intercommunity interactions in the chimpanzee population of the Gombe National Park. Pages 13–53 in D. A. Hamburg and E. R. McCown, eds. *The Great Apes*. Benjamin Cummings, Menlo Park, CA.

Gordh, G. 1981. The phenomenon of insect hyperparasitism and its taxonomic occurence in the Insecta. Pages 10–18 in D. Rosen, ed. *The Role of Hyperparasitism in Biological Control: A Symposium*. University of California, Berkeley.

Gordon, D. 1999. *Ants at Work: How an Insect Society Is Organized*. The Free Press, New York.

Goudet, J., M. Raymond, T. de Meeüs, and F. Rousset. 1996. Testing differentiation in diploid populations. *Genetics* 144: 1933–1940.

Grant, B. R., and P. R. Grant. 1993. Evolution of Darwin's finches caused by a rare climatic event. *Proceedings of the Royal Society of London Series B* 251: 111–117.

———. 1997. Genetics and the origin of bird species. *Proceedings of the National Academy of Sciences USA* 94: 7768–7775.

Grant, P. R. 1986. *Ecology and Evolution of Darwin's Finches*. Princeton University Press, Princeton, NJ.

———. 1999. *Ecology and Evolution of Darwin's Finches* [reprinting with new Afterword]. Princeton University Press, Princeton, NJ.

Guilford, T. 1988. The evolution of conspicuous coloration. *The American Naturalist* 131: S7–S21.

Guppy, C. S., and A. I. Fischer. 2001. Garry oak ecosystems rare/endangered butterflies inventory—2001 report.

Guppy, C. S., and J. H. Shepard. 2001. *Butterflies of British Columbia*. UBC Press, Vancouver, BC.

Gurney, W. S. C., and R. M. Nisbet. 1978. Single species fluctuations in patchy environments. *The American Naturalist* 112: 1075–1090.

Gyllenberg, M., and I. Hanski. 1992. Single-species metapopulation dynamics: A structured model. *Theoretical Population Biology* 72: 35–61.

Gyllenberg, M., and D. S. Silvestrov. 1994. Quasi-stationary distribution of a stochastic metapopulation model. *Journal of Mathematical Biology* 33: 35–70.

Gyllenberg, M., G. Söderbacka, and S. Ericsson. 1993. Does migration stabilize local population dynamics? Analysis of a discrete metapopulation model. *Mathematical Biosciences* 118: 25–49.

Haag, C., J. Hottinger, M. Riek, and D. Ebert. 2002. Strong inbreeding depression in a *Daphnia* metapopulation. *Evolution* 56: 518–526.

Haddad, N. M. 1999a. Corridor and distance effects on interpatch movements: A landscape experiment with butterflies. *Ecological Applications* 9: 612–622.

———. 1999b. Corridor use predicted from behaviors at habitat boundries. *The American Naturalist* 153: 215–227.

Haddad, N. M., and K. Baum. 1999. An experimental test of corridor effects on butterfly densities. *Ecological Applications* 9: 623–633.

Haikola, S., W. Fortelius, R. B. O'Hara, M. Kuussaari, N. Wahlberg, I. J. Saccheri, M. C. Singer, and I. Hanski. 2001. Inbreeding depression and the maintenance of genetic load in *Melitaea cinxia* metapopulations. *Conservation Genetics* 2: 325–335.

Haikola, S., M. C. Singer, and I. Pen. 2004. Has inbreeding depression led to avoidance of sib mating in the Glanville fritillary butterfly (*Melitaea cinxia*)? Manuscript.

Haldane, J. B. S. 1931. A mathematical theory of natural selection VI Isolation. *Proceedings of the Cambridge Philosophical Society* 26: 220–230.

———. 1937. The effect of variation on fitness. *The American Naturalist* 71: 337–349.

Hämet-Ahti, L., J. Suominen, T. Ulvinen, and P. Uotila, eds. 1998. *Retkeilykasvio* [*Field Flora of Finland*]. Yliopistopaino, Helsinki.

Hamilton, W. D., and R. M. May. 1977. Dispersal in stable habitats. *Nature* 269: 578–581.

Hammond, P. C., and D. V. McCorkle. 1983. The decline and extinction of *Speyeria* populations resulting from human environmental disturbances (Nymphalidae: Argynninae). *Journal of Research on Lepidoptera* 22: 217–228.

Hanski, I. 1976. Breeding experiments with carrion flies (Diptera) in natural conditions. *Annales Entomologici Fennici* 42: 113–121.

———. 1981. Coexistence of competitors in patchy environment with and without predation. *Oikos* 37: 306–312.

———. 1982. Dynamics of regional distribution:

The core and satellite species hypothesis. *Oikos* 38: 210–221.
———. 1983. Coexistence of competitors in patchy environment. *Ecology* 64: 493–500.
———. 1985. Single-species spatial dynamics may contribute to long-term rarity and commonness. *Ecology* 66: 335–343.
———. 1987. Carrion fly community dynamics: Patchiness, seasonality and coexistence. *Ecological Entomology* 12: 257–266.
———. 1990. Density dependence, regulation and variability in animal populations. *Philosophical Transactions of the Royal Society of London B* 330: 141–150.
———. 1991. Single-species metapopulation dynamics: Concepts, models and observations. *Biological Journal of the Linnean Society* 42: 17–38.
———. 1992. Inferences from ecological incidence functions. *The American Naturalist* 139: 657–662.
———. 1994a. Patch-occupancy dynamics in fragmented landscapes. *Trends in Ecology and Evolution* 9: 131–135.
———. 1994b. A practical model of metapopulation dynamics. *Journal of Animal Ecology* 63: 151–162.
———. 1998a. Connecting the parameters of local extinction and metapopulation dynamics. *Oikos* 83: 390–396.
———. 1998b. Metapopulation dynamics. *Nature* 396: 41–49.
———. 1999a. Habitat connectivity, habitat continuity, and metapopulations in dynamic landscapes. *Oikos* 87: 209–219.
———. 1999b. *Metapopulation Ecology*. Oxford University Press, Oxford.
———. 2001. Spatially realistic theory of metapopulation ecology. *Naturwissenschaften* 88: 372–381.
———. 2002a. In the midst of ecology, conservation, and competing interests in the society. *Annales Zoologici Fennici* 39: 183–186.
———. 2002b. Metapopulations of animals in highly fragmented landscapes and population viability analysis. Pages 86–108 in S. R. Beissinger and D. R. McCullough, eds. *Population Viability Analysis*. Chicago University Press, Chicago.
———. 2003. Biology of extinctions in butterfly metapopulations. Pages 577–602 in C. L. Boggs, W. B. Watt, and P. R. Ehrlich, eds. *Butterflies: Ecology and Evolution Taking Flight*. University of Chicago Press, Chicago.
Hanski, I., J. Alho, and A. Moilanen. 2000. Estimating the parameters of survival and migration of individuals in metapopulations. *Ecology* 81: 239–251.
Hanski, I., C. J. Breuker, K. Schöps, R. Setchfield, and M. Nieminen. 2002. Population history and life history influence the migration rate of female Glanville fritillary butterflies. *Oikos* 98: 87–97.
Hanski, I., P. Foley, and M. Hassell. 1996a. Random walks in a metapopulation: How much density dependence is necessary for long-term persistence? *Journal of Animal Ecology* 65: 274–282.
Hanski, I., and M. Gilpin. 1991. Metapopulation dynamics: Brief history and conceptual domain. Pages 3–16 in M. Gilpin and I. Hanski, eds. *Metapopulation Dynamics: Empirical and Theoretical Investigations*. Academic Press, London.
Hanski, I., and M. Gyllenberg. 1993. Two general metapopulation models and the core-satellite species hypothesis. *The American Naturalist* 142: 17–41.
Hanski, I., and P. Hammond. 1995. Biodiversity in boreal forests. *Trends in Ecology and Evolution* 10: 5–6.
Hanski, I., L. Hansson, and H. Henttonen. 1991. Specialist predators, generalist predators, and the microtine rodent cycle. *Journal of Animal Ecology* 60: 353–367.
Hanski, I., and M. Heino. 2003. Metapopulation-level adaptation of insect host plant preference and extinction-colonization dynamics in heterogeneous landscapes. *Theoretical Population Biology* 64: 281–290.
Hanski, I., and H. Henttonen. 2002. Population cycles of small rodents in Fennoscandia. Pages 44–68 in A. Berryman, ed. *Population Cycles: The case for Trophic Interactions*. Oxford University Press, Oxford.
Hanski, I., H. Henttonen, E. Korpimäki, L. Oksanen, and P. Turchin. 2001. Small-rodent dynamics and predation. *Ecology* 82: 1505–1520.
Hanski, I., and S. Kuusela. 1977. An experiment on competition and diversity in the carrion fly community. *Annales Entomologici Fennici* 43: 108–115.
———. 1980. The structure of carrion fly communities: Differences in breeding seasons. *Annales Zoologici Fennici* 17: 185–190.
Hanski, I., and M. Kuussaari. 1995. Butterfly metapopulation dynamics. Pages 149–171 in N. Capuccino and P. W. Price, eds. *Population Dynamics: New Approaches and Synthesis*. Academic Press, London.
Hanski, I., M. Kuussaari, and M. Nieminen. 1994. Metapopulation structure and migration in the butterfly *Melitaea cinxia*. *Ecology* 75: 747–762.
Hanski, I., A. Moilanen, and M. Gyllenberg.

1996b. Minimum viable metapopulation size. *The American Naturalist* 147: 527–541.
Hanski, I., A. Moilanen, T. Pakkala, and M. Kuussaari. 1996c. The quantitative incidence function model and persistence of an endangered butterfly metapopulation. *Conservation Biology* 10: 578–590.
Hanski, I., and O. Ovaskainen. 2000. The metapopulation capacity of a fragmented landscape. *Nature* 404: 755–758.
———. 2002. Extinction debt at extinction threshold. *Conservation Biology* 16: 666–673.
———. 2003. Metapopulation theory for heterogeneous fragmented landscapes. *Theoretical Population Biology* 64: 119–127.
Hanski, I., T. Pakkala, M. Kuussaari, and G. Lei. 1995a. Metapopulation persistence of an endangered butterfly in a fragmented landscape. *Oikos* 72: 21–28.
Hanski, I., J. Pöyry, T. Pakkala, and M. Kuussaari. 1995b. Multiple equilibria in metapopulation dynamics. *Nature* 377: 618–621.
Hanski, I., and E. Ranta. 1983. Coexistence in a patchy environment: Three species of *Daphnia* in rock pools. *Journal of Animal Ecology* 52: 263–280.
Hanski, I., and D. Simberloff. 1997. The metapopulation approach, its history, conceptual domain and application to conservation. Pages 5–26 in I. Hanski and M. E. Gilpin, eds. *Metapopulation Biology*. Academic Press, San Diego.
Hanski, I., and M. Singer. 2001. Extinction-colonization dynamics and host-plant choice in butterfly metapopulations. *The American Naturalist* 158: 341–353.
Hanski, I., and C. D. Thomas. 1994. Metapopulation dynamics and conservation: A spatially explicit model applied to butterflies. *Biological Conservation* 68: 167–180.
Hansson, L. 1991. Dispersal and connectivity in metapopulations. *Biological Journal of the Linnean Society* 42: 89–103.
Hansson, L., and H. Henttonen. 1985. Gradients in density variations of small rodents: The importance of latitude and snow cover. *Oecologia* 67: 394–402.
Harrison, R. G. 1998. Linking evolutionary pattern and process: The relevance of species concepts for the study of speciation. Pages 19–31 in D. J. Howard and S. H. Berlocher, eds. *Endless Forms: Species and Speciation*. Oxford University Press, New York.
Harrison, S. 1989. Long-distance dispersal and colonization in the Bay checkerspot butterfly, *Euphydryas editha bayensis*. *Ecology* 70: 1236–1243.
———. 1991. Local extinction in a metapopulation context: An empirical evaluation. *Biological Journal of the Linnean Society* 42: 73–88.
———. 1994. Metapopulations and conservation. Pages 111–128 in P. J. Edwards, R. M. May, and N. R. Webb, eds. *Large-scale Ecology and Conservation Biology*. Blackwell Scientific, Oxford.
———. 1997. Persistent, localized outbreaks in the western tussock moth (*Orgyia vetusta*); roles or resource quality, predation and poor dispersal. *Ecological Entomology* 22: 158–166.
Harrison, S., and E. Bruna. 1999. Habitat fragmentation and large-scale conservation: What do we know for sure? *Ecography* 22: 225–232.
Harrison, S., D. D. Murphy, and P. R. Ehrlich. 1988. Distribution of the bay checkerspot butterfly, *Euphydryas editha bayensis*: Evidence for a metapopulation model. *The American Naturalist* 132: 360–382.
Harrison, S., J. F. Quinn, J. F. Baughman, D. D. Murphy, et al. 1991. Estimating the effects of scientific study on two butterfly populations. *The American Naturalist* 137: 227–243.
Harrison, S., K. Rice, and J. Maron. 2001. Habitat patchiness promotes invasion by alien grasses on serpentine soil. *Biological Conservation* 100: 45–53.
Hartl, D. L., and A. G. Clark. 1997. *Principles of Population Genetics*, 3rd ed. Sinauer Associates, Sunderland, MA.
Hartung, T. G., and D. A. Dewsbury. 1978. A comparative analysis of copulatory plugs in muroid rodents and their relationship to copulatory behavior. *Journal of Mammalogy* 59: 717–723.
Harvey, D. J. 1991. Higher classification of the Nymphalidae. Pages 255–273 in H. F. Nijhout, ed. *The Development an Evolution of Butterfly Wing Patterns*. Smithsonian Institution Press, Washington, DC.
Harvey, P. H. 1996. Phylogenies for ecologists. *Journal of Animal Ecology* 65: 255–263.
Harvey, P. H., and S. Nee. 1997. The phylogenetic foundations of behavioural ecology. Pages 334–349 in J. R. Krebs and N. B. Davies, eds. *Behavioural Ecology: An Evolutionary Approach*. Blackwell Scientific, Oxford.
Harvey, P. H., and M. D. Pagel. 1991. *The Comparative Method in Evolutionary Biology*. Oxford University Press, Oxford.
Hassell, M. P. 1969. A study of the mortality factors acting upon *Cyzenis albicans* (Fall.), a tachinid parasite of the winter moth

(*Operophtera brumata* (L.)). *Journal of Animal Ecology* 38: 329–339.
———. 1998. *Insect Parasitoid Population Dynamics*. Oxford University Press, Oxford.
———. 2000a. Host-parasitoid dynamics. *Journal of Animal Ecology* 69: 543–566.
———. 2000b. *The Spatial and Temporal Dynamics of Host–Parasitoid Interactions*. Oxford University Press, London.
Hassell, M. P., H. N. Comins, and R. M. May. 1991. Spatial structure and chaos in insect population dynamics. *Nature* 353: 255–258.
———. 1994. Species coexistence and self-organizing spatial dynamics. *Nature* 370: 290–292.
Hassell, M. P., and T. R. E. Southwood. 1978. Foraging strategies of insects. *Annual Reviews of Ecology and Systematics* 9: 75–98.
Hastings, A. 1983. Can spatial variation alone lead to selection for dispersal? *Theoretical Population Biology* 24: 244–251.
———. 2000. Parasitoid spread: Lessons for and from invasion biology. Pages 70–82 in M. E. Hochberg and A. R. Ives, eds. *Parasitoid Population Biology*. Princeton University Press, Princeton, NJ.
Haukioja, E. 1980. On the role of plant defences in the fluctuation of herbivore populations. *Oikos* 35: 202–213.
———. 1993. Effects of food and predation on population dynamics. Pages 425–447 in N. E. Stamp and T. M. Casey, eds. *Caterpillars: Ecological and Evolutionary Constraints on Foraging*. Chapman & Hall, New York.
Haukioja, E., K. Kapiainen, P. Niemelä, and J. Tuomi. 1983. Plant availabiltiy hypothesis and other explanations of herbivore cycles: Complementary or exclusive alternatives. *Oikos* 40: 419–432.
Hawkins, B. A. 1994. *Patterns and Process in Host-Parasitoid Interactions*. Cambridge University Press, Cambridge.
———. 2000. Species coexistance in parasitoid communities: Does competition matter? Pages 198–214 in M. E. Hochberg and A. R. Ives, eds. *Parasitoid Population Biology*. Princeton University Press, Princeton, NJ.
Hawkins, B. A., N. J. Mills, M. A. Jervis, and P. W. Price. 1999. Is the biological control of insects a natural phenomenon? *Oikos* 86: 493–506.
Hawkins, B. A., and W. Sheehan. 1994. *Parasitoid Community Ecology*. Oxford University Press, Oxford.
Hawthorne, D. J., and S. Via. 2001. Genetic linkage of ecological specialization and reproductive isolation in pea aphids. *Nature* 412: 904–907.
Heath, J. 1981. *Threatened Rhopalocera (Butterflies) in Europe*. European Committee for the Conservation of Nature and Natural Resources, Nature and Environment Series 23. Strasbourg.
Heath, J., E. Pollard, and J. Thomas. 1984. *Atlas of Butterflies in Britain and Ireland*. Viking, New York.
Hebert, P. D. N. 1983. Egg dispersal patterns and adult feeding behaviour in the Lepidoptera. *The Canadian Entomologist* 115: 1477–1481.
Hedrick, P. W. 1994. Purging inbreeding depression and the probability of extinction: Full-sib mating. *Heredity* 73: 363–372.
Hedrick, P. W., and S. T. Kalinowski. 2000. Inbreeding depression in conservation biology. *Annual Review of Ecology and Systematics* 31: 139–162.
Heino, M., and I. Hanski. 2001. Evolution of migration rate in a spatially realistic metapopulation model. *The American Naturalist* 157: 495–511.
Heinrich, B. 1979. *Bumblebee Economics*. Harvard University Press, Cambridge, MA.
———. 1993. *The Hot Blooded Insects*. Springer-Verlag, Berlin.
Hellmann, J. J. 2000. The role of environmental variation in the dynamics of an insect-plant interaction. PhD dissertation, Department of Biological Sciences, Stanford University, Stanford, CA.
———. The effect of an environmental change on mobile butterfly larvae and the nutritional quality of their hosts. *Journal of Animal Ecology* 70: 925–936.
———. 2002b. Butterflies as model systems for understanding and predicting climate change. Pages 93–126 in S. H. Schenider and T. L. Root, eds. *Wildlife Responses to Climate Change: North American Case Studies*. Island Press, Washington, DC.
———. 2002c. The effect of an environmental change on mobile butterfly larvae and the nutritional quality of their hosts. *Journal of Animal Ecology* 70: 925–936.
Hellmann, J. J., P. R. Armsworth, and J. Roughgarden. 2004. Resource availability, environmental change, and the abundance of an insect herbivore. Manuscript.
Hellmann, J. J., S. B. Weiss, J. F. McLaughlin, C. L. Boggs, P. R. Ehrlich, A. E. Launer, and D. D. Murphy. 2003. Testing short-term hypotheses with a long-term study of a butterfly population. *Ecological Entomology* 28: 74–84.
Hendricks, P. 1986. Avian predation of alpine butterflies. *Journal of the Lepidopterists' Society* 40: 129.

Henter, H. J. 1995. The potential for coevolution in a host-parasitoid system. II. Genetic variation within a population of wasps in the ability to parasitize an aphid host. *Evolution* 49: 439–445.

Herben, T., H. J. During, and R. Law. 2000. Spatio-temporal patterns in grassland communities. Pages 48–64 in U. Dickmann, R. Law, and J. A. J. Metz, eds. *The Geometry of Ecological Interactions*. Cambridge University Press, Cambridge.

Heske, E. J., J. H. Brown, and S. Mistry. 1994. Long-term experimental study of a desert rodent community: 13 years of competition. *Ecology* 75: 438–445.

Hestbeck, J. B., J. D. Nichols, and R. A. Malecki. 1991. Estimates of movement and site fidelity using mark-resight data of wintering Canada geese. *Ecology* 72: 523–533.

Hewitt, G. M. 1996. Some genetic consequences of ice ages, and their role in divergence and speciation. *Biological Journal of the Linnean Society* 58: 247–276.

———. 1999. Post-glacial re-colonization of European biota. *Biological Journal of the Linnean Society* 68: 87–112.

———. Speciation, hybrid zones and phylogeography—or seeing genes in space and time. *Molecular Ecology* 10: 537–549.

Hey, J. 2001. *Genes, Categories, and Species*. Oxford University Press, New York.

Heywood, V. H., ed. 1995. *Global Biodiversity Assessment*. Cambridge University Press, Cambridge.

Hickman, J. C., ed. 1993. *The Jepson Manual of Higher Plants in California*. University of California Press, Bekeley.

Higashiura, Y., M. Ishihara, and P. W. Schaefer. 1999. Sex ratio distortion and severe inbreeeding depression in the gypsy moth *Lymantria dispar* L. in Hokkaido, Japan. *Heredity* 83: 290–297.

Higgins, K., and M. Lynch. 2000. Metapopulation extinction caused by mutation accumulation. *Proceedings of the National Academy of Sciences USA* 98: 2928–2933.

Higgins, L. G. 1941. An illustrated catalogue of the Palearctic *Melitaea* (Lep. Rhopalocera). *Transactions of the Royal Entomological Society of London* 91: 175–365.

———. 1978. A revision of the genus *Euphydryas* scudder (Lepidoptera: Nymphalidae). *Entomologist's Gazette* 29: 109–115.

———. 1981. A revision of *Phyciodes* Huebner and related genera, with a review of the classification of the Melitaeinae (Lepidoptera: Nymphalidae). *Bulletin of the British Museum of Natural History* 43: 177–243.

Higgins, L. G., and N. D. Riley. 1983. *A Field Guide to the Butterflies of Britain and Europe*, 5th ed. Collins, London.

Hilborn, R. 1975. Similarities in dispersal tendency among siblings in four species of voles (*Microtus*). *Ecology* 56: 1221–1225.

———. 1991. Modeling the stability of fish schools: Exchange of individual fish between schools of skipjack tuna (*Katsuwonus pelamis*). *Canadian Journal of Fisheries and Aquatic Sciences* 48: 1081–1091.

Hill, J. K., C. D. Thomas, and D. S. Blakeley. 1999a. Evolution of flight in a butterfly that has recently expanded its geographic range. *Oecologia* 121: 165–170.

Hill, J. K., C. D. Thomas, and B. Huntley. 1999b. Climate and habitat availability determine 20th century changes in a butterfly's range margin. *Proceedings of the Royal Society of London, Series B* 266: 1197–1206.

———. 2002. Climate and recent range changes in butterflies. Pages 77–88 in G.-R. Walther, C. A. Burga, and P. J. Edwards, eds. *Fingerprints of Climate Change: Adapted Behavior and Shifting Species Ranges*. Kluwer Academic, London.

———. 2003. Modeling present and potential future ranges of European butterflies using climate response surfaces. Pages 149–167 in C. L. Boggs, W. B. Watt, and P. R. Ehrlich, eds. *Butterflies: Ecology and Evolution Taking Flight*. University of Chicago Press, Chicago.

Hill, J. K., C. D. Thomas, B. Huntley, and R. Fox. 2001. Analysing and modeling range changes in UK butterflies. Pages 415–442 in I. P. Woiwood, D. Reynolds, and C. D. Thomas, eds. *Insect Movement: Mechanisms and Consequences*. CABI, London.

Hill, J. K., C. D. Thomas, and O. T. Lewis. 1996. Effects of habitat patch size and isolation on dispersal by *Hesperia comma* butterflies: Implications for metapopulation structure. *Journal of Animal Ecology* 65: 725–735.

———. 1999c. Flight morphology in fragmented populations of a rare British butterfly, *Hesperia comma*. *Biological Conservation* 87: 227–283.

Hill, M. F., and H. Caswell. 2001. The effects of habitat destruction in finite landscapes: A chain-binomial. *Oikos* 93: 321–331.

Hilty, S. 1994. *Birds of Tropical America*. Chapters, Shelburne, VT.

Hobbs, R. J., and H. A. Mooney. 1995. Spatial and temporal variability in California annual grassland: Results from a long-term study. *Journal of Vegetation Science* 6: 43–56.

Hochberg, M. E. 2000. What, conserve parasitoids? Pages 266–277 in M. E. Hochberg and A. R. Ives, eds. *Parasitoid Population Biology*. Princeton University Press, Princeton, NJ.

Hochberg, M. E., and B. A. Hawkins. 1994. The implications of population dynamics theory to parasitoid diversity and biological control. Pages 451–471 in B. A. Hawkins and W. Sheehan, eds. *Parasitoid Community Ecology*. Oxford University Press, New York.

Hochberg, M. E., and A. R. Ives, eds. 2000. *Parasitoid Population Biology*. Princeton University Press, Princeton, NJ.

Hoffman, A. A. 1985. Effects of experience on oviposition and attraction in Drosophila: comparing apples and oranges. *The American Naturalist* 126: 41–51.

Holdren, C. E., and P. R. Ehrlich. 1981. Long range dispersal in checkerspot butterflies: Transplant experiments with *Euphydryas gillettii*. *Oecologia* 50: 125–129.

———. 1982. Ecological determinants of food plant choice in the checkerspot butterfly *Euphydryas editha* in Colorado. *Oecologia* 52: 417–423.

Holdren, J. P., and P. R. Ehrlich. 1974. Human population and the global environment. *American Scientist* 62: 282–292.

Holl, K. D. 1995. Nectar resources and their influence on butterfly communities on reclaimed coal surface mines. *Restoration Ecology* 3: 76–85.

Hölldobler, B., and E. O. Wilson. 1990. *The Ants*. Harvard University Press, Cambridge, MA.

Holt, R. D. 1977. Predation, apparent competition and the structure of prey communities. *Theoretical Population Ecology* 12: 197–229.

———. 1995. Demographic constraints in evolution: Towards unifying the evolutionary theories of senescence and the niche conservatism. *Evolutionary Ecology* 10: 1–11.

———. 1997. From metapopulation dynamics to community structure. Pages pp. 149–165 in I. Hanski and M. E. Gilpin, eds. *Metapopulation Biology*. Academic Press, San Diego, CA.

———. 2002. Food webs in space: On the interplay of dynamic instability and spatial processes. *Ecological Research* 17: 261–273.

Holt, R. D., and M. S. Gaines. 1992. Analysis of adaptation in heterogeneous landscapes: Implications for the evolution of fundamental niches. *Evolutionary Ecology* 6: 433–447.

Hopper, K. 1984. The effects of host-finding and colonization rates on abundance of parasitoids of a gall midge. *Ecology* 65: 20–27.

Horn, H. S., and R. H. MacArthur. 1972. Competition among fugitive species in a harlequin environment. *Ecology* 53: 749–752.

Horner-Devine, C., G. C. Daily, P. R. Ehrlich, and C. L. Boggs. 2002. Countryside biogeography of tropical butterflies. *Conservation Biology* 17: 168–177.

Howard, L. O., and W. F. Fiske. 1911. The importation into the United States of the parasites of the gipsy moth and the browntail moth. *Bulletin of the United States Bureau of Entomology* 91: 1–344.

Hufbauer, R. A. 2002. Aphid population dynamics: Does resistance to parasitim influence population size? *Ecological Entomology* 27: 25–32.

Huffaker, C. B. 1958. Experimental studies on predation: Dispersion factors and predator-prey oscillations. *Hilgardia* 27: 343–383.

Hughes, J. B., G. C. Daily, and P. R. Ehrlich. 1997. Population diversity: Its extent and extinction. *Science* 278: 689–692.

———. 1998. The loss of population diversity and why it matters. Pages 71–83 in P. H. Raven, ed. *Nature and Human Society*. National Academy Press, Washington, DC.

———. 2002. Conservation of tropical forest birds in countryside habitats. *Ecology Letters* 5: 121–129.

Hughes, L. 2000. Biological consequences of global warming: Is the signal already apparent? *Trends in Ecology and Evolution* 15: 56–61.

Hull, D. 1976. Are species really individuals? *Systematic Zoology* 25: 174–191.

Hunter, A. F. 1993. Gypsy moth population sizes and the window of opportunity in spring. *Oikos* 68: 531–538.

———. 2000. Gregariousness and repellent defences in the survival fo phytophagous insects. *Oikos* 91: 213–224.

Huntsman, A. G. 1937. The cause of periodic scarcity in Atlantic salmon. *Transactions of the Royal Society of Canada* 31: 17–27.

Hutchinson, D. W., and A. R. Templeton. 1999. Correlation of pairwise genetic and geographic distance measures: Inferring the relative influences of gene flow and drift on the distribution of genetic variability. *Evolution* 53: 1898–1914.

Idyll, C. P. 1973. The anchovy crisis. *Scientific American* 22–29.

Imbert, E. 2001. Capitulum characters in a seed heteromorphic plant, *Crepis sancta* (Asteraceae): Variance partitioning and inference for the evolution of dispersal rate. *Heredity* 86: 78–86.

Ims, R. A., and N. G. Yoccoz. 1997. Studying

transfer processes in metapopulations: Emigration, migration and colonization. Pages 247–266 in I. Hanski and M. E. Gilpin, eds. *Metapopulation Biology*. Academic Press, San Diego, CA.

Ingvarsson, P. K., and M. C. Whitlock. 2000. Heterosis increases the effective migration rate. *Proceedings of the Royal Society: Biological Sciences* 267: 1321–1326.

Ivlev, V. S. 1961. *Experimental Ecology of the Feeding of Fishes*. Yale University Press, New Haven, CT.

Iwasa, Y., F. J. Odendaal, D. D. Murphy, P. R. Ehrlich, et al. 1983. Emergence patterns in male butterflies: A hypothesis and a test. *Theoretical Population Biology* 23: 363–379.

Jackel, A. K., and P. Poschlod. 1996. Why are some plant species of fragmented continental dry grasslands frequent and some rare? Pages 194–203 in J. Settele, C. Margules, P. Poschlod, and K. Henle, eds. *Species Survival in Fragmented Landscapes*. Kluwer Academic, Dordrecht.

Jaenike, J. 1990. Host specialization in phytophagous insects. *Annual Review of Ecology and Systematics* 21: 243–274.

Janz, N., and S. Nylin. 1997. The role of female search behaviour in determining host plant range in plant feeding insects: A test of the information processing hypothesis. *Proceedings of the Royal Society of London, Biology* 264: 701–707.

———. 1998. Butterflies and plants: A phylogenetic study. *Evolution* 52: 486–502.

Jennions, M. D., and M. Petrie. 2000. Why do females mate multiply? A review of the genetic benefits. *Biological Reviews of the Cambridge Philosophical Society* 75: 21–64.

Jensen, S. R. 1991. Plant iridoids, their biosynthesis and distribution in angiosperms. Pages 133–158 in J. B. Harborne and F. A. Tomas-Barberan, eds. *Ecological Chemistry and Biochemistry of Plant Terpenoids*. Clarendon Press, Oxford.

Jensen, S. R., B. J. Nielsen, and R. Dahlgren. 1975. Iridoid compounds, their occurrence and systematic importance in the angiosperms. *Botanisk Notiser* 128: 148–180.

Jermy, T. 1976. Insect-host-plant relationship—co-evolution or sequential evolution? *Symposia Biologica Hungarica* 16: 109–113.

———. 1984. Evolution of insect/host plant relationships. *The American Naturalist* 124: 609–630.

———. 1993. Evolution of insect-plant relationships—a devil's advocate approach. *Entomologia Experimentalis et Applicata* 66: 3–12.

Jiménez, J. A., K. A. Hughes, G. Alaks, L. Graham, and R. C. Lacy. 1994. En experimental study of inbreeding depression in a natural habitat. *Science* 266: 271–273.

Johannesen, J., M. Veith, and A. Seitz. 1996. Population genetic structure of the butterfly *Melitaea didyma* (Nymphalidae) along a northern distribution range border. *Molecular Ecology* 5: 259–267.

Johannesson, K. 2001. Parallel speciation: A key to sympatric divergence. *Trends in Ecology and Evolution* 16: 148–153.

Johnson, C. G. 1969. *Migration and Dispersal of Insects by Flight*. Methuen, London.

Jolly, G. M. 1965. Explicit estimates from capture-recapture data with both death and immigration—stochastic model. *Biometrika* 52: 225–247.

Jones, C. G., R. S. Ostfeld, M. P. Richard, E. M. Schauber, and J. O. Wolff. 1998. Chain reactions linking acorns to gypsy moth outbreaks and lyme disease risk. *Science* 179: 1023–1026.

Jones, K. N., F. J. Odendaal, and P. R. Ehrlich. 1986. Evidence against the spermatophore as paternal investment in checkerspot butterflies (Euphydryas: Nymphalidae). *The American Midland Naturalist* 116: 1–6.

Jones, R. E. 1977. Movement patterns and egg distribution in cabbage butterflies. *Journal of Animal Ecology* 46: 195–212.

Judd, W. S., C. S. Campbell, E. A. Kellogg, and P. F. Stevens. 1999. *Plant Systematics: A Phylogenetic Approach*. Sinauer Associates, Sunderland, MA.

Kadmon, R. 1993. Population dynamic consequences of habitat heterogeneity: An experimental study. *Ecology* 74: 816–825.

Kaitala, A., and C. Wiklund. 1995. Female mate choice and mating costs in the polyandrous butterfly *Pieris napi* (Lepidoptera: Pieridae). *Journal of Insect Behavior* 8: 355–363.

Kaitala, V., and W. M. Getz. 1992. Sex ratio genetics and the competitiveness of parasitic wasps. *Bulletin of Mathematical Biology* 54: 295–312.

Kamata, N. 2000. Population dynamics of the beech caterpillar, *Syntypistis punctatella* and biotic and abiotic factors. *Population Ecology* 42: 267–278.

Kaplan, N. L., R. R. Hudson, and C. H. Langley. 1989. The "hitchhiking effect" revisited. *Genetics* 123: 887–899.

Kareiva, P. 1983. Influence of vegetation texture on herbivore populations: resource concentration and herbivore movements. Pages 259–290 in R. F. Denno and M. S. McClure, eds. *Variable Plants and Herbivores in Natural and Managed Systems*. Academic Press, New York.

Karowe, D. N. 1990. Predicting host range evolution: colonization of *Coronilla varia* by *Colias philodice* (Lepidoptera: Pieridae). *Evolution* 44: 1637–1647.

Keller, L. F. 1998. Inbreeding and its fitness effects in an insular population of song sparrows (*Melospiza melodia*). *Evolution* 52: 240–250.

Keller, L. F., P. Arcese, J. N. M. Smith, W. M. Hochachka, and S. C. Stearns. 1994. Selection against inbred song sparrows during a natural population bottleneck. *Nature* 372: 356–357.

Keller, L. F., P. R. Grant, B. R. Grant, and K. Petren. 2002. Environmental conditions affect the magnitude of inbreeding depression in survival of Darwin's finches. *Evolution* 56: 1229–1239.

Keller, L. F., and D. M. Waller. 2002. Inbreeding effects in wild populations. *Trends in Ecology and Evolution* 17: 230–241.

Kendall, B. E., C. J. Briggs, W. W. Murdoch, P. Turchin, S. P. Ellner, E. McCauley, R. M. Nisbet, and S. N. Wood. 1999. Inferring the causes of population cycles: A synthesis of statistical and mechanistic modeling approaches. *Ecology* 80: 1789–1805.

Kennedy, J. S. 1961. A turning-point in the study of insect migration. *Nature* 189: 785–791.

Kennedy, J. S., and C. O. Booth. 1963. Free flight of aphids in the laboratory. *Journal of Experimental Biology* 40: 67–85.

Kester, K. M., and P. Barbosa. 1994. Behavioral responses to host food plants of two populations of the insect parasitoid *Cotesia congregata* (Say). *Oecologia* 99: 151–157.

Kettlewell, B. 1955. Selection experiments on industrial melanism in the Lepidoptera. *Heredity* 10: 287–301.

———. 1973. *The Evolution of Melanism: The Study of a Recurring Necessity*. Clarendon Press, Oxford.

Keyghobadi, N., J. Roland, S. Fownes, and C. Strobeck. 2003. Ink marks and molecular markers: Examining effects of landscape on dispersal using both mark-recapture and molecular methods. Pages 149–167 in C. L. Boggs, W. B. Watt, and P. R. Ehrlich, eds. *Butterflies: Ecology and Evolution Taking Flight*. University of Chicago Press, Chicago.

Kimura, M. 1968. Evolutionary rate at the molecular level. *Nature* 217: 624–662.

Kimura, M., and T. Maruyama. 1971. Pattern of neutral polymorphism in a geographically structured population. *Genetical Research* 18: 125–131.

Kimura, M., T. Maruyama, and J. F. Crow. 1963. The mutation load in small populations. *Genetics* 48: 1303–1312.

Kimura, M., and T. Ohta. 1971a. Protein polymorphism as a phase of molecular evolution. *Nature* 229: 467–469.

———. 1971b. *Theoretical Aspects of Population Genetics*. Princeton University Press, Princeton, NJ.

Kindvall, O. 1995. The impact of extreme weather on habitat preference and survival in a metapopulation of the bush cricket *Metrioptera bicolor* in Sweden. *Biological Conservation* 73: 51–58.

———. 1996. Habitat heterogeneity and survival in a bush cricket metapopulation. *Ecology* 77: 207–214.

King, C. E., and P. S. Dawson. 1972. Population biology and the *Tribolium* model. *Evolutionary Biology* 5: 133–227.

King, J. L., and T. H. Jukes. 1969. Non-Darwinian evolution. *Science* 164: 788–798.

King, R. B. 1993. Color pattern variation in Lake Erie water snakes: Prediction and measurement of selection. *Evolution* 47: 1819–1833.

King, R. B., and R. Lawson. 1995. Color pattern variation in Lake Erie water snakes: The role of gene flow. *Evolution* 49: 885–896.

Kingsolver, J. G. 1983. Ecological significance of flight activity in *Colias* butterflies: Implications for reproductive strategy and population structure. *Ecology* 64: 546–551.

———. 1985. Thermoregulatory significance of wing melanization in *Pieris* butterflies (Lepidoptera: Pieridae): Physics, posture, and pattern. *Oecologia* 66: 546–553.

Kingsolver, J. G., and W. B. Watt. 1983. Thermoregulatory strategies in *Colias* butterflies: Thermal stress and the limits to adaptation in temporally varying environments. *The American Naturalist* 121: 32–55.

Klemetti, T., and N. Wahlberg. 1997. The ecology and population structure of the marsh fritillary (*Euphydryas aurinia*) in Finland [in Finnish with English summary]. *Baptria* 22: 87–93.

Knapton, R. W. 1985. Lek structure and territoriality in the chryxus arctic butterfly, *Oeneis chryxus* (Satyridae). *Behavioral Ecology and Sociobiology* 17: 389–395.

Koenig, W. D., F. A. Pitelka, W. J. Carmen, R. L. Mumme, and M. T. Stanback. 1992. The evolution of delayed dispersal in cooperative breeders. *Quarterly Review of Biology* 67: 111–150.

Komonen, A. 1997. The parasitoid complexes attacking Finninsh populations of two threatened butterflies, *Euphydryas maturna* and *E. aurinia*. *Baptria* 22: 105–109.

———. 1998. Host species used by parasitoids of Melitaeini in southern France. *Baptria* 22: 87–93.

Konopka, R. J., and S. Benzer. 1971. Clock mutants of *Drosophila melanogaster*. *Proceedings of the National Academy of Sciences USA* 68: 2112–2116.

Kons Jr., H. L. 2000. Phylogenetic studies of the Melitaeini (Lepidoptera: Nymphalidae: Nymphalinae) and a revision of the genus *Chlosyne* Butler. PhD dissertation, Department of Entomology and Nematology, University of Florida, Gainesville.

Kooiman, P. 1972. The occurrence or iridoid glycosides in the Scrophulariaceae. *Acta Botanica Neerlandica* 19: 329–340.

Kornfield, I., and P. F. Smith. 2000. African cichlid fishes: Model systems for evolutionary biology. *Annual Review of Ecology and Systematics* 31: 163–196.

Korshunov, J. P., and P. J. Gorbunov. 1995. *Dnevnie Babochki Aziatskoi Chasti Rossii: Spravochnik*. Izdatelstvo Urap, Ekaterinburg.

Koskinen, M. T., T. O. Haugen, and C. R. Primmer. 2002. Contemporary Fisherian life-history evolution in small salmonid populations. *Nature* 419: 826–830.

Kouki, J., and I. Hanski. 1995. Population aggregation facilitates coexistence of many competing carrion fly species. *Oikos* 72: 223–227.

Krebs, C. J., S. Boutin, R. Oonstra, A. R. E. Sinclair, J. N. M. Smith, M. R. T. Dale, K. Martin, and R. Turkington. 1995. Impact of food and predation on the snowshoe hare cycle. *Science* 269: 1112–1115.

Krebs, C. J., and J. H. Myers. 1974. Population cycles in small mammals. *Advances in Ecological Research* 8: 268–400.

Kreitman, M., and H. Akashi. 1995. Molecular evidence for natural selection. *Annual Review of Ecology and Systematics* 26: 403–422.

Kremen, C. 1994. Biological inventory using target taxa: A case study of the butterflies of Madagascar. *Ecological Applications* 4: 407–422.

Krischik, V. A., and R. F. Denno. 1983. Individual, population, and geographic patterns in plant defense. Pages 463–512 in R. F. Denno and M. S. McClure, eds. *Variable Plants and Herbivores in Natural and Managed Systems*. Academic Press, New York.

Kristensen, N. P. 1976. Remarks on the family-level phylogeny of butterflies (Insecta, Lepidoptera, Rhopalocera). *Systematic Zoology* 14: 25–33.

Krogh, A. 1929. Progress in physiology. *American Journal of Physiology* 90: 243–251.

Kukkonen, I. 1992. Tähkätädyke ja idäntie. *Lutukka* 8: 103–111.

Kuussaari, M. 1998. Biology of the Glanville fritillary butterfly (*Melitaea cinxia*). PhD thesis, Department of Ecology and Systematics, University of Helsinki.

Kuussaari, M., J. Heliölä, J. Salminen, and I. Niininen. 2001. Maatalousympäristön päiväperhosseurannan vuoden 2000 tulokset [Results of the butterfly monitoring scheme in Finnish agricultural landscapes for the year 2000]. *Baptria* 26: 69–80.

Kuussaari, M., and M. Nieminen. 2004. Transfers in and between populations. In J. Settele et al., eds. *Ecology of Butterflies in Europe*. In press.

Kuussaari, M., M. Nieminen, and I. Hanski. 1996. An experimental study of migration in the Glanville fritillary butterfly *Melitaea cinxia*. *Journal of Animal Ecology* 65: 791–801.

Kuussaari, M., M. Nieminen, J. Pöyry, and I. Hanski. 1995. Täpläverkkoperhosen (*Melitaea cinxia*) elinkierto ja esiintyminen Suomessa [Life history and distribution of Glanville fritillary *Melitaea cinxia* (Nymphalidae) in Finland]. *Baptria* 20: 167–180.

Kuussaari, M., I. Saccheri, M. Camara, and I. Hanski. 1998. Allee effect and population dynamics in the Glanville fritillary butterfly. *Oikos* 82: 384–392.

Kuussaari, M., M. C. Singer, and I. Hanski. 2000. Local specialization and landscape-level influence on host use in an herbivorous insect. *Ecology* 81: 2177–2187.

Labine, P. A. 1964. Population biology of the butterfly, *Euphydryas editha*. I. Barriers to multiple inseminations. *Evolution* 18: 335–336.

———. 1966a. The population biology of the butterfly, *Euphydryas editha*. IV. Sperm precedence—a preliminary report. *Evolution* 20: 580–586.

———. 1966b. The reproductive biology of the checkerspot butterfly, *Euphydryas editha*. PhD dissertation, Department of Biological Sciences, Stanford University, Stanford, CA.

———. 1968. The population biology of the butterfly, *Euphydryas editha*. VIII. Oviposition and its relation to patterns of oviposition in other butterflies. *Evolution* 22: 799–805.

Lack, D. 1947. *Darwin's Finches*. Cambridge University Press, Cambridge, UK.

Lande, R. 1987. Extinction thresholds in demographic models of territorial populations. *The American Naturalist* 130: 624–635.

———. 1993. Risks of population extinction from demographic and environmental stochasticity and random catastrophies. *The American Naturalist* 142: 911–927.

———. 1995. Mutation and conservation. *Conservation Biology* 9: 782–791.

Landis, D. A., and F. D. Menalled. 1998. Ecological considerations in the conservation of effective parasitoid communities in agricultural systems. Pages 101–122 in P. Barbosa, ed. *Conservation Biological Control*. Academic Press, San Diego, CA.

Landweber, L. F., and A. P. Dobson, eds. 1999. *Genetics and the Extinction of Species*. Princeton University Press, Princeton, NJ.

Lappalainen, I. 1998. *Suomen Luonnon Monimuotoisuus*. Edita, Helsinki.

Larsen, T. B. 1974. *Butterflies of Lebanon*. National Council for Scientic Research, Beirut, Lebanon.

Launer, A. E., D. D. Murphy, C. L. Boggs, J. F. Baughman, S. B. Weiss, and P. R. Ehrlich. 1993 [1996]. Puddling behavior by Bay checkerspot butterflies (*Euphydryas editha bayensis*). *Journal of Research on Lepidoptera* 32: 45–52.

Lavery, T. A. 1993. A review of the distribution, ecology and status of the Marsh Fritillary *Euphydryas aurinia* Rottemburg, 1775 (Lepidoptera: Nymphalidae) in Ireland. *Irish Naturalist's Journal* 24: 192–199.

Lawes, M. J., P. E. Mealin, and S. E. Piper. 2000. Patch occupancy and potential metapopulation dynamics of three forest mammals in fragmented afromontane forest in South Africa. *Conservation Biology* 14: 1088–1098.

Lawrence, W. S. 1990. The effects of group size and host species on development and survivorship of a gregarious caterpillar *Halisidota caryae* (Lepidoptera: Arctiidae). *Ecological Entomology* 15: 53–62.

Lawton, J. H., and S. MacNeill. 1979. Between the devil and the deep blue sea: On the problem of being a herbivore. Pages 223–244 in R. M. Anderson, B. D. Turner, and L. R. Taylor, eds. *Population Dynamics*. Blackwell Scientific, Oxford.

Lawton, J. H., and R. M. May, eds. 1995. *Extinction Rates*. Oxford University Press, Oxford.

Lederhouse, R. C., M. P. Ayres, and J. M. Scriber. 1990. Adult nutrition affects male virility in *Papilio glaucus* L. *Functional Ecology* 4: 743–752.

Lei, G. C. 1997. Metapopulation dynamics of host-parasitoid interactions. PhD dissertation, Department of Ecology and Systematics, University of Helsinki.

Lei, G., and M. D. Camara. 1999. Behaviour of a specialist parasitoid, *Cotesia melitaearum*: From individual behaviour to metapopulation processes. *Ecological Entomology* 24: 59–72.

Lei, G. C., and I. Hanski. 1997. Metapopulation structure of *Cotesia melitaearum*, a specialist parasitoid of the butterfly *Melitaea cinxia*. *Oikos* 78: 91–100.

———. 1998. Spatial dynamics of two competing specialist parasitoids in a host metapopulation. *Journal of Animal Ecology* 67: 422–433.

Lei, G. C., V. Vikberg, M. Nieminen, and M. Kuussaari. 1997. The parasitoid complex attacking the Finnish populations of Glanville fritillary *Melitaea cinxia* (Lep: Nymphalidae), an endangered butterfly. *Journal of Natural History* 31: 635–648.

Leimar, O., and U. Norberg. 1997. Metapopulation extinction and genetic variation in dispersal-related traits. *Oikos* 80: 448–458.

Leirs, H., N. C. Stenseth, J. D. Nichols, J. E. Hines, R. Verhagen, and W. Verheyen. 1997. Stochastic seasonality and non-linear density-dependent factors regulate population size in an African rodent. *Nature* 389: 176–180.

L'Empereur, K. M., and F. R. Stermitz. 1990a. Iridoid glycoside content of *Euphydryas anicia* (Lepidoptera: Nymphalidae) and its major host plant, *Besseya plantaginea* (Scrophulariaceae), at a high plains Colorado site. *Journal of Chemical Ecology* 16: 1495–1506.

———. 1990b. Iridoid glycoside metabolism and sequestration by *Poladryas minuta* (Lepidoptera: Nymphalidae) feeding on *Penstemon virgatus* (Scrophulariaceae). *Journal of Chemical Ecology* 16: 1495–1506.

León-Cortés, J. L., M. J. R. Cowley, and C. D. Thomas. 1999. Detecting decline in a formerly widespread species: How common is the common blue butterfly *Polyommatus icarus*. *Ecography* 22: 643–650.

Letourneau, D. K. 1998. Conservation biology: Lessons for conserving natural enemies. Pages 9–38 in P. Barbosa, ed. *Conservation Biological Control*. Academic Press, San Diego, CA.

Levins, R. 1968. *Evolution in Changing Environments*. Princeton University Press, Princeton, NJ.

———. 1969. Some demographic and genetic consequences of environmental heterogeneity for biological control. *Bulletin of the Entomological Society of America* 15: 237–240.

———. 1970. Extinction. Pages 77–107 in M. Gerstenhaber, ed. *Some Mathematical Problems in Biology*. American Mathematical Society, Providence, RI.

Levins, R., and D. Culver. 1971. Regional

coexistence of species and competition between rare species. *Proceedings of the National Academy of Sciences USA* 68: 1246–1248.

Lewellen, R. H., and S. H. Vessey. 1998. The effect of density dependence and weather on population size of a polyvoltine species. *Ecological Monographs* 68: 571–594.

Lewis, A. C. 1982. Leaf wilting alters a plant species ranking by the grasshopper *Melanoplus differentialis*. *Ecological Entomology* 7: 391–395.

Lewis, O. T., and C. Hurford. 1997. Assessing the status of the marsh fritillary butterfly (*Eurodryas aurinia*): An example from Glamorgan, UK. *Journal of Insect Conservation* 1: 159–166.

Lewis, O. T., and C. D. Thomas. 2001. Adaptations to captivity in the butterfly *Pieris brassicae* (L.) And the implications for *ex situ* conservation. *Journal of Insect Conservation* 5: 55–63.

Lewis, O. T., C. D. Thomas, J. K. Hill, M. I. Brookes, T. P. R. Crane, Y. A. Graneau, et al. 1997. Three ways of assessing the metapopulation structure in the butterfly *Plebejus argus*. *Ecological Entomology* 22: 283–293.

Lewontin, R. C. 1974. *The Genetic Basis of Evolutionary Change*. Columbia University Press, New York.

Lewontin, R. C., and J. L. Hubby. 1966. A molecular approach to the study of genic heterozygosity in natural populations. II. Amount of variation and degree of heterozygosity in natural populations of *Drosophila pseudoobscura*. *Genetics* 54: 595–609.

Lewontin, R. C., and J. Krakauer. 1973. Distribution of gene frequency as a test of the theory of the selective neutrality of polymorphisms. *Genetics* 74: 175–195.

Liebhold, A., J. Elkinton, D. Williams, and R.-M. Muzika. 2000. What causes outbreaks of the gypsy moth in North America? *Population Ecology* 42: 257–266.

Lincoln, D. E. 1980. Leaf resin flavonoids of *Diplacus aurantiacus*. *Biochemical Systematics and Ecology* 8: 397–400.

———. 1985. Host-plant protein and phenolic resin effects on larval growth and survival of a butterfly. *Journal of Chemical Ecology* 11: 1459–1467.

Lincoln, D. E., T. S. Newton, P. R. Ehrlich, and K. S. Williams. 1982. Coevolution of the checkerspot butterfly *Euphydryas chalcedona* and its larval food plant *Diplacus aurantiacus*: Larval response to protein and leaf resin. *Oecologia* 52: 216–223.

Linnaeus, C. 1758. *Systema Naturae. Regnum Animale*, 10th ed. Engelmann, Leipzig.

Long, J. L. 1981. *Introduced Birds of the World*. Universe Books, New York.

Losos, J. B. 1998. Ecological and evolutionsary determinants of the species-area relationship in Caribbean anoline lizards. Pages 210–224 in P. R. Grant, ed. *Evolution on Islands*. Oxford University Press, Oxford.

Losos, J. B., T. R. Jackman, A. Larson, K. De Queiroz, and L. Rodriguez-Schettino. 1998. Contingency and determinism in replicated adaptive radiations of island lizards. *Science* 279: 2115–2118.

Lotka, A. J. 1925. *Elements of Physical Biology*. Williams & Wilkins, Baltimore, MD.

Louda, S. M., D. Kendall, J. Connor, and D. Simberloff. 1997. Ecological effects of an insect introduced for the biological control of weeds. *Science* 277: 1088–1090.

Lowther, P. E., and C. L. Cink. 1992. House sparrow. *The Birds of North America* 12: 1–19.

Lubchenco, J., and B. A. Menge. 1978. Community development and persistence in a low rocky intertidal zone. *Ecological Monographs* 48: 67–94.

Luck, G., G. C. Daily, and P. R. Ehrlich. 2003. Population diversity and ecosystem services. *Trends in Ecology and Evolution* 18: 331–336.

Luckens, C. J. 1985. *Hypodryas intermedia* Menetries in Europe: An account of the life history. *Entomologist's Record and Journal of Variation* 97: 37–45.

Ludwig, D. 1996. The distribution of population survival times. *The American Naturalist* 147: 506–526.

Ludwig, D., R. Hilborn, and C. Walters. 1993. Uncertainty, resource exploitation, and conservation: Lessons from history. *Science* 260: 17–36?

Lynch, M., J. Conery, and R. Buerger. 1995. Mutation accumulation and the extinction of small populations. *The American Naturalist* 146: 489–514.

MacArthur, R. H., and E. O. Wilson. 1963. An equilibrium theory of insular zoogeography. *Evolution* 17: 373–387.

———. 1967. *The Theory of Island Biogeography*. Princeton University Press, Princeton, NJ.

Mackay, D. A. 1985. Prealighting search behavior and host plant selection by ovipositing *Euphydryas editha* butterflies. *Ecology* 66: 142–151.

Mackay, D. A., and M. C. Singer. 1982. The basis of an apparent preference for isolated plants by ovipositing *Euptychia libye* butterflies. *Ecological Entomology* 7: 299–303.

MacLulich, D. A. 1937. Fluctuations in the numbers of the varying hare (*Lepus americanus*). PhD dissertation, University of Toronto, Toronto, Canada.

Mac Nally, R., and E. Fleishman. 2002. Using 'indicator' species to model species richness: Model development and predictions for Great Basin butterfly assemblages. *Ecological Applications* 12: 79–92.

Macy, R. W., and H. H. Shepard. 1941. *Butterflies: A Handbook of the Butterflies of the United States, Complete for the Region North of the Potomac and Ohio Rivers and East of the Dakotas*. University of Minnesota Press, Minneapolis.

Maes, D., and H. Van Dyck. 2001. Butterfly diversity loss in Flanders (north Belgium): Europe's worst case scenario? *Biological Conservation* 99: 263–276.

Majerus, M. E. N. 1998. *Melanism: Evolution in Action*. Oxford University Press, Oxford.

Mallet, J. 1995. A species definition for the modern synthesis. *Trends in Ecology and Evolution* 10: 294–299.

Mallet, J., and L. E. Gilbert. 1995. Why are there so many mimicry rings? Correlations between habitat, behaviour and mimicry in Heliconius butterflies. *Biological Journal of the Linnean Society* 55: 159–180.

Mangel, M. 1989. An evolutionary interpretation of the motivation to oviposit. *Journal of Evolutionary Biology* 2: 157–172.

Manly, B. F. J. 1977. The determination of key factors from life table data. *Oecologia* 31: 111–117.

Marak, H. B., A. Biere, and J. M. M. Van Damme. 2000. Direct and correlated responses to selection on iridoid glycosides in *Plantago lanceolata* L. *Journal of Evolutionary Biology* 13: 985–996.

Margules, C. R., I. D. Cresswell, and A. O. Nicholls. 1994. A scientific basis for establishing networks of protected areas. *Systematics and Conservation Evaluation* 50: 327–350.

Marino, P. C., and D. A. Landis. 1996. Effect of landscape structure on parasitoid diversity and parasitism in agroecosystems. *Ecological Applications* 6: 276–284.

Marquis, R. J. 1992. Selective impact of herbivores. Pages 301–325 in R. S. Fritz and E. L. Simms, eds. *Plant Resistance to Herbivores and Pathogens: Ecology, Evolution and Genetics*. University of Chicago Press, Chicago.

Marsh, P. 1979. Braconidae. Pages 144–295 in K. V. Krombein, P. D. Hurd, D. R. Smith, and B. D. Burk, eds. *Catalog of Hymenoptera in America North of Mexico, vol. 1, Symphyta and Apocrita (Parasitica)*. Smithsonian Institution, Washington, DC.

Martan, J., and B. A. Shepard. 1976. The role of the copulatory plug in reproduction of the guinea pig. *Journal of Experimental Zoology* 196: 79–84.

Martin, J. A., and D. P. Pashley. 1992. Molecular systematic analysis of butterfly family and subfamily relationships (Lepidoptera: Papilionoidea). *Annals of the Entomological Society of America* 85: 127–139.

Martin, J.-F., A. Gilles, and H. Descimon. 2000. Molecular phylogeny and evolutionary patterns of the European satyrids (Lepidoptera: Satyridae) as revealed by mitochondrial gene sequences. *Molecular Phylogenetics and Evolution* 15: 70–82.

Martin, T. E. 1998. Are microhabitat preferences of coexisting species under selection and adaptive? *Ecology* 79: 656–670.

———. 2001. Abiotic vs. biotic influences on habitat selection of coexisting species: Climate change impacts? *Ecology* 82: 175–188.

Marttila, O., T. Haahtela, H. Aarnio, and P. Ojalainen. 1990. *Suomen Päiväperhoset*. Kirjayhtymä, Helsinki.

Mason, L. G., P. R. Ehrlich, and T. C. Emmel. 1967. The population biology of the butterfly, *Euphydryas editha*. V. Character clusters and asymmetry. *Evolution* 21: 85–91.

Mathias, A., E. , E. Kisdi, and I. Olivieri. 2001. Divergent evolution of dispersal in a heterogeneous landscape. *Evolution* 55: 246–259.

Matson, P., P. Vitousek, J. Ewel, M. Mazzarino, and G. Robertson. 1987. Nitrogen transformations following tropical forest felling and burning on a volcanic soil. *Ecology* 68: 491–502.

Mattoni, R., G. F. Pratt, T. R. Longcore, J. F. Emmel, and J. N. George. 1997. The endangered quino checkerspot butterfly, *Euphydryas editha quino* (Lepidoptera: Nymphalidae). *Journal of Research on the Lepidoptera* 34: 99–118.

Mauricio, R., and M. D. Rausher. 1997. Experimental manipulation of putative selective agents provides evidence for the role of natural enemies in the evolution of plant defenses. *Evolution* 51: 1435–1444.

May, R. M. 1973. *Stability and Complexity in Model Ecosystems*. Princeton University Press, Princeton, NJ.

———. 1979. Simple models for single populations. *Fortschritte Zoologie* 25: 95–107.

———. 1988. How many species are there on Earth? *Science* 241: 1441–1149.

Maynard Smith, J. 1974. *Models in Ecology*. Cambridge University Press, Cambridge.

Mayr, E. 1942. *Systematics and the Origin of Species*. Columbia University Press, New York.

———. 1982. Adaptation and selection. *Biologisches Zentralblatt* 101: 161–174.

Mayr, E., E. G. Linsley, and R. L. Usinger. 1953. *Methods and Principles of Systematic Zoology*. McGraw-Hill, New York.

Mazel, R. 1986. Structure et evolution du peuplement d'Euphydras aurinia (Lepidoptera) dans le sud-ouest europeen. *Vie Milieu* 36: 205–225.

McCarty, J. P. 2001. Ecological consequences of recent climatic change. *Conservation Biology* 15: 320–331.

McCauley, D. E. 1993. Evolution in metapopulations with frequent local extinction and recolonization. *Oxford Surveys of Evolutionary Biology* 10: 109–134.

McCauley, D. E., J. Raveill, and J. Antonovics. 1995. Local founding events as determinants of genetic structure in a plant metapopulation. *Heredity* 75: 630–636.

McEvoy, P. B. 1996. Host specificity and biological pest control. *BioScience* 46: 401–405.

McFadyen, R. E. C. 1998. Biological control of weeds. *Annual Review of Entomology* 43: 369–393.

McGarrahan, E. 1997. Much-studied butterfly winks out on Stanford preserve. *Science* 275: 479–480.

McKechnie, S. W., P. R. Ehrlich, and R. R. White. 1975. Population genetics of *Euphydryas* butterflies. I. Genetic variation and the neutrality hypothesis. *Genetics* 81: 571–594.

McLaughlin, J. F., J. J. Hellmann, C. L. Boggs, and P. R. Ehrlich. 2002a. Climate change hastens population extinctions. *Proceedings of the National Academy of Sciences USA* 99: 6070–6074.

———. 2002b. The route to extinction: Population dynamics of a threatened butterfly. *Oecologia* 132: 538–548.

McNeely, C., and M. C. Singer. 2001. Contrasting the roles of learning in butterflies foraging for nectar and oviposition sites. *Animal Behaviour* 61: 847–852.

Mead, E. W., T. A. Foderaro, D. R. Gardner, and F. R. Stermitz. 1993. Iridoid glycoside sequestration by *Thessalia leanira* (Lepidoptera: Nymphalidae) feeding on *Castilleja integra* (Scrophulariaceae). *Journal of Chemical Ecology* 19: 1155–1166.

Meagher, S., D. J. Penn, and W. K. Potts. 2000. Male-male competition magnifies inbreeding depression in wild house mice. *Proceedings of the National Academy of Sciences USA* 97: 3324–3329.

Meffe, G. K., C. R. Carroll, and contributors. 1997. *Principles of Conservation Biology*, 2nd ed. Sinauer Associates, Sunderland, MA.

Meier, N. F. 1968. *Keys to the parasitic hymenoptera (Family Ichneumonidae) of the USSR and adjacent countries*. Israel Program for Scientific Translations, Jerusalem.

Menge, B. A. 2000. Top-down and bottom-up community regulation in marine rocky intertidal habitats. *Journal of Experimental Marine Biology and Ecology* 250: 257–289.

Menges, E. S. 1988. Conservation biology of Furbish's lousewort. Final report to region 5. U.S. Fish and Wildlife Service, Holcomb Research Institute report no. 126. Butler University, Indianapolis, Indiana.

———. 1990. Population viability analysis for an endangered plant. *Conservation Biology* 4: 52–62.

Merriam, G. 1991. Corridors and connectivity: Animal populations in heterogeneous environments. Pages 133–142 in D. A. Saunders and R. J. Hobbs, eds. *Nature Conservation 2: The Role of Corridors*. Surrey Beatty & Sons, Chipping Norton, NSW.

Michener, C. D. 1974. *The Social Behavior of the Bees: A Comparative Study*. Harvard University Press, Cambridge, MA.

———. 2000. *The Bees of the World*. Johns Hopkins University Press, Baltimore, MD.

Middleton, D. A. J., A. R. Veitch, and R. M. Nesbit. 1995. The effect of an upper limit to population size on persistence time. *Theoretical Population Biology* 48: 277–305.

Mikio, Y., T. Zhijian, and H. Hirose. 2001. Possible horizontal transfer of a transposable element from host to parasitoid. *Molecular Biology and Evolution* 18: 1952–1958.

Mischler, B. D., and M. J. Donoghue. 1982. Species concepts: A case for pluralism. *Systematic Zoology* 31: 491–503.

Mitchell-Olds, T. 2001. *Arabidopsis thaliana* and its wild relatives: A model system for ecology and evolution. *Trends in Ecology and Evolution* 16: 693–700.

Mitter, C., and B. Farrell. 1991. Macroevolutionary aspects of insect-plant relationships. Pages 35–78 in E. Bernays, ed. *Insect-Plant Interactions*. CRC Press, Boca Raton, FL.

Mitton, J. B. 1997. *Selection in Natural Populations*. Oxford University Press, Oxford.

Moilanen, A. 1999. Patch occupancy models of metapopulation dynamics: Efficient parameter estimation using implicit statistical inference. *Ecology* 80: 1031–1043.

———. 2000. The equilibrium assumption in estimating the parameters of metapopulation models. *Journal of Animal Ecology* 69: 143–153.

———. 2002. Implications of empirical data quality to metapopulation model parameter estimation and application. *Oikos* 96: 516–530.

Moilanen, A., and M. Cabeza. 2002. Single-species dynamic site selection. *Ecological Applications* 12: 913–926.

Moilanen, A., and I. Hanski. 1998. Metapopulation dynamics: Effects of habitat quality and landscape structure. *Ecology* 79: 2503–2515.

———. 2001. On the use of connectivity measures in spatial ecology. *Oikos* 95: 147–151.

Moilanen, A., and M. Nieminen. 2002. Simple connectivity measures for spatial ecology. *Ecology* 84: 1131–1145.

Moilanen, A., A. T. Smith, and I. Hanski. 1998. Long-term dynamics in a metapopulation of the American pika. *The American Naturalist* 152: 530–542.

Mönkkönen, M., R. Hardling, J. T. Forsman, and J. Tuomi. 1999. Evolution of heterospecific attraction: Using other species as cues in habitat selection. *Evolutionary Ecology* 13: 91–104.

Monteiro, A., and N. E. Pierce. 2001. Phylogeny of *Bicyclus* (Lepidoptera: Nymphalidae) inferred from COI, COII, and EF-1alpha gene sequences. *Molecular Phylogenetics and Evolution* 18: 264–281.

Montllor, C. B., and E. A. Bernays. 1993. Invertebrate predators and caterpillar foraging. Pages 179–202 in N. E. Stamp and T. M. Casey, eds. *Caterpillars: Ecological and Evolutionary Constraints on Foraging*. Chapman & Hall, London.

Mooney, H. A., P. R. Ehrlich, D. E. Lincoln, and K. S. Williams. 1980. Environmental controls on the seasonality of a drought deciduous shrub, *Diplacus aurantiacus* and its predator, the checkerspot butterfly, *Euphydryas chalcedona*. *Oecologia* 45: 143–146.

Mooney, H. A., K. S. Williams, D. E. Lincoln, and P. R. Ehrlich. 1981. Temporal and spatial variability in the interaction between the checkerspot butterfly, *Euphydryas chalcedona* and its principal food source, the Californian Shrub, *Diplacus aurantiacus*. *Oecologia* 50: 195–198.

Moore, R. A., and M. C. Singer. 1987. Effects of maternal age and adult diet on egg weight in the butterfly *Euphydryas editha*. *Ecological Entomology* 12: 401–408.

———. ND. Age-specific components of reproductive strategies. Manuscript.

Moore, S. D. 1987. Male-biased mortality in the butterfly *Euphydryas editha*: A novel cost of mate acquisition. *The American Naturalist* 130: 306–309.

———. 1989a. Patterns of juvenile mortality within an oligophagous insect population. *Ecology* 70: 1726–1737.

———. 1989b. Regulation of host diapause by an insect parasitoid. *Ecological Entomology* 14: 93–98.

Mopper, S. 1998. Local adaptation and stochastic events in an oak leafminer population. Pages 139–155 in S. Mopper and S. Y. Strauss, eds. *Genetic Structure and Local Adaptation in Natural Insect Populations*. Chapman & Hall, New York.

Morbey, Y. E., and R. C. Ydenberg. 2001. Protandrous arrival timing to breeding areas: A review. *Ecology Letters* 4: 663–673.

Morgan, T. H. 1911. The origins of five mutations in eye color in *Drosophila* and their modes of inheritance. *Science* 33: 534–537.

Morinaka, S., N. Minaka, M. Sekiguchi, Erniwati, S. N. Prijono, I. K. Ginarsa, T. Miyata, and T. Hidaka. 2000. Molecular phylogeny of birdwing butterflies of the tribe Troidini (Lepidoptera: Papilionidae): Using all species of the genus *Ornithoptera*. *Biogeography* 2: 103–111.

Moritz, C. 1994. Defining 'evolutionarily signficant units' for conservation. *Trends in Ecology and Evolution* 9: 373–375.

Morris, R. F. 1959. Single-factor analysis in population dynamics. *Ecology* 40: 580–588.

———. 1972a. Predation by wasps, birds, and mammals on *Hyphantria cunea*. *The Canadian Entomologist* 104: 1581–1591.

———. 1972b. Predation by insects and spiders inhabiting colonial webs of *Hyphantria cunea*. *The Canadian Entomologist* 104: 1197–1207.

———. 1976. Relation of parasitic attack to the colonial habit of *Hyphantria cunea*. *The Canadian Entomologist* 108: 833–836.

Morrison, G., and D. R. Strong. 1980. Spatial variation in host density and the intensity of parasitism: Some empirical examples. *Environmental Entomology* 9: 149–152.

Morton, A. C. 1984. The effects of marking and capture on recapture frequencies of butterflies. Thesis. University of Southampton, Southampton, UK.

Morton, N. E., J. F. Crow, and J. J. Muller. 1956. An estimate of the mutational damage in man from data on consanguineous marriages. *Proceedings of the National Academy of Sciences USA* 42: 855–863.

Mueller, L. D. 1985. The evolutionary ecology of *Drosophila*. *Evolutionary Biology* 19: 37–98.

———. 1997. Theoretical and empirical examination of density-dependent selection. *Annual Review of Ecology and Systematics* 28: 269–288.

Mueller, L. D., and A. Joshi. 2000. *Stability in Model Populations*. Princeton University Press, Princeton, NJ.

Mueller, L. D., B. A. Wilcox, P. R. Ehrlich, D. G. Heckel, and D. D. Murphy. 1985. A direct assessment of the role of genetic drift in determining allele frequency variation in populations of *Euphydryas editha*. *Genetics* 110: 495–511.

Müller, C. B., and H. C. J. Godfray. 1998. The response of aphid secondary parasitoids to different patch densities of their host. *Biocontrol* 43: 129–139.

Muller, C. H. 1969. The "co" in coevolution. *Science* 164: 197–198.

Muller, H. J. 1950. Our load of mutations. *American Journal of Human Genetics* 2: 111–176.

Munguira, M. L., J. Martin, E. G. Barros, and L. V. Jose. 1997. Use of space and resources in a Mediterranean population of the butterfly *Euphydryas aurinia*. *Acta Oecologica*: 597–612.

Murdock, W. W., R. F. Luck, S. J. Walde, J. D. Reeve, and D. S. Yu. 1989. A refuge for red scale under control by *Aphytis*: Structural aspects. *Ecology* 70: 1707–1714.

Murdock, W. W., J. D. Reeve, C. B. Huffaker, and C. E. Kennett. 1984. Biological control of olive scale and its relevance to ecological theory. *The American Naturalist* 123: 371–392.

Murphy, D. D. 1983. Nectar sources as constraints on the distribution of egg masses by the checkerspot butterfly, *Euphydryas chalcedona* (Lepidoptera: Nymphalidae). *Environmental Entomology* 12: 463–466.

———. 1984. Butterflies and their nectar plants: The role of the checkerspot butterfly *Euphydryas editha* as a pollen vector. *Oikos* 43: 113–117.

Murphy, D. D., and P. R. Ehrlich. 1983. Crows, bobs, tits, elfs and pixies: The phoney "common name" phenomenon. *The Journal of Research on the Lepidoptera* 22: 154–158.

———. 1989. Conservation biology of California's remnant native grasslands. Pages 201–211 in L. F. Huenneke and H. Mooney, eds. *Grassland Structure and Function: California Annual Grassland*. Kluwer Academic, Dordrecht.

Murphy, D. D., K. E. Freas, and S. B. Weiss. 1990. An environment-metapopulation approach to population viability analysis for a threatened invertebrate. *Conservation Biology* 4: 41–51.

Murphy, D. D., A. E. Launer, and P. R. Ehrlich. 1983. The role of nectar feeding in egg production and population dynamics of the checkerspot butterfly *Euphydryas editha*. *Oecologia* 56: 257–263.

Murphy, D. D., M. S. Menninger, and P. R. Ehrlich. 1984. Nectar source distribution as a determinant of oviposition host species in *Euphydryas chalcedona*. *Oecologia* 62: 269–271.

Murphy, D. D., M. S. Menninger, P. R. Ehrlich, and B. A. Wilcox. 1986. Local population dynamics of adult butterflies and the conservation status of two closely related species. *Biological Conservation* 37: 201–223.

Murphy, D. D., and S. B. Weiss. 1988a. Ecological studies and the conservation of the bay checkerspot butterfly, *Euphydryas editha bayensis*. *Biological Conservation* 46: 183–200.

———. 1988b. A long-term monitoring plan for a threatened butterfly. *Conservation Biology* 2: 367–374.

———. 1992. Effects of climate change on biological diversity in western North America: Species losses and mechanisms. Pages 355–368 in R. L. Peters and T. E. Lovejoy, eds. *Global Warming and Biological Diversity*. Yale University Press, New Haven, CT.

Murphy, D. D., and R. R. White. 1984. Rainfall, resources, and dispersal in southern populations of *Euphydryas editha* (Lepidoptera: Nymphalidae). *Pan-Pacific Entomologist* 60: 350–354.

Myers, J. 1969. Distribution of foodplant chemoreceptors on the female Florida queen butterfly, *Danaus gilippus berenice* (Nymphalidae). *Journal of the Lepidopterists' Society* 23: 196–198.

———. 1993. Population outbreaks in forest Lepidoptera. *American Scientist* 81: 240–251.

———. 2000. Population fluctuations of the western tent caterpillar in southwestern British Columbia. *Population Ecology* 42: 231–241.

Myers, J. H., and L. D. Rothman. 1995. Field experiments to study regulation of fluctuating populations. Pages 229–250 in N. Cappuccino and P. W. Price, eds. *Population Dynamics: New Approaches and Synthesis*. Academic Press, San Diego, CA.

Myers, N. 1979. *The Sinking Ark*. Pergamon Press, New York.

Nachman, G. 1987. Systems analysis of acarine predator-prey interactions. I. A stochastic simulation model of spatial processes. *Journal of Animal Ecology* 56: 247–265.

———. 1988. Regional persistence of locally unstable predator-prey interactions. *Experimental and Applied Acarology* 5: 293–318.

Naef-Daenzer, B. 2000. Patch time allocation and patch sampling by foraging great and blue tits. *Animal Behaviour* 59: 989–999.

National Climatic Data Center (NCDC). 2001. Digital weather data.: Available at http://ftp.ncdc.noaa.gov/.

Neck, R. W. 1973. Foodplant ecology of the butterfly *Chlosyne lacinia* (Geyer) (Nymphalidae). I. Larval foodplants. *Journal of the Lepidopterists' Society* 27: 22–33.

Nee, S., and R. M. May. 1992. Dynamics of metapopulations: Habitat destruction and competitive coexistence. *Journal of Animal Ecology* 61: 37–40.

Nee, S., R. M. May, and M. P. Hassell. 1997. Two-species metapopulation models. Pages 123–148 in I. Hanski and M. E. Gilpin, eds. *Metapopulation Biology: Ecology, Genetics, and Evolution*. Academic Press, San Diego.

NERC Centre for Population Biology, I. C. 1999. The global population dynamics database. Natural Environment Research Council, Silwood Park, Imperia College, London.

Nève, G., B. Barascud, H. Descimon, and M. Baguette. 2000. Genetic structure of *Proclossiana eunomia* populations at the regional scale (Lepidoptera, Nymphalidae). *Heredity* 84: 657–666.

Nève, G., B. Barascud, R. Hughes, J. Aubert, H. Descimon, P. Lebrun, and M. Baguette. 1996. Dispersal, colonisation power and metapopulation structure in the vulnerable butterfly *Proclossiana eunomia* (Lepidoptera: Nymphalidae). *Journal of Applied Ecology* 33: 14–22.

Nève, G., and E. Meglécz. 2000. Microsatellite frequencies in different taxa. *Trends in Ecology and Evolution* 15: 376–377.

New, T. R., R. M. Pyle, J. A. Thomas, C. D. Thomas, and P. C. Hammond. 1995. Butterfly conservation management. *Annual Review of Entomology* 40: 57–83.

Newman, D., and D. Pilson. 1997. Increased probability of extinction due to decreased effective population size: Experimental populations of *Clarkia pulchella*. *Evolution* 51: 354–362.

Ng, D. 1988. A novel level of interactions in plant-insect systems. *Nature* 334: 611–612.

Nice, M. M. 1937. Studies in the life history of the song sparrow. I. *Transactions of the Linnean Society of New York* 4: 1–246.

———. 1943. Studies in the life history of the song sparrow. II. *Transactions of the Linnean Society of New York* 6: 1–328.

Nichols, R. A., M. W. Bruford, and J. J. Groombridge. 2001. Sustaining genetic variation in a small population: Evidence from the Mauritius kestrel. *Molecular Ecology* 10: 593–602.

Nicholson, A. J. 1933. The balance of animal populations. *Journal of Animal Ecology* 2: 132–178.

———. 1954. An outline of the dynamics of animal populations. *Australian Journal of Zoology* 2: 9–65.

———. 1957. The self-adjustment of populations to change. *Cold Springer Harbor Symposium of Quantitative Biology* 22: 153–173.

Nicholson, A. J., and V. A. Bailey. 1935. The balance of animal populations. Part I. *Proceedings of the Zoological Society of London* 1935: 551–598.

Niemelä, J., D. Langor, and J. R. Spence. 1993. Effects of clear-cut harvesting on boreal ground-beetle assemblages (Coleoptera: Carabidae) in western Canada. *Conservation Biology* 7: 551–561.

Nieminen, M., M. C. Singer, W. Fortelius, K. Schöps, and I. Hanski. 2001. Experimental confirmation that inbreeding depression increases extinction risk in butterfly populations. *The American Naturalist* 157: 237–244.

Nieminen, M., J. Suomi, S. van Nouhuys, P. Sauri, and M.-L. Riekkola. 2003. Effect of iridoid glycoside content on oviposition host plant choice and parasitism in a specialist herbivore. *Journal of Chemical Ecology* 29: 823–844.

Nijhout, H. F. 1991. *The Development and Evolution of Butterfly Wing Patterns*. Smithsonian Institution Press, Washington, DC.

———. 1994. Genes on the wing. *Science* 265: 44–45.

Nishida, R. 2002. Sequestration of defensive substances from plants by Lepidotera. *Annual Review of Entomology* 47: 57–92.

Nixon, K. C., and Q. D. Wheeler. 1990. An amplification of the phylogenetic species concept. *Cladistics* 6: 211–223.

Norberg, U., K. Enfjäll, and O. Leimar. 2002. Habitat exploration in butterflies—an outdoor cage experiment. *Evolutionary Ecology* 16: 1–14.

Nothnagle, P. J., and J. C. Schultz. 1987. What is a forest pest? Pages 59–80 in J. C. Schultz and P. Barbosa, eds. *Insect Outbreaks*. Academic Press, New York.

Nowak, R. M. 1991. *Walker's Mammals of the World*, vol. 2 Johns Hopkins University Press, Baltimore, MD.

Nylin, S., K. Nyblom, F. Ronquist, N. Janz, J. Belicek, and M. Källersjö. 2001. Phylogeny of *Polygonia, Nymphalis* and related butterflies (Lepidoptera: Nymphalidae): A total-evidence analysis. *Zoological Journal of the Linnean Society* 132: 441–468.

Oaks, S. C., Jr., V. S. Mitchell, G. W. Pearson, and C. C. J. Carpenter, eds. 1991. *Malaria: Obstacles and Opportunities*. National Academy Press, Washington, DC.

Oates, M. R., and M. S. Warren. 1990. *A Review of Butterfly Introductions in Britain and Ireland*. JCCBI/WWF, Godalming.

Obara, Y. 1982. Mate refusal hormone in the cabbage white butterfly? *Naturwissenschaften* 69: 551.

O'Brien, D. M., C. L. Boggs, and M. Fogel. 2004. Making eggs from nectar: Connections between butterfly life history and the importance of nectar carbon in reproduction. *Oikos*, in press.

O'Brien, D. M., M. L. Fogel, and C. L. Boggs. 2002. Renewable and non-renewable resources: Amino acid turnover and allocation to reproduction in Lepidoptera. *Proceedings of the National Academy of Sciences USA* 99: 4413–4418.

O'Brien, D. M., D. P. Schrag, and C. M. del Rio. 2000. Allocation to reproduction in a hawkmoth: A quantitative analysis using stable carbon isotopes. *Ecology* 81: 2822–2832.

Odendaal, F. J., Y. Iwasa, and P. R. Ehrlich. 1985. Duration of female availability and its effect on butterfly mating systems. *The American Naturalist* 125: 673–678.

Odendaal, F. J., P. Turchin, and F. R. Stermitz. 1988. An incidental-effect hypothesis explaining aggregation of males in a population of *Euphydryas anicia. The American Naturalist* 132: 735–749.

———. 1989. Influence of host-plant density and male harassment on the distribution of female *Euphydryas anicia* (Nymphalidae). *Oecologia* 78: 283–288.

O'Hara, R. B., E. Arjas, H. Toivonen, and I. Hanski. 2002. Bayesian analysis of metapopulation data. *Ecology* 83: 2408–2415.

Ohasaki, N., and Y. Sato. 1999. The role of parasitoids in evolution of habitat and larval food plant preference by three Pieris butterflies. *Research in Population Ecology* 41: 107–119.

Ohta, T. 1992. The nearly neutral theory of molecular evolution. *Annual Review of Ecology and Systematics* 23: 263–286.

Okubo, A. 1980. *Diffusion and Ecological Problems: Mathematical Models*. Springler-Verlag, Berlin.

Oliviera, E., R. B. Srygley, and R. Dudley. 1998. Do neotropical migrant butterflies navigate using a solar compass? *Journal of Experimental Biology* 201: 3317–3331.

Olivieri, I., and P. H. Gouyon. 1997. Evolution of migration rate and other traits: The metapopulation effect. Pages 293–324 in I. A. Hanski and M. E. Gilpin, eds. *Metapopulation Biology*. Academic Press, San Diego, CA.

Olivieri, I., M. Swan, and P.-H. Gouyon. 1983. Reproductive system and colonizing strategy of two species of *Carduus* (Compositae). *Oecologia* 60: 114–117.

Olmstead, R. G., B. Bremer, K. M. Scott, and J. D. Palmer. 1993. A parsimony analysis of the Asteridae sensu lato based on *rbc*L sequences. *Annals of the Missouri Botanical Gardens* 80: 700–722.

Olmstead, R. G., C. W. dePamphilis, A. D. Wolfe, N. D. Young, W. J. Elisons, and P. A. Reeves. 2001. Disintegration of the Scrophulariaceae. *American Journal of Botany* 88: 348–361.

Olmstead, R. G., K. J. Kim, R. K. Jansen, and S. J. Wagstaff. 2000. The phylogeny of the Asteridae sensu latu based on chloroplast ndhF gene sequences. *Molecular Phylogenetics and Evolution* 16: 96–112.

Oosting, H. J. 1956. *The Study of Plant Communities: An Introduction to Plant Ecology*. 2nd ed. W.H. Freeman, San Francisco, CA.

Opler, P., and G. O. Krizek. 1984. *Butterflies: East of the Great Plains*. Johns Hopkins University Press, Baltimore, MD.

Opler, P., H. Pavulaan, and R. E. Stanford, eds. 1995. *Butterflies of North America*. Northern Prairie Wildlife Research Center Home Page, http://www.npwrc.usgs.gov/resource/distr/lepid/bflyusa.htm.

Orive, M. E., and J. F. Baughman. 1989. Effects of handling on *Euphydryas editha* (Nymphalidae). *Journal of the Lepidopterists' Society* 43: 244–247.

Orr, A. G. 1988. Mate conflict and the evolution of the sphragis in butterflies. PhD dissertation, Griffith University, Australia.

———. 2002. The sphragis of *Heteronympha penelope* Waterhouse (Lepidoptera: Satyridae): Its structure, formation and role in sperm guarding. *Journal of Natural History* 36: 185–196.

Orr, A. G., and R. L. Rutowski. 1991. The function of the sphragis in *Cressida cressida* (Fab) (Lepidoptera, Papilionidae): A visual deterrent to copulation attempts. *Journal of Natural History* 25: 703–710.

Ouborg, N. J., and R. Vantreuren. 1994. The significance of genetic erosion in the process of extinction. 4. Inbreeding load and heterosis in relation to population size in the mint *Salvia pratensis*. *Evolution* 48: 996–1008.

Ovaskainen, O. 2002a. The effective size of a metapopulation living in a heterogeneous patch network. *The American Naturalist* 160: 612–628.

———. 2002b. Long-term persistence of species and the SLOSS problem. *Journal of Theoretical Biology* 218: 419–433.

———. 2003. Habitat destruction, habitat restoration and eigenvector-eigenvalue relations. *Mathematical Biosciences* 181: 165–176.

Ovaskainen, O., and B. Grenfell. 2003. Mathematical tools for planning effective intervention scenarios for sexually transmitted diseases. *Sexually Transmitted Diseases* 30: 388–394.

Ovaskainen, O., and I. Hanski. 2001. Spatially structured metapopulation models: Global and local assessment of metapopulation capacity. *Journal of Theoretical Biology* 60: 281–304.

———. 2002. Transient dynamics in metapopulation response to perturbation. *Journal of Theoretical Biology* 61: 285–295.

———. 2003. How much does an individual habitat fragment contribute to metapopulation dynamics and persistence? *Theoretical Population Biology*, in press.

Owen, D. F. 1997. Natural selection and evolution in moths: Homage to J. W. Tutt. *Oikos* 78: 177–181.

Pacala, S. W., and J. A. Silander Jr. 1990. Field tests of neighborhood population dynamic models of two annual weed species. *Ecological Monographs* 60: 113–134.

Paine, R. T. 1974. Intertidal community structure: Experimental studies on the relationship between a dominant competitor and its principal predator. *Oecologia* 15: 93–120.

Parker, G. A. 1970. Sperm competition and its evolutionary consequences in the insects. *Biological Reviews of the Cambridge Philosophical Society* 45: 525–567.

Parker, G. A., and S. P. Courtney. 1984. Models of clutch size in insect oviposition. *Theoretical Population Biology* 26: 27–48.

Parmesan, C. 1991. Evidence against plant "apparency" as a constraint on evolution of insect search efficiency (Lepidoptera: Nymphalidae). *Journal of Insect Behavior* 4: 417–430.

———. 1995. Ecological and behavioral aspects of the interaction between host-plants and checkerspot butterfly. PhD dissertation, University of Texas, Austin.

———. 1996. Climate and species' range. *Nature* 382: 765–766.

———. 2000. Unexpected density-dependent effects of herbivory in a wild populaiton of the annual *Collinsia torreyi*. *Journal of Ecology* 88: 392–400.

Parmesan, C., T. L. Root, and M. R. Willig. 2000. Impacts of extreme weather and climate on terrestrial biota. *Bulletin of the American Meterological Society* 81: 443–450.

Parmesan, C., N. Ryrholm, C. Stefanescu, J. K. Hill, C. D. Thomas, H. Descimon, B. Huntley, L. Kaila, J. Kullberg, T. Tammaru, W. J. Tennent, J. A. Thomas, and M. Warren. 1999. Poleward shifts in geographical ranges of butterfly species associated with regional warming. *Nature* 399: 579–583.

Parmesan, C., M. Singer, and I. Harris. 1995. Absence of adaptive learning from the oviposition foraging behaviour of a checkerspot butterfly. *Animal Behaviour* 50: 161–175.

Parmesan, C., and G. Yohe. 2003. A globally coherent fingerprint of climate change impacts across natural systems. *Nature* 421: 37–42.

Parsons, M. 1999. *The Butterflies of Papua New Guinea: Their Systematics and Biology*. Academic Press, London.

Parvinen, K., U. Dieckmann, M. Gyllenberg, and J. A. J. Metz. 2000. Evolution of dispersal in metapopulations with local density dependence and demographic stochasticity. *Interim Report*. IIASA, Laxenburg, Austria.

Patterson, H. E. H. 1985. The recognition concept of species. Pages 21–29 in E. S. Vrba, ed. *Species and Speciation*. Transvaal Museum, Pretoria, South Africa.

Pearson, D. L., and F. Cassola. 1992. World-wide species richness patterns of tiger beetles (Coleoptera: Cicindelidae): Indicator taxon for biodiversity and conservation studies. *Conservation Biology* 6: 376–391.

Pellmyr, O., and C. J. Huth. 1994. Evolutionary stability of mutualism between yuccas and yucca moths. *Nature* 372: 257–260.

Pennisi, E. 2001. Linnaeus's last stand? *Science* 291: 2304–2307.

Penz, C. M. 1999. Higher level phylogeny for the passion-vine butterflies (Nymphalidae, Heliconiinae) based on early stage and adult morphology. *Zoological Journal of the Linnean Society* 127: 277–344.

Pereyra, P. C., and M. D. Bowers. 1988. Iridoid

glycosides as oviposition stimulants for the buckeye butterfly, *Junonia coenia* (Nymphalidae). *Journal of Chemical Ecology* 14: 917–928.

Perez, S. M., O. R. Taylor, and R. Jander. 1997. A sun compass in monarch butterflies. *Nature* 387: 29.

Perry, D. A. 1995. Status of forest habitat of the marbled murrelet. Pages 381–383 in C. J. Ralph, J. G. L. Hunt, M. G. Raphael, and J. F. Piatt, eds. *Ecology and Conservation of the Marbled Murrelet*. U.S. Forest Service, Pacific Southwest Research Station, Albany, CA.

Peterson, S. C., N. D. Johnson, and J. L. LeGuyader. 1987. Defensive regurgitation of allelochemicals derived from host cyanogenesis by eastern tent caterpillars. *Ecology* 68: 1268–1272.

Petit, S., A. Moilanen, I. Hanski, and M. Baguette. 2001. Metapopulation dynamics of the bog fritillary butterfly: Movements between habitat patches. *Oikos* 92: 491–500.

Petren, K., B. R. Grant, and P. R. Grant. 1999. A phylogeny of Darwin's finches based on microsatellite DNA length variation. *Proceedings of the Royal Society of London B, Biological Sciences* 266: 321–330.

Pfadt, R. E. 1985. *Fundamentals of Applied Entomology*, 4th ed. Macmillan, New York.

Pfisterer, A. B., and B. Schmid. 2002. Diversity-dependent productivity can decrease the stability of ecosystem functioning. *Nature* 416: 84–86.

Pilson, D., and M. D. Rausher. 1988. Clucth size adjustment by a swallowtail butterfly. *Nature* 333: 361–363.

Pimm, S. L. 1991. *The Balance of Nature: Ecological Issues in the Conservation of Species and Communities*. University of Chicago Press, Chicago.

Pimm, S. L., and A. Redfearn. 1988. The variation of population densities. *Nature* 334: 613–614.

Pivnick, K. A., and J. N. McNeil. 1987. Puddling in butterflies: Sodium affects reproductive success in *Thymelicus lineola*. *Physiological Entomology* 12: 461–472.

Pollard, E., and T. J. Yates. 1992. The extinction and foundation of local butterfly populations in relation to populaiton variability and other factors. *Ecological Entomology* 17: 249–254.

———. 1993. *Monitoring Butterflies for Ecology and Conservation*. Chapman & Hall, London.

Pollock, D. D., W. B. Watt, V. K. Rashbrook, and E. V. Iyengar. 1998. Molecular phylogeny for *Colias* butterflies and their relatives (Lepidoptera: Pieridae). *Annals of the Entomological Society of America* 91: 524–531.

Porter, K. 1981. The population dynamics of small colonies of the butterfly *Euphydryas aurinia*. PhD disseration, University of Oxford.

———. 1982. Basking behaviour in larvae of the butterfly *Euphydryas aurinia*. *Oikos* 38: 308–312.

———. 1983. Multivoltinism in *Apanteles bignellii* and the influence of weather on synchronisation with its host *Euphydryas aurinia*. *Experimental and Applied Entomology* 34: 155–162.

———. 1984. Sunshine, sex ratio and behavior of *Euphydryas aurinia* larvae. Pages 309–311 in R. T. Vane-Wright and P. R. Ackery, eds. *The Biology of Butterflies: Symposium of the Royal Entomological Society of London*. Academic Press, London.

Possingham, H. P., I. Ball, and S. Andelman. 2000. Mathematical methods for reserve system design. Pages 291–306 in S. Ferson and M. Burgman, eds. *Quantitative Methods for Conservation Biology*. Springer-Verlag, New York.

Possingham, H. P., D. B. Lindenmayer, and G. N. Tuck. 2002. Decision theory for population viability analysis. Pages 470–489 in S. R. Beissinger and D. R. McCullough, eds. *Population Viability Analysis*. The University of Chicago Press, Chicago.

Pratt, G. F., E. W. Hein, and K. D. M. 2001. Newly discovered populations and food plants extend the range of the endangered quino checkerspot butterfly, *Euphydryas editha quino* (Nymphalidae) in southern California. *Journal of the Lepidopterists' Society* 55: 169–171.

Prendergast, J. R. 1997. Species richness covariation in higher taxa—empirical tests of the biodiversity indicator concept. *Ecography* 20: 210–216.

Prendergast, J. R., R. M. Quinn, J. H. Lawton, B. C. Eversham, and D. W. Gibbons. 1993. Rare species, the coincidence of diversity hotspots and conservation strategies. *Nature* 365: 335–337.

Pressey, R. L., and A. O. Nicholls. 1989. Efficiency in conservation evaluation: Scoring versus iterative approaches. *Biological Conservation* 50: 199–218.

Price, P. W. 1984. *Insect Ecology*. Wiley & Sons, New York.

———. 1992. Evolution and ecology of gall-induced sawflies. Pages 208–224 in J. D. Shorthouse and O. D. Rohfritsch, eds.

Biology of Insect-induced Galls. Oxford University Press, Oxford.

Price, T., and D. Schluter. 1991. On the low heritability of life history traits. *Evolution* 45: 853–861.

Prinkkilä, M.-L., and I. Hanski. 1995. Complex competitive interactions in four species of *Lucilia* blowflies. *Ecological Entomology* 20: 261–272.

Pulliam, H. R. 1988. Sources, sinks, and population regulation. *The American Naturalist* 132: 652–661.

———. 1996. Sources and sinks: Empirical evidence and population consequences. Pages 45–70 in O. E. Rhodes Jr., R. K. Chester, and M. H. Smith, eds. *Population Dynamics in Ecological Space and Time*. University of Chicago Press, Chicago.

Pullin, A. S., ed. 1995. *Ecology and Conservation of Butterflies*. Chapman & Hall, London.

———. 1996. Restoration of butterfly populations in Britain. *Restoration Ecology* 4: 71–80.

Pullin, A. S., I. F. G. McLean, and W. M.R. 1995. Ecology and conservation of *Lycaena dispar*: British and European perspectives. Pages 150–164 in A. S. Pullin, ed. *Ecology and Conservation of Butterflies*. Chapman & Hall, London.

Puttick, G. M., and M.D. Bowers. 1988. Effect of qualitative and quantitative variation in allelochemicals on a generalist insect: Iridoid glycosides and the southern armyworm. *Journal of Chemical Ecology* 14: 335–351.

Radtkey, R. R., and M. C. Singer. 1995. Repeated reversals of host preference evolution in a specialist insect herbivore. *Evolution* 49: 351–359.

Raffa, K. F. 1991. Where next for plant-insect interactions? *Bulletin of the Ecological Society of America* 72: 127–130.

Ralls, K., J. D. Ballou, and A. Templeton. 1988. Estimates of lethal equivalents and the cost of inbreeding in mammals. *Conservation Biology* 2: 185–193.

Rand, D. B., A. Heath, T. Suderman, and N. E. Pierce. 2000. Phylogeny and life history evolution of the genus *Chrysotis* within Aphnaeini (Lepidoptera: Lycaenidae), inferred from mitochondrial *cytochrome oxidase I* sequences. *Molecular Phylogenetics and Evolution* 17: 85–96.

Rankin, M. A., and C. A. Burchsted. 1992. The cost of migration in insects. *Annual Reviews of Entomology* 37: 533–559.

Rassi, P., A. Alanen, T. Kanerva, and I. Mannerkoski, eds. 2001. *Suomen lajien uhanalaisuus 2000*. Ympäristöministeriö & Suomen ympäristökeskus, Helsinki.

Ratte, H. T. 1984. Temperature and insect development. Pages 33–66 in K. H. Hoffmann, ed. *Environmental Physiology and Biochemistry of Insects*. Springer Verlag, Berlin.

Rausher, M. D. 1978. Search image for leaf shape in a butterfly. *Science* 200: 1071–1073.

———. 1982. Population differentiation in *Euphydryas editha* butterflies: Larval adaptation to different hosts. *Evolution* 36: 581–590.

———. 1988. Is coevolution dead? *Ecology* 69: 898–901.

Rausher, M. D., D. A. Mackay, and M. C. Singer. 1981. Pre- and post-alighting host discrimination by *Euphydryas editha* butterflies: The behavioural mechanisms causing clumped distributions of egg clusters. *Animal Behaviour* 29: 1220–1228.

Reed, J. M., and A. P. Dobson. 1993. Behavioral constraints and conservation biology: conspecific attraction and recruitment. *Trends in Ecology and Evolution* 8: 253–255.

Rees, M., R. Condit, M. Crawley, S. Pacala, and D. Tilman. 2001. Long-term studies of vegetation dynamics. *Science* 293: 650–655.

Rhoades, D. F., and R. G. Cates. 1976. A general theory of plant antiherbivore chemistry. Pages 168–213 in J. W. Wallace and R. L. Mansell, eds. *Recent Advances in Phytochemistry: Biochemical Interaction between Plants and Insects*. Plenum Press, New York.

Rhodes O. E. Jr., R. K. Chesser, and M. H. Smith, eds. 1996. *Population Dynamics in Ecological Space and Time*. University of Chicago Press, Chicago.

Richards, C. M. 2000. Inbreeding depression and genetic rescue in a plant metapopulation. *The American Naturalist* 155: 383–394.

Richards, C. M., S. Church, and D. E. McCauley. 1999. The influence of population size and isolation on gene flow by pollen in *Silene alba*. *Evolution* 53: 63–73.

Richards, L. J., and J. H. Myers. 1980. Maternal influences on size and emergence time of the cinnabar moth. *Canadian Journal Zoology* 58: 1452–1457.

Richards, O. W. 1940. The biology of the small white butterfly (*Pieris rapae*), with special reference to the factors controlling its abundance. *Journal of Animal Ecology* 9: 243–288.

Ricketts, T. H. 2001. The matrix matters: Effective isolation in fragmented landscapes. *The American Naturalist* 158: 87–99.

Ricketts, T. H., G. C. Daily, P. R. Ehrlich, and J. P. Fay. 2001. Countryside biogeography of moths in a fragmented landscape: Biodiversity in native and agricultural habitats. *Conservation Biology* 15: 378–388.

Ridley, M. 1996. *Evolution.* Blackwell Scientific, Cambridge, MA.

Ries, L., and D. M. Debinski. 2001. Butterfly responses to habitat edges in the highly fragmented prairies of Central Iowa. *Journal of Animal Ecology* 70: 840–852.

Ries, L., D. M. Debinski, and M. L. Wieland. 2001. Conservation value of roadside prairie restoration to butterfly communities. *Conservation Biology* 15: 401–411.

Robbins, R. K. 1988. Male foretarsal variation in Lycaenidae and Riodinidae, and the systematic placement of *Styx infernalis* (Lepidoptera). *Proceeding of the Entomological Society of Washington* 90: 356–368.

Roff, D. A. 1986. The evolution of wing dimorphism in insects. *Evolution* 40: 1009–1020.

———. 1994. Habitat persistence and the evolution of wing dimorphism in insects. *The American Naturalist* 144: 772–798.

Roff, D. A., and D. J. Fairbairn. 2001. The genetic basis of dispersal and migration, and its consequences for the evolution of correlated traits. Pages 191–202 in J. Clobert, A. A. Dhondt, E. Danchin, and J. Nichols, eds. *Dispersal—Causes, Consequences and Mechanisms of Dispersal at the Individial, Population and Community Level.* Oxford University Press, Oxford.

Rohani, P., R. M. May, and M. P. Hassell. 1996. Metapopulations and local stability: The effects of spatial structures. *Journal of Theoretical Biology* 181: 97–109.

Roininen, A., P. W. Price, and J. Tahvanainen. 1996. Bottom-up and top-down influences in the trophic system of a willow, a galling sawfly, parasitoids and inquilines. *Oikos* 77: 44–50.

Roland, J. 1982. Melanism and the diel activity of alpine *Colias* (Lepidoptera: Pieridae). *Oecologia* 53: 214–221.

———. 1993. Large-scale forest fragmentation increases the duration of tent caterpillar outbreaks. *Oecologia* 93: 25–30.

———. 1994. After the decline: What maintains low winter moth density after successful biological control? *Journal of Animal Ecology* 63: 392–398.

Roland, J., N. Keyghobadi, and S. Fownes. 1999. Alpine *Parnassius* butterfly dispersal: Effects of landscape and population size. *Ecology* 81: 1642–1653.

Roland, J., and P. D. Taylor. 1995. Herbivore-natural enemy interactions in fragmented and continuous forests. Pages pp. 195–208 in N. Cappuccino and P. Price, eds. *Population Dynamics: New Approaches and Synthesis.* Academic Press, San Diego, CA.

———. 1997. Insect parasitoid species respond to forest structure at different spatial scales. *Nature* 386: 710–713.

Rolstad, J., and P. Wegge. 1987. Distribution and size of capercaillie leks in relation to old forest fragmentation. *Oecologia* 72: 389–394.

Ronce, O., and M. Kirkpatrick. 2001. When sources become sinks: Migrational meltdown in heterogeneous habitats. *Evolution* 55: 1520–1531.

Ronce, O., F. Perret, and I. Olivieri. 2000. Evolutionarily stable dispersal rates do not always increase with local extinction rates. *The American Naturalist* 155: 485–496.

Root, K. V. 1998. Evaluating the effects of habitat quality, connectivity, and catastrophies on a threatened species. *Ecological Applications* 8: 854–865.

Root, T. L. 1988. Energy constraints on avian distributions and abundances. *Ecology* 69: 330–339.

Root, T. L., J. T. Price, K. R. Hall, S. H. Schneider, C. Rosenzweig, and A. Pounds. 2003. "Fingerprints" of global warming on wild animals and plants. *Nature* 421: 57–60.

Rose, M. R., and G. V. Lauder, eds. 1996. *Adaptation.* Academic Press, San Diego, CA.

Rosenheim, J. A., L. R. Wilhoit, and C. A. Armer. 1993. Influence of intraguild predation among generalist insect predators on the supression of an herbivore populations. *Oecologia* 96: 439–449.

Rosenzweig, M. L. 1995. *Species Diversity in Space and Time.* Cambridge University Press, Cambridge.

Ross, P., Jr., and D. Crews. 1977. Influence of the seminal plug on mating behaviour in the garter snake. *Nature* 267: 344–345.

Roughgarden, J. 1979. *Theory of Population Genetics and Evolutionary Ecology: An Introduction.* Macmillan, New York.

———. 1995. *Anolis Lizards of the Caribbean.* Oxford University Press, Oxford.

Rousset, F. 1997. Genetic differentiation and estimation of gene flow from *F*-statistics under isolation by distance. *Genetics* 145: 1219–1228.

Roy, D. B., and T. H. Sparks. 2000. Phenology of British butterflies and climate change. *Global Change Biology* 6: 407–416.

Royama, T. 1996. A fundamental problem in key factor analysis. *Ecology* 77: 87–93.

Rubinoff, D. 2001. Evaluating the California Gnatcatcher as an umbrella species for conservation of southern California coastal sage scrub. *Conservation Biology* 15: 1374–1383.

Rundle, H. D., L. Nagel, J. W. Boughman, and D. Schluter. 2000. Natural selection and parallel speciation in sticklebacks. *Science* 287: 306–308.

Ruse, M. 1987. Biological species: Natural kinds, individuals, or what? *British Journal for the Philosophy of Science* 38: 225–242.

Rutowski, R. L. 1979. Courtship behavior of the checkered white, *Pieris protodice* (Pieridae). *Journal of the Lepidopterists' Society* 33: 42–49.

———. 1982a. Epigamic selection by males as evidenced by courtship partner preferences in the checkered white butterfly (*Pieris protodice*). *Animal Behaviour* 30: 108–112.

———. 1982b. Mate choice and lepidopteran mating behavior. *Florida Entomologist* 65: 72–82.

———. 1984. Sexual selection and the evolution of butterfly mating behavior. *Journal of Research on the Lepidoptera* 23: 125–142.

———. 1985. Evidence for mate choice in a sulfur butterfly (*Colias eurytheme*). *Zeitschrift für Tierpsychologie* 70: 103–114.

———. 1988. Male mate-locating behavior in *Euphydryas chalcedona* (Lepidoptera: Nymphalidae) related to pupation site preferences. *Journal of Insect Behavior* 1: 277–289.

———. 1991. The evolution of male mate-locating behavior in butterflies. *The American Naturalist* 138: 1121–1139.

Rutowski, R. L., J. L. Dickinson, and B. Terkanian. 1989. The structure of the mating plug in the checkerspot butterfly *Euphydryas chalcedona*. *Psyche* 96: 279–286.

Rutowski, R. L., and G. W. Gilchrist. 1987. Courtship, copulation and oviposition in the chalcedon checkerspot, *Euphydryas chalcedona* (Lepidoptera: Nymphalidae). *Journal of Natural History* 21: 1109–1117.

Rutowski, R. L., M. Newton, and S. J. 1983. Interspecific variation in the size of the nutrient investment made by male butterflies during copulation. *Evolution* 37: 708–713.

Ruxton, G. D. 1993. Linked populations can still be chaotic. *Oikos* 68: 347–348.

Ruxton, G. D., J. L. Gonzales-Andujar, and J. N. Perry. 1997. Mortality during dispersal stabilizes local population fluctuations. *Journal of Animal Ecology* 66: 289–292.

Ryder, O. A. 1986. Species conservation and systematics: The dilemma of subspecies. *Trends in Ecology and Evolution* 1: 9–10.

Saari, L., J. Aberg, and J. E. Swenson. 1998. Factors influencing the dynamics of occurrence of the Hazel Grouse in a fine-grained managed landscape. *Conservation Biology* 12: 586–592.

Saccheri, I. J., and P. M. Brakefield. 2002. Rapid spread of immigrant genomes into inbred populations. *Proceedings of the Royal Society B* 269: 1073–1078.

Saccheri, I. J., P. M. Brakefield, and R. A. Nichols. 1996. Severe inbreeding depression and rapid fitness rebound in the butterfly *Bicyclus anynana* (Satyridae). *Evolution* 50: 2000–2013.

Saccheri, I. J., M. Kuussaari, P. Vikman, W. Fortelius, and I. Hanski. 1998. Inbreeding and extinction in a butterfly metapopulation. *Nature* 392: 491–494.

Saccheri, I. J., H. Lloyd, and P. M. Brakefield. ND. Sex-specific inbreeding depression in a butterfly is inconsistent with dominance-fitness relationship. Manuscript.

Sappington, T. W., and O. R. Taylor. 1990. Disruptive sexual selection in *Colias eurytheme* butterflies. *Proceedings of the National Academy of Sciences USA* 87: 6132–6135.

Sato, A., C. O'Huigin, F. Figueroa, P. R. Grant, B. R. Grant, H. Techy, and J. Klein. 1999. Phylogeny of Darwin's Finches as revealed by mitochondrial DNA sequences. *Proceedings of the National Academy of Sciences USA* 96: 5101–5106.

Savalli, U. M., M. E. Czesak, and C. W. Fox. 2000. Paternal investment in the seed beetle *Callosobruchus maculatus* (Coleoptera: Bruchidae): Variation among populations. *Annals of the Entomological Society of America* 93: 1173–1178.

Schaeffer, S. W., and E. L. Miller. 1993. Estimates of linkage disequilibrium and the recombination parameter determined from segregating nucleotide sites in the alcohol dehydrogenase region of *Drosophila pseudoobscura*. *Genetics* 135: 541–552.

Schluter, C. 2001. Ecology and the origin of species. *Trends in Ecology and Evolution* 16: 372–380.

Schluter, D. 1994. Experimental evidence that competition promotes divergence in adaptive radiation. *Science* 266: 798–801.

———. 2000. *The Ecology of Adaptive Radiations*. Oxford University Press, Oxford.

Schluter, D., T. D. Price, and P. R. Grant. 1985. Ecological character displacement in Darwin's finches. *Science* 227: 1056–1059.

Schneirla, T. C. 1971. *Army Ants: A Study in Social Organization*. W.H. Freeman, San Francisco, CA.

Schoener, T. W., and A. Schoener. 1983. Distribution of vertebrates on some very small islands.

I. Occurrence sequences of individual species. *Journal of Animal Ecology* 52: 209–235.

Schoener, T. W., and D. A. Spiller. 1992. Is extinction rate related to temporal variability in population size? An empirical answer for orb spiders. *The American Naturalist* 139: 1176–1207.

Schoonhoven, L. M., T. Jermy, and J. J. A. van Loon. 1998. *Insect-Plant Biology: From Physiology to Evolution*. Chapman & Hall, London.

Schöps, K., and I. Hanski. 2001. Correlation between pre-alighting and post-alighting host plant preference in the Glanville fritillary butterfly. *Ecological Entomology* 26: 517–524.

Schrier, R. D., M. J. Cullenward, P. R. Ehrlich, and R. R. White. 1976. The structure and genetics of a montane population of the checkerspot butterfly, *Chlosyne palla*. *Oecologia* 25: 279–289.

Schultz, C., and K. Dlugosch. 1999. Nectar and hostplant scarcity limits populations of an endangered butterfly. *Oecologia* 119: 231–238.

Schultz, C. B. 1998. Dispersal behavior and its implications for reserve design in a rare Oregon butterfly. *Conservation Biology* 12: 284–292.

Schwarz, C. H., and A. F. Seber. 1999. Estimating animal abundance: Review III. *Statistical Science* 14: 427–456.

Schwarz, M., and M. R. Shaw. 1999. Western Palaearctic Cryptinae (Hymenoptera: Ichneumonidae) in the National Museums of Scotland, with nomenclatural changes, taxonomic notes, rearing records and special reference to the British check list. Part 2. Genus *Gelis* Thunberg (Phygadeuontini: Gelina). *Entomologist's Gazette* 50: 117–125.

Schwerdtfeger, F. 1941. Über die Ursachen des Massenwechsels der Insekten. *Zeitschrift für angewandt Entomologie* 28: 254–303.

Scoble, M. J. 1986. The structure and affinities of the Hedyloidea: A new concept of the butterflies. *Bulletin of the British Museum (Natural History)* 53: 251–286.

———. 1992. *The Lepidoptera: Form, Function and Diversity*. Oxford University Press, Oxford.

Scott, J. A. 1968. Hilltopping as a mating mechanism to aid the survival of low density species. *Journal of Research on Lepidoptera* 7: 191–204.

———. 1973. Convergence of population biology and adult behaviour in two sympatric butterflies, *Neominois ridingsii* (Papilionoidea: Nymphalidae) and *Amblyscirtes simius* (Hesperioidea: Hesperiidae). *Journal of Animal Ecology* 42: 663–672.

———. 1974a. Adult behavior and population biology of *Poladryas minuta*, and the relationship of the Texas and Colorado populations. *Pan-Pacific Entomologist* 50: 9–22.

———. 1974b. Mate-locating behavior of butterflies. *The American Midland Naturalist* 91: 103–117.

———. 1978. A survey of valvae of *Euphydryas chalcedona, E. c. colon, and E. c. anicia*. *Journal of Research on the Lepidoptera* 17: 245–252.

———. 1986. *The Butterflies of North America: A Natural History and Field Guide*. Stanford University Press, Stanford, CA.

Scriber, J. M. 2003. Hybrid zone ecology and tiger swallowtail trait clines in North America. Pages 367–391 in C. L. Boggs, W. B. Watt, and P. R. Ehrlich, ed. *Butterflies: Ecology and Evolution Taking Flight*. University of Chicago Press, Chicago.

Scriber, J. M., R. C. Lederhouse, and R. Hagen. 1991. Food plants and evolution within the *Papilio glaucus* and *Papilio troilus* species groups (Lepidoptera: Papilionidae) in P. W. Price, T. M. Lewinsohn, G. W. Fernandes, and W. W. Benson, eds. *Plant-animal Interactions in Tropical and Temperate Regions*. Wiley, New York.

Scriber, J. M., R. L. Lindroth, and J. Nitao. 1989. Differential toxicity of a phenolic glycoside from quaking aspen leaves by *Papilio glaucus* subspecies, their hybrids, and backcrosses. *Oecologia* 81: 186–191.

Scriber, J. M., Y. Tsubaki, and R. C. Lederhouse, eds. 1995. *Swallowtail butterflies: Their ecology and evolutionary biology*. Scientific Publications, Gainsville, FL.

Scudder, S. H. 1889. *The Butterflies of the Eastern United States and Canada with Special Reference to New England*. Author, Cambridge, MA.

Sculley, C. E., and C. L. Boggs. 1996. Mating systems and sexual division of foraging effort affect puddling behaviour by butterflies. *Ecological Entomology* 21: 193–197.

Seber, G. A. F. 1982. *The Estimation of Animal Abundance and Related Parameters*. Macmillan, New York.

Seigler, D. S. 1998. *Plant Secondary Metabolism*. Kluwer, Norwell, MA.

Shapiro, A. M. 1976. Beau Geste? *The American Naturalist* 110: 900–902.

———. 1981. The pierid red-egg syndrome. *The American Naturalist* 117: 276–294.

Sharpe, P. J. H., and D. W. DeMichele. 1977. Reaction kinetics of poikilotherm development. *Journal of Theoretical Biology* 64: 649–670.

Shaw, M. R. 1994. Parasitoid host ranges. Pages 111–144 in B. A. Hawkins and W. Sheehan, eds. *Parasitoid Community Ecology*. Oxford University Press, Oxford.

———. 2002. Experimental confirmation that *Pteromalus apum* (Retzius) (Hym., Pteromalidae) parasitizes both leaf-cutter bees (Hym. Megachilidae) and Fritillary butterflies (Lep., Nymphalidae). *Entomologist's Monthly Magazine* 138: 37–41.

Shaw, M. R., and M. E. Hochberg. 2001. The neglect of parasitic Hymenoptera in insect conservation strategies: The British fauna as a prime example. *Journal of Insect Conservation* 5: 253–263.

Shaw, W. T. 1925. The seasonal differences of north and south slopes in controlling the activities of the Columbian ground squirrel. *Ecology* 6: 157–162.

Sheppard, P. M. 1951. Fluctuations in the selective value of certain phenotypes in the polymorphic land snail *Cepaea nemoralis* (L.). *Heredity* 5: 125–134.

———. 1952. Natural selection in two colonies of the polymorphic land snail *Cepaea nemoralis*. *Heredity* 6: 233–238.

———. 1953. Polymorphism and population studies. *Symposia of the Society of Experimental Biology* 7: 274–289.

———. 1956. Ecology and its bearing on population genetics. *Proceedings of the Royal Society of London B* 145: 308–315.

Sheppe, W. 1965. Island populations and gene flow in the deer mouse, *Peromyscus leucopus*. *Evolution* 19: 480–495.

Shields, O. 1967. Hilltopping. *Journal of Research on Lepidoptera* 6: 69–178.

Shiga, M. 1976. A quantitative study of food consumption and growth of the tent caterpillar *Malacosoma neustria testacea* Motschulsky (Lepidoptera: Lasiocampidae). *Bulletin of the Fruit Tree Research Station A* 3: 67–86.

Sillén-Tullberg, B. 1988. Evolution of gregariousness in aposematic butterfly larvae: A phylogenetic analysis. *Evolution* 42: 293–305.

Simberloff, D. 1988. The contribution of population and community biology to conservation science. *Annual Review of Ecology and Systematics* 19: 473–511.

———. 1994. The ecology of extinction. *Acta Palaeontologica Polonica* 38: 159–174.

Simberloff, D., and P. Stiling. 1996. How risky is biological control? *Ecology* 77: 1965–1974.

Simmons, L. W. 1992. Quantification of role reversal in relative parental investment in a bush cricket. *Nature* 358: 61–63.

———. 1994. Reproductive energetics of the role reversing bush-cricket, *Kawanaphila nartee* (Orthoptera, Tettigoniidae, Zaprochilinae). *Journal of Evolutionary Biology* 7: 189–200.

Sinclair, A. R. E. 1989. Population regulation in animals. Pages 197–241 in J. M. Cherrett, ed. *Ecological Concepts*. Blackwell Scientific, Oxford.

Sinclair, A. R. E., and M. Norton-Griffiths. 1979. *Serengeti: Dynamics of an Ecosystem*. University of Chicago Press, Chicago.

Singer, M. C. 1971a. Ecological studies on the butterfly *Euphydryas editha*. PhD disseration, Department of Biological Sciences, Stanford University, Stanford, CA.

———. 1971b. Evolution of food-plant preference in the butterfly *Euphydryas editha*. *Evolution* 25: 383–389.

———. 1972. Complex components of habitat suitability within a butterfly colony. *Science* 176: 75–77.

———. 1982. Quantification of host preference by manipulation of oviposition behavior in the butterfly *Euphydryas editha*. *Oecologia* 52: 224–229.

———. 1983. Determinants of multiple host use by a phytophagous insect population. *Evolution* 37: 389–403.

———. 1984. Butterfly-hostplant relationships: Host quality, adult choice and larval success. Pages 81–88 in R. I. Vane-Wright and P. R. Ackery, eds. *The Biology of Butterflies*. Academic Press, London.

———. 1986. The definition and measurement of oviposition preference in plant-fedding insects. Pages 65–94 in J. Miller and T. A. Miller, eds. *Plant-insect Interactions*. Springer-Verlag, Berlin.

———. 1994. Behavioural constraints on the evolutionary expansion of insect diet: A case history from checkerspot butterflies. Pages 279–296 in L. Real, ed. *Behavioural Mechanisms in Evolutionary Ecology*. University of Chicago Press, Chicago.

———. 2000. Reducing ambiguity in describing plant-insect interaction: "preference", "acceptability" and "electivity". *Ecology Letters* 3: 159–162.

———. 2003. Spatial and temporal patterns of checkerspot butterfly-host plant association: The diverse roles of oviposition preference. Pages 207–228 in C. L. Boggs, W. B. Watt, and P. R. Ehrlich, eds. *Butterflies: Ecology and Evolution Taking Flight*. University of Chicago Press, Chicago.

Singer, M. C., and P. R. Ehrlich. 1979. Population dynamics of the checkerspot butterfly *Euphydryas editha*. *Fortschritte der Zoologie* 25: 53–60.

Singer, M. C., P. R. Ehrlich, and L. E. Gilbert. 1971. Butterfly feeding on Lycopsid. *Science* 172: 1341–1342.

Singer, M. C., and J. R. Lee. 2000. Discrimination within and between host species by a butterfly: Implications for design of preference experiments. *Ecology Letters* 3: 101–105.

Singer, M. C., D. Ng, and R. A. Moore. 1991. Genetic variation in oviposition preference between butterfly populations. *Journal of Insect Behavior* 4: 531–535.

Singer, M. C., D. Ng, and C. D. Thomas. 1988. Heritability of oviposition preference and its relationship to offspring performance within a single insect population. *Evolution* 42: 977–985.

Singer, M. C., D. Ng, D. Vasco, and C. D. Thomas. 1992a. Rapidly evolving associations among oviposition preferences fail to constrain evolution of insect diet. *The American Naturalist* 139: 9–20.

Singer, M. C., and C. Parmesan. 1993. Sources of variation in patterns of plant-insect association. *Nature* 361: 251–253.

Singer, M. C., C. Stefanescu, and I. Pen. 2002. When random sampling does not work: standard design falsely indicates maladative host preferences in a butterfly. *Ecology Letters* 5: 1–6.

Singer, M. C., and C. D. Thomas. 1992. The difficulty of deducing behavior from resource use: An example from hilltopping in checkerspot butterflies. *The American Naturalist* 140: 654–664.

———. 1996. Evolutionary responses of a butterfly metapopulation to human- and climate-caused environmental variation. *The American Naturalist* 148: S9–S39.

Singer, M. C., C. D. Thomas, H. L. Billington, and C. Parmesan. 1989. Variation among conspecific insect populations in the mechanistic basis of diet breadth. *Animal Behaviour* 37: 751–759.

———. 1994. Correlates of speed of evolution of host preferences in a set of twelve populations of butterfly *Euphydryas editha*. *Ecoscience* 1: 107–114.

Singer, M. C., C. D. Thomas, and C. Parmesan. 1993. Rapid human-induced evolution of insect diet. *Nature* 366: 681–683.

Singer, M. C., D. Vasco, C. Parmesan, C. D. Thomas, and D. Ng. 1992b. Distinguishing between 'preference' and 'motivation' in food choice: An example from insect oviposition. *Animal Behaviour* 44: 463–471.

Singer, M. C., and P. Wedlake. 1981. Capture does affect probability of recapture in a butterfly species. *Ecological Entomology* 6: 215–216.

Sisk, T. D., D. D. Murphy, S. B. Weiss, P. R. Ehrlich, and A. E. Launer. ND. Movements of checkerspot butterflies across varied landscapes: implications for invertebrate conservation. Manuscript.

Skellam, J. G. 1951. Random dispersal in theoretical populations. *Biometrika* 38: 196–218.

Slatkin, M. 1974. Competition and regional coexistence. *Ecology* 55: 128–134.

———. 1985. Gene flow in natural populations. *Annual Review of Ecology and Systematics* 16: 393–430.

———. 1987. Gene flow and the geographic structure of natural populations. *Science* 236: 787–792.

Smith, A. T. 1980. Temporal changes in insular populations of the pika *Ochotona princeps*. *Ecology* 61: 8–13.

Smith, H. S. 1935. The role of biotic factors in the determination of population densities. *Journal of Economic Entomology* 28: 873–898.

Smith, K. G. V., ed. 1973. *Insects and Other Arthropods of Medical Importance*. British Museum, Natural History, London.

Snyder, W. E., and A. R. Ives. 2001. Generalist predators disrupt biological control by a specialist parasitoid. *Ecology* 82: 705–716.

Sokal, R. R., and T. J. Crovello. 1970. The biological species concept: A critical evaluation. *The American Naturalist* 104: 127–153.

Solbreck, C., and D. B. Andersen. 1989. Wing reduction: its control and consequences in a lygaeid bug, *Spilostethus pandurus*. *Hereditas* 111: 1–6.

Sonenshine, D. E. 1991. *Biology of Ticks*. Oxford University Press, New York.

Soulé, M. E., ed. 1986. *Conservation Biology: The Science of Scarcity and Diversity*. Sinauer Associates, Sunderland, MA.

———. 1987. *Viable Populations for Conservation*. Cambridge University Press, New York.

Soulé, M. E., D. T. Bolger, A. C. Alberts, J. Wright, M. Sorice, and S. Hill. 1988. Reconstructed dynamics of rapid extinctions of chaparral-requiring birds in urban habitat island. *Conservation Biology* 2: 75–92.

Southwood, T. R. E. 1962. Migration of terrestrial arthropods in relation to habitat. *Biology Review* 37: 171–214.

———. 1977. Habitat, the templet for ecological strategies? *Journal of Animal Ecology* 46: 337–366.

Sperlich, D., H. Feuerbach-Mravlag, P. Lange, A. Michaelidis, and A. Pentzos-Daponte.

1977. Genetic load and viability distribution in central and marginal populations of *Drosophila subobscura. Genetics* 86: 835–848.

Sperling, F. A. H. 2003. Buterfly molecular systematics: From species definitions to higher-level phylogenies. Pages 431–458 in C. L. Boggs, W. B. Watt, and P. R. Ehrlich, ed. *Butterflies: Ecology and Evolution Taking Flight*. University of Chicago Press, Chicago.

Spiess, E. 1977. *Genes in Populations*. Wiley, New York.

Srygley, R. B. 2001. Compensation for fluctuations in crosswind drift without stationary landmarks in butterflies migrating over seas. *Animal Behaviour* 61: 191–203.

Srygley, R. B., and E. G. Oliveira. 2001. Orientation mechanisms and migration strategies within the flight boundary layer. Pages 183–206 in I. P. Woiwod, D. R. Reynolds, and C. D. Thomas, eds. *Insect Movement: Mechanisms and Consequences*. CABI publishing, Wallingford, UK.

Stamp, N. E. 1980. Egg deposition patterns in butterflies: Why do some species cluster their eggs rather than deposit them singly? *The American Naturalist* 115: 367–380.

———. 1981a. Effect of group size on parasitism in a natural population of the Baltimore checkerspot *Euphydryas phaeton. Oecologia* 49: 201–206.

———. 1981b. Parasitism of single and multiple egg clusters of *Euphydryas phaeton* (Nymphalidae). *New York Entomological Society* 89: 89–97.

———. 1982a. Aggregation behavior in Baltimore checkerspot caterpillars, *Euphydryas phaeton* (Nymphalidae). *Journal of Lepidopterists' Society* 36: 31–41.

———. 1982b. Behavioral interactions of parasitoids and Baltimore Checkerspot caterpillars (*Euphydryas phaeton*). *Environmental Entomology* 11: 100–104.

———. 1982c. Searching behaviour of parasitoids for web-making caterpillars: A test of optimal searching theory. *Journal of Animal Ecology* 52: 387–395.

———. 1982d. Selection of oviposition sites by the Baltimore checkerspot, *Euphydryas phaeton* (Nymphalidae). *Journal of the Lepidopterists' Society* 36: 290–302.

———. 1984. Interactions of parasitoids and checkerspot caterpillars *Euphydryas* spp. (Nymphalidae). *Journal of Research on the Lepidoptera* 23: 2–18.

———. 1992. Relative susceptibility to predation of two species of caterpillars on plantain. *Oecologia* 92: 124–129.

———. 2001. Enemy-free space via host plant chemistry and dispersion: Assessing the influence of tri-trophic interactions. *Oecologia* 128: 153–163.

Stamp, N. E., and M. D. Bowers. 1990. Body temperature, behavior and growth of early-spring caterpillars (*Hemileuca lucina*: Saturniidae). *Journal of Lepidopterists' Society* 44: 143–155.

———. 1992. Foraging behavior of specialist and generalist caterpillars on plantain (*Plantago lanceolata*). *Oecologia* 92: 596–602.

———. 1994. Effect of temperature and leaf age on growth versus moulting time of a generalist caterpillar fed plantain (*Plantago lanceolata*). *Ecological Entomology* 19: 199–206.

———. 2000a. Do enemies of herbivores influence plant growth and chemistry? Evidence from a seminatural experiment. *Journal of Chemical Ecology*: 2367–2386.

———. 2000b. Foraging behaviour of caterpillars given a choice of plant genotypes in the presence of insect predators. *Ecological Entomology*: 486–492.

Stamp, N. E., and T. M. Casey, eds. 1993. *Caterpillars: Ecological and evolutionary constraints on foraging*. Chapman & Hall, New York.

Stamps, J. A. 1987. Conspecifics as cues to territory quality. I. A preference for previously-used territories by juvenile lizards (*Anolis aeneus*). *The American Naturalist* 129: 629–642.

Stefanescu, C. 2000. El butterfly monitoring scheme en Cataluña: Los primeros cinco años. *Treb. Soc. Cat. Lep.* 15: 5–48.

Stefanescu, C. B. Pintureau, H-P Tschorsnig, J. Pujade-Villar. 2003. The parasitoid complex of the butterfly *Iphiclides podalirius feisthamelii* (Lepidoptera: Papilionidae) in north-east Spain. *Journal of Natural History* 37: 379–396.

Sterelny, K., and P. E. Griffiths. 1999. *Sex and Death: An Introduction to the Philosophy of Biology*. University of Chicago Press, Chicago.

Stermitz, F. R., G. N. Belofsky, D. Ng, and M. C. Singer. 1989. Quinolizidine alkaloids obtained by *Pedicularis semibarbata* (Scrophulariaceae) from *Lupinus fulcratus* (Leguminosae) fail to influence the specialist herbivore *Euphydryas editha* (Lepidoptera). *Journal of Chemical Ecology* 15: 2521–2530.

Stermitz, F. R., D. R. Gardner, F. J. Odendaal, and P. R. Ehrlich. 1986. *Euphydryas anicia* (Lepidoptera: Nymphalidae) utilization of iridoid glycosides from *Castilleja* and

Besseya (Scrophulariaceae) host plants. *Journal of Chemical Ecology* 12: 1459–1468.

Stermitz, F. R., M. S. A. Kader, T. A. Foderaro, and M. Pomeroy. 1994. Iridoid glycosides from some butterflies and their larval food plants. *Phytochemistry* 37: 997–999.

Stern, V. M., and R. F. Smith. 1960. Factors affecting egg production and oviposition in populations of *Colias philodice eurytheme* Boisduval (Lepidoptera: Pieridae). *Hilgardia* 29: 411–454.

Sternberg, D. W. 1990. Genetic control of cell type and pattern formation in *Caenorhabditis elegans*. *Advances in Genetics* 27: 63.

Stiles, F. G., and A. F. Skutch. 1989. *A Guide to the Birds of Costa Rica*. Cornell University Press, Ithaca, NY.

Stiling, P. 1988. Density-dependent processes and key factors in insect populations. *Journal of Animal Ecology* 57: 581–593.

Stiling, P., and A. M. Rossi. 1994. The window of parasitoid vulnerability to hyperparasitism: Template from parasitoid complex structure. Pages 228–244 in B. A. Hawkins and W. Sheehan, eds. *Parasitoid Community Ecology*. Oxford University Press, Oxford.

Stjernholm, F., and B. Karlsson. 2000. Nuptial gifts and the use of body resources for reproduction in the green-veined white butterfly *Pieris napi*. *Proceedings of the Royal Society of London B* 267: 807–811.

Strauss, S. Y., and R. Karban. 1998. The stregth of selection: intraspecific variation in host-plant quality and the fitness of herbivores. Pages 156–180 in S. Mopper and S. Y. Strauss, eds. *Genetic Structure and Local Adaptation in Natural Insect Populations*. Chapman & Hall, London.

Strong, D. R. 1989. Density independence in space and inconsistent temporal relationships for host mortality caused by a fairyfly parasitoid. *Journal of Animal Ecology* 58: 1065–1076.

Sugawara, T. 1979. Stretch reception in the bursa copulatrix of the butterfly, *Pieris rapae crucivora*, and its role in behavior. *Journal of Comparative Physiology* 130: 191–199.

———. 1981. Fine-structure of the stretch receptor in the bursa copulatrix of the butterfly, *Pieris rapae crucivora*. *Cell and Tissue Research* 217: 23–36.

Sugihara, G., and R. M. May. 1990. Nonlinear forecasting as a way of distinguishing chaos from measurement error in time series. *Nature* 344: 734–741.

Suomi, J., H. Sirén, S. K. Weidmer, and M. Riekkola. 2001. Isolation of aucubin and catalpol from *Melitaea cinxia* larvae and quantification by micellar electrokinetic capillary chromatography. *Analytica Chimica Acta* 429: 91–99.

Suomi, J., S. K. Wiedmer, M. Jussila, and M.-L. Riekkola. 2002. Analysis of eleven iridoid glycosides by micellar electrokinetic capillary chromatography and screening of plant samples by partial filling micellar electrokinetic capillary chromatography-electrospray ionisation mass spectrometry. *Journal of Chromotography A* 970: 287–296.

Svärd, L., and C. Wiklund. 1989. Mass and production rate of ejaculates in relation to monandry/polyandry in butterflies. *Behavioral Ecology and Sociobiology* 24: 395–402.

Swengel, A. B. 1996. Effects of fire and hay management on abundance of prairie butterflies. *Biological Conservation* 76: 73–85.

Swengel, A. B., and S. R. Swengel. 2001. Effects of prairie and barrens management on butterfly faunal composition. *Biodiversity and Conservation* 10: 1757–1785.

Tabashnik, B. E. 1980. Population structure of pierid butterflies. III. Pest populations of *Colias philodice eriphyle*. *Oecologia* 47: 175–183.

———. 1983. Host range evolution: the shift from native legume hosts to alfalfa by the butterfly *Colias philodice eriphele*. *Evolution* 37: 150–162.

Tadros, T. M. 1957. Evidence of the presence of an edapho-biotic factor in the problem of serpentine tolerance. *Ecology* 38: 14–23.

Takekawa, J. E., and S. R. Beissinger. 1989. Cyclic drought, dispersal, and conservation of the snail kite in Florida: Lessons in critical habitat. *Conservation Biology* 3: 302–311.

Taylor, A. D. 1988. Large-scale spatial structure and population dynamics in arthropod predator-prey systems. *Annales Zoologici Fennici* 25: 63–74.

Taylor, L. R. 1974. Monitoring change in the distribution and abundance of insects. *Report of the Rothamsted Experimental Station for 1973* 2: 202–239.

———. 1986. Synoptic dynamics, migration and the Rothamsted Insect Survey. *Journal of Animal Ecology* 55: 1–38.

Telfer, W. H., and L. D. Rutberg. 1960. The effects of blood protein depletion on the growth of oocytes in the cecropia moth. *Biological Bulletin* 118: 352–366.

Templeton, A. R. 1986. Coadaptation and outbreeding depression. Pages 105–116 in M. E. Soulé, ed. *Conservation Biology: The*

Science of Scarcity and Diversity. Sinauer Associates, Sunderland, MA.

———. 1989. The meaning of species and speciation: A genetic perspective. Pages 3–27 in D. Otte and J. A. Endler, eds. *Speciation and Its Consequences*. Sinauer Associates, Sunderland, MA.

Tennent, J. 1996. *The Butterflies of Morocco, Algeria and Tunisia*. Gem Publishing, Wallingford, UK.

ter Braak, J. F., I. Hanski, and J. Verboom. 1998. The incidence function approach to modeling of metapopulation dynamics. Pages 167–188 in J. Bascompte and R. V. Solé, eds. *Modeling Spatiotemporal Dynamics in Ecology*. Springer-Verlag, Berlin.

Tewksbury, J. J., D. J. Levey, N. M. Haddad, S. Sargent, J. L. Orrock, A. Weldon, B. J. Danielson, J. Brinkerhoff, E. I. Damschen, and P. Townsend. 2002. Corridors affect plants, animals, and their interactions in fragmented landscapes. *Proceedings of the National Academy of Sciences USA* 99: 12923–12926.

Theodoratus, D. H., and D. Bowers. 1999. Effects of sequestered iridoid glycosides on prey choice of the prairie wolf spider, *Lycosa carolinensis*. *Journal of Chemical Ecology* 25: 283–295.

Thomas, C. D. 1987. Behavioural determination of diet breadth in insect herbivores: The effect of leaf age on choice of host species by beetles feeding on Passiflora vines. *Oikos* 48: 211–216.

———. 1994a. Extinction, colonization and metapopulations: Environmental tracking by rare species. *Conservation Biology* 8: 373–378.

———. 1994b. Local extinctions, colonizations and distributions: Habitat tracking by British butterflies. Pages 319–336 in S. R. Leather, A. D. Watt, N. J. Mills, and K. F. A. Walters, eds. *Individuals, Populations and Patterns in Ecology*. Intercept, Andover.

Thomas, C. D., E. J. Bodsworth, R. J. Wilson, A. D. Simmons, Z. G. Davies, M. Musche, and L. Conradt. 2001a. Ecological and evolutionary processes at expanding range margins. *Nature* 411: 577–581.

Thomas, C. D., and I. Hanski. 1997. Butterfly metapopulations. Pages 359–386 in I. Hanski and M. E. Gilpin, eds. *Metapopulation Biology*. Academic Press, San Diego, CA.

Thomas, C. D., and S. Harrison. 1992. Spatial dynamics of a patchily distributed butterfly species. *Journal of Animal Ecology* 61: 437–446.

Thomas, C. D., J. K. Hill, and O. T. Lewis. 1998. Evolutionary consequences of habitat fragmentation in a localized butterfly. *Journal of Animal Ecology* 67: 485–497.

Thomas, C. D., and J. J. Lennon. 1999. Birds extend their ranges northwards. *Nature* 399: 213.

Thomas, C. D., D. Ng, M. C. Singer, J. L. B. Mallet, C. Parmesan, and H. L. Billington. 1987. Incorporation of a European weed into the diet of a North American herbivore. *Evolution* 41: 892–901.

Thomas, C. D., and M. C. Singer. 1987. Variation in host preference affects movement patterns within a butterfly population. *Ecology* 68: 1262–1267.

———. 1998. Scale-dependent evolution of specialization in a checkerspot butterfly: From individuals to metapopulations and ecotypes. Pages 343–374 in S. Mopper and S. Y. Strauss, eds. *Genetic Structure and Local Adaptation in Natural Insect Populations*. Chapman & Hall, New York.

Thomas, C. D., M. C. Singer, and D. A. Boughton. 1996. Catastrophic extinction of population sources in a butterfly metapopulation. *The American Naturalist* 148: 957–975.

Thomas, C. D., J. A. Thomas, and M. S. Warren. 1992. Distributions of occupied and vacant butterfly habitats in fragmented landscapes. *Oecologia* 92: 563–567.

Thomas, J. A. 1980. Why did the large blue become extinct in Britain. *Oryx* 15: 243–247.

———. 1984. The conservation of butterflies in temperate countries: Past efforts and lessons for the future. Pages 333–353 in R. I. Vane-Wright and P. R. Ackery, eds. *The Biology of Butterflies*. Princeton University Press, Princeton, NJ.

———. 1989. The return of the large blue butterfly. *British Wildlife* 1: 2–13.

———. 1991. Rare species conservation: Case studies of European butterflies. Pages 149–197 in I. F. Spellerberg, F. B. Goldsmith, and M. G. Morris, eds. *The Scientific Management of Temperate Communities for Conservation*. Blackwell Scientific, Oxford.

———. 1995a. The conservation of declining butterfly populations in Britain and Europe: Priorities, problems and successes. *Biological Journal of the Linnean Society* 56 (suppl.): 55–72.

———. 1995b. The ecology and conservation of *Maculinea arion* and other European species of large blue butterfly. Pages 180–197 in A. S. Pullin, ed. *Ecology and Conservation of Butterflies*. Chapman & Hall, London.

Thomas, J. A., N. A. D. Bourn, R. T. Clarke, K. E. Stewart, D. J. Simcox, G. S. Pearman, R. Curtis, and B. Goodger. 2001. The quality and isolation of habitat patches both determine where butterflies persist in fragmented landscapes. *Proceedings of the Royal Society London B* 268: 1791–1796.

Thomas, J. A., and D. J. Simcox. 1982. A quick method for estimating larval populations of *Melitaea cinxia* L. during surveys. *Biological Conservation* 22: 315–322.

Thomas, J., and N. Webb. 1984. *Butterflies of Dorset*. The Dorset Natural History and Archaeological Society, Dorchester, UK.

Thompson, J. N. 1988. Evolutionary genetics of oviposition preference in swallowtail butterflies. *Evolution* 42: 1223–1234.

———. 1993. Preference hierarchies and the origin of geographic specialization in host use in swallowtail butterflies. *Evolution* 47: 1585–1594.

———. 1994. *The Coevolutionary Process*. University of Chicago Press, Chicago.

———. 1995. The origins of host shifts in swallowtail butterflies versus other insects. Pages 195–204 in J. M. Scriber, Y. Tsubaki, and R. C. Lederhouse, eds. *Swallowtail Butterflies: Their Ecology and Evolutionary Biology*. Scientific Publishers, Gainesville, FL.

———. 1996. Trade-offs in larval performance on normal and novel hosts. *Entomologia Experimentalis et Applicata* 80: 133–139.

———. 1998. The evolution of diet breadth: Monophagy and polyphagy in swallowtail butterflies. *Journal of Evolutionary Biology* 11: 563–578.

Thompson, W. R. 1939. Biological control and the theories of the interactions of populations. *Parasitology* 31: 299–388.

Thornhill, N. W. 1993. *The Natural History of Inbreeding and Outbreeding*. University of Chicago Press, Chicago.

Thornhill, R., and J. Alcock. 1983. *The Evolution of Insect Mating Systems*. Harvard University Press, Cambridge, MA.

Thorpe, K. W., and P. Barbosa. 1986. Effects of consumption of high and low nicotine tobbacco by *Manduca sexta* (Lepidoptera: Sphyngidae) on the survival of gregarious endoparasitoid *Cotesia congregata*. *Journal of Chemical Ecology* 12: 1329–1337.

Thrall, P. H., and J. Antonovics. 1995. Theoretical and empirical studies of metapopulations: Population and genetic dynamics of the Silene-Ustilago system. *Canadian Journal of Botany* 73: 1249–1258.

Tilman, D. 1982. *Resource Competition and Community Structure*. Princeton University Press, Princeton, NJ.

———. 1988. *Plant Strategies and the Dynamics and Structure of Plant Communities*. Princeton University Press, Princeton, NJ.

———. 1994. Competition and biodiversity in spatially structured habitats. *Ecology* 75: 2–16.

Tilman, D., and P. Kareiva. 1997. *Spatial Ecology*. Princeton University Press, Princeton, NJ.

Tischendorf, L., and L. Fahrig. 2000. On the usage of landscape connectivity. *Oikos* 90: 7–19.

Tolman, T., and R. Lewington. 1997. *Butterflies of Britain and Europe*. Harper Collins, London.

Travis, J., D. J. Murrell, and C. Dytham. 1999. The Evolution of density-dependent dispersal. *Proceedings of the Royal Society, B* 266: 1837–1842.

Tscharntke, T., and B. A. Hawkins. 2002. *Multitrophic Level Interactions: An Introduction*. Cambridge University Press, New York.

Tschorsnig, H. P., and B. Herting. 1994. Die Raupenfliegen (Diptera: Tachinidae) Mitteleuropas: Bestimmungstabellen und Angaben zur Verbreitung und ökologie der einzeinen Arten. *Stuttgarter Beiträge zur Naturkunde Serie A Biologie* 506: 1–170.

Tucker, R. P. 2000. *Insatiable Appetite: The United States and the Ecological Degradation of the Tropical World*. University of California Press, Berkeley.

Turchin, P. 1991. Translating foraging movements in heterogeneous environments into the spatial distribution of foragers. *Ecology* 72: 1253—1266.

———. 1995. Population regulation: Old arguments and a new synthesis. Pages 19–40 in N. Cappuccino and P. W. Price, eds. *Population Dynamics: New Approaches and Synthesis*. Academic Press, London.

———. 1996. Nonlinear time-series modeling of vole popualtion fluctuations. *Research in Population Ecology* 38: 121–132.

———. 1998. *Quantitative Analysis of Movement: Measuring and Modelling Population Redistribution in Animals and Plants*. Sinaeur Associates, Sunderland, MA.

———. 2002. *Complex Population Dynamics*. Princeton University Press, Princeton, NJ.

Turchin, P., and A. D. Taylor. 1992. Complex dynamics in ecological time series. *Ecology* 73: 289–305.

Turlings, T. C. J., J. H. Loughrin, P. J. McCall, U. S. R. Rose, W. J. Lewis, and J. H. Tumlinson. 1995. How caterpillar -damaged plants protect themselves by attracting parasitic wasps. *Proceedings of the National Academy of Sciences USA* 92: 4169–4174.

Turner, M. G. 1989. Landscape ecology: The effect of pattern on process. *Annual Review of Ecology and Systematics* 20: 171–197.

Turner, M. G., R. H. Gardner, and R. V. O'Neill. 2001. *Landscape Ecology in Theory and Practice*. Springer, New York.

Tuzov, V. K., and S. Churkin. 2000. Genus *Melitaea*. Pages 59–76 in V. k. Tuzov, P. V. Bogdanov, S. V. Churkin, A. V. Dantchenko, A. L. Devyatkin, V. S. Murzin, G. D. Samodurov, and A. B. Zhdanko, eds. *Guide to the Butterflies of Russia and Adjacent Territories (Lepidoptera, Rhopalocera)*. Pensoft, Sofia, Bulgaria.

Vahed, K. 1998. The function of nuptial feeding in insects: Review of empirical studies. *Biological Reviews of the Cambridge Philosophical Society* 73: 43–78.

van der Meijden, E., and C. A. M. van der Veen-van Wijk. 1997. Tritrophic metapopulation dynamics: A case study of ragwort, the Cinnabar moth and the parasitoid *Cotesia popularis*. Pages 387–406 in I. Hanski and M. E. Gilpin, eds. *Metapopulation Biology: Ecology, Genetics, and Evolution*. Academic Press, San Diego, CA.

van Dyck, H., and E. Matthysen. 1999. Habitat fragmentation and insect flight: A changing 'design' in a changing landscape? *Trends in Ecology and Evolution* 14: 172–174.

Van Horne, B. 1983. Density as a misleading indicator of habitat quality. *Journal of Wildlife Management* 47: 893–901.

Van Klinken, R. D., and O. R. Edwards. 2002. Is host-specificty of weed biological control agents likely to evolve raplidly following establishment? *Ecology Letters* 5: 590–596.

van Nouhuys, S., and J. Ehrnsten. 2004. Spatial learning enhances stability of a host-parasite system. *Journal of Behavioral Ecology*, in press.

van Nouhuys, S., and I. Hanski. 1999. Host diet affects extinctions and colonizations in a parasitoid metapopulation. *Journal of Animal Ecology* 68: 1248–1258.

———. 2000. Apparent competition between parasitoids mediated by a shared hyper-parasitoid. *Ecology Letters* 3: 82–84.

———. 2002a. Colonization rates and distances of a host butterfly and two specific parasitoids in a fragmented landscape. *Journal of Animal Ecology* 71: 630–650.

———. 2002b. Multitrophic interactions in space: Metacommunity dynamics in fragmented landscapes. Pages 124–147 in T. Tscharntke and B. A. Hawkins, eds. *Multitrophic Level Interactions*. Cambridge University Press, Cambridge.

van Nouhuys, S., M. C. Singer, and M. Nieminen. 2003. Spatial and temporal patterns of caterpillar performance and the suitability of two host plant species. *Ecological Entomology* 28: 193–202.

van Nouhuys, S., and W. T. Tay. 2001. Causes and consequences of mortality in small populations of a parasitoid wasp in a fragmented landscape. *Oecologia* 128: 126–133.

van Nouhuys, S., and S. Via. 1999. Natural selection and genetic differentiation of behavior between parasitoids from wild and cultivated habitats. *Heredity* 83: 127–137.

van Oosterhout, C., W. G. Zijlstra, M. K. van Heuven, and P. M. Brakefield. 2000. Inbreeding depression and genetic load in laboratory metapopulations of the butterfly *Bicyclus anynana*. *Evolution* 54: 218–225.

van Swaay, C. A. M. 1990. An assesment of the changes in the butterfly abundance in The Netherlands during the 20th century. *Biological Conservation* 52: 287–302.

van Swaay, C. A. M., D. Maes, and C. Plate. 1997. Monitoring butterflies in the Netherlands and Flanders: The first results. *Journal of Insect Conservation* 1: 81–87.

van Swaay, C., and M. Warren. 1999. *Red Data Book of European Butterflies (Rhopalocera)*. Council of Europe Publishing, Strasbourg.

van Valen, L. 1971. Group selection and the evolution of dispersal. *Evolution* 25: 591–598.

Varley, G. C. 1947. The natural control of population balance in the Knapweed gall fly (*Urophaga jaceana*). *Journal of Animal Ecology* 16: 139–187.

Varley, G. C., and G. R. Gradwell. 1960. Key factors in population studies. *Journal of Animal Ecology* 29: 399–401.

Varley, G. C., G. R. Gradwell, and M. P. Hassell. 1973a. *Insect Population Ecology*. Blackwell Scientific, Oxford.

———. 1973b. *Insect Population Ecology: An Analytical Approach*. Blackwell Scientific, Oxford.

Veltman, C. J., S. Nee, and M. J. Crawley. 1996. Correlates of introduction success in exotic New Zealand birds. *The American Naturalist* 147: 542–557.

Venable, D. L. 1979. The demographic consequences of achene polymorphism in *Heterotheca latifolia* Buckl. (Compositae): Germination, survivorship, fecundity and dispersal. PhD dissertation, University of Texas, Austin.

Venable, D. L., and L. Lawlor. 1980. Delayed germination and dispersal in desert annuals: escape in space and time. *Oecologia* 46: 272–282.

Vepsäläinen, K. 1973. The distribution and habitats of Gerris Fabr. Species (Heteroptera:Gerridae) in Finland. *Annales Zoolologici Fennici* 10: 419–444.

Verboom, J., A. Schotman, P. Opdam, and J. A. J. Metz. 1991. European nuthatch metapopulations in a fragmented agricultural landscape. *Oikos* 61: 149–156.

Vet, L. E. M., and M. Dicke. 1992. Ecology of infochemicals used by natural enemies in a tritrophic context. *Annual Review of Entomology* 37: 141–172.

Via, S. 1999. Reproductive isolation between sympatric races of pea aphids. I. Gene flow restriction and habitat choice. *Evolution* 53: 1446–1457.

———. 2001. Sympatric speciation in animals: The ugly duckling grows up. *Trends in Ecology and Evolution* 16: 381–390.

Vitousek, P. M. 2002. Oceanic islands as model systems for ecological studies. *Journal of Biogeography* 29: 573–582.

———. 2003. *The Hawaiian Islands as a Model Ecosystem*. Princeton University Press, Princeton, NJ.

Vogel, K., and J. Johannesen. 1996. Research on population viability of *Melitaea didyma* (Esper, 1779) (Lepidoptera, Nymphalidae). Pages 262–267 in J. Settele, C. Margules, P. Poschold, and K. Henle, eds. *Species Survival in Fragmented Landscapes*. Kluwer, Dordrecht.

Vogler, A. P., and R. DeSalle. 1994. Diagnosing units of conservation management. *Conservation Biology* 8: 354–363.

Volterra, V. 1926. Variations and fluctuations of the numbers of individuals in animal species living together. *Journal du Conseil permanent international pour l'Exploration de la Mer* 3: 3–51.

Voss, R. 1979. Male accessory glands and the evolution of copulatory plugs in rodents. *Occasional Papers of the Museum of Zoology, University of Michigan* 689: 1–27.

Vrana, P., and W. Wheeler. 1992. Individual organisms as terminal entities: Laying the species problem to rest. *Cladistics* 8: 67–72.

Vulinec, K. 1990. Collective security: Aggregation by insects as a defense. Pages 251–288 in D. Evans and J. O. Schmidt, eds. *Insect Defenses*. State University of New York Press, New York.

Wahlberg, N. 1995. One day in the life of a butterfly. A study of the biology of the Glanville fritillary *Melitaea cinxia*. MSc thesis, University of Helsinki.

———. 1997a. The life history and ecology of *Melitaea diamina* (Nymphalidae) in Finland. *Nota Lepidoptera* 20: 70–81.

———. 1997b. Ratamoverkkoperhosen (*Mellicta athalia*) elinkierto Etelä-Suomessa [The life history of the heath fritillary (*Mellicta athalia*) in southern Finland]. *Baptria* 22: 149–153.

———. 1998. The life history and ecology of *Euphydryas maturna* (Nymphalidae: Melitaeini) in Finland. *Nota Lepidoptera* 21: 154–169.

———. 2000a. Comparative descriptions of the immature stages and ecology of five Finnish melitaeine butterfly species (Lepidoptera: Nymphalidae). *Entomologica Fennica* 11: 167–174.

———. 2000b. The ecology and evolution of melitaeine butterflies. University of Helsinki, Helsinki.

———. 2001a. On the status of the scarce fritillary *Euphydryas maturna* (Lepidoptera: Nymphalidae) in Finland. *Entomologica Fennica* 12: 244–250.

———. 2001b. The phylogenetics and biochemistry of host plant specialization in melitaeine butterflies (Lepidoptera: Nymphalidae). *Evolution* 55: 522–537.

Wahlberg, N., T. Klemetti, and I. Hanski. 2002a. Dynamic populations in a dynamic landscape: The metapopulation structure of the marsh fritillary butterfly. *Ecography* 25: 224–232.

Wahlberg, N., T. Klemetti, V. Selonen, and I. Hanski. 2002b. Metapopulation structure and movements in five species of checkerspot butterflies. *Oecologia* 130: 33–43.

Wahlberg, N., J. Kullberg, and I. Hanski. 2001. Natural history of some Siberian melitaeine butterfly species (Nymphalidae: Melitaeini) and their parasitoids. *Entomologica Fennica* 12: 72–77.

Wahlberg, N., A. Moilanen, and I. Hanski. 1996. Predicting the occurrence of endangered species in fragmented landscapes. *Science* 273: 1536–1538.

Wahlberg, N., and M. Zimmermann. 2000. Pattern of phylogenetic relationships among members of the tribe Melitaeini (Lepidoptera: Nymphalidae) inferred from mtDNA sequences. *Cladistics* 16: 347–363.

Wahlström, E., E.-L. Hallanaro, and S. Manninen. 1996. *Suomen Ympäristön Tulevaisuus*. Edita, Helsinki.

Walde, S. J., and W. W. Murdock. 1988. Spatial density dependence in parasitoids. *Annual Review of Entomology* 33: 441–466.

Walker, R. B. 1954. The ecology of serpentine soils. II. Factors affecting plant growth on serpentine soils. *Ecology* 35: 259–266.

Walling, L., and C. L. Boggs. ND. Family-specific

nutrient preferences of temperate puddling butterflies. Manuscript.
Wang, J. 2000. Effects of population structures and selection strategies on the purging of inbreeding depression due to deleterious mutations. *Genetical Research* 76: 75–86.
Wang, J., W. G. Hill, D. Charlesworth, and B. Charlesworth. 1999. Dynamics of inbreeding depression due to deleterious mutations in small populations: Mutation parameters and inbreeding rate. *Genetical Research* 74: 165–178.
Wanntorp, H.-E., D. R. Brooks, T. Nilsson, S. Nylin, F. Ronquist, S. C. Stearns, and N. Wedell. 1990. Phylogenetic approaches in ecology. *Oikos* 57: 119–132.
Warren, M. S. 1987a. The ecology and conservation of the heath fritillary butterfly, *Mellicta athalia*. I. Host selection and phenology. *Journal of Applied Ecology* 24: 467–482.
———. 1987b. The ecology and conservation of the heath fritillary butterfly, *Mellicta athalia*. II. Adult population structure and mobility. *Journal of Applied Ecology* 24: 483–498.
———. 1987c. The ecology and conservation of the heath fritillary butterfly, *Mellicta athalia*. III. Population dynamics and the effect of habitat management. *Journal of Applied Ecology* 24: 499–513.
———. 1992. The conservation of British butterflies. Pages 246–274 in R. L. H. Dennis, ed. *The Ecology of Butterflies in Britain*. Oxford University Press, Oxford.
———. 1994. The UK status and suspected metapopulation structure of a threatened European butterfly, the marsh fritillary *Eurodryas aurinia*. *Biological Conservation* 67: 239–249.
Warren, M. S., J. K. Hill, J. A. Thomas, J. Asher, R. Fox, B. Huntley, D. B. Roy, M. G. Telfer, S. Jeffcoate, P. Harding, G. Jeffcoate, S. G. Willis, J. N. Greatorex-Davies, D. Moss, and C. D. Thomas. 2001. Rapid responses of British butterflies to opposing forces of climate and habitat change. *Nature* 414: 65–69.
Waser, P. M., L. F. Elliott, N. M. Creel, and S. R. Creel. 1995. Habitat variation and mongoose demography. Pages 421–447 in A. R. E. Sinclair and P. Arcese, eds. *Serengeti II. Dynamics, Management, and Conservation of an Ecosystem*. University of Chicago Press, Chicago.
Watkinson, A. R., and W. J. Sutherland. 1995. Sources, sinks, and pseudosinks. *Journal of Animal Ecology* 64: 126–130.
Watson, R. W. 1969. Notes on *Melitaea cinxia* L. 1945–1968. *Entomologist's Record and Journal of Variation* 81: 18–20.
Watt, W. B. 1968. Adaptive significance of pigment polymorphisms in *Colias* butterflies. I. Variation of melanin pigment in relation to thermoregulation. *Evolution* 22: 437–458.
———. 1977. Adaptation at specific loci. I. Natural selection on phosphoglucose isomerase of *Colias* butterflies: Biochemical and population aspects. *Genetics* 87: 177–184.
———. 1992. Eggs, enzymes, and evolution: Natural genetic variants change insect fecundity. *Proceedings of the National Academy of Sciences USA* 89: 10608–10612.
———. 1994. Allozymes in evolutionary genetics: Self-imposed burden or extraordinary tool? *Genetics* 136: 11–16.
———. 2003. Mechanistic studies of butterfly adaptations. Pages 603–613 in C. Boggs, W. B. Watt, and P. R. Ehrlich, eds. *Evolution and Ecology Taking Flight: Butterflies as Model Systems*. University of Chicago Press, Chicago.
Watt, W. B., and C. L. Boggs. 1987. Allelic enzymes as probes of the evolution of metabolic organization. *Isozymes* 15: 27–47.
———. 2003. Butterflies as model systems in ecology and evolution: Present and future. Pages 603–613 in C. L. Boggs, W. B. Watt, and P. R. Ehrlich, eds. *Butterflies: Ecology and Evolution taking Flight*. University of Chicago Press, Chicago.
Watt, W. B., P. A. Carter, and K. Donohue. 1986. An insect mating system promotes the choice of "good genotypes" as mates. *Science* 233: 1187–1190.
Watt, W. B., R. C. Cassin, and M. B. Swann. 1983. Adaptation at specific loci. III. Field behavior and survivorship differences among *Colias* PGI genotypes are predictable from *in vitro* biochemistry. *Genetics* 103: 725–739.
Watt, W. B., F. S. Chew, L. R. G. Snyder, A. G. Watt, and D. E. Rothschild. 1977. Population structures of pierid butterflies 1. Numbers and movements in some montane *Colias* species. *Oecologia* 27: 1–22.
Watt, W. B., and A. M. Dean. 2000. Molecular-functional studies of adaptive genetic variation in procaryotes and eucaryotes. *Annual Review of Genetics* 34: 593–622.
Watt, W. B., K. Donohue, and P. A. Carter. 1996. Adaptation at specific loci. VI. Divergence vs. parallelism of polymorphic allozymes in molecular function and fitness-component effects among *Colias* species (Lepidoptera, Pieridae). *Molecular Biology and Evolution* 13: 699–709.
Watt, W. B., C. W. Wheat, E. H. Meyer, and

J.-F. Martin. 2003. Adaptation at specific loci. VII. Natural selection, dispersal, and the diversity of molecular-functional variation patterns among butterfly species complexes (*Colias*: Lepidoptera, Pieridae). *Molecular Ecology*, 12: 1265–1275.

Wedell, N., C. Wiklund, and P. A. Cook. 2002. Monandry and polyandry as alternative lifestyles in a butterfly. *Behavioral Ecology* 13: 450–455.

Wedin, D., and D. Tilman. 1993. Competition among grasses along a nitrogen gradient: Initial conditions and mechanisms of competition. *Ecological Monographs* 63: 199–229.

Wehling, W. F., and J. N. Thompson. 1997. Evolutionary conservatism of oviposition preference in a widespread polyphagous insect herbivore, *Papilio zelicaon*. *Oecologia* 111: 209–215.

Weibull, A. C., J. Bengtsson, and E. Nohlgren. 2000. Diversity of butterflies in the agricultural landscape: The role of farming system and landscape heterogeneity. *Ecography* 23: 743–750.

Weidemann, H. J. 1988. *Tagfalter, Band 2. Biologie-Ökologie-Biotopschutz*. Neumann-Neudamm, Melsungen.

Weiss, S. B. 1999. Cars, cows, and checkerspot butterflies: Nitrogen deposition and management of nutrient poor grasslands for a threatened species. *Conservation Biology* 13: 1476–1486.

Weiss, S. B., and D. D. Murphy. 1988a. Fractal geometry and caterpillar dispersal: Or how many inches can inchworms inch? *Journal of Functional Ecology* 2: 116–118.

———. 1988b. Landscapes, topoclimate, and conservation. *Endangered Species UPDATE* 5: 10.

———. 1990. Thermal microenvironments and the restoration of rare butterfly habitat. Pages 50–60 in J. J. Berger, ed. *Environmental Restoration: Science and Strategies for Restoring the Earth*. Island Press, Washington, DC.

———. 1993. Climatic considerations in reserve design and ecological restoration. Pages 89–107 in D. A. Saunders, R. J. Hobbs, and P. R. Ehrlich, eds. *Nature Conservation 3. Reconstruction of Fragmented Ecosystems*. Surrey Beatty & Sons, Chipping Norton, New South Wales.

Weiss, S. B., D. D. Murphy, P. R. Ehrlich, and C. F. Metzler. 1993. Adult emergence phenology in checkerspot butterflies: The effects of macroclimate, topoclimate, and population history. *Oecologia* 96: 261–270.

Weiss, S. B., D. D. Murphy, and R. R. White. 1988. Sun, slope, and butterflies: Topographic determinants of habitat quality for *Euphydryas editha*. *Ecology* 69: 1486–1496.

Weiss, S. B., and A. Weiss. 1998. Landscape-level phenology of a threatened butterfly: a GIS-based modeling approach. *Ecosystems* 1: 299–309.

Weiss, S. B., R. R. White, D. D. Murphy, and P. R. Ehrlich. 1987. Growth and dispersal of larvae of the checkerspot butterfly *Euphydryas editha*. *Oikos* 50: 161–166.

Weller, S. J., D. P. Pashley, and J. A. Martin. 1996. Reassessment of butterfly family relationships using independent genes and morphology. *Annals of the Entomological Society of America* 89: 184–192.

Weseloh, R. M. 1993. Potential effects of parasitoids on the evolution of caterpillar foraging behavior. Pages 203–223 in N. E. Stamp and T. M. Casey, eds. *Caterpillars: Ecological and Evolutionary Constraints on Foraging*. Chapman & Hall, London.

Wheye, D., and P. R. Ehrlich. 1985. The use of fluorescent pigments to study insect behavior: Investigating mating patterns in a butterfly population. *Ecological Entomology* 10: 231–234.

White, R. R. 1973. Community relationships of the butterfly, *Euphydryas editha*. Ph.D dissertation, Department of Biological Sciences, Stanford University, Stanford, CA.

———. 1974. Food plant defoliation and larval starvation of *Euphydryas editha*. *Oecologia* 14: 307–315.

———. 1979. Foodplant of alpine *Euphydryas anicia* (Nymphalidae). *Journal of the Lepidopterists' Society* 33: 170–173.

———. 1980. Inter-peak dispersal in alpine checkerspot butterflies (Nymphalidae). *Journal of the Lepidopterists' Society* 34: 353–362.

———. 1986. Pupal mortality in the Bay checkerspot butterfly (Lepidoptera: Nymphalidae). *Journal of Research on the Lepidoptera* 25: 52–62.

White, R. R., and M. P. Levin. 1981. Temporal variation in vagility: Implications for evolutionary studies. *The American Midland Naturalist* 105: 348–357.

White, R. R., and M. C. Singer. 1974. Geographical distribution of hostplant choice in *Euphydryas editha* (Nymphalidae). *Journal of the Lepidopterists' Society* 28: 103–107.

Whitlock, M. C. 1992. Nonequilibrium population structure in forked fungus beetles: Extinction, colonization, and the genetic variance among populations. *The American Naturalist* 139: 952–970.

———. 2002. Selection, load, and inbreeding depression in a large metapopulation. *Genetics* 160: 1191–1202.

Whitlock, M. C., and N. H. Barton. 1997. The effective size of a subdivided population. *Genetics* 146: 427–441.

Whitlock, M. C., P. K. Ingvarsson, and T. Hatfield. 2000. Local drift load and the heterosis of interconnected populations. *Heredity* 84: 452–457.

Whitlock, M. C., and D. E. McCauley. 1990. Some population genetic consequences of colony formation and extinction: Genetic correlations within founding groups. *Evolution* 44: 1717–1724.

———. 1999. Indirect measures of gene flow and migration: F_{ST} not equal $1/(4Nm+1)$. *Heredity* 82: 117–125.

Wickman, P.-O. 1988. Dynamics of mate-searching behaviour in a hilltopping butterfly, *Lasiommata megera* (L.): The effects of weather and male density. *Zoological Journal of the Linnean Society* 93: 357–377.

Wiens, J. A., N. C. Stenseth, B. Van Horne, and R. A. Ims. 1993. Ecological mechanisms and landscape ecology. *Oikos* 66: 369–380.

Wiernasz, D. C. 1989. Female choice and sexual selection of male wing melanin pattern in *Pieris occidentalis* (Lepidoptera). *Evolution* 43: 1672–1682.

———. 1995. Male choice on the basis of female melanin pattern in *Pieris* butterflies. *Animal Behaviour* 49: 45–51.

Wijngaarden, P. J. 2000. Quantitative and endocrine genetics of reaction norms for wing pattern in the butterfly *Bicyclus anynana*. PhD disseration, Leiden University, Leiden, the Netherlands.

Wiklund, C. 1975. The evolutionary relationship between adult oviposition preferences and larval host plant range in *Papilio machaon*. *Oecologia* 18: 185–197.

———. 1982. Generalist versus specialist utilization of host plants among butterflies. *Proceedings of the Fifth International Symposium on Insect-Plant Relationships*: 181–191.

Wiklund, C., and T. Fagerström. 1977. Why do males emerge before females? A hypothesis to explain the incidence of protandry in butterflies. *Oecologia* 31: 153–158.

Wiklund, C., and G. J. Forsberg. 1991. Sexual size dimorphism in relation to female polygamy and protandry in butterflies: A comparative study of Swedish Pieridae and Satyridae. *Oikos* 60: 373–381.

Wiklund, C., and B. Karlsson. 1984. Egg size variation in satyrid butterflies: Adaptive versus historical, "Bauplan", and mechanistic explanations. *Oikos* 43: 391–400.

Wilcove, D. S., D. Rothstein, J. Dubow, A. Phillips, and E. Losos. 1998. Quantifying threats to imperiled species in the United States. *BioScience* 48: 607–615.

Wilkinson, G. 1907. Habits and habitats of *Melitaea aurinia*. *Entomologist's Record and Journal of Variation* 19: 273–275.

Williams, C. B. 1930. *The Migration of Butterflies*. Oliver & Boyd, Edinburgh.

Williams, E. H., C. E. Holdren, and P. R. Ehrlich. 1984. The life history and ecology of *Euphydryas gillettii* Barnes (Nymphalidae). *Journal of the Lepidopterists' Society* 38: 1–12.

Williams, G. C. 1975. *Sex and Evolution*. Princeton University Press, Princeton, NJ.

Williams, K. S. 1983. The coevolution of *Euphydryas chalcedona* butterflies and their larval host plants: III. Oviposition behavior and host plant quality. *Oecologia* 56: 336–340.

Williams, K. S., and L. E. Gilbert. 1981. Insects as selective agents on plant vegetative morphology: Egg mimicry reduces egg laying by butterflies. *Science* 212: 467–469.

Williams, K. S., D. E. Lincoln, and P. R. Ehrlich. 1983a. The coevolution of *Euphydryas chalcedona* butterflies and their larval host plants. I. Larval feeding behavior and host plant chemistry. *Oecologia* 56: 323–329.

———. 1983b. The coevolution of *Euphydryas chalcedona* butterflies and their larval host plants: II. Maternal and host plant effects on larval growth, development, and food-use efficiency. *Oecologia* 56: 330–335.

Williams, P. 2000. Complementarity. Pages 813–829 in S. A. Levin, ed. *Encyclopedia of Biodiversity*. Academic Press, London.

Williamson, J., and S. Harrison. 2002. Biotic and abiotic limits to the spread of exotic revegetation species. *Ecological Applications* 12: 40–51.

Wilson, D. S. 1989. The diversification of single gene pools by density- and frequency-dependent selection. Pages 366–385 in D. Otte and J. A. Endler, eds. *Speciation and Its Consequences*. Sinauer Associates, Sunderland, MA.

Wilson, E. O., ed. 1988. *Biodiversity*. National Academy Press, Washington, DC.

———. 1992. *The Diversity of Life*. Harvard University Press, Cambridge, MA.

Wilson, E. O., and W. L. Brown. 1953. The subspecies concept and its taxonomic application. *Systematic Zoology* 2: 97–111.

Woiwood, I. P., and I. Hanski. 1992. Patterns of density dependence in moths and aphids. *Journal of Animal Ecology* 61: 619–629.

Wood, T. K., and K. J. Tilmon. 1999. The role of host-plant fidelity in initiating insect race formation. *Evolutionary Ecology Research* 1: 317–332.

Wood, W. B., ed. 1988. *The Nematode Caenorhabditis elegans*. Cold Spring Harbor Laboratory, Cold Spring Harbor, NY.

Woolfenden, G. E., and J. W. Fitzpatrick. 1984. *The Florida Scrub Jay: Demography of a Cooperative Breeding Bird*. Princeton University Press, Princeton, NJ.

Wooton, J. T. 1997. Estimates and tests of per capita interaction strength: Diet, abundance, and impact of intertidally foraging birds. *Ecological Monographs* 67: 45–64.

Wright, S. 1931. Evolution in Mendelian populations. *Genetics* 16: 97–159.

———. 1940. Breeding structure of populations in relation to speciation. *The American Naturalist* 74: 232–248.

———. 1948. On the roles of directed and random changes in gene frequency in the genetics of populations. *Evolution* 2: 279–294.

———. 1951. The genetical structure of populations. *Annals of Eugenics* 15: 323–354.

Yogi, S., ed. 1996. *Highway 99: A Literary Journey Through California's Great Central Valley*. Heyday Books, Berkeley, CA.

Young, A. M. 1973. Notes on the biology of *Phyciodes (Eresia) eutropia* (Lepidoptera: Nymphalidae) in a Costa Rican mountain forest. *Journal of the New York Entomological Society* 81: 87–100.

Young, A. M., and M. W. Moffett. 1979. Studies on the population biology of the tropical butterfly *Mechanitis isthmia* in Costa Rica. *The American Midland Naturalist* 101: 309–319.

Zalucki, M. P., A. R. Clarke, and S. B. Malcolm. 2002. Ecology and behavior of first instar larval Lepidoptera. *Annual Review of Entomology* 47: 361–393.

Zimmermann, M., N. Wahlberg, and H. Descimon. 2000. A phylogeny of *Euphydryas* checkerspot butterflies (Lepidoptera; Nymphalidae) based on mitochondrial DNA sequence data. *Annals of the Entomological Society of America* 93: 347–355.

Zwölfer, H., and P. Harris. 1971. Host specificity determination of insects for biological control of weeds. *Annual Review of Entomology* 16: 159–178.

Index

Note: page numbers followed by *f* and *t* indicate figures and tables